Raymond W. Luckenbaugh 1991
DuPont
Agricultural Products

Managing Resistance
to Agrochemicals

ACS SYMPOSIUM SERIES **421**

Managing Resistance to Agrochemicals

From Fundamental Research to Practical Strategies

Maurice B. Green, EDITOR
Independent Consultant

Homer M. LeBaron, EDITOR
Ciba-Geigy Corporation

William K. Moberg, EDITOR
E. I. du Pont de Nemours and Company

Developed from a symposium sponsored
by the Division of Agrochemicals
at the 196th National Meeting
of the American Chemical Society,
Los Angeles, California,
September 25–30, 1988

American Chemical Society, Washington, DC 1990

CHEM
SB
957
.M35
1990

Library of Congress Cataloging-in-Publication Data

Managing resistance to agrochemicals: from fundamental research to practical strategies
  Maurice B. Green, editor; Homer M. LeBaron, editor; William K. Moberg, editor.

  p.   cm.—(ACS Symposium Series, ISSN 0097–6156; 421)

  "Developed from a symposium sponsored by the Division of Agrochemicals at the 196th National Meeting of the American Chemical Society, Los Angeles, California, September 25–30, 1988."

  Includes bibliographical references.

  ISBN 0-8412-1741-6

  1. Pesticide resistance—Congresses.   I. Green, Maurice B. (Maurice Berkeley) II. LeBaron, Homer M. III. Moberg, William K., 1948–   . IV. American Chemical Society. Division of Agrochemicals. V. American Chemical Society. Meeting (196th: 1988: Los Angeles, Calif.) VI. Series.

SB957.M35 1990
623'.95—dc20   89–77580
                                                       CIP

The paper used in this publication meets the minimum requirements of American National Standard for Information Sciences—Permanence of Paper for Printed Library Materials, ANSI Z39.48–1984.

Copyright © 1990

American Chemical Society

All Rights Reserved. The appearance of the code at the bottom of the first page of each chapter in this volume indicates the copyright owner's consent that reprographic copies of the chapter may be made for personal or internal use or for the personal or internal use of specific clients. This consent is given on the condition, however, that the copier pay the stated per-copy fee through the Copyright Clearance Center, Inc., 27 Congress Street, Salem, MA 01970, for copying beyond that permitted by Sections 107 or 108 of the U.S. Copyright Law. This consent does not extend to copying or transmission by any means—graphic or electronic—for any other purpose, such as for general distribution, for advertising or promotional purposes, for creating a new collective work, for resale, or for information storage and retrieval systems. The copying fee for each chapter is indicated in the code at the bottom of the first page of the chapter.

The citation of trade names and/or names of manufacturers in this publication is not to be construed as an endorsement or as approval by ACS of the commercial products or services referenced herein; nor should the mere reference herein to any drawing, specification, chemical process, or other data be regarded as a license or as a conveyance of any right or permission to the holder, reader, or any other person or corporation, to manufacture, reproduce, use, or sell any patented invention or copyrighted work that may in any way be related thereto. Registered names, trademarks, etc., used in this publication, even without specific indication thereof, are not to be considered unprotected by law.

PRINTED IN THE UNITED STATES OF AMERICA

ACS Symposium Series

M. Joan Comstock, *Series Editor*

1990 ACS Books Advisory Board

Paul S. Anderson
Merck Sharp & Dohme Research
  Laboratories

V. Dean Adams
Tennessee Technological
  University

Alexis T. Bell
University of California—
  Berkeley

Malcolm H. Chisholm
Indiana University

Natalie Foster
Lehigh University

G. Wayne Ivie
U.S. Department of Agriculture,
  Agricultural Research Service

Mary A. Kaiser
E. I. du Pont de Nemours and
  Company

Michael R. Ladisch
Purdue University

John L. Massingill
Dow Chemical Company

Robert McGorrin
Kraft General Foods

Daniel M. Quinn
University of Iowa

Elsa Reichmanis
AT&T Bell Laboratories

C. M. Roland
U.S. Naval Research Laboratory

Stephen A. Szabo
Conoco Inc.

Wendy A. Warr
Imperial Chemical Industries

Robert A. Weiss
University of Connecticut

Foreword

The ACS SYMPOSIUM SERIES was founded in 1974 to provide a medium for publishing symposia quickly in book form. The format of the Series parallels that of the continuing ADVANCES IN CHEMISTRY SERIES except that, in order to save time, the papers are not typeset but are reproduced as they are submitted by the authors in camera-ready form. Papers are reviewed under the supervision of the Editors with the assistance of the Series Advisory Board and are selected to maintain the integrity of the symposia; however, verbatim reproductions of previously published papers are not accepted. Both reviews and reports of research are acceptable, because symposia may embrace both types of presentation.

Contents

Preface xi

1. Understanding and Combating Agrochemical Resistance:
   A Chemist's Perspective on an Interdisciplinary Challenge 1
   William K. Moberg

INSECTICIDES

2. Overview of Insecticide Resistance 18
   George P. Georghiou

3. Resistance Mechanisms to Carbamate and Organophosphate
   Insecticides 42
   L. B. Brattsten

4. Biochemical and Genetic Mechanisms of Insecticide
   Resistance 61
   Thomas M. Brown

5. Characterization and Exploitation of Instability of Spider
   Mite Resistance to Acaricides 77
   T. J. Dennehy, J. P. Nyrop, and T. E. Martinson

6. Laboratory Evaluation and Empirical Modeling
   of Resistance-Countering Strategies 92
   I. Denholm, M. Rowland, A. W. Farnham, and
   R. M. Sawicki

7. Targeting Insecticide-Resistant Markets: New Developments
   in Microbial-Based Products 105
   Wendy D. Gelernter

8. Strategies for Managing Resistance to Insecticides in *Heliothis* Pests of Cotton 118
   D. L. Bull and J. J. Menn

9. Pyrethroid Resistance in *Heliothis* spp.: Current Monitoring and Management Programs 134
   S. L. Riley

10. Management of Pesticide Resistance in Arthropod Pests: Research and Policy Issues 149
    B. A. Croft

FUNGICIDES

11. Antiresistance Strategies: Design and Implementation in Practice 170
    F. J. Schwinn and H. V. Morton

12. Sterol Biosynthesis Inhibitors: Model Studies with Respect to Modes of Action and Resistance 184
    D. Berg, K.-H. Büchel, G. Holmwood, W. Krämer, and R. Pontzen

13. Resistance to Sterol Biosynthesis-Inhibiting Fungicides: Current Status and Biochemical Basis 199
    D. W. Hollomon, J. A. Butters, and J. A. Hargreaves

14. Biochemical Basis of Resistance to Phenylamide Fungicides 215
    L. C. Davidse

15. Mechanism of Action of *N*–Phenylcarbamates in Benzimidazole-Resistant *Neurospora* Strains 224
    Makoto Fujimura, Kenji Oeda, Hirokazu Inoue, and Toshiro Kato

16. Binding of Cellular Protein from *Venturia nashicola* Isolates to Carbendazim: Its Relationship with Sensitivity to *N*–Phenylcarbamates, *N*–Phenylformamidoximes, and Rhizoxin 237
    H. Ishii, S. Iwasaki, Z. Sato, and I. Inoue

17. Edifenphos Resistance in *Pyricularia oryzae* and *Drechslera oryzae*: In Vitro Techniques for Detection and Biochemical Studies..249
    D. Lalithakumari and P. Annamalai

18. Management of Fungicide Resistance by Using Computer Simulation..264
    Phil A. Arneson

19. Population Biology and Management of Fungicide Resistance........275
    W. E. Fry and M. G. Milgroom

20. Impact of Fungicide Resistance on Citrus Fruit Decay Control..286
    Joseph W. Eckert

21. Predicting the Evolution of Fungicide Resistance.....................303
    K. J. Brent, D. W. Hollomon, and M. W. Shaw

22. The Fungicide Resistance Action Committee: An Update on Goals, Strategies, and North American Initiatives...............320
    M. Wade and C. J. Delp

HERBICIDES

23. Herbicide Resistance in Weeds and Crops: An Overview and Prognosis..336
    Homer M. LeBaron and Janis McFarland

24. Fate of Herbicide Resistance Genes in Weeds..........................353
    H. Darmency and J. Gasquez

25. Structural and Biochemical Characterization of Dinitroaniline-Resistant *Eleusine*....................................364
    Kevin C. Vaughn and Martin A. Vaughan

26. Herbicide Resistance in *Alopecurus myosuroides*.....................376
    Malcolm S. Kemp, Stephen R. Moss, and Tudor H. Thomas

27. Herbicide Cross-Resistance in Annual Ryegrass *(Lolium rigidum Gaud)*: The Search for a Mechanism.................394
    S. B. Powles, J. A. M. Holtum, J. M. Matthews, and D. R. Liljegren

28. Peroxidizing Herbicides: Some Aspects on Tolerance 407
    Gerhard Sandmann and Peter Böger

29. Fitness and Ecological Adaptability of Herbicide-Resistant
    Biotypes 419
    Jodie S. Holt

30. Herbicide Rotations and Mixtures: Effective Strategies
    to Delay Resistance 430
    J. Gressel and L. A. Segel

31. Herbicide-Resistant Plants Carrying Mutated Acetolactate
    Synthase Genes 459
    Mary E. Hartnett, Chok-Fun Chui, C. Jeffry Mauvais,
    Raymond E. McDevitt, Susan Knowlton, Julie K. Smith,
    S. Carl Falco, and Barbara J. Mazur

32. Genetic Modification of Crop Responses to Imidazolinone
    Herbicides 474
    K. E. Newhouse, D. L. Shaner, T. Wang, and R. Fincher

Author Index 482

Affiliation Index 483

Subject Index 483

Preface

RESISTANCE OF PESTS TO AGROCHEMICALS is a problem of growing importance and international scope. The consequences include serious crop damage, loss of highly effective compounds soon after their introduction, and abandonment of certain crops in areas where pests have become insensitive to all available chemical controls. Every class of agrochemical is affected. Resistance is a major threat which, if ignored or improperly managed, will significantly reduce worldwide agricultural production and public health. It is the greatest challenge that the agrochemical industry faces.

This book reflects the conviction of the editors and authors that resistance can be understood and combated effectively. It describes the current situation: distribution and seriousness of pest resistance to insecticides, fungicides, and herbicides, and what is known about the nature and mechanisms of resistance. It goes on to discuss what can be done with this knowledge and what is needed for continued progress.

Though the first instance of insecticide resistance was recorded as long ago as 1914, resistance was not serious or widespread until potent, highly specific organic compounds such as DDT and organochlorines were introduced in the 1940s. Almost three decades passed before fungicides were widely affected, but when protectants were replaced by more active systemic compounds in the early 1970s, resistance developed within a few years. More recently, herbicide resistance has evolved rapidly from localized and isolated cases, such as those involving triazines, to a worldwide acknowledgment of the problem, even with the newer, low-rate herbicides.

Meanwhile, our chemical arsenal has been hard-pressed to keep pace. The traditional response to resistance, switching to new compounds, has become less practical due to substantial increases in the time and expense of agrochemical discovery and development. Simultaneously, older products, including many that are valuable for resistance management, are subjected to increasing regulatory scrutiny and possible cancellation.

Biotechnology and biological control methods show promise in selected situations but are unlikely to have a major effect on agriculture in the near future. Nor should they be considered as replacements for, or in competition with, chemicals. Rather, we anticipate and encourage their complementary use. Experience with pests makes it clear that the maximum number of control options must be maintained; otherwise, biological and genetic controls will likely be overcome by resistant pests, just as many chemical controls have been.

Fortunately, the situation is far from hopeless. Rapid progress is being made as ignorance and controversy are replaced by interdisciplinary research, communication, and cooperation, especially within the international agrochemical industry, to limit the use of specific toxophores. Chemists should play a role in this process, but until now much of the discussion on understanding and combating resistance has been taking place in biologically oriented scientific societies. Yet even here relatively few coordinated efforts have brought together international expertise, science from all disciplines, and basic and applied approaches.

With all this in mind, one of us (Maurice B. Green) proposed the symposium on which this book is based with the following objectives:

- to update the rapidly advancing field of resistance management;
- to inform and involve chemists in combating resistance; and
- to offer a broad, integrated perspective by bringing together researchers from various scientific disciplines, agronomic focuses, and geographic locations.

The result is a comprehensive, thought-provoking overview of the current status and future directions of resistance management. The extensive treatment of herbicide resistance, previously accorded little space in unified works on resistance management, is particularly valuable and reflects the rapid development of this area in recent years. The authors represent 10 countries, and many have experience on a worldwide level. These participants represent diverse professional experience in academic, industrial, and government laboratories working on fundamental and applied research. Together, their contributions chart a clear course for the future: interdisciplinary research that cuts across agronomic and geographic lines; communication among all parties affected; and proactive, cooperative implementation of practical measures based on the best scientific evidence available.

The editors began with the conviction that resistance can be managed. The papers and discussions at the symposium reinforced this conviction. We believe that this volume will do the same for its readers, especially those in the chemical community; that it will also provide them with the latest findings in a dynamic field of critical importance;

and that it will inspire many readers to meet the challenge of pest resistance in the future.

We gratefully acknowledge that a program of this scope, especially one involving many speakers from overseas, could not have been presented without generous financial support from the following companies: in the United States, BASF, Ciba-Geigy, Dow, Du Pont, Monsanto, NOR AM, Pioneer International, Sandoz Zoecon Division, and Uniroyal; in the United Kingdom, BASF, Ciba-Geigy, Cyanamid International, Du Pont, ICI, Pan Brittanica Industries, and Shell; in the Netherlands, Duphar; and in the Federal Republic of Germany, Schering. Their contributions underscore the commitment of industry around the world to support the science needed for effective resistance management.

Finally, we acknowledge with pleasure the efforts of each scientist who participated in the symposium and contributed a chapter to the book. It is to their credit that we have met our objectives.

MAURICE B. GREEN
79/81 Woodcote Road
Wallington
Surrey SM6 OPZ
England

HOMER M. LeBARON
Ciba-Geigy Corporation
Greensboro, NC 27419

WILLIAM K. MOBERG
E. I. du Pont de Nemours and Company
Newark, DE 19714

November 3, 1989

Chapter 1

Understanding and Combating Agrochemical Resistance

A Chemist's Perspective on an Interdisciplinary Challenge

William K. Moberg

Agricultural Products Department, E. I. du Pont de Nemours and Company, Stine-Haskell Research Center 300–312C, P.O. Box 30, Newark, DE 19714

> Resistance of agricultural pests to chemical control is a problem of growing importance, international scope, and potentially serious consequences. Although resistance-induced crises are not yet the norm, enough have occurred, and with sufficient severity, to make clear the need for action. It is also clear that effective action can only result from interdisciplinary research and cooperation, in which chemists should play an important role. The intent of this chapter is to provide readers, particularly chemists, with background information on the history and current status of pesticide resistance, its genetic basis, and definitions of terms relating to it. Succeeding chapters of this volume support an optimistic view that we are entering an era when most cases of resistance can be managed, provided awareness is followed by a sequence of properly focused fundamental research, communication, consensus building, and proactive implementation of practical measures based on the best scientific information available.

Resistance of agricultural pests to chemical control is now recognized as one of the most important challenges to sustaining worldwide production of food and fiber. The amount and quality of food produced today are unique in recorded history, arising from less land and labor than ever before. For the first time, man has the ability to eliminate hunger if only the barriers to effective distribution of food and agricultural technology can be overcome. However, at least for the foreseeable future, this abundance cannot be sustained, much less expanded, without the continuing contributions of agrochemicals. A similar situation exists in public

0097–6156/90/0421–0001$06.00/0
© 1990 American Chemical Society

health, where goals such as malaria eradication, once tantalizingly close, are threatened by insecticide resistance. Agrochemicals, and pest populations susceptible to them, are a critical national resource for developed and developing countries alike.

Objectives of the Current Publication

This book arose from the editors' conviction that resistance must be managed effectively, and that the progress currently being made is bringing this goal within reach in most cases. It brings together experts from the three major agrochemical areas -- insecticides, fungicides, and herbicides -- with the following objectives:

- to provide an update on the current status of resistance, recent advances in understanding and managing it, and needs for future progress;

- to integrate, in one convenient source, fundamental and applied research, a broad range of scientific disciplines, and a worldwide perspective; and

- to inform chemists and other scientists, and stimulate their participation in resistance research.

Authors of succeeding chapters have admirably accomplished the first objective, describing the latest progress across a broad front of fundamental studies to understand resistance and practical measures to combat it. It should be noted that two very important papers presented in the symposium -- Neuropharmacology and Molecular Genetics of Insect Nerve Insensitivity to Pyrethroids, and Genetic Systems for Engineering Plants for Resistance to Glyphosate -- are not included in the book because of publications that were simultaneously in press and have now been published (pyrethroids, 1,2; glyphosate, 3,4).

The second and third objectives deserve brief elaboration. Until now, most discussions of resistance have taken place in biologically oriented scientific societies, often with a strong disciplinary, national, or regional focus. This can insulate the various scientific disciplines from each other, so that lessons learned by one group may not be fully appreciated or applied by others.

It may also have tended to isolate chemists from resistance research, even though they have much to contribute. Chemists are an integral part of new compound discovery, they form a natural bridge across agronomic and biological lines, and they bring a unique scientific perspective to questions of enzyme inhibition, metabolism, and chemical transport in crops and pests. These are all germane to resistance research, yet most chemists concentrate solely on understanding and optimizing activity in sensitive species. And if they find an interesting new compound, they typically 'throw it over the wall' to biologists, who are expected to cope with resistance on their own. To the extent this attitude exists, it deprives resistance research of a potentially important component. We hope that this book can play a role in making chemists more active partners in combating resistance.

Overview Objectives

Each section of the volume has its own overview, written by scientists of international stature (insecticides, Georghiou; fungicides, Schwinn and Morton; herbicides, LeBaron and McFarland). Thus the objective of this chapter is to provide general information, applicable across all three sections.

The Literature of Resistance Research

Previous Literature. As the importance of resistance has become apparent, a steadily increasing amount of research has been done to understand and combat it. Excellent summaries of earlier work, with extensive references to the original literature, are available in recent books and review articles, some dealing with all areas of agrochemistry (5-8), and others addressing specific classes of pests (insects, 9-11; fungi, 12,13; weeds, 14). Information on resistance to rodenticides, antibiotics, and anticoccidial drugs, which are not covered in this volume, is available in two of the references (7,8).

Each of these works may be recommended for its scientific quality. However, one stands out as having refocused the subject, broadened its constituency, and helped crystallize information into action. In 1984, the energy building around resistance management was channeled into consensus by a National Academy of Sciences (NAS) symposium and workshop, the proceedings of which were published in 1986 (7). Most notably, this was the first meeting of its kind to integrate science and public policy across the whole agrochemical spectrum. What had been confined to the scientific and commercial realm was expanded to include governmental functions, ranging from registration to antitrust legislation. Aside from making this important connection, and emphasizing its potential for implementing a truly integrated response to resistance, the symposium and publication also brought resistance management to the attention of a broader audience, and provided new impetus for action. Areas needing further research and cooperation were identified, and the important point was made that resistance management should be a factor in regulatory decision making for new and existing agrochemicals. The NAS book remains invaluable reading today, and we hope the current volume, in which public policy issues are treated by several authors, will serve to update and expand upon it.

Current Awareness. Scanning the Literature Cited in subsequent chapters quickly identifies journals to be monitored in each field. However, the the volume of primary literature, reviews, and symposia challenges even the most diligent observer.

A promising new vehicle for current awareness is a biannual newsletter, published by Michigan State University in cooperation with the Western Regional Communication Committee on Pesticide

Resistance and Resistance Management (WRCC-60). This publication, which first appeared in January, 1989 (15), features concise reports on original research, distribution of resistant pests, symposia announcements and proceedings, legislative activity, reviews, funding opportunities, and other information on pesticide resistance, drawn from around the world. Subscription information may be obtained from Ms. Rosie Bickert, Pesticide Research Center, Michigan State University, East Lansing, MI 48824-1311, U.S.A.

Defining the Challenge

Scope. The extent of resistance was summarized succinctly by Georghiou in 1986: "Whereas the presence of resistance was a rare phenomenon during the early 1950s, it is the fully susceptible population that is rare in the 1980s". He goes on to cite reports of resistance to 447 species of insects and mites, 100 species of plant pathogens, and 48 species of weeds, world wide (16). In this volume, he raises the number for insects and mites to 504 species, noting that this is doubtless a lower limit, and LeBaron and McFarland raise the number of weed species to 78.

Consequences. The impact of resistance ranges from increased costs at the least, to catastrophic control failures and complete loss of valuable chemical tools at the worst.

In any cropping system, certain agrochemicals become the control method of choice because they offer improvements in cost effectiveness, worker safety, or environmental impact, relative to other available products. Simple logic indicates that their loss to resistance degrades one or more of these factors.

Safety and environmental issues can be relatively clear cut, as in the negative consequences of replacing pyrethroids with organophosphorus insecticides (see Riley, this volume). However, the economics of resistance can only be estimated, since "detailed estimates of crop losses, with and without pesticide use, are still surprisingly difficult to obtain" (17). Available data suggest that costs can increase dramatically. Pimentel et al. estimated added costs from insecticide resistance in the U.S. alone at $133 million in the late 1970s (18); estimates were also made of the economics of losing agrochemicals completely (19). Steiner documented an eventual 6-fold rise in the expense of German apple production as mites became resistant to a series of acaricides over time (20). Metcalf has summarized resistance-related costs for controlling mosquito vectors of malaria, showing that expenditures can increase up to 20-fold -- in countries least able to afford them -- with the end result of increased malaria incidence (21). Clearly these represent but a small fraction of resistance costs, and in the case of malaria control one sees that there can be a significant toll in human terms as well. Many more such analyses are needed to provide better inputs for decision making by growers, industry, and public policy makers.

One should also recognize that these are the 'better' cases, where alternative chemicals are still available. The potential consequences of resistance are perhaps best understood by considering instances where all agrochemicals have lost effectiveness. For example, resistance of the tobacco budworm to all available insecticides led to the virtual abandonment of cotton production in southern Texas and northeastern Mexico during the early 1970s, with devastating consequences to the local economies (22,23). Other cases where resistance has reached crisis proportions include Colorado potato beetle on Long Island, New York (9) and diamondback moth in Southeast Asia (24).

So far, the worst cases have been confined to insecticides, where the history of resistance has been longest, selection pressure greatest, and the modes of action available fewest. Fungicide and herbicide resistance can create short-term hardships -- for example, control failures of benomyl on apple scab in Michigan (25) and on grape gray mold in France and peanut leafspots in the Southeastern U.S. (26), localized distributions of triazine-resistant weed biotypes world wide (14; LeBaron and McFarland, Gressel, this volume) -- but so far serious crises have been held at bay by the availability of alternate compounds. That this may not persist, however, is illustrated by the recent discovery of multiple resistance to several classes of herbicides in biotypes of ryegrass in Australia (Powles et al., LeBaron and McFarland, this volume) and blackgrass in the United Kingdom (Kemp et al., this volume).

Prospects. Several conclusions can be drawn from the foregoing. First, resistance is steadily spreading, in terms of pest species, classes of agrochemicals, and geographic location. Second, costs, though difficult to estimate with precision, are undoubtedly great, and can extend beyond mere economics to real human suffering. Third, crisis situations are, fortunately, not yet the norm. In most instances, there is still time to plan and to act. Fourth, delay could be costly, and even disastrous. Crises can and do occur when all available control methods fail.

In a carefully reasoned look to the future (27), Conway concluded that "At the world level, there is little to support the assertion that ... there is a threat now or in the foreseeable future to global food supplies ... Of greater likelihood ... is the build up of a [localized] resistance problem to the point at which national and regional food production is threatened." Potential trouble spots were identified as wheat, corn, or soybean production in the U.S.; wheat or root crops in Europe; wheat or rice in India; and rice or vegetables in southeast Asia, China or Japan. Any one "could severely affect regional food production and have extensive repercussions on the global food markets." Success or failure depends upon "smoothness of substitution" -- are multiple control alternatives, chemical and non-chemical, available; if so, will they be used in an integrated fashion, and will enough growers cooperate?

In this connection, Conway does point to one "tenable global threat": agrochemical discovery and manufacture is increasingly concentrated in a few companies, and should one or more decide to abandon insecticide research in favor of "more profitable pesticides" the threat to food production would be serious (27).

The same could be said of fungicide research, which historically has been concentrated in even fewer companies than insecticide research.

Taken together, the scope, consequences, and future outlook for resistance are a clear call to action. Fortunately, succeeding chapters show that this call is being heeded, in creative and cooperative ways. Moreover, the robust agrochemical patent literature, which demonstrates continuing research directed toward new insecticides, acaricides and fungicides, and the growing commitment of industry to resistance management (see below), show no evidence of declining interest in these critical areas. Industrial consolidation has admittedly placed strategic decision making in fewer hands, but it has also generally been undertaken with the intent of establishing the broadest possible product lines. One can even argue that the resulting research organizations, having greater resources than their smaller predecessors, will be better able and more willing to retain all areas of agrochemical discovery, with the intent of maintaining a broad offering to their customers.

The Nature of Resistance

To create the means -- and indeed, the will -- to combat resistance, one must first understand it. Though the details of each specific situation are complex, the general nature of resistance can be understood in terms of genetics and natural selection.

A Natural Phenomenon. Each organism in the complex natural world around us is a survivor. Through evolutionary processes extending over millennia, every species of plant and animal has adapted to cope with the physical conditions and compete with the other organisms in its ecological niche, so that reproduction of sufficient individuals, and thus survival of the species, is assured. The attributes needed are supplied by the genetic diversity of large populations: provided even a small number of individuals with the required characteristics are present, they can survive and pass these traits on to progeny, which can in time become common in the population.

Agricultural pests are no exception. In fact, they are characterized by very large populations and frequent reproduction, and are often even more adaptable to adverse conditions than other organisms. From this perspective, chemicals used to control them should be seen as only the latest in a long history of threats to their survival. Therefore, it should come as no surprise that many pests possess means to resist the lethal effects of agrochemicals. Indeed, without claiming that resistance will occur in every case, it seems logical that the wisest approach to any pest-agrochemical interaction should not be to ask _whether_ resistance will develop, but _when, and in what form_. The same must be expected for genetic, biological, or even mechanical control measures.

A Manageable Challenge. At the same time, there are positive aspects in the realization that resistance is part of a larger, ongoing biological process. As a natural phenomenon, controlled by genetics and expressed in specific biochemical processes, resistance can be studied, understood, and combated. It is neither mysterious nor totally beyond our control. This is particularly true if strategies and tactics to avoid or delay resistance are implemented proactively, before resistant pests become serious problems; i.e. before the survival characteristics become common in the population.

Moreover, resistance is not something new and sinister, unleashed on the world by use of agrochemicals. It is a fact of nature that must be integrated into the discovery and development of new products and the use of existing ones.

Definitions

One might expect that terms relating to resistance would be standardized by now. However, they are still being debated and refined, as can be seen by considering the word 'resistance' itself.

'Resistance'. For many years this word meant different things to different people, and was used interchangeably with 'tolerance' and 'insensitivity'. In 1960, Crow emphasized the importance of genetics by defining resistance as "a genetic change in response to selection" (28). In 1971, the World Health Organization proposed a standardized definition for resistance to insecticides: "The development of an ability in a strain of insects to tolerate doses of toxicant which would prove lethal to the majority of individuals in a normal population of the same species" (29). Subsequently, the United Nations Food and Agriculture Organization (FAO) proposed a similar definition for fungicides and bactericides, adding the genetic qualifier that resistance be limited to "hereditable changes" (30). 'Tolerance' or 'insensitivity' were discouraged as ambiguous. Some confusion persists, however, since 'resistance' is also used by plant pathologists to denote the ability of plants to resist attack by pathogens. In weed science, multiple terms are still used; for example, in the definitions of Gressel: "Tolerance is defined as any decrease in susceptibility, compared with the wild type. Resistance is complete tolerance to agriculturally used levels of a herbicide" (31).

In general, it appears that variants of the FAO definition are now gaining ground. However, two somewhat interrelated problems remain.

First, what is the 'normal' reference population, against which decreases in sensitivity should be measured? Initially, 'normal' meant pest populations never subjected to toxicant pressure and maintained in the laboratory (32). However, there is now growing sentiment that meaningful results require using field-collected 'normal' populations with a 'normal' history of agrochemical exposure for the area in question; these may already be resistant to the toxicant in question, or may display cross resistance in the case of new compounds (33). The key difficulty here is proper monitoring to establish meaningful baseline data (34; see also Bull, Riley, Croft, Brent, Delp and Wade, this volume).

Second, how should laboratory and field studies of resistance be correlated? Laboratory studies necessarily involve small collections of individuals, whereas field resistance is dependent on the distribution and stability of resistance levels across a whole field population; moreover, 'discriminating doses' used in these studies may bear little relation to residues resulting from recommended field application rates. Thus laboratory studies can err in either direction, causing unwarranted concern in some cases or failing to indicate real risks in others (see Schwinn and Morton, this volume, for an example involving phenylamide fungicides). At the same time, field 'resistance' can sometimes be traced to poor application techniques or timing, or even to low-quality products (34). Thus researchers usually differentiate 'field' resistance from 'laboratory' or 'physiological' resistance, a distinction which we favor.

Recently, a definition that tries to take these factors into account has been offered by Sawicki: "Resistance marks a genetic change in response to selection by toxicants that may impair control in the field" (34). One might well add "at recommended application rates." In any event, the very length of the discussion above shows that more conformity is required on this simple but important issue.

'Cross Resistance'. This seemingly simple term is also subject to various interpretations. It is probably best defined as resistance of an organism to two or more toxicants (with the implication that the 'normal' population is sensitive to both). Recently the added qualifier 'positive' or 'positively correlated' cross resistance has been suggested, since there are instances where resistance to one toxicant is accompanied by _increased_ sensitivity to another, which are then termed 'negative' or 'negatively correlated' cross resistance. (For an example, see Fujimura, Ishii, this volume.) This distinction, though useful, seems cumbersome and possibly confusing as presently worded.

Moreover, there are subclasses of cross resistance, defined according to their genetic and mechanistic basis. Cross resistance _mediated by a single gene_ ('monogenic' or 'monofactorial') may be due either to a single defense mechanism operating against both toxicants (this is sometimes given as the sole definition of cross resistance), or to multiple mechanisms that may not act equally on different toxicants ('pleiotropic resistance'). The latter can sometimes be traced to a regulatory gene influencing several others (35).

Cross resistance _mediated by multiple genes_ ('polygenic' or 'polyfactorial') may, similarly, result from the same overall mechanism(s) acting against both toxicants, or from different mechanisms acting against each. The latter case is sometimes called 'multiple resistance', but this seems ambiguous. The broader definition of 'multiple resistance' or 'multiresistance' as "having a genotype [total genetic constitution] conferring resistance to a wide range of pesticide groups" (36) would seem preferable.

The reader who finds this confusing cannot be blamed. Communication inside and outside the community of resistance researchers would benefit from better standardization of terms.

Related Genetic Terms. Though an extensive discussion is beyond the scope of this chapter, it may be helpful to define a few terms commonly appearing in resistance literature (adapted from 36):

- Chromosomes. DNA polymers that carry genes in linear arrangement. Except for specialized cells formed for sexual reproduction, chromosomes occur as matched ('homologous') pairs in each cell of an organism.
- Gene. A short length of a chromosome influencing a particular set of the organism's characteristics. This set of characteristics is sometimes called a 'trait'.
- Alleles. Different forms of a gene (i.e., coding for variants of the same trait).
- Homozygous. Having identical alleles of a given gene in both homologous chromosomes. The individual is a 'homozygote'.
- Heterozygous. Having different alleles of a given gene in each chromosome of the homologous pair. The individual is a 'heterozygote'.
- Dominant Gene. An allele that partially or wholly overrides the effect of its paired allele in a heterozygote.
- Recessive Gene. An allele having no effect on the organism unless homozygous.

Historical Perspective

Exceptionally comprehensive, thoughtful coverage of the history of resistance may be found in a paper of Georghiou and Mellon ("Pesticide Resistance in Time and Space", 37) and in a more recent update by Georghiou (38). This section offers a much abbreviated summary, followed by speculations on why resistance developed as it did and observations on the response of industry.

The Stepwise Progression of Resistance. Resistance was first noted for insects, when Melander reported in 1914 that San Jose scale on apples in Washington State had become insensitive to lime-sulfur (39). This was greeted with some skepticism, and was followed by scattered reports, limited to insects, over the next 30 years. By 1946, resistance was documented for only 11 species (40).

The picture changed rapidly after introduction of the potent and persistent compound, DDT. DDT resistance appeared quickly, worldwide, and in several species, following the first report in 1947 (41). New insecticides were introduced steadily, providing better control of both resistant and susceptible insects, but most new products eventually suffered the same fate. The number of species showing resistance to one or more toxicant doubled about every six years between 1948 and 1983 (37).

Yet even as this happened, the prevailing opinion was that other agrochemicals would encounter resistance rarely, if at all. Indeed, the story for fungicides parallels that for insecticides remarkably. Though resistance was first reported for Penicillium spp. to biphenyl in 1940 by Farkas and Aman (42), 27 years later Georgopoulos and Zaracovitis noted that "The reported cases of tolerance to agricultural fungicides are very few, and the knowledge accumulated hardly justifies a review" (43). However, in the early 1970s new, highly active systemic compounds such as benomyl did for fungicides what DDT had done for insecticides. Resistance appeared in a matter of years, and some uses were lost that have not been recovered to this day (26). Some degree of resistance has now become the norm for all classes of fungicides, with the exception of a few protectant compounds.

The final stage -- herbicide resistance -- is now upon us. Herbicides were long thought safe from resistance because weeds lack the mobility and rapid reproduction of insects and fungi, and because a few susceptible individuals surviving treatment can produce large numbers of seeds that can infest the soil seed bank and germinate over many years. Field resistance of common groundsel to triazines in Washington State was first reported in 1970 (44), but resistance was slow to spread, and 13 years later Radosevich wrote that "... there are few examples of formerly susceptible weed species that have developed resistance" (45). This has changed dramatically in recent years, as is apparent throughout the herbicide chapters of this book. Of particular note are the surprising speed of resistance development to acetolactate synthase inhibitors (Mazur et al., Gressel, this volume), and the emergence of multiply resistant ryegrass and blackgrass biotypes noted earlier.

A Common Thread. Though each stage in the history of resistance took many by surprise, hindsight shows common characteristics. As long as a class of agrochemical was dominated by relatively inefficient, short residual compounds, selection pressure was low, and resistance was slow to develop. This was especially true for the earliest insecticides and fungicides, which interfered with multiple biochemical processes at the expense of selectivity, or had limited persistence (e.g., botanical insecticides). But as soon as compounds offering high effective kill, long residual, and selectivity through action at a single biochemical target site appeared, selection pressure was high, and resistance soon followed.

Ironically, the seeds of resistance were carried in the very properties that made many products seem more desirable -- in some cases, strikingly so -- than their predecessors. Moreover, the efficacy of such products often led to the unwise practice of relying on a single compound or class of compounds over one or more seasons, exacerbating the problem through increased selection pressure. We must learn and apply the basic principles of this history if we are to improve upon the poor performance outlined above, and this topic forms an important part of many of the chapters that follow. One point deserving reemphasis here is the importance of maintaining the maximum number of control options for a given crop-pest complex, developing new options, and using all options in an integrated manner to give each the maximum effective life. A particular corollary for chemists and other discovery scientists is the importance of compounds with new modes of action.

Industry Involvement. Agrochemical companies did not escape being caught off guard by the relentless progression of resistance, nor was their response either coordinated or uniformly effective at first. In a candid assessment of the early days of resistance management, Delp wrote that "Industry's response to fungicide resistance problems varies from denial to active participation in the solution" (46).

Happily, this stance has evolved rapidly to a coordinated, positive response. At first, the 'solution' to resistance was seen as new compounds, but the increasing expense of discovery and the accelerating pace of resistance development soon led to the realization that good product stewardship must include resistance management. It was also clear that, because of the various forms of cross resistance, no one company could act in isolation.

The result was industry cooperation in wide variety of programs, including governmental intiatives such as the Australian Wheat Board Working Party on Grain Protectants in the early 1970s (47), inter-company working groups such as the Pyrethroid Efficacy Group (PEG, 1979), and scientific society-based groups such as the Herbicide Resistant Weeds Committee of the Weed Science Society of America (1986). Meanwhile, a broad framework for inter-company cooperation was provided within the International Group of National Associations of Agrochemical Manufacturers (generally known as GIFAP, the acronym for the French version of its name). A Fungicide Resistance Action Committee (FRAC) was formed in 1981, followed by similar groups for insecticides (IRAC, 1984) and herbicides (HRAC, 1989).

Despite constraints of commercial rivalry, antitrust laws, and intra-company differences between research and marketing staffs (47), these groups have taken a leadership role in their respective areas and have tried to foster both inter-company cooperation and communication of industry with universities and government agencies (Riley, Delp and Wade, this volume; LeBaron, H. M., Ciba-Geigy, 1989, personal communication). Progress has been slow but steady, and these efforts bode well for the future.

Conclusions

Reactions to resistance have included surprise, denial, panic, and resignation. The brief history presented above is not a bright one: in some cases we have been slow to respond to the challenge, and in others we have failed to apply lessons learned in one area to actual or potential resistance problems in another. Today, none of this is warranted.

On the one hand, resistance is clearly a serious challenge. It is a natural phenomenon that will not go away, and its consequences can be severe for world agricultural markets, public health, and worldwide standards of living.

On the other hand, resistance can be understood and combated. A discussion of the technologies available to do this is beyond the scope of this chapter, but looking forward to succeeding chapters one sees that the science we need is either in hand, or within our grasp if research is properly focused and funded. Many gaps remain to be filled, but the foundations have been laid and the future needs have been clarified.

What are needed most -- and what have often been most difficult to achieve -- are communication and cooperation. These two factors are interrelated, and essential at both fundamental and practical levels.

For existing resistance problems, no individual grower, researcher, or company can design and implement effective solutions in isolation. The pests are too mobile and resilient, and the chemical tools available are too few in number and too limited in modes of action for anything but a coordinated effort to succeed. We must get the most from what we have, basing practical measures on the best knowledge available. And we must look beyond short-term profits or other considerations of narrow self interest, to the long-term viability of crop protection chemicals and the industry that provides them.

For the future, we need to stimulate and focus basic research, making sure that all disciplines are brought to bear in a coordinated way. We must build understanding in growers, whose cooperation is essential for practical success, and who may have to settle for less than perfect pest control to realize the long-term benefit of preserving agrochemical efficacy. The same is true for public policy makers, who need to understand the real value of agrochemicals, the importance of maintaining the maximum number of pest control options, and the importance of resistance management in benefit analysis. Finally, it will be critical to incorporate anti-resistance strategies both in new compound discovery and in the earliest stages of new product development and introduction.

Fortunately, there are clear signs that communication can lead to cooperation and well-targeted research, and that these in turn can lead to successful resistance management. In the chapters ahead, examples include proactive approaches to acaricide resistance (Croft), government-industry cooperation in pyrethroid resistance management (Bull, Riley), genetic engineering to provide new herbicide options in a variety of crops (Mazur et al., Newhouse et al.; see also 3,4) and new insecticides targeting resistant populations (Gelernter), successful management of phenylamide fungicide resistance (Schwinn and Morton), the development of short-residual compounds to manage sulfonylurea herbicide resistance (Mazur et al.), and the first practical demonstration of the utility of negatively correlated cross resistance (Fujimura, Ishii). Other chapters describe basic research and cooperative initiatives that will form the basis for future successes.

Acknowledgments

The initiator of the symposium on which this book is based was Dr. Maurice B. Green. Its successful implementation and publication are due in large part to his efforts and those of Dr. Homer M. LeBaron, both of whom also made important contributions to this chapter. In addition, Dr. Thomas M. Brown provided valuable advice on both topics and participants for the symposium. I also acknowledge with gratitude the debt this chapter owes to the earlier reviews cited above (5-14), and to the outstanding contributions that form the rest of this volume. It is a pleasure to express my respect to their authors and editors for making the vast resistance literature accessible to the scientific community. Finally, thanks are due to Ms. Cheryl Shanks and Ms. Robin Giroux of ACS Books for support throughout the publication process, and to Ms. Florine Clarke for preparation of this manuscript.

Literature Cited

1. Soderlund, D. M.; Knipple, D. C. In Neurotox '88: Molecular Basis of Drug and Pesticide Action; Lunt, G. G., Ed.; Elsevier: Amsterdam, 1988, pp 553-61.
2. Soderlund, D. M.; Bloomquist, J. R.; Wong, F.; Payne, L. L.; Knipple, D. C. Pestic. Sci. 1989, 26, 359-74.
3. Kishore, G. M.; Shah, D. M. Ann. Rev. Biochem. 1988, 57, 627-63.
4. Padgette, S.; della-Cioppa, G.; Shaw, D. M.; Fraley, R. T.; Kishore, G. M. In Cell Culture and Somatic Cell Genetics of Plants; Constabel, F.; Vasil, I. K., Eds.; Academic: New York, 1989; Vol. 6, pp 441-76.
5. Pesticide Resistance and World Food Production; Conway, G. R., Ed.; Imperial College Centre for Environmental Toxicology: London, 1982.
6. Pest Resistance to Pesticides; Georghiou, G. P.; Saito, T., Eds.; Plenum Press: New York, 1983.
7. Pesticide Resistance: Strategies and Tactics for Management; Glass, E. H., Ed.; National Academy Press: Washington, DC, 1986.
8. Combating Resistance to Xenobiotics: Biological and Chemical Approaches; Ford, M. G.; Holloman, D. W.; Khambay, B. P. S.; Sawicki, R. M., Eds.; Ellis Horwood: Chichester, England, 1987.
9. Forgash, A. J. Pestic. Biochem. Physiol. 1984, 22, 178.
10. Brattsten, L. B.; Holyoke, C. W., Jr.; Leeper, J. R.; Raffa, K. F. Science 1986, 231, 1255.
11. Rutgers Entomology Centennial Symposium, published as a dedicated issue in Pestic. Sci. 1989, 26, 329-441.
12. Fungicide Resistance in Crop Protection; Dekker, J.; Georgopoulos, S. G., Eds.; Centre for Agricultural Publishing and Documentation: Wageningen, The Netherlands, 1982.
13. Fungicide Resistance in North America; Delp, C. J., Ed.; APS Press: St. Paul, 1988.

14. Herbicide Resistance in Plants; LeBaron, H. M.; Gressel, J., Eds.; Wiley: New York, 1982.
15. Pesticide Resistance Management; Whalon, M. E.; Hollingworth, R. M., Eds.; Michigan State University Pesticide Research Center: East Lansing, 1989; Vol. 1, No. 1, 24 pp.
16. Georghiou, G. P. In Pest Resistance to Pesticides; Georghiou, G. P.; Saito, T., Eds.; Plenum Press: New York, 1983; p 14.
17. Craig, I.; Conway, G. R.; Norton, G. A. In Pesticide Resistance and World Food Production; Conway, G. R., Ed.; Imperial College Centre for Environmental Toxicology: London, 1982; p 57.
18. Pimentel, D.; Andow, D.; Gallahan, D.; Schreiner, I.; Thompson, T.; Dyson-Hudson; R.; Jacobson, S. Irish, M.; Kroop, S.; Moss, A.; Shepard, M.; Vinzant, B. In Pest Control: Cultural and Environmental Aspects; Pimentel, D.; Perkind, J. J., Eds.; Westview: Boulder, 1979; pp 99-158.
19. Pimentel, D.; Krummel, J.; Gallahan, D.; Hough, J.; Merrill, A.; Schreiner, I.; Vittum, P.; Koziol, F.; Back, E.; Yen, D.; Fiance, S. Bioscience 1978, 28, 772-84.
20. Steiner, H. OEPP/EPPO Bulletin 1973, 3, 27-36.
21. Metcalf, R. L. In Pest Resistance to Pesticides; Georghiou, G. P.; Saito, T., Eds.; Plenum Press: New York, 1983; pp 712-8.
22. Craig, I.; Conway, G. R.; Norton, G. A. In Pesticide Resistance and World Food Production; Conway, G. R., Ed.; Imperial College Centre for Environmental Toxicology: London, 1982; pp 54-5.
23. Adkisson, P. L. In Agricultural Chemicals - Harmony or Discord for Food, People, and the Environment; Swift, J. E., Ed.; University of California Division of Agricultural Sciences: Sacramento, 1971; pp 43-51.
24. Conway, G. R. In Pesticide Resistance and World Food Production; Conway, G. R., Ed.; Imperial College Centre for Environmental Toxicology: London, 1982; pp 80-1.
25. Dover, M. J.; Croft, B. A. Getting Tough: Public Policy and the Management of Pesticide Resistance; World Resources Institute: Washington, DC, 1984; p 8.
26. Smith, C. M. In Fungicide Resistance in North America; Delp, C. J., Ed.; APS Press: St. Paul, 1988; pp 23-4.
27. Conway, G. R. In Pesticide Resistance and World Food Production; Conway, G. R., Ed.; Imperial College Centre for Environmental Toxicology: London, 1982; pp 77-90.
28. Crow, J. F. Miscellaneous Publication of the Entomological Society of America 1960, 2, pp 69-74.
29. Brown, A. W. A.; Pal, K. In World Health Organization Monograph Series; World Health Organization: Geneva, 1971; No. 38, p 491.
30. FAO Panel of Experts on Pest Resistance to Pesticides, Report of Meeting 28 August-1 September, 1978; as summarized by Dekker, J. In Fungicide Resistance in Crop Protection; Dekker, J.; Georgopoulos, S. G., Eds.; Centre for Agricultural Publishing and Documentation: Wageningen, The Netherlands, 1982; p 2.

31. Gressel, J. In *Pesticide Resistance: Strategies and Tactics for Management*; Glass, E. H., Ed.; National Academy Press: Washington, DC, 1986; p 55.
32. Muggleton, J. *Proc. Brit. Crop Prot. Conf. - Pests and Diseases*, 1984, p 585.
33. Brattsten, L. B. *Pestic. Sci.* 1989, 26, 329.
34. Sawicki, R. M. In *Combating Resistance to Xenobiotics: Biological and Chemical Approaches*; Ford, M. G.; Holloman, D. W.; Khambay, B. P. S.; Sawicki, R. M., Eds.; Ellis Horwood: Chichester, England, 1987.; pp 105-17.
35. Plapp, F. W. In *Pesticide Resistance: Strategies and Tactics for Management*; Glass, E. H., Ed.; National Academy Press: Washington, DC, 1986; pp 74-86.
36. Glossary of Genetic Terms In *Pesticide Resistance and World Food Production*; Conway, G. R., Ed.; Imperial College Centre for Environmental Toxicology: London, 1982; pp 127-33.
37. Georghiou, G. P.; Mellon, R. B. In *Pest Resistance to Pesticides*; Georghiou, G. P.; Saito, T., Eds.; Plenum Press: New York, 1983; pp 1-46.
38. Georghiou, G. P. In *Pesticide Resistance: Strategies and Tactics for Management*; Glass, E. H., Ed.; National Academy Press: Washington, DC, 1986; pp 14-43.
39. Melander, A. L. *J. Econ. Entomol.* 1914, 7, 167.
40. Metcalf, R. L. *Pestic. Sci.* 1989, 26, 334.
41. Weismann, R. *Mitt. Schweiz. Entomol. Ges.* 1947, 20, 484.
42. Farkas, A.; Aman, J. *Palest. J. Bot. Jerusalem* 1940, Ser. 2, 38.
43. Georgopoulos, S. G.; Zaracovitis, C. *Ann. Rev. Phytopathol.* 1967, 5, 109.
44. Ryan, G. E. *Weed Sci.* 1970, 18, 614.
45. Radosevich, S. R. In *Pest Resistance to Pesticides*; Georghiou, G. P.; Saito, T., Eds.; Plenum Press: New York, 1983; p 453.
46. Delp, C. J. *Crop Prot.* 1984, 3, 2-8.
47. Davies, R. A. H. *Proc. Brit. Crop Prot. Conf. - Pests and Diseases*, 1984, pp 595-7.

RECEIVED November 21, 1989

INSECTICIDES

Chapter 2

Overview of Insecticide Resistance

George P. Georghiou

Department of Entomology, University of California, Riverside, CA 92521

>The incidence of insecticide resistance continues to increase, having been recorded in at least 504 species of insects and mites, a 13% increase since 1984. In a number of species, the presence of multiple resistance has created a crisis situation due to lack of alternative, effective and affordable insecticides. Most of these cases concern pests in high intensity cropping systems or vectors of human disease that are subjected to organized control programs. Notable progress has been achieved in research on the biochemistry, molecular genetics and dynamics of resistance. These advances are discussed as they provide improved opportunities for designing control strategies aimed at forestalling or delaying the evolution of resistance.

The doubtful distinction that insects have earned as the first pests to have developed resistance to a pesticide has continued to keep them in the spotlight as resistance has progressively expanded to all types of commercialized insecticides. The questions being asked today on how to cope with resistance are no less challenging than those asked when resistance first appeared. It is clear, however, that the huge amount of research conducted during the ensuing three quarters of a century on the mechanisms, genetics, and dynamics of resistance is an invaluable asset as we begin to move from passive counteraction of resistance to preventive resistance management.
Resistance to insecticides has been discussed recently in two comprehensive symposia volumes ($\underline{1},\underline{2}$) and in several reviews ($\underline{3-13}$) dealing with the biochemistry, dynamics, management, or economics of the problem. My

brief overview cannot include the multitude of topics that qualify for discussion. Many of these will be covered in other papers of this symposium. I shall limit myself to developments that I feel are of special significance as we contemplate future directions. I shall be covering
- A status report on the incidence of resistance with emphasis on "critical" cases;
- Recent developments in resistance to pyrethroids;
- The prospects for resistance to biopesticides;
- Progress in research on resistance mechanisms with emphasis on gene amplification; and
- Diagnostic tests for resistance monitoring.

Status of Resistance

In view of the evolutionary nature of resistance, additional cases are bound to arise as new insecticides are being introduced and selection pressure remains high. Figure 1, based on a data bank that we are maintaining for the Food and Agriculture Organization of the United Nations (FAO), indicates the progressive increase in the number of species of insects and mites world-wide, that have developed resistance to one or more insecticides. The total of 504 species is probably an understatement of the problem since many cases of pest control failure undoubtedly remain uninvestigated or unreported.

When the various cases of resistance are grouped according to the decade of first occurrence, it is evident that the number of new resistant species recorded in the present decade is smaller than those recorded in the previous decade, i.e. 90 vs. 190 species (Table I). This is because many of the recent reports concern expansion of the resistance spectrum in species which have previously been reported as resistant to other compounds. It would thus appear that most of the economically important species have by now developed resistance to at least one insecticide.

Table I. Chronological Increase in Number of Insects and Mites with Documented Cases of Resistance to Pesticides

Period	Number of Species	
	New in Decade	Total
1908	1	1
1909-1918	2	3
1919-1928	2	5
1929-1938	2	7
1939-1948	7	14
1949-1957	62	76
1958-1967	148	224
1968-1978	190	414
1979-1988	90	504

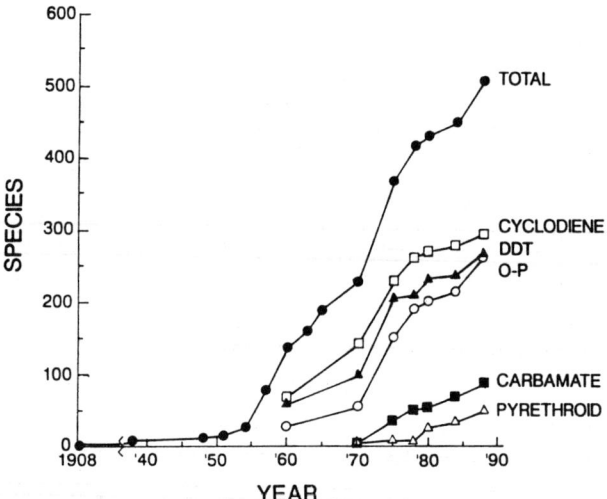

Figure 1. Chronological increase in the number of insect and mite species resistant to at least one type of insecticide (total), and species resistant to each of the 5 principal classes of insecticides.

Cyclodiene insecticide resistance is found in more species than any other type of resistance: 291 species are able to resist this class of insecticide, compared to smaller numbers for DDT, OPs, carbamates or pyrethroids (Table II). The preponderance of resistance to the cyclodienes is almost certainly due to the long persistence of these chemicals in the environment.

Table II. The Occurrence of Resistance in Insects and Mites According to Pesticide Chemical Group

Chemical Group	Number of Species	Percent of Total
Cyclodiene	291	57.7
DDT	263	52.2
Organophosphate	260	51.6
Carbamate	85	16.9
Pyrethroid	48	9.5
Fumigant	12	2.4
Other	40	7.9

The majority of resistant species (56.1%) are of agricultural importance, but a substantial proportion (39.3%) represents insects of medical importance, especially mosquitoes and flies. Only 4.6% are beneficial species (predators, parasites, or pollinators) (Table III). For a detailed index of cases of resistance see Georghiou, G. P.; Lagunes, A., The Occurrence of Resistance to Pesticides in Arthropods, 2nd ed., F.A.O.: Rome, in press.

Table III. Resistant Insect and Mite Species According to Economic Importance

Economic Importance	Number of Species	Percent
Agricultural	283	56.1
Medical/Veterinary	198	39.3
Beneficial	23	4.6

Critical Cases of Resistance

There are a number of especially critical cases of resistance in which nearly all of the affordable, previously effective insecticides have been depleted. Most of these cases concern high intensity cropping systems, or vectors of disease that are subjected to organized control programs. A partial list of such species follows:

    Diamondback moth   *(Plutella xylostella)*
    Whitefly   *(Bemisia tabaci)*
    Green peach aphid   *(Myzus persicae)*
    Leafminer   *(Liriomyza trifolii)*
    Budworms   *(Heliothis virescens)*
                                     (H. armigera)

Colorado potato beetle	*(Leptinotarsa decemlineata)*
Twospotted spider mite	*(Tetranychus urticae)*
European red mite	*(Panonychus ulmi)*
Malaria mosquitoes	*(Anopheles* spp.)
Housefly	*(Musca domestica)*
German cockroach	*(Blattella germanica)*
Blackfly	*(Simulium damnosum s.l.)*

Diamondback Moth. A major pest of cruciferous crops, the diamondback moth has demonstrated one of the highest propensities for development of resistance, especially in southeast Asia, and the Far East (13-15). In Thailand, where nearly every available insecticide has been used and eventually failed, this insect is highly resistant to a variety of organophosphates (OP), pyrethroids, and chitin inhibitors, but continues to be susceptible to *Bacillus thuringiensis* (BT) and to Avermectin (16). Cases of resistance in this species have also been noted recently in the U.S. and Central America, involving OP, pyrethroid and carbamate insecticides (Shelton, A. M., N. Y. State Agr. Exp. Station, Geneva, personal communication, 1988). Several mechanisms of resistance have been identified in this species, including carboxylesterase, general esterases, glutathione-*S*-transferase, insensitive acetylcholinesterase (AChE) (14-17) and kdr (pyrethroid target site insensitivity) (14).

Whitefly. Pests of high value crops in greenhouses or in the field, are often subjected to strong selection pressure by a variety of insecticides, resulting in high levels of broad spectrum resistance. The whitefly, *Bemisia tabaci*, is a classic example of such resistance, which at times is accompanied by massive population explosions.

As described for the Sudan by Dittrich et al. (18), this species became a primary pest of cotton in the late 1970's, superseding the bollworm *Heliothis armigera*. Subsequent applications of DDT/dimethoate to control whitefly, bollworm and jassids, resulted in failure of control by dimethoate, monocrotophos and other OPs by 1980/81 and a "tremendous flare-up" of whitefly populations. More recent studies have shown higher resistance to several OPs, certain pyrethroids, as well as to endosulfan and methamidophos (Dittrich, V., Ciba-Geigy Ltd., Basel, unpublished data, 1988). Similar patterns are evident in Turkey and especially in Guatemala, where extremely high levels of resistance (300x - 2000x) toward pyrethroids (deltamethrin, cypermethrin, fenpropathrin, biphenthrin, cyhalothrin) have been observed (Dittrich, V., unpublished data, 1988).

Aphids. Resistance in various species of aphids, especially those transmitting plant virus diseases, remains a serious problem. There are more cases of resistance reported in the green peach aphid, *Myzus persicae*, than in any other species of agricultural importance. Such resistance has been reported from 31 countries involving a total of at least 69 different insecticides, representing organochlorines, OPs, carbamates, and pyrethroids (Georghiou, G. P.; Lagunes, A., F.A.O.: Rome, in press).

Resistance in this species has been the subject of long-standing, productive investigations at Rothamsted, U.K. Resistance is due to a carboxylesterase (E4) which confers broad spectrum resistance involving not only OPs but also carbamates and certain pyrethroids ([19]). Three classes of resistant aphids have been identified (R_1, R_2, R_3) based on the intensity of total esterase activity. Earlier surveys had shown aphids with a moderate (R_1) level of resistance to be widespread in the U.K. ([20]), but those with higher resistance (R_2) to be localized in northern England, Scotland, and less commonly in East Anglia, while those with the highest level (R_3) were associated with greenhouse crops ([20-24]). More recent surveys, based on immunoassay of E4 ([25]), rather than total esterase activity, indicate an increased frequency of very resistant R_2 and R_3 aphids in untreated fields in southern and eastern England ([26]). Field treatments also revealed that all three major classes of aphicides (OP, carbamate and pyrethroid) select for highly resistant individuals, eliminating those with susceptible levels of esterase activity ([27], [28]).

Leafminers. A relatively recent case of resistance with serious implications for the floricultural and vegetable crops industries concerns the leafminer, *Liriomyza trifolii*. Over the past ten years there has been a dramatic rise in the importance of *Liriomyza* species as major pests of numerous ornamental and agricultural crops ([29]) and a world-wide extension of the geographic distribution of some of these insects, mainly through infested plant material.

L. trifolii is believed to have been introduced into California from Florida during the late 1970s on chrysanthemum cuttings and celery transplants. It is now recognized as the primary pest of chrysanthemum and a major pest of celery and tomato ([30]). During the past 10 years, the average effective field life of an insecticide used against this species in Florida has been less than three years ([31]). In the confined environment of greenhouses, in which chrysanthemum crops are primarily grown in California, the species has developed resistance to pyrethroids within two years ([32]). This is perhaps not surprising in view of the extremely high frequency of insecticide applications made to ensure that ornamental plants and cut flowers are free of insect blemishes, as demanded by the market. In a some southern California greenhouse operations, insecticides are applied with a frequency of one or two per week throughout the year (31). This case clearly illustrates the high resistance potential of pests in greenhouse environments, as well as the risks for the spread of resistance with the distribution of infested plant material.

Mosquitoes. Insecticide resistance in mosquitoes continues to be of concern, especially as it involves disease vector species of *Anopheles, Aedes* and *Culex*. Resistance continues to impede the effectiveness of well-established malaria control programs, such as those of Pakistan, India, Mexico, and Central America, and the initiation of large scale control programs in poorer countries, as in Africa. According to data of the World Health Organization, insecticide resistance in *Anopheles* is found in 50 species, 14 of which show multiple resistance to 3 or 4 chemical groups. At least 11 of the 50 resistant species are important vectors of malaria (33). The quantities and succession of insecticides used in the malaria control program of Pakistan, the chronology of first evidence of resistance in the vector *Anopheles stephensi* to each of the insecticides employed, and the changes in the incidence of malaria are illustrated in Figure 2. There is strong evidence from a number of areas, especially in Central America, Turkey, the Sudan, India and elsewhere, that heavy use of insecticides on certain crops, such as cotton, aggravates the mosquito resistance problem by imposing additional selection pressure on mosquitoes that breed in the same environment (34-36).

Colorado Potato Beetle. Resistance to insecticides in the Colorado potato beetle, *Leptinotarsa decemlineata*, in the intensive potato growing region of Long Island, N.Y., has attracted considerable attention because of its economic importance, broad spectrum, and rapidity with which it has evolved (7, 37). Following the development of resistance to aldicarb (7) and the banning of this insecticide due to the appearance of residues in underground water, growers resorted to repeated applications of fenvalerate in combination with the synergist piperonyl butoxide (p.b.), the botanical insecticide rotenone + p.b., and combinations of oxamyl and endosulfan. The cost of insecticides alone was estimated at $300 to $700/ha (38). Resistance has recently led to the replacement of these insecticides by the biopesticide BT subsp. *tenebrionis* and the inorganic insecticide cryolite (aluminum trisodium fluoride). Rotenone and cryolite are pre-DDT insecticides that were abandoned following the introduction of the more effective, synthetic organic insecticides. Neither BT nor cryolite are especially effective against later-instar larvae or adults of this species.

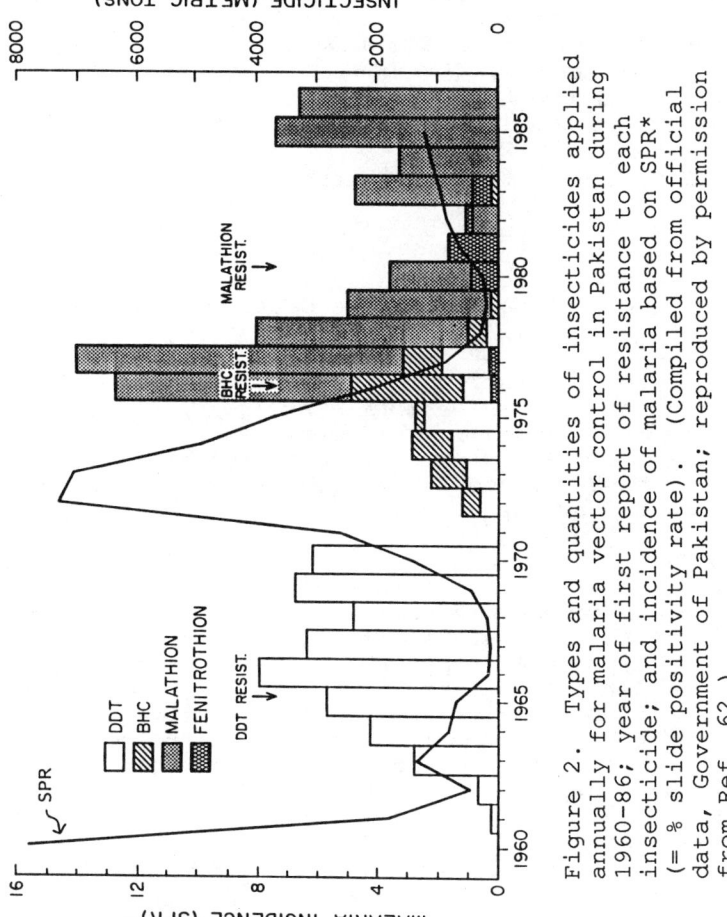

Figure 2. Types and quantities of insecticides applied annually for malaria vector control in Pakistan during 1960-86; year of first report of resistance to each insecticide; and incidence of malaria based on SPR* (= % slide positivity rate). (Compiled from official data, Government of Pakistan; reproduced by permission from Ref. 62.)

The Long Island case is a classic illustration of many of the factors known to enhance the development of resistance, including intensive commercial production of the insect's principal hosts (potato, tomato, eggplant), a very small number of wild hosts (*Solanum rostratum, S. dulcamara*) offering only minimal opportunities for avoidance of selection on untreated refugia, and a relatively isolated island location.

Organophosphate, carbamate, and in certain cases pyrethroid resistance have been reported more recently from a number of northeastern U.S. locations (7, 37, 39) and eastern Canada (40, 41) as well as from Michigan and Wisconsin (Hollingworth, R. M., Michigan State University, East Lansing, personal communication, 1988).

Developments in Resistance to Pyrethroids

In view of the importance attached to the synthetic pyrethroids as "end of the line" conventional insecticides, some recent significant developments in resistance to this group are considered separately in this section.

Pests of Cotton. In several countries in which principal pests of cotton such as *Heliothis armigera, H. virescens* and *Spodoptera littoralis* have developed resistance to many of the chemicals previously used for control, pyrethroids have assumed major importance. In the U.S., the annual cost of insecticides on cotton is estimated at $200 million, of which 46% is spent on pyrethroids (42). The importance attached to the preservation of pyrethroid efficacy is apparent in the introduction of pyrethroid resistance management programs in a number of countries. The first of these, a state-directed program, initiated in Egypt in 1978, restricts pyrethroid use to a single application per year against *S. littoralis*, on cotton only, and includes various supplementary, integrated pest management (IPM) measures against this and other pests. This strategy apparently continues to function successfully (43).

An Australian program (44) was initiated in 1983 after the discovery of emerging pyrethroid resistance in *H. armigera* at Emerald, Central Queensland (45). The program limits the use of pyrethroids to 3 applications over a 42-day mid-season period, limits endosulfan to early and mid-season sprays only, and includes additional IPM measures. Grower participation, though voluntary, appears to be excellent (43). This strategy, which is described as "a delicate balance between pyrethroid resistance and endosulfan resistance", has been running successfully for 5 crop seasons (8, 46). It is feared, however, that without new integrated technologies, the strategy will probably fail, due to contamination of the refugia to the extent that they will no longer be useful as a source of dilution by susceptible individuals (8, 46).

In U.S. cotton growing areas, control problems with pyrethroids against *H. virescens* were first detected in California in 1979 (47,48) and subsequently in Texas (49). Surveys conducted in 1986 using a vial bioassay test on adults, revealed a significant decline in susceptibility to pyrethroids in the Brazos River Valley and the Uvalde area of Texas, and lower levels of decline in portions of Mississippi, Arkansas and Louisiana (49-51). Since 1986, because of concern over emerging resistance, a pyrethroid resistance management program has gradually evolved, which recommends avoiding early and late season use of pyrethroids (50). As in Australia the program is voluntary, but less vigorous in detail, and at present lacks a central resistance monitoring organization. Grower compliance appears to be low but improving (52).

The consequences of lack of grower compliance are nowhere as evident as in Thailand where, despite official recommendations for moderation in the use of pyrethroids, a large number of treatments are applied annually on cotton. It has been estimated that at least 34% of the total production cost is spent on insecticides for bollworm (*H. armigera*) control. Pyrethroids were introduced for bollworm control in 1976 and rapidly became popular, but evidence of resistance was detected as early as 1979 (53). By 1986, broad spectrum pyrethroid resistance involving fenvalerate, cypermethrin, cyfluthrin and deltamethrin, occurred widely (Wangboonkong, S., Thailand Department of Agriculture, Bangkok, personal communication, 1987). Available data on control efficacy reveal a textbook-like pattern of decline of effectiveness of pyrethroids as dosages were increased in apparently unsuccessful attempts to cope with resistance (Figure 3).

Cockroaches. Insecticide resistance in cockroaches, though a long-standing problem, has assumed greater prominence in recent years because of increases in the numbers of eating establishments and single-occupant apartments, which provide greater opportunities for increases in cockroach populations. Many of these sites are under routine, commercial treatment by insecticides, which enhances the risk for resistance, especially in the fast-breeding German cockroach, *Blattella germanica*. A countrywide survey in the U.S. during 1981-87 involving 45 collections, and using a minimum LT50 knockdown threshold as criterion for resistance, has revealed widespread resistance to the carbamates propoxur and bendiocarb in 50% and 75% of the strains, respectively, and resistance to the organophosphates chlorpyrifos, diazinon and malathion in 7%, 13% and 80% of the strains, respectively (54). Also of significance is the evidence for resistance to pyrethroids. It was shown as early as 1956 that some strains of *B. germanica* were resistant to natural

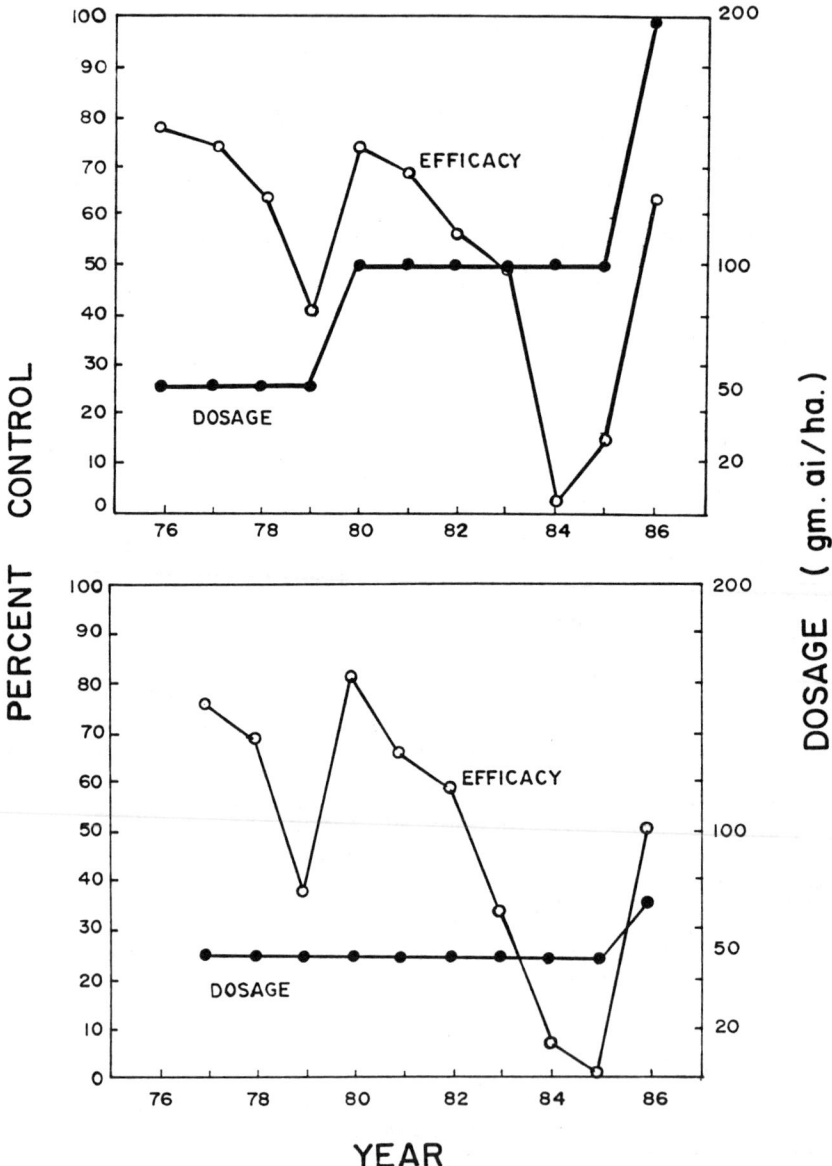

Figure 3. Decline in fenvalerate (top) and cypermethrin (below) efficacy against *Heliothis armiger* on cotton, and dosages of insecticide applied, 1976-86. (Compiled from data provided by Wangboonkong, S., Thailand Department of Agriculture, Bangkok, 1987.)

pyrethrins (55, 56). The present survey has shown resistance to pyrethrins in 51% of the collections, to allethrin in 18%, and varying levels and frequencies of resistance to pyrethroids (e.g. phenothrin in 13% of the strains, permethrin 9%, fenvalerate 9%, and cyfluthrin 2%).

Although the mechanisms of resistance to pyrethroids in these strains has not been investigated, it is important to note that in certain strains high resistance to pyrethrins (>100x) and allethrin (>190x) includes only minimal resistance to pyrethroids (permethrin (1.9x) and phenothrin (2.5x). High resistance to permethrin (>120x) and phenothrin (>140x), however, also involves low to moderate resistance to other pyrethroids (fenvalerate 7.1x, cyfluthrin 5.4x) (54). These variations in resistance spectrum would suggest that metabolic, as well as non-metabolic (kdr) mechanisms have been selected. The involvement of kdr resistance is more clearly evident in a strain of *B. germanica* from Osaka, Japan, that has been under field selection by permethrin. The strain has shown an extensive spectrum of cross resistance involving all seven pyrethroids tested and has yielded electro-physiological evidence of nerve insensitivity to the action of pyrethroids (57).

The U.S. survey provides empirical support of the importance of population isolation as a factor enhancing the effectiveness of selection for resistance: pyrethroid resistance was evident mainly in populations obtained from naval vessels and a mess hall that were under intensive pyrethroid treatment (Cochran, D.G., Virginia Polytechnic Inst. and State Univ., Blacksburg, personal communication, 1988).

Horn fly. The rapidity of toxic action of pyrethroids, complemented by a degree of repellency, has led to the introduction of pyrethroid-impregnated ear tags for control of horn flies (*Haematobia irritans*) on cattle. It was hoped that such topical use of pyrethroids would hinder development of resistance by allowing some insects to remain unexposed. However, the widespread use of this simple method, and the season-long effectiveness of the ear tags (58), equivalent to 10-14 generations (59), led to selection of resistant populations within the relatively brief period of two years, first in Florida in 1982, and in 14 other states by 1985 (59).

The wide spectrum of pyrethroid resistance in these populations, which involves permethrin, fenvalerate, flucythrinate, cypermethrin, deltamethrin, and cyhalothrin, the relative lack of synergism by p. b. (60) or DEF, and the presence of DDT as well as methoxychlor resistance (59), suggest that this resistance is due to the site insensitivity mechanism *kdr*. Interestingly, some evidence of behavioral resistance was also detected. It was observed that pyrethroid-resistant flies tended to frequent the untreated ventral and posterior region of tagged cattle, a behavior that tended to persist even in the absence of the insecticide stimulus (61).

Developments in Resistance to Biopesticides

Biopesticides, as represented by insecticidally active toxins produced by various bacteria, e.g. BT and other organisms, have been in commercial use for nearly two decades; however, no information is available concerning the mechanisms of potential resistance in target populations. The available reports reveal no consistent pattern of response to selection and may suggest that several mechanisms, some of them unique, may determine the risk for resistance in any insect-species/Bt-strain combination (9). Laboratory selections of house flies and *Drosophila melanogaster* by BT β-exotoxin produced resistance slowly and of low levels (<14x), while selection of the mosquito *Culex quinquefasciatus* by BT subsp. *israelensis* (BTI) δ-endotoxin produced 16.5x resistance within 46 generations (62). The results obtained with *Culex* are consistent with the continued absence of operationally limiting resistance in these mosquitoes under field conditions. Against Lepidoptera, selections in the laboratory with BT δ-endotoxin for less than 10 generations had no effect on susceptibility in *Anagasta kuehniella*, *Plutella xylostella* or *Spodoptera littoralis* (9), but longer-term selection of *Cadra cautella* (24 generations) produced ca.7x resistance (63). These are relatively low levels compared to resistance that usually develops toward conventional insecticides (Figure 4).

While these results are encouraging, recent studies on the Indian mealmoth, *Plodia interpunctella*, and tobacco budworm, *H. virescens*, discussed below, have shown that under certain conditions the risk for resistance remains high.

Laboratory selection of a field strain of the Indianmeal moth by BT, rapidly produced significant levels of resistance (97x within 5 generations), which by F35 attained a plateau in the 250-310x range (63, 64). It should be pointed out that earlier data (65) had shown variation of up to 42x in the intrinsic susceptibility of different strains of this insect, prior to exposure to commercial BT applications. These results indicate that certain natural populations of this species may be only "marginally" susceptible and thus more prone to respond promptly to selection pressure. Toxicological, immunological, or other criteria for identifying such cases would enable more accurate choice of the appropriate BT strain or construct against a given pest.

The second case of resistance involves larvae of the tobacco budworm selected with a genetically engineered strain of the epiphytic bacterium *Pseudomonas fluorescens* expressing the 130 kDa endotoxin protein of BT subsp. *kurstaki*, strain HD-1 (66). Selection involved incorporation of lyophilized powder of transformed *P. fluorescens* into the diet of neonate *H. virescens* larvae. These selections produced 24x resistance by generation F7, which subsequently fluctuated between 13x and 20x (67). It should be stressed that these selections utilized a single toxin protein (130 kDa), whereas BT spore crystal preparations contain at least four different toxin proteins (68, 69) some of which interact synergistically (70, 71). Interestingly, the selected insect strain was much less resistant (only 4x) toward Dipel (67), a commercial product of BT HD-1, which contains at least 4 toxin proteins and from which the single gene transferred to *P. fluorescens* was obtained.

There is increasing interest in the development of transgenic plants containing BT toxin genes for controlling insect pests (72). It would appear from these limited data, that transgenic plants containing several toxin genes of BT might have a better chance of delaying the selection of resistance than transgenic plants in which only a single gene was introduced. This expectation is consistent with numerous field observations concerning the rapidity of development of resistance to single-site acting pesticides (e.g. benzimidazole fungicides) as opposed to multi-site acting pesticides (copper, mercury, sulfur, cryolite, etc.) (73). This view is also supported by the results of mathematical analysis and simulations that indicate that simultaneous selection by two or more insecticides is more likely to delay the evolution of resistance than tandem selection by these chemicals (73; Taylor, C. E., Univ. of Calif., Los Angeles, CA, personal communication, 1988).

Progress on Resistance Mechanisms

Recent work in several laboratories has significantly enriched the available knowledge on the biochemistry, genetics and dynamics of resistance (6, 75-78). Exciting advances have been made by the use of molecular biology techniques, indicating the tremendous benefits that can be derived through these technologies. Of special interest is the cloning of insecticide-sequestering/detoxifying esterase genes from the mosquito *C. quinquefasciatus* (79) and the aphid *M. persicae* (80), of cytochrome P-450 genes from the house fly (81), and progress achieved on the molecular genetics of AChE (82-84). These advances open new horizons in the understanding of resistance, including the development of more specific diagnostic tests for detecting and monitoring of resistance in populations, as well as the transfer of resistance genes to beneficial organisms (85, 86).

Esterases. Esterases are a major mechanism of resistance to organophosphates and in certain cases may also contribute resistance toward carbamates and certain pyrethroids. The role of esterases in resistance in mosquitoes and aphids has been the subject of especially fruitful research during the past few years. In collaborative studies between our laboratory and French laboratories in Montpellier, Antibes and Pau, several electrophoretic forms of esterases A and B have been identified in mosquitoes (Figure 5). Esterases A1, A2, A4, B1, B2, and B4 are found in the *Culex pipiens* complex, i.e. *C. pipiens* and *C. quinquefasciatus*. Two other forms, esterases A3 and B3, are present in *C. tarsalis*, and still others (not yet named) in *Aedes aegypti* and *Ae. nigromaculis* from California, *C. tritaeniorhynchus* from Japan (87) and various *Culex* species from Mexico.

These esterases are classified as types A or B on the basis of their preferential hydrolysis of α- or β- naphthyl acetate, respectively, in the presence of equal amounts of both substrates, after electrophoretic separation (88). To the extent studied, these esterases appear to have distinct geographic distributions, B1 being found in N. America and the Far East, A1 in Europe and the Middle East, A2B2 mainly in Africa and Asia, and A4B4 in the Mediterranean Basin. We now have evidence of the presence of A2B2 in the U.S., mainly near major ports (Oakland, Los Angeles, Houston, New Orleans, Paramus NJ), suggesting the possible introduction of these esterases through ocean trade routes. In this case, the ability to identify resistance esterases has provided evidence of transport of resistant strains across oceans.

Organophosphate resistance in *M. persicae* is due to the production of large amounts of carboxylesterase E4 that degrade as well as sequester these insecticides (19). This is believed to be the only biochemical resistance mechanism in this species in many countries, including the UK., continental Europe, Japan, and Australia (25). E4 additionally confers low levels of resistance to carbamates and to (1S)-*trans*-permethrin (19). This esterase has been characterized extensively by toxicological, biochemical, immunological and molecular studies (6, 25, 80, 89, 90).

Of considerable significance is the discovery of amplification of genes that encode esterases responsible for OP resistance in *Myzus* and *Culex*. In these cases, gene duplication was first proposed on the basis of indirect evidence from work on variant lines (for *Myzus*) (89), or offspring of single pair matings (for *Culex*) (91, 92) in which a stepwise increase in levels of esterase activity was accompanied by corresponding increases in levels of resistance to organophosphates. Subsequently, the gene encoding esterase B1 of *C. quinquefasciatus* was cloned in λgt11 phage, and a cDNA fragment of the gene was used to demonstrate that adult mosquitoes of the highly

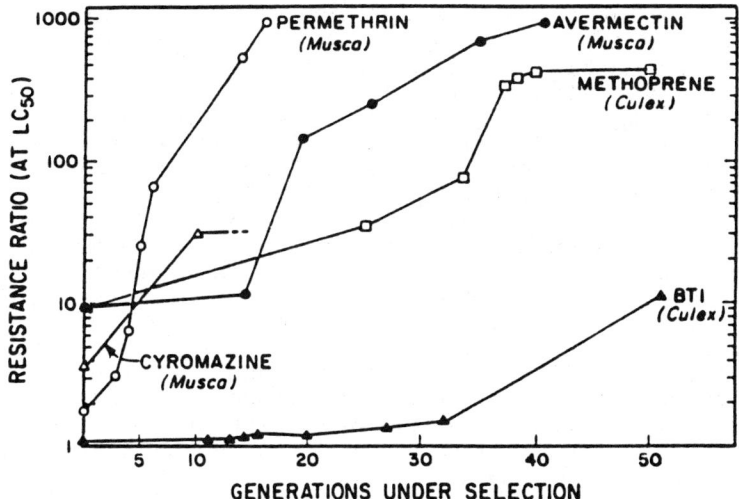

Figure 4. Rates and levels of resistance developed in house flies (*Musca*) and mosquitoes (*Culex*) through selection with various insecticides in the laboratory. (Reproduced with permission from Ref. 62.)

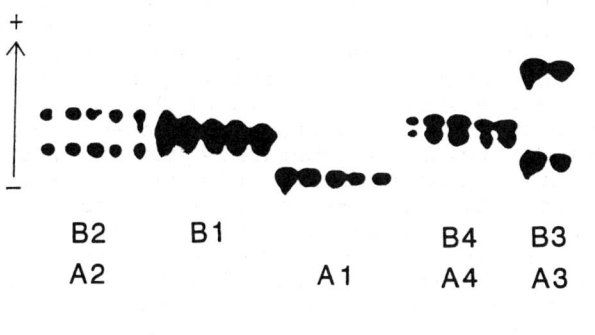

Figure 5. Relative electrophoretic mobility of esterases involved in organophosphate insecticide resistance in *Culex* mosquitoes.

resistant Tem-R strain possess at least 250 times more copies of the B1 gene than adults of a susceptible strain (79). Through the use of this probe it was also shown that esterases B2 and B3 are also amplified (18x and 250x, respectively) and that amplification is also present in natural populations of the species (Raymond, M. et al. Biochem. Genet., in press).

Overproduction of OP-detoxifying esterases in mosquitoes was also demonstrated by an immunoblot method. Esterase B1 of *C. quinquefasciatus* was shown to be overproduced by a factor of at least 500x, approximating the level of resistance to the OP insecticide chlorpyrifos (ca. 800x) as determined by bioassay (93).

In the aphid *M. persicae*, molecular evidence that insecticide resistance results from amplification of an esterase (E4) gene was provided following successful isolation and use of cDNA clones for the esterase. The degree of amplification of the structural gene was correlated with the activity of the esterase and the level of resistance. Quantitative differences between restriction patterns in different clones of resistant aphids were found to be correlated with the presence or absence of a specific chromosome translocation and with production of the esterase (80).

The presence of gene amplification in insects, although so far investigated only in mosquitoes and aphids, emphasizes the importance of this genetic process as a survival mechanism under high insecticidal selection pressure. The biochemical cost to the insect of such amplification of a defence mechanism may be considerable. In *Culex*, for example, the overproduced esterase amounts to 6%-12% of the total protein of the insect (94) and in the aphid 3% (19). This is a large diversion of resources into production of an enzyme that is apparently of little or no use in the absence of insecticide, but its consequences to the insect in terms of "fitness" are not yet adequately known. In *Culex*, resistance in newly colonized field strains is moderately unstable, but becomes relatively stable after prolonged rearing of the strain under insecticidal selection in the laboratory.

Progress has also been made in molecular studies of cytochrome P-450 monooxygenases that are known to play an important role in insecticide resistance. The cytochrome P-450 gene from the house fly has been cloned, which should enable studies of the genetic regulation of insect cytochrome P-450 and further elucidation of its role in insecticide resistance (81).

Resistance Detection and Monitoring Tests

There are several approaches to detecting esterases, each with different degrees of feasibility, practicality and costs. The conventional approach of using bioassay to determine the complete dose-response relationship of a population toward a given chemical continues to be essential, at least on a periodic basis, since it is the only means by which the combined effects of all barriers and rate-limiting steps to insecticide activation and detoxication can be measured.

The progress accomplished in the study of the biochemical mechanisms of resistance, aided by improvements in microassay techniques, has facilitated work on the development of simple biochemical diagnostic tests for the detection of individual mechanisms of resistance in single insects, and therefore, the monitoring of the frequency of these mechanisms in populations (3, 95). The importance of detection and monitoring of resistance is self evident: early detection provides advanced warning of impending dangers; subsequent monitoring of changes in the frequency of resistant individuals provides the means for judging the effectiveness of measures that are applied for management of resistance.

Biochemical tests are especially useful where the type of mechanism responsible for resistance has already been identified. Significant progress has been achieved in devising biochemical tests aimed at detecting esterases of high metabolic activity in single insects, using model substrates such as naphthyl acetate (96-99), or immunological reactions (25, 28, 100). Advances have also been made in the development of practical tests for detection of reduced sensitivity of AChE to OPs and carbamates in single insects (101-105).

In general, testing in most of these cases is done either in microtiter plates in which several homogenates at various dilutions can be evaluated simultaneously, or on filter paper on which the homogenates are deposited. In the case of esterases, the homogenates are exposed to an appropriate substrate such as napthyl acetate and subsequently to a staining solution of Fast Garnet GBC salt. The intensity of the resulting color reveals the presence of the esterase and the level of its activity (97). The simplicity of the filter paper procedure makes it a desirable test for use under both laboratory and field conditions. The qualitative and quantitative accuracy of these procedures and their complexity may vary, so the choice of a test would depend on the precision required and the facilities and skills available.

There is general agreement that as these tests are further improved with respect to their specificity and quantification accuracy, they will become essential tools in resistance management programs.

Conclusion

It must be obvious from this overview that the stage is now set for fruitful research on tactics and strategies for the management of resistance in pest populations.
 Principles for resistance management have already been proposed under three categories, e.g. "moderation", "saturation" and "multiple attack" (106) based on mode of insecticide usage. Management by moderation refers to application of insecticides at lower rates, lower frequency and less thoroughly, and aims at delaying or forestalling resistance by allowing a portion of susceptible individuals to survive. This approach would be attractive where supplementary, non-chemical control measures are also available and feasible. Progress in integrated pest management makes this an attractive strategy on certain crops.
 Management by saturation aims at preventing the evolution of resistance by rendering resistance genes functionally recessive, through exposure to dosages that are lethal to heterozygous-resistant individuals. This can be accomplished in certain cases without increasing the rate of insecticide per unit area, through microencapsulation of appropriately high dosages, combination of insecticide with a synergist, or with attractants (pheromones, phagostimulants, etc.).
 Finally, management by multiple attack involves the use of two (or more) insecticides in rotation or in combination. This strategy requires the availability of pairs of insecticides with non-overlapping cross-resistance spectra. The strategy takes advantage of the initially rare frequency of genes for resistance to new types of insecticides, possible interactive effects between insecticides (synergism, negative cross-resistance), or low fitness in resistant phenotypes.
 Many of the principles outlined above have been examined by several investigators through modelling (5, 74, 107) and in a small number of cases have been tested in laboratory population cages (108-110). The initial successes of resistance management in *Heliothis* and *Spodoptera* in Australia and Egypt, respectively, provide encouragement for further comprehensive research on resistance management in other pests.

Literature Cited

1. Georghiou, G. P.; Saito, T. Eds. Pest Resistance to Pesticides; Plenum, NY, 1983; p 809.
2. National Research Council, Pesticide Resistance: Strategies and Tactics for Management; Nat. Acad. Press: Washington, DC, 1986; p 471.
3. Brown, T. M.; Brogdon, W. G. Ann. Rev. Entomol. 1987, 32, 145-62.

4. Brown, A. W. A. In *Integrated Mosquito Control Methodologies*; Academic: New York, 1983; p 161.
5. Curtis, C. F. In *Combatting Resistance to Xenobiotics*; Ellis Horwood Ltd.: Chichester, U.K., 1987; Chapter 13.
6. Devonshire, A. L. In *Combatting Resistance to Xenobiotics*; Ellis Horwood Ltd.: Chichester, U.K., 1987; Chapter 20.
7. Forgash, A. J. In *Proc. Symp. Colorado Potato Beetle, 17th Intern. Congr. Entomol.*; Ferro, D. N.; Voss, R. H., Eds.; U. Mass: Amherst, 1985, p 33.
8. Forrester, N. W.; Cahill, M. In *Combatting Resistance to Xenobiotics*; Ellis Horwood Ltd.: Chichester, U.K., 1987; Chapter 11.
9. Georghiou, G. P. In *Proc. Conf. Biotech., Biological Pesticides and Novel Plant Pest Resistance for Insect Pest Management*; Roberts, D. W.; Granados, R. R., Eds.; Boyce Thompson Inst.: Ithaca, NY, 1988, p 137.
10. Ford, M. G.; Holloman, D. W.; Khambay, B. P. S.; Sawicki, R. M., Eds.; *Combatting Resistance to Xenobiotics*; Ellis Horwood, Ltd.: Chichester, U.K., 1987; p 320.
11. Knight, A. L.; Norton, G. W. *Ann. Rev. Entomol.* 1989, 34, 293-13.
12. Roush, R. T.; McKenzie, J. A. *Ann. Rev. Entomol.* 1987, 32, 361-80.
13. Talekar, N. S.; Griggs, T. D., Eds.; *Diamondback Moth Management; Proc. 1st Int. Workshop, Tainan, Taiwan*; 1986, p 471.
14. Sun, C. N.; Wu, T. K.; Chen, J. S.; Lee, W. T. In *Diamondback Moth Management; Proc. 1st Int. Workshop, Tainan, Taiwan*; Talekar, N. S.; Griggs, T. D., Eds.; Asian Veg. Res. Dev. Center: Taiwan, 1986, p 359.
15. Perng, F. S.; Yao, M. C.; Hung, C. F.; Sun, C. N. *J. Econ. Entomol.* 1988, 81, 1277-82.
16. Rushtapakornchai, W.; Vattanatangum, A. In *Diamondback Moth Management; Proc. 1st Int. Workshop, Tainan, Taiwan*; Talekar, N. S.; Griggs, T. D., Eds.; Asian Veg. Res. Dev. Center: Taiwan, 1986, p 359.
17. Miyata, T.; Saito, T.; Noppun, V. In *Diamondback Moth Management; Proc. 1st Int. Workshop, Tainan, Taiwan*; Talekar, N. S.; Griggs, T. D., Eds.; Asian Veg. Res. Dev. Center: Taiwan, 1986, p 359.
18. Dittrich, V.; Hassan, S. O.; Ernst, G. H. *Crop Protection* 1985, 4, 161-76.
19. Devonshire, A. L.; Sawicki, R. M. *Nature* 1979, 280, 140-41.
20. Sawicki, R. M.; Devonshire, A. L.; Rice, A. D.; Moores, G. D.; Petzing, S. M.; Cameron, A. *Pestic. Sci.* 1978, 9, 189-01.
21. Sykes, G. B. *Plant Path.* 1977, 26, 91-93.

22. Devonshire, A. L.; Foster, G. N.; Sawicki, R. M. Plant Path., 1977, 22, 60-62.
23. Sawicki, R. M.; Rice, A. D.; Gibson, R. W. Aspects Appl. Biol. 2, Pests, Diseases, Weeds; AAB: Wellesbourne, 1983, p 29.
24. Dunn, J. A.; Kempton, D. P. H. Ann. Appl. Biol. 1977, 85, 175-79.
25. Devonshire, A. L.; Moores, G. D.; Ffrench-Constant, R. H. Bull. Entomol. Res. 1986, 76, 97-107.
26. Ffrench-Constant, R. H.; Devonshire, A. L. Aspects Appl. Biol. 13, Crop Protection of Sugar Beet; AAB: Wellesbourne, 1986, p 115.
27. Ffrench-Constant, R. H.; Devonshire, A. L.; Clark, S. J. Bull. Entomol. Res. 1987, 77, 227-38.
28. Ffrench-Constant, R. H.; Devonshire, A. L. Bull. Entomol. Res. 1988, 78, 163-71.
29. Parrella, M. P. Ann. Rev. Entomol. 1987, 32, 201-24.
30. Parrella, M. P.; Robb, K. L.; Morishita, P. J.Econ. Entomol. 1982, 75, 1104-08.
31. Parrella, M. P.; Keil, C. B.; Morse, J. G. Calif. Agric. 1984, 38(1-2), 22-23.
32. Keil, C. B.; Parrella, M. P.; Morse, J. G. J. Econ. Entomol.1985, 78, 419-22.
33. W.H.O. Resistance of Vectors and Reservoirs of Disease to Pesticides; Tech. Report Ser. 655, World Health Organization: Geneva, Switzerland, 1986, p 87.
34. Georghiou, G. P. Environ. Entomol. 1978, 2, 369-74.
35. Georghiou, G. P. In Resistance to Insecticides Used in Public Health and Agriculture; Proc. Int. Workshop; Nat. Sci. Council: Colombo, Sri Lanka, 1982; p 95.
36. Lines, J. D. Parasitol. Today 1988, 4, S17-S20.
37. Forgash, A. J. In Advances in Potato Pest Management; Lashoumd, J. H.; Casagrande, R., Eds.; Hutchinson Ross: Stausburg, PA, 1981; p 34.
38. Ferro, D. N. In Proc. Symp. Colorado Potato Beetle, 17th Int. Congr. Entomol.; Ferro, D. N.; Voss, R. H., Eds.; U. of Mass.: Amherst, 1985, p 1.
39. Hare, J. D. J. Econ. Entomol. 1980, 73, 230-31.
40. Harris, C. R.; Svec, H. J. J. Econ. Entomol. 1981, 74, 421-24.
41. Boiteau, G.; Parry, R. H.; Harris, C. R. Can. Entomol. 1987, 119, 459-63.
42. Ferguson, J. Agr. Consultant August 1988, p 13.
43. Sawicki, R. M. Agribusines Worldwide 1986, June, 20-25.
44. Forrester, N. W. Australian Cotton Grower Nov. 1984, p 45.
45. Gunning, R. V.; Easton, C. S.; Greenup, L. R.; Edge, V. E. J. Econ. Entomol. 1984, 77, 1283-87.
46. Forrester, N. W. Cotton Grower 1989, 25, p 49.
47. Twine, P. H.; Reynolds, H. T. J. Econ. Entomol. 1980, 73, 239-42.

48. Martinez-Carrillo, J. L.; Reynolds, H. T. *J. Econ. Entomol*. 1983, *76*, 983-86.
49. Plapp, F. W., Jr.; Campanhola, C. *Proc. 1986 Beltwide Cotton Production Research Conf*., 1986, p 167.
50. Watkinson, I. *Cotton Grower* 1989, *25*, p 54.
51. Campanhola, C.; Plapp, F. W., Jr. *J. Econ. Entomol*. 1989, *82*, 22-28.
52. Certain, G. *Agr. Consultant* 1988, May, p 7.
53. Wangboonkong, S. *Trop. Pest Management* 1981, *27*, 495-500.
54. Cochran, D. G. *J. Econ. Entomol*. 1989, *82*, 336-41.
55. Keller, J. C.; Clark, P. H.; Lofgren, C. S. *Pest Contr*. 1956, *24(11)*, 14-15, 30.
56. Cochran, D. G. *J. Econ. Entomol*. 1973, *66*, 27-30.
57. Umeda, C.; Yano, T.; Hirano, M. *Appl. Entomol. Zool*. 1988, *23*, 373-80.
58. Miller, J. A.; Oehler, D. D.; Kunz, S. E. *J. Econ. Entomol*. 1983, *76*, 1335-40.
59. Byford, R. L.; Sparks, T. C. In *Combatting Resistance to Xenobiotics*; Ellis Horwood Ltd.: Chichester, U.K., 1987; Chapter 15.
60. Byford, R. L.; Lockwood, J. A.; Smith, S. M.; Sparks, T. C.; Luther, D. G. *J. Econ. Entomol*. 1987, *80*, 111-16.
61. Byford, R. L.; Lockwood, J. A.; Smith, S. M.; Franke, D. E. *Environ. Entomol*. 1987, *16*, 467-70.
62. Georghiou, G. P. *Insecticides and Pest Resistance: The Consequences of Abuse, 36th Annu. Faculty Res. Lecture*. Academic Senate, Univ. of Calif., Riverside, CA 1987; p 27.
63. McGaughey, W. H.; Beeman, R. W. *J. Econ. Entomol*. 1988, *81*, 28-33.
64. McGaughey, W. H. *Science* 1985, *229*, 193-95.
65. Kinsinger, R. A.; McGaughey, W. H. *J. Econ. Entomol*. 1979, *72*, 346-49.
66. Watrud, L. S.; Perlak, F. J.; Tran, M.-T.; Kusano, K.; Mayer, E. J.; Miller-Wideman, M. A.; Obukowicz, M. G.; Nelson, D. R.; Kreitenger, J. P.; Kaufman, R. J. In *Engineered Organisms in the Environment*; Halverson, H. O.; Pramer, D.; Rogul, M., Eds.; American Society for Microbiology: Washington, DC, 1985; p 40.
67. Stone, T. B.; Sims, S. R.; Marrone, P. G. *J. Invert. Pathol*. 1989, *53*, 228-34.
68. Hofte, H.; Whiteley, H. R. *Microbiol. Rev*. 1989, 242-55.
69. Widner, W. R.; Whiteley, H. R. *J. Bacteriol*. 1989, *171*, 965-74.
70. Chilcott, C. N.; Ellar, D. J. *J. Gen. Microbiol*. 1988, *134*, 2551-58.
71. Wu, D.; Chang, F. N. *FEBS Letters* 1985, *190*, 232-36.

72. Vaeck, M.; Reynaerts, A.; Hoften, H.; Jansens, S.; De Beuckeleer, M.; Dean, C.; Zabeau, M.; Van Montagu, M.; Leemans, J. Nature 1987, 328, 33-37.
73. Georghiou, G. P.; Mellon, R. In Pest Resistance to Pesticides; Plenum: New York, 1983; p 1.
74. Mani, G. S. Genetics 1985, 109, 761-83.
75. Soderlund, D. M.; Hessney, C. W.; Jiang, M. J. Agr. Food Chem. 1987, 35, 100-05.
76. Sunseth, S. S.; Kennel, S. J.; Waters, L. C. Pestic. Biochem. Physiol. 1989, 33, 176-88.
77. Scott, J. G.; Georghiou, G. P.; Pestic. Sci. 1986, 17, 195-06.
78. Farnham, A. W.; Murray, A. W. A.; Sawicki, R. M.; Denholm, I.; White, J. C. Pestic. Sci. 1987, 197, 209-20.
79. Mouchès, C.; Pasteur, N.; Bergé, J. B.; Hyrien, O.; Raymond, M.; de Saint Vincent, B. R.; de Silvestri, M.; Georghiou, G. P. Science, 1986, 223, 778-80.
80. Field, L. M.; Devonshire, A. L.; Forde, B. G. Biochem. J. 1988, 251, 309-12.
81. Feyereisen, R.; Loener, J. F.; Farnsworth, D. E.; Nebert, D. W. Proc. Natl. Acad. Sci. USA 1989, 86, 1465-69.
82. Fournier, D.; Cuany, A.; Bride, J. M.; Bergé, J. B. J. Neurochem. 1987, 49, 1455-61.
83. Fournier, D.; Bergé, J. B.; Cardoso de Almeida, M.-L.; Bordier, C. J. Neurochem. 1988, 50, 1158-63.
84. Fournier, D.; Bride, J.-M.; Karch, F.; Bergé, J.-B. FEBS Letters 1988, 238, 333-37.
85. Miller, L. H.; Sakai, R. K.; Romans, P.; Gwadz, R. W.; Kantoff, P.; Coon, H. G. Science 1987, 237, 779-81.
86. Morris, A. C.; Eggleston, P.; Crampton, J. M. Med. Vet. Entomol. 1989, 3, 1-7.
87. Takahashi, M.; Yasutomi, K. J. Med. Entomol. 1987, 24, 595-03.
88. Georghiou, G. P.; Pasteur, N. J. Econ. Entomol. 1978, 71, 201-05.
89. Devonshire, A. L.; Sawicki, R. M. Nature 1979, 280, 140-41.
90. Devonshire, A. L.; Searle, M.; Moores, G. D. Insect Biochem. 1986, 16, 659-65.
91. Pasteur, N.; Georghiou, G. P.; Ranasinghe, L. E. Proc. Ann. Conf. Calif. Mosq. Contr. Assoc. 1980, p 69.
92. Pasteur, N.; Georghiou, G. P.; Iseki, A. Genet. Sel. Evol. 1984, 16, 271-84.
93. Mouchès, C.; Magnin, M.; Bergé, J. B.; de Silvestri, M.; Beyssat, V.; Pasteur, N.; Georghiou, G. P. Proc. Nat. Acad. Sci. USA. 1987, 84, 2113-16.
94. Fournier, D.; Bride, J. M.; Mouchès, C.; Raymond, M.; Magnin, M.; Bergé, J. B.; Pasteur, N.; Georghiou, G. P. Pestic. Biochem. Physiol. 1987, 27, 211-17.
95. Brogdon, W. G. Parasitol. Today 1989, 5, 56-60.

96. Pasteur, N. P.; Georghiou, G. P. *J. Econ. Entomol.* 1989, **82**, 347-53.
97. Pasteur, N.; Georghiou, G. P. *Mosquito News* 1981, **41**, 181-83.
98. Rees, A. T.; Field, W. N.; Hitchen, J. M. *J. Amer. Mosq. Contr. Assoc.* 1985, **1**, 23-27.
99. Brogdon, W. G.; Hobbs, J. H.; St. Jean, Y.; Lacques, J. R.; Charles, L. B. *J. Amer. Mosq. Contr. Assoc.* 1988, **4**, 152-58.
100. Beyssat-Arnaouty, V.; Mouchès, C.; Georghiou, G. P.; Pasteur, N. *J. Amer. Mosq. Contr. Assoc.* 1989, **5**, 196-200.
101. Ffrench-Constant, R. H.; Bonning, B. C. *Med. Vet. Entomol.* 1989, **3**, 9-16.
102. Devonshire, A. L.; Moores, G. D. *Pestic. Biochem. Physiol.* 1984, **21**, 341-48.
103. Raymond, M.; Fournier, D.; Bergé, J.; Cuany, A.; Bride, J.-M.; Pasteur, N. *J. Amer. Mosq. Contr. Assoc.* 1985, **1**, 425-27.
104. Hemingway, J.; Smith, C.; Jayawardena, K. G. I.; Herath, P. R. J. *Bull. Entomol. Res.* 1986, **76**, 559-65.
105. Brogdon, W. G. *Comp. Biochem. Physiol.* 1988, **90C**, 145-50.
106. Georghiou, G. P. In *Pest Resistance to Pesticides*; Plenum: New York, 1983; p 769.
107. Taylor, C. E.; Georghiou, G. P. *J. Econ. Entomol.* 1979, **72**, 105-09.
108. Taylor, C. E.; Georghiou, G. P. *Environ. Entomol.* 1982, **11**, 746-50.
109. Taylor, C. E.; Quaglia, F.; Georghiou, G. P. *J. Ecol. Entomol.* 1983, **76**, 704-07.
110. Georghiou, G. P.; Lagunes, A.; Baker, J. D. *Proc. 5th Int. Congr. Pestic. Chem.* 1983, p 183.

RECEIVED September 1, 1989

Chapter 3

Resistance Mechanisms to Carbamate and Organophosphate Insecticides

L. B. Brattsten

Department of Entomology, Rutgers University, J. B. Smith Hall, New Brunswick, NJ 08903-0231

Organophosphates (OPs), introduced in 1944, and carbamates, introduced in 1956, remain widely used and effective insecticides although not free from resistance problems. Metabolic resistance to OPs was reported 14 years after their introduction, compared to only 7 years for DDT and 5 for the carbamates. The complex metabolic fate of the OPs, including attack by cytochrome P-450 leading either to activation or detoxification, as well as by glutathione transferases and esterases, may play a role in this delay. Carbamates are not bioactivated; they are detoxified by cytochrome P-450.

In the presence of a continued selection pressure, metabolic resistance may facilitate the evolution of other defenses such as target site resistance, reported for the OPs and carbamates 6 and 10 years after metabolic resistance. Target site resistance to OPs and carbamates resides in modified forms of acetyl-cholinesterases (AChEs) with reduced affinity for the insecticides. AChE-based target site resistance does not necessarily confer cross resistance to all other OPs and carbamates and may be unstable in the absence of a selection pressure.

Historical Perspective

The organophosphorous (OP) insecticides were among the first synthetic insecticides to be used on a large scale; they are still widely used today. The first organophosphate to be used as an insecticide was TEPP (tetraethyl pyrophosphate), introduced in 1944. This compound was soon discontinued, however, because of its very high mammalian toxicity. Parathion was the first successful OP but also suffered from very high mammalian toxicity. Parathion was introduced in 1947 and quickly followed by a large number of other OPs, some of which have improved selectivity between insects and mammals because of small changes in the molecule. For instance, methyl parathion, introduced in 1949, is less toxic to mammals because of the phosphoester methyl substituents, and fenitrothion is a further improvement thanks to the methyl substituent on the phenolic ring (Table 1) (1-3).

Table 1. Comparison of acute toxicities (LD_{50}, mg/kg) of OPs and carbamates to mammals (rat, oral) and insects (house fly, topical). The data are compiled from several sources (1-3)

Insecticide	Mammal	Insect
Parathion	3.6	0.9
Methyl parathion	24	1.2
Fenitrothion	250	2.3
Malathion	885	26.5
Carbaryl	500	100
Propoxur	95	26
Carbofuran	8	4.6

Because of their combination of high acute mammalian toxicity and very high efficiency in controlling insects, the OPs are credited with having given a strong impetus for neurophysiological and neurobiochemical studies in both mammalian and insect systems, prompting the development of what are now highly sophisticated areas in science. The introduction of malathion in 1950 increased the acceptance of this type of compounds for commercial field use, as malathion was the first OP insecticide with remarkable mammalian safety (Table 1).

The development of the N-methylcarbamate insecticides and the successful introduction of carbaryl in 1956 seemed to solve the problems with the high mammalian toxicity of the OPs, but many subsequently developed carbamates are also highly toxic to vertebrates (Table 1).

In fact, only two new classes of insecticides have been developed for commercial use in the last 30 years. Both the synthetic pyrethroids and the avermectins have excellent mammalian saftety. However, both are encumbered by previously evolved target site resistance, selected by over-use of DDT and cyclodiene insecticides, respectively. Thus, the remaining importance of the OPs and the carbamates is obvious.

Structures

As shown in Figure 1, the basic structural skeleton of both OPs and carbamates allows a very large variety of substituents to be attached without loss of toxicity. With the exception of the oxime carbamates, e. g. aldicarb and methomyl, the successful carbamates have aromatic or heteroaromatic substituents. There are three categories of OPs: the aliphatic OPs, exemplified by malathion; the aromatic OPs, exemplified by methyl parathion; and the heteroaromatic OPs, exemplified by chlorpyrifos. The last category, containing the least biodegradable compounds of the three (4), is the one most widely used in agricultural applications. Recently however, aliphatic OPs are used increasingly as systemic agricultural insecticides as replacements for aldicarb, which has caused ground water contamination problems. The basic structures also show that, depending on the nature of the substituents (R), the OP and carbamate molecules are susceptible to enzymatic attack in not just one, but several places. The microsomal cytochrome P-450 dependent polysubstrate monooxygenases (EC 1.14.14.1) (PSMOs), carboxylesterases (EC 3.1.1.1 or 3.1.1.2) with phosphatase activity, and glutathione transferases (EC 2.5.1.18) all metabolize the OPs, whereas the carbamates are detoxified with sufficient speed only by cytochrome P-450.

Resistance

Resistance to the OPs was not reported until 14 years after their introduction. This is twice the time it took insects to evolve resistance to DDT. Resistance to the carbamates appeared more quickly, 5 years after their introduction, and was probably conditioned by previous use of chlorinated hydrocarbon and OP insecticides, both of which acted as selection agents of the major detoxification mechanism, cytochrome P-450, for the carbamates.

Metabolic resistance mechanisms are still the most widely encountered causes for OP and carbamate resistance. Yet, after the first few cases of target site insensitivity were reported in the early 1970's in mites and ticks, many such cases have been found also in insects. Resistance can also be enhanced by a decreased rate of penetration through the integument. This resistance mechanism by itself is of minor importance, but provides an increased opportunity for detoxification. A 50-fold increase in resistance to carbaryl was seen in a house fly strain, in which a gene for reduced penetration had been combined (by selective breeding) with a gene for increased detoxification (5).

Target site resistance is probably always combined with some other form of resistance; even a relatively insensitive target site would eventually be overwhelmed unless there were also an effective way to eliminate the insecticide. Combinations of resistance mechanisms are becoming increasingly frequent in field populations of insects under persistent selection pressure. In insect populations that are specialized to feed on toxic plants, the occurrence of several or many resistance mechanisms appears to be a rule with extremely few exceptions (6).

Metabolic Resistance to Carbamates

Oxidation. Any and all biotransformations of carbamate insecticides result in their loss of anti-AChE activity. Oxidation is the most important metabolic fate of the carbamates and cytochrome P-450 is the most important catalyst. One or more isoenzymes of cytochrome P-450 can oxidize a carbamate molecule in many different places, depending on its structure. In Figure 2, arrows indicate places susceptible to oxidation in selected carbamate molecules.

Cytochrome P-450 removes the N-methyl group by direct hydroxylation of the carbon atom adjacent to the nitrogen; the aromatic substituent can be hydroxylated at specific carbons or undergo epoxidation at electron dense sites, followed by conversion to the dihydrodiol either spontaneously or catalyzed by epoxide hydrolases (7). Early experiments with synergized carbamates led to the realization that insects are well endowed with "drug-metabolizing enzymes" and that these, in particular cytochrome P-450, are involved in insecticide detoxification. Therapeutically used drug extenders, such as SKF 525A and Lilly 18947, were shown to be synergists for carbamates by virtue of inhibiting detoxification (8-10).

As in vitro methods for studying cytochrome P-450 in insects became available (11-13), it soon became clear that insects with high cytochrome P-450 activities were resistant to carbamates and most other insecticides. This phenomenon is termed metabolic cross resistance and derives from the characteristic of cytochrome P-450 of accepting a very wide range of molecular structures as substrates; the cytochrome binds the substrate very loosely by a lipophilic interaction and rapidly oxidizes it by an oxygen free radical-mediated reaction, a very powerful combination. Moreover, the cytochrome occurs in several or many different isoenzymic forms with broadly overlapping substrate preferences. A normally infrequent form may be selectively induced by allelochemicals in the crop plants (14), and if the induced form has survival value in the presence of an insecticide, it could be selected to dominate in the exposed population (15).

Selection of laboratory strains of insects with insecticides has repeatedly resulted in populations with considerably higher cytochrome P-450 activities than the original ones (16, 17). The role of cytochrome P-450 in the detoxification and metabolism of the carbamates has been studied extensively (18). High cytochrome P-450 activities are associated with virtually all cases of carbamate resistance.

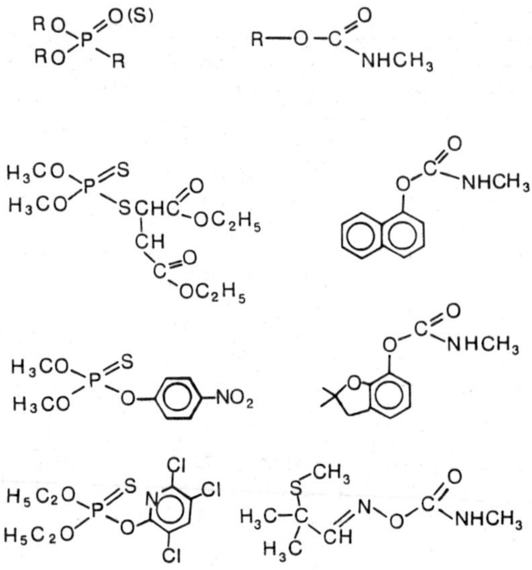

Figure 1. Basic structures of OPs and carbamates and examples of different substituents (R).

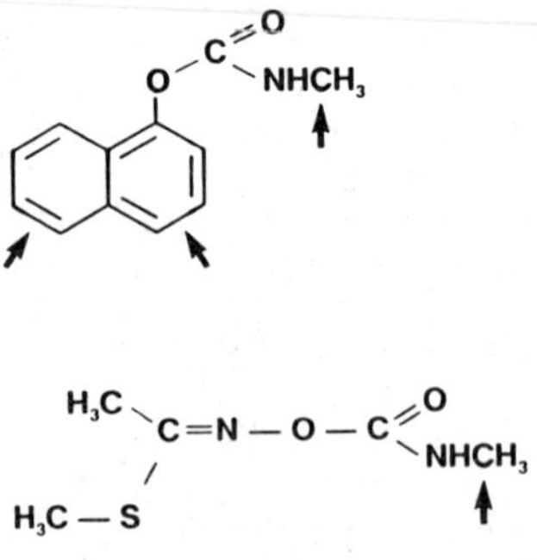

Figure 2. Carbamates with arrows pointing to sites in the molecules that can be attacked by cytochrome P-450.

Hydrolysis. Although carbamate molecules are detoxified by cytochrome P-450-catalyzed oxidation, they contain a carbamic ester bond, susceptible to hydrolysis. Insects have traditionally not had a carboxylesterase capable of hydrolyzing this bond with toxicologically relevant speed. Hydrolysis products are recovered after longer holding times, 18-24 hours in *in vivo* metabolism experiments, showing that either the insects have the appropriate esterase but it works very slowly, or that, if ingested, the carbamates undergo slow spontaneous hydrolysis in the alkaline gut of insects (19, 20).

Mixtures of insecticides. Laboratory selection experiments with mixtures of cytochrome P-450 inhibiting synergists such as piperonyl butoxide show that resistance either does not evolve at all for the duration of the experiment or that it is greatly delayed (21), compared to the rate of resistance evolution when a single compound is used. This has led to the idea of controlling insects, even susceptible populations, with insecticide-synergist mixtures. This is based on the rational consideration that the synergist will eliminate the advantage of high detoxifying ability and give the insecticide the opportunity to kill individuals that would otherwise be selected (22). (It could also, in some cases, help the survival of beneficial insects.) This tactic may delay evolution of resistance to carbamates temporarily, because they are detoxified by one single enzyme system, the cytochrome P-450-dependent PSMOs (23). It will not delay evolution of resistance to any insecticides that are detoxified by more than one enzyme system, such as the OPs. In the latter case, blocking detoxification by one enzyme would only result in selection of another detoxifying enzyme.

This strategy is problematic even with the carbamates because of the potential for the evolution other resistance mechanisms. These could include selection of an esterase which can rapidly hydrolyze the carbamic ester bond, or overproduction of an already existing esterase capable of binding and sequestering them. The green peach aphid has an esterase, E4, that eliminates the toxicity of carbamates, pyrethroids, and OPs not by hydrolysis but by having a very high binding affinity for these compounds (24) and, thus, functioning as a storage protein. Armyworms have esterases with similar properties (15). Lipophorins and arylphorins, large hemolymph storage proteins that bind xenobiotics with intermediate lipophilicity characteristics, have been found in several orders of insects (25).

Any insecticide-synergist mixture, as, indeed, any other kinds of mixtures, constitutes *per se* a selection pressure, although different from that exerted by a single insecticide. Thus, mixtures, synergistic or otherwise, can best be used in strategies of rotation with other insecticides or mixtures to avoid selecting new, and possibly harder to overcome, resistance mechanisms. Despite results of laboratory selection experiments with mixtures, it is predictable, that mixtures with few components will be nearly as effective selecting agents as single compounds. To slow resistance evolution to a pace comparable to that in natural insect-plant associations, will require multi-component mixtures such as those employed by plants. A lupine in Colorado effectively protects its inflorescences against an otherwise specialized caterpillar by a mixture of alkaloids containing up to 18 different compounds (26).

Metabolic Resistance to OPs

Oxidation. OPs are attacked by no less than three enzyme systems, the PSMOs, carboxylesterases with phosphatase activity, and glutathione transferases. The latter two invariably detoxify the OPs by splitting off alkyl or other substituents. However, the action of cytochrome P-450 can result in detoxification if the carbon of one of the small alkyl substituents is oxidized, or in activation of phosphorothioates if the P=S is converted to P=O by oxidative desulfuration (Figure 3). Except in a very few cases, the thiophosphate (P=S) is not an AChE inhibitor. It is usually the phosphate (P=O) form of the OPs that inhibits AChE, because the P=S bond does not polarize the phosphorous atom sufficiently for strong binding to the esteratic subsite. The oxidative desulfuration reaction is unusual for cytochrome P-450, which typically oxidizes carbon atoms. It is thought to occur in analogy with the epoxidation reaction of aromatic carbons, as outlined by Nakatsugawa and Morelli (27) (Figure 4).

Hydrolysis. Carboxylesterases are frequently one of the major factors in OP resistance. In some insects, for instance the house fly (28), there are highly substrate specific esterases which attack only one or a very few molecules. "Malathionase", the prominent esterase responsible for many cases of malathion resistance, is highly specific for malathion. It cleaves one or both of the ethyl ester groups leaving malathion mono- or diacid (29). This enzyme is a true serine carboxylesterase that is inhibited by malaoxon (28) and does not hydrolyze any of the phosphoester bonds. In Anopheles stephensi from Pakistan, the malathion resistance decreased with adult age, but there was no concommittant decrease in general esterase activity as measured with 1- and 2-naphthylacetate as model substrates (30). Other mosquitoes have a carboxylesterase with broad substrate specificity that is associated with resistance (31-33). As mentioned above, the green peach aphid has a carboxylesterase, E4, with broad substrate specificity that sequesters toxicants (24).

Since OPs inhibit serine hydrolases such as AChE by phosphorylating them, it follows that OPs are potential inhibitors of all other serine hydrolases. This means that OPs can be used as synergists, notably for the pyrethroids that are detoxified by ester hydrolysis, and OP-containing mixture have proven successful in several cases (34). The usefulness of these mixtures can be prolonged by using them in rotations with other mixtures and/or insecticides.

Glutathione conjugation. The involvement of glutathione transferases in OP metabolism was realized in the early 1960's (35, 36). It was difficult to establish this fact because of similarities between glutathione transferase- and carboxylesterase-produced metabolites. Induction of glutathione transferase activity in the fall armyworm caused a 2- to 3-fold decrease in the toxicity of diazinon, methamidophos, and methyl parathion (37). This shows indirectly the importance of glutathione transferase activity in the detoxification of these OPs.
Glutathione transferases catalyze a substitution reaction at electrophilic centers of molecules. They are also binding proteins in analogy with the E4 esterase; a mammalian form, called ligandin, binds with high affinity to a broad spectrum of compounds but does not catalyze the subsequent substitution reaction (38). The role of the transferases in the catalytic reaction is thought to be to provide close proximity between the xenobiotic and the reduced glutathione anion, GS^-.

There are, as seems to be the rule with enzymes involved in the metabolism of xenobiotics, multiple isoenzymes with glutathione transferase activity. This fact combined with their low substrate specificity is considered to account for the variety of OP metabolites produced as shown in Figure 5 (39). There are at least three different isoenzymes in the house fly (40). One of the house fly glutathione transferases is identical to the enzyme responsible for DDT dehydrochlorination (41). This form is rather substrate specific, and is not known to confer metabolic cross resistance between DDT and any OP.

There are few highly specific inhibitors of the glutathione transferases, but organisms and tissues can be depleted of endogenous glutathione by diethylmaleate and other compounds, with which it reacts spontaneously. The herbicide synergist tridiphane also synergized diazinon toxicity towards house flies by acting as an alternative substrate (42). The house fly glutathione transferase conjugated tridiphane 22 times faster than diazinon.

Evolution of metabolic resistance to the OPs. The following is an attempt to understand why it took such an unusually long time for insects to evolve metabolic resistance to the OPs: Clearly, the selection pressure is diluted by having to act on three different, genetically unrelated enzyme systems, all of which detoxify the com-

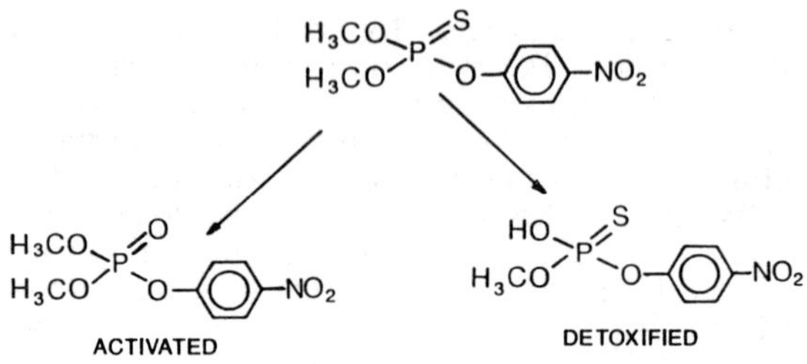

Figure 3. Detoxification and activation of a thiophosphate insecticide by cytochrome P-450.

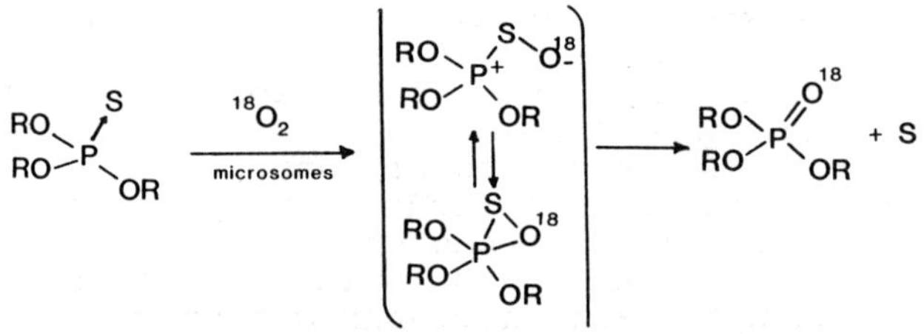

Figure 4. Hypothetic mechanism for the P=S to P=O conversion showing its analogy with an aromatic epoxidation (27).

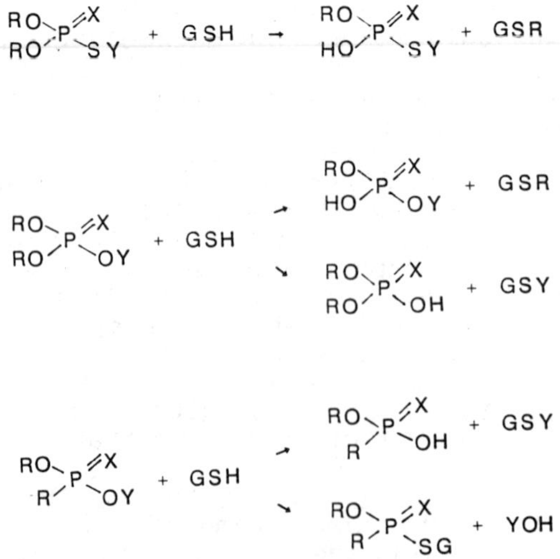

Figure 5. Metabolism of OPs by glutathione transferases. The scheme is adapted from (39).

pounds. But the situation may be further complicated by the dual involvement of cytochrome P-450. It is likely that the OP-activating cytochrome P-450 isoenzyme does not catalyze the reactions leading to their detoxification. If so, conflicting selection pressures then act on different isoenzymic forms, which, however, are very closely genetically related. Such a situation can conceivably result in slower evolution than if conflicting selection pressures act on genetically unrelated enzyme systems. A hypothetical example of the latter sort may be where an organism evolves predominance of a transferase activity that would detoxify at the expense of a hydrolase activity that would activate. This could happen in response to selection pressure by a glycosidic toxicant. The genetic unrelatedness of the enzymes may allow more rapid evolution. Thus, from the point of view of finding enduring chemical crop protection agents with low potential for precipitating resistance, the OPs are nearly ideal. The fact that there is, indeed, resistance to these compounds, is most likely a consequence of their unwise use.

Reviews on detoxification. All the enzyme systems involved in carbamate and OP metabolism and detoxification have been described in detail by several recent reviewers (7, 43-45). For an in-depth discussion of the genetics of these resistance mechanisms, see Oppenoorth (28).

Target Site Interactions

Both organophosphates and N-methylcarbamates are inhibitors of acetylcholinesterase (EC 3.1.1.7) (AChE), an enzyme of critical importance in synaptic nerve impulse transmission. This enzyme hydrolyzes the neurotransmitter, acetylcholine, so that its concentration near the receptor on the post-synaptic membrane is below the threshold for initiating the post-synaptic nerve impulse, except when a pre-synaptic nerve impulse has caused the release of a pulse of transmitter. The reaction mechanism of AChE is a straightforward hydrolysis mediated by water molecules and not requiring any high energy cofactor input. Synapses of insects appear to contain very much higher concentrations of AChE than mammalian synapses, on the order of 100-fold more (46, 47).

The AChE active site has two well defined subsites. The esteratic subsite contains a serine residue to which the acetyl group of the transmitter, the phosphorous of the OPs, and the carbamoyl carbon of the carbamates all bind. The anionic subsite contains an ionized carboxylic acid that binds the charged trimethylammonium group of the neurotransmitter. There is no clearly indicated interaction between this subsite and any part of the inhibitor molecules. Their leaving groups may bind to a different region in the active site from the substrate (48), or a charge-transfer complex may be involved in the

transient attachment of the leaving group to this subsite (49) but the evidence for this is not firm.
The AChE reaction is usually represented as follows (50, 51):

$$E\text{-}OH + RX \underset{k_{-1}}{\overset{k_1}{\rightleftarrows}} E\text{-}OH\text{-}RX \xrightarrow{k_2} E\text{-}OR \xrightarrow{k_3} E\text{-}OH + ROH$$
$$\downarrow$$
$$HX$$

where E-OH stands for the esteratic site of the AChE; the R in RX is the acetyl group of the transmitter, the phosphorous of an OP, or the carbamoyl carbon; and X of the RX is the choline of the transmitter or the leaving group of the inhibitor. First, a (theoretically) reversible complex forms, which as the leaving group leaves, results in an acetylated, phosphorylated, or carbamylated enzyme (E-OR). The latter two conditions represent an inhibited enzyme because of the relatively slow rate of k_3.

The interaction of OPs and carbamates with the esteratic subsite is, thus, analogous with the interaction by the neurotransmitter. However, whereas the transmitter is easily and very quickly released from the esteratic active subsite (in about 40 microseconds), the carbamates are released more slowly, over several minutes. The k_3 for a monomethyl carbamylated house fly brain AChE was 0.01 $\min^{-1}$, $brain^{-1}$ (25 C, pH 7.4) (52). Since the carbamoyl residue on all carbamylated AChEs is the same,

$$-C(O)-NHCH_3$$

all should be released with the same rate constant. That they are not reflects the existence of differences in AChEs between species and populations. The phosphorylated enzyme is hydrolyzed even more slowly, in several hours to days, or not at all.

Target Site Resistance

Target site resistance to OPs was first reported in 1964, 20 years after their introduction, in a spider mite (53) and subsequently in several insect species including the green rice leaf hopper (54), mosquitoes (55), house flies (56-58), and an armyworm (59). Target site resistance to the carbamates was first observed in 1971 (60). It is, today, a common form of resistance to these insecticides and always combined with metabolic resistance.

Target site resistance can be diagnosed in vivo by the use of synergists, which block detoxification but have no effect if the target site is insensitive or protected. When target site resistance is combined with metabolic resistance, the use of synergists can partially but not fully reverse the resistance. Alternatively, target site resistance can be established by direct in vitro studies of the enzyme-inhibitor interactions. Practically, it is common to measure residual activity after preincubation with the insecticide. The results can then be expressed in terms of an I_{50} value. Under certain circumstances, the I_{50} relates to the overall bimolecular rate constant, k_i, which equals k_2/K_d. K_d is the ratio of k_{-1} and k_1. I_{50} values can be used to compare the inhibitory potency of a series of insecticides to the overall AChE population in the insect population and thereby outline a target site resistance profile.

However, I_{50} measurements do not provide information about the biochemical characteristics of the enzyme-inhibitor interaction and what specific feature of the interaction may have changed and so caused the resistance. Also, I_{50} measurements need to be done with pooled tissues of several to many insects, depending on their size. To study the frequency of the resistant target site in a population, measurements must be done in individual insects. This can and has been done in house flies (<u>58, 61</u>), an armyworm (<u>59</u>), and plant hoppers and leaf hoppers (<u>62</u>). Detailed inhibition kinetics studies will reveal if a mutation has occurred (<u>58</u>). A mutation causing the enzyme to have decreased affinity for the inhibitor is most clearly reflected in K_d measurements. Usually, k_i values are reported. Because of the inverse relationship between K_i and k_d, k_i is smaller in resistant insects with insensitive AChE. This is illustrated in Table 2.

Table 2. Bimolecular rate constants for acetylcholinesterase inhibition by organophosphates and a carbamate in resistant (R) and susceptible (S) insects

Insect	Compound	k_i* R	k_i* S	Ref.
House fly	paraoxon	578	139	(<u>63</u>)
	dimethoxon	20	2	(<u>63</u>)
Anopheles albomanus	paraoxon	780	2.1	(<u>55</u>)
	propoxur	295	0.02	(<u>55</u>)
Spodoptera littoralis	dichlorvos	430	28	(<u>59</u>)
	monocrotophos	130	0.61	(<u>59</u>)

*($mM^{-1} min^{-1}$)

Kinetic studies have also revealed the presence of multiple AChE forms in house fly heads (<u>56, 61, 64</u>) and thoraces (<u>56</u>), and in a fruit fly (<u>65</u>), possibly resulting from mutations that were unrelated to insecticide exposure.

The existence of multiple AChE isoenzymes has several consequences. First, it increases the chances of an insect having one that is, or by a minor genetic change can be rendered, insensitive. The molecular redundancy combined with a selection pressure in the form of persistent insecticide applications would facilitate target site resistance development. Second, it could be a factor in the frequent lack of target site cross resistance between OPs and carbamates, and even between different OPs. Third, it would facilitate the disappearance of the insensitive form(s) in the absence of a selection pressure. This would especially easily explain observed instability of resistance if the form(s) with decreased affinity for the inhibitors also have decreased affinity for the neurotransmitter. Insensitivity to the inhibitor may be accompanied by a reduced rate of neurotransmitter hydrolysis (56, 28), but this is not always the case. It seems that the reduced rate of neurotransmitter hydrolysis does not impair survival, at least in laboratory cultures of insects. It is unclear what impact such reduced rates have in field populations.

In practical terms, target site resistance is commonly considered the most serious form of resistance for several reasons. First, it can not be counteracted by synergists. Second, once aquired it is a "built in" defense that does not require any extra energy expenditure and is available at all times (66), at least as long as the selection pressure prevails. Third, it usually means that the insect is defended against any other compounds with the same mode of action, as in the case of DDT, the pyrethrins, and the type I pyrethroids. In other words, an insect population with target site resistance often has target site cross resistance to other insecticides with the same mode of action.

However, the multiplicity of AChEs and the existence of several different insensitive isoenzymes exclude the automatic assumption that an OP resistant insect population has target site cross resistant to all other OPs and all carbamates. There is, in fact, a case where decreased affinity for one type of compounds is combined with increased sensitivity to inhibition by another type. In a population of the green rice leaf hopper resistant to N-methyl carbamates, the AChE has a 40-fold increased sensitivity to inhibition by N-propyl carbamates (67). This is a preciously rare case of negative target site cross resistance to insecticides, and it has not been commercially exploited.

Implications of Resistance

Cost of Resistance. Even though insect defenses against plant allelochemicals carry energy costs (68, 69), there are few cases known where resistance impairs the biotic potential of the insects. There was no correlation between resistance and the fitness in diazinon resistant house flies (70), Australian sheep blow flies (71), malathion resistant rice weevils (72), or OP resistant aphids (73). But a malathion resistant, laboratory-selected population of Drosophila melanogaster had increased microsomal oxidase activities combined with decreased reproduction and reduced growth (74). A temephos resistant strain of Culex quinquefasciatus had reduced growth rate (75), and the fecundity and fitness of an azinphos methyl resistant population of the Colorado potato beetle were reduced (76).

Few studies have been undertaken of the energetic cost of resistance, the resistance mechanism is often not specified in these studies, and the parameters chosen for measurements vary between studies. Comprehensive information is therefore not available. Still, this kind of information is important for understanding the evolution of resistance and its stability. Most new insecticides, it appears, have a mode of action identical to or extremely similar to one that was used previously. The successful use of new insecticides therefore depends on what kinds and frequencies of resistance genes are already present in insect populations to be controlled. It can be expected that resistance mechanisms that affect fitness will be less stable than those that do not. There is, thus, a need for more studies of the relationships between resistance mechanisms and biotic potential in insects.

Use of Chemicals and Evolution of Resistance. It is logical to assume that target site resistance is more difficult to aquire than metabolic resistance because it involves changes in a crucially important process that is, presumably, optimized through evolution and can be disturbed by even small changes. Metabolic resistance, on the other hand, involves changes in a system that is already primarily designed for defense against toxic chemicals. This idea is supported by the appearance of metabolic resistance to all major classes of insecticides before the appearance of target site resistance (77). Because it is in the interest of an organism to be as well protected as possible within the constraints of energetics, metabolic resistance, once acquired, may facilitate the evolution of target site resistance by simple opportunity. Mutations leading to target site resistance have a better chance of being spread in a population if they occur in individuals that survive insecticide exposure by virtue of already having metabolic resistance.

As mentioned before, in the few cases of natural, chemical-mediated insect-plant associations where the resistance mechanisms of the insect have been studied in detail, the insects, with rare exceptions, rely on more than one defense. The tobacco hornworm is a good example. This species is highly specialized to feed on alkaloid-containing plants and has at least five resistance mechanisms against nicotine (15). Insects with a feeding specialty for cardenolide-containing plants, for instance monarch caterpillars and the large milkweed bug, have an insensitive target site, metabolic defenses, and ability to store the compounds in special structures or throughout their body tissues (see 77 for a review of several other cases). It seems clear that the evolutionary endpoint for defensive adaptations against toxic chemicals is a condition of several or multiple genetically heritable defenses. The thus far poorly studied "side effects" of insecticides, interference with biochemical and physiological processes less critical for survival than the target sites in the nervous system, may account for some of this apparent "striving" for as many defenses as possible.

The obvious implication for insecticide resistance is that persistent selection pressures with relatively few chemicals will result in fully resistant insects, a guild of insects specialized to feed on insecticide-treated crop plants (78, 79). Still, chemicals are among the primary defenses of plants against attack by insects, diseases, and other competition. The use of chemicals, synthetic or otherwise, is therefore obviously the most natural method for crop protection. A major problem in using chemicals is the mistaken notion that they can be used according to present-day economic marketing ideas; toxic chemicals must obviously rather be used in accordance with biological and evolutionary processes, of which we have inadequate knowledge.

Structure of chemicals and Evolution of Resistance. Resistance to the carbamates develops readily when exposure is persistent. In contrast, there are still many insects that are not resistant to the OPs, for instance, the boll weevil is still well controlled with azinphos-methyl, methyl parathion and other OPs and the German cockroach can still be controlled with chlorpyrifos and other OPs. There may be several reasons for this. First, the OPs are relatively degradable in the environment which avoids lengthy exposure to diminishing residues. Second, there are at least three enzyme systems involved in their detoxification, which spreads the selection pressure over several mechanisms instead of concentrating and intensifying it on one single mechanism. And third, the target site is quite flexible even though it constitutes a crucial factor in nerve impulse transmission. It accomodates several different isoenzymes with diminished affinity for inhibitors, implying that across-the-board cross resistance is not obligatory. The frequently seen combination of reduced affinity to inhibitors and the normal substrate may contribute to instability of target site resistance in the absence of a selection pressure.

The major difference between OPs and carbamates is in their metabolic fate. This strongly indicates that metabolism and detoxification are the most important factors in resistance evolution. The current emphasis on target site interactions and their associated resistance mechanisms reflects their perceived practical importance, the importance of their study for the development of new insecticidal molecules, and the fundability of such research. But detailed studies of the biochemical characteristics and physiological behavior of the enzymes involved in insecticide metabolism and detoxification are still very important.

Because of their relatively short residual activity, the OPs can be useful as components in integrated pest management programs. Because of their inhibitory action on carboxylesterases that detoxify pyrethroids, they can be employed in strategies of rotating mixtures. It seems quite possible that the structural diversity of the OPs has not yet been exhausted and that new OPs can be developed. This is also true for the methylcarbamates, and attempts are being made to develop proinsecticidal compounds of this type (80), which like the phosphothioates would need enzyme-catalyzed activation and so introduce ambiguity in the selection pressure.

Acknowledgments

I thank C. W. Holyoke, Jr., J. G. Hollingshaus, I. M. McDougall, and W. K. Moberg for discussions and helpful comments. This is New Jersey Agricultural Experiment Station Publication No. D-08111-32-88, supported by state funds and U. S. Hatch Act funds.

Literature Cited

1. Hollingworth, R. M. In Insecticide Biochemistry and Physiology, C. F. Wilkinson, Ed., Plenum: New York, 1976, pp. 431-506.
2. Ware, G. W. Pesticides - Theory and Application, Freeman and Co.: New York, 1983.
3. El-Sebae, A. H.; Metcalf, R. L.; Fukuto, T. R. J. Econ. Entomol. 1964, 57, 478-482.
4. Cremlyn, R. Pesticides - Preparation and Mode of Action, Wiley: New York, 1978.
5. Georghiou, G. P., Ann. Rev. Ecol. Syst. 1972, 3, 133-163.
6. Brattsten, L. B., J. Chem. Ecol. 1988, 14, 1919-1939.
7. Ahmad, S.; Brattsten, L. B.; Mullin, C. A.; Yu, S. J. In Molecular Aspects of Insect-Plant Associations L. B. Brattsten and S. Ahmad, Eds, Plenum: New York, 1986, pp. 73-151.

8. Moorefield, H. H. Contrib. Boyce Thompson Inst. 1958, **19**, 501-507.
9. Hodgson, E.; Casida, J. E. Biochem. Pharmacol. 1961, **8**, 179-191.
10. Metcalf, R. L.; Fukuto, T. R. J. Agr. Food Chem. 1965, **13**, 220-231.
11. Ray, J. W. Biochem. Pharmacol. 1967, **16**, 99-107.
12. Lewis, S. E.; Wilkinson, C. F.; Ray, J. W. Biochem. Pharmacol. 1967, **16**, 1195-1210.
13. Krieger, R. I.; Wilkinson, C. F. Biochem. Pharmacol. 1969, **18**, 1403-1415.
14. Yu, S. J. In Molecular Aspects of Insect-Plant Associations, L. B. Brattsten and S. Ahmad, Eds, Plenum: New York, 1986, pp. 153-174.
15. Brattsten, L. B. In Novel Aspects of Insect-Plant Interactions, P. Barbosa and D. K. Letourneau, Eds, Wiley: New York, 1988, pp. 313-348.
16. Georghiou, G. P.; Metcalf, R. L.; Gidden, F. E. Bull. World Hlth Organiz. 1966, **35**, 691-708.
17. Terriere, L. C. Ann. Rev. Entomol. 1968, **13**, 75-98.
18. Kuhr, R. J.; Dorough, H. W. Carbamate Insecticides: Chemistry, Biochemistry, and Toxicology CRC Press: Cleveland, 1976.
19. Fukuto, T. R.; Fahmy, M. A. H.; Metcalf, R. L. J. Agr. Food Chem. 1967, **15**, 273-281.
20. Berenbaum, M. R. Amer. Nat. 1980, **115**, 138-146.
21. Ranasinghe, L. E.; Georghiou, G. P. Pestic. Science, 1979, **10**, 502-508.
22. Raffa, K. F.; Priester, T. M. J. Agric. Entomol. 1985, **2**, 27-45.
23. Wilkinson, C. F. In Agricultural Chemicals of the Future, J. L. Hilton, Ed., Rowman and Allanheldublishers: Totowa, N. J., pp. 311-325.
24. Devonshire, A. L.; Moores, G. D. Pestic. Biochem. Physiol. 1982, **18**, 235-246.
25. Helling, D. J.; Brownie, C.; Guthrie, F. E. Pestic. Biochem. Physiol. 1986, **25**, 125-132.
26. Dolinger, P.M.; Ehrlich, P.R.; Fitch, W.L.; Breedlove, D.E. Oecologia (Berl.), 1973, **13**, 191-204.
27. Nakatsugawa, T.; Morelli, M. A. In Insecticide Biochemistry and Physiology, C. F. Wilkinson, Ed., Plenum: New York, 1976, pp. 61-114.
28. Oppenoorth, F. J. In Comprehensive Insect Physiology, Biochemistry, and Pharmacology, G. A. Kerkut and L. I. Gilbert, Eds, Pergamon: Oxford, 1985, **12**, 731-773.
29. Matsumura, F.; Hogendijk, C. J. J. Agr. Food Chem. 1964, **12**, 447-452.
30. Rowland, M.; Hemingway, J. Pestic. Biochem. Physiol. 1987, **28**, 239-247.
31. Georghiou, G. P.; Pasteur, N. J. Econ. Entomol. 1978, **71**, 201-205.
32. Georghiou, G. P.; Pasteur, N. J. Econ. Entomol. 1980, **73**, 489-492.

33. Georghiou, G. P.; Pasteur, N.; Hawley, M. K. J. Econ. Entomol. 1980, **73**, 301-305.
34. Asher, K. R. S.; Eliyahu, M.; Ishaaya, I.; Zur, M.; Ben-Moshe, E. Phytoparasitica, 1986, **14**, 101-110.
35. Hodgson, E.; Casida, J. E. J. Agr. Food Chem. 1962, **10**, 208-214.
36. Shishido, T.; Fukami, J. Botyu-Kaguku, 1963, **28**, 69-76.
37. Yu, S. J. Pestic. Biochem. Physiol. 1982, **18**, 101-106.
38. Litwack, G.; Ketterer, B.; Arias, I. M. Nature, 1971, **234**, 466-467.
39. Motoyama, N.; Dauterman, W. C. Revs Biochem. Toxicol. 1980, **2**, 49-69.
40. Clark, A. G.; Shaaman, N. A.; Dauterman, W. C.; Hayaoka, T. Pestic. Biochem. Physiol. 1984, **22**, 51-59.
41. Clark, A. G.; Shaaman, N. A. Pestic. Biochem. Physiol. 1984, **22**, 249-261.
42. Lamoreux, G. L.; Rusness, D. G. Pestic. Biochem. Physiol. 1987, **27**, 318-329.
43. Wilkinson, C. F. In Foreign Compound Metabolism, J. Caldwell and G. D. Paulson, Eds, Taylor and Francis: New York, 1984, pp. 133-147.
44. Hodgson, E. In Comprehensive Insect Physiology, Biochemistry, and Pharmacology, G. A. Kerkut and L. I. Gilbert, Eds, Pergamon: Oxford, 1985, **11**, 206-321.
45. Agosin, M. In Op. Cit. **12**, 647-712.
46. Breer, H., J. Comp. Physiol. 1981, **141**, 271-275.
47. Pichon, Y.; Manaranche, R. In Comprehensive Insect Physiology, Biochemistry, and Pharmacology, G. A. Kerkut and L. I. Gilbert, Eds, Pergamon: Oxford, 1985, **10**, 417-450.
48. Jarv, J.; Aaviksaar, A.; Godovikov, N.; Lobanov, D., Biochem. J. 1977, **167**, 823-825.
49. Hetnarski, B.; O'Brien, R. D., Biochemistry 1973, **12**, 3883-3887.
50. Corbett, J. R.; Wright, K.; Baillie, A. C. The Biochemical Mode of Action of Pesticides, Academic Press: New York, 1984, pp. 99-140.
51. O'Brien, R. D. In Insecticide Biochemistry and Physiology, C. F. Wilkinson, Ed. Plenum: New York, 1976, pp. 271-296.
52. Fuchs, R. A.; Schroder, R. In Chemistry of Pesticides, K. H. Buchel, Ed., Wiley: New York, 1983, pp. 9-226.
53. Smissaert, H. R., Science 1964, **143**, 129-131.
54. Hama, H.; Iwata, T., Appl. Entomol. Zool. 1971, **6**, 183-191.
55. Ayad, A.; Georghiou, G. P., J. Econ. Entomol. 1975, **68**, 295-297.
56. Tripathi, R. K.; O'Brien, R. D. Pestic. Biochem. Physiol. 1973, **3**, 495-498.
57. Devonshire, A. L. Biochem. J. 1975, **149**, 463-470.

58. Oppenoorth, F. J.; Smissaert, H. R.; Welling, W.; van der Pas, L. J. T.; Hitman, K. T. Pestic. Biochem Physiol. 1977, 7, 34-47.
59. Voss, G. J. Econ. Entomol. 1980, 73, 189-192.
60. Brown, A. W. A.; Pal, R., World Health Organization Monograph Series No. 38, Geneva, 1971.
61. Devonshire, A. L.; Moores, G. D. Pestic. Biochem. Physiol. 1984, 21, 341-348.
62. Hama, H. In Pesticide Chemistry - Human Welfare and the Environment, 1982, J. Miyamoto and P. C. Kearney, Eds, Pergamon: New York, 1983, 3, 203-208.
63. Sawicki, R. M; Denholm, I. 1984 Ciba Found. Symp. 102, 152-166.
64. Devonshire, A. L.; Moores, G. D. Pestic. Biochem. Physiol. 1984, 21, 336-340.
65. Duday, Y. Dros. Inform. Serv. 1977, 52, 65-66.
66. Berenbaum, M. R. In Molecular Aspects of Insect-Plant Associations, L. B. Brattsten and S. Ahmad, Eds, Plenum: New York, 1986, pp. 257-272.
67. Yamamoto, I.; Takahashi, Y; Kyomura, N. In Pest Resistance to Pesticides, G. P. Georghiou and T. Saito, Eds, Plenum: New York, 1983, pp. 579-594.
68. Tallamy, D. W. In Molecular Aspects of Insect-Plant Associations, L. B. Brattsten and S. Ahmad, Eds, Plenum: New York, 1986, pp 273-300.
69. Weaver Jr., J. B.; Reddy, M. S. J. Econ. Entomol. 1977, 70, 283-285.
70. Roush, R. T.; Plapp, Jr., W. F. J. Econ. Entomol. 1982, 75, 708-713.
71. McKenzie, J. A.; Whitten, M. J.; Adena, M. A., Heredity, 1982, 49, 1-9.
72. Heather, N. W., Queensl. J. Agr. Anim. Sci. 1982, 39, 61-68.
73. Pozarowska, B. J. Bull. ent. Res. 1987, 77, 123-134.
74. Halpern, M. E.; Morton, R. A. Pestic. Biochem. Physiol. 1987, 28, 44-56.
75. El-Khatib, Z. I.; Georghiou, G. P. J. Econ. Entomol. 1985, 78, 1023-1029.
76. Argentine, J. A.; Clark, J. M.; Ferro, D. N. Environ. Entomol. 1989, 18, 705-710.
77. Brattsten, L. B. In Molecular Aspect of Insect-Plant Associations, L. B. Brattsten and S. Ahmad, Eds, Plenum: New York, 1986, pp. 211-256.
78. Atsatt, P. R.; O'Dowd, D. J. Science, 1976, 193, 24-29.
79. Brattsten, L. B. Pestic. Sci. 1989, 26, in press.
80. Fahmy, M. A.; Fukuto. T. R. In Pesticide Chemistry - Human Welfare and the Environment, J. Miyamoto and P. C. Kearney, Eds, Pergamon: New York, 1983, 1, 193-200.

RECEIVED October 27, 1989

Chapter 4

Biochemical and Genetic Mechanisms of Insecticide Resistance

Thomas M. Brown

Department of Entomology, Clemson University, Clemson, SC 29634

> The biochemical mechanisms of insecticide resistance are categorized as affecting insecticide pharmacokinetics or pharmacodynamics. The enzymes and targets involved are reviewed from the perspective of understanding the genetic basis for each mechanism. This perspective is important since insecticide resistance is a problem in population genetics and each case of resistance must be managed according to the particular resistance gene, or combination of genes, present.

The source of the insecticide resistance problem is traced ultimately from biochemical mechanisms to changes in nucleic acid chemistry of genes which confer resistance. Advances in genetics and recombinant DNA have provided the opportunity for detailed study of genes that produce biochemically resistant populations.

Insecticide Resistance Mechanisms. These can be catagorized as behavioral avoidance and physiological changes which allow survival upon contact with the insecticide. Major physiological mechanisms are categorized by whether they influence insecticide pharmacokinetics (the penetration, distribution, metabolism, and elimination of a drug or other xenobiotic) or insecticide pharmacodynamics (the interaction of a drug or other xenobiotic with its site of action). Consideration will be given to the recent advances in the molecular genetics of resistance mechanisms known or suspected to be important (Table I).

This review will focus upon detoxication as a pharmacodynamic mechanism and also pharmacodynamic mechanisms in which various targets of insecticides, often proteins in the nervous system, become less sensitive to poisoning.

Besides detoxication, another pharmacokinetic mechanism of resistance, retarded penetration, has been identified and mapped to linkage group III of the house fly where it is known as *Pen* (1) or *tin* (2). This gene imparts low resistance; however, its combination with other mechanisms can be synergistic in producing higher than expected resistance. Because the biochemical basis of this mechanism is unknown, it will not be discussed further in this paper.

The genetic basis of insecticide resistance is not well understood. Many genes for resistance traits have been mapped to chromosomes of the house fly, *Musca domestica* (3,4), *Drosophila melanogaster* (78), and mosquitoes, *Aedes aeqypti* and *Culex quinquefasciatus* (5,6). In some cases, the expression of the biochemical mechanism, such as increased activity of enzyme or decreased sensitivity of target, has also been mapped to the same locus as resistance; however, few studies have established the genetic mechanism responsible for resistance by mapping the biochemical polymorphism or analyzing the gene sequence.

Table I. Principal Physiological Mechanisms of Insecticide Resistance

Pharmacokinetic mechanisms	Pharmacodynamic mechanisms (reduced sensitivity of target)
Decreased penetration	Acetylcholinesterase
	Sodium ion channel
Enhanced detoxication	γ-Aminobutyric acid (GABA) receptor
Monooxygenase	
Arylester hydrolase	
Carboxylester hydrolase	
Catalytic hydrolysis	
Insecticide sequestering	
Glutathione S-transferase	

Beyond these examples, there is very little known of the genetics of insecticide resistance in many important insects of agriculture which are the targets for much of the insecticide used in the world; e. g., boll weevil, tobacco budworm, codling moth, and Colorado potato beetle are practically unknown genetically. The present challenge is to apply new techniques of molecular genetics to gain a better understanding of resistance in many of the pests of greatest economic impact for the future development of insecticide resistance management.

A case in point is the unraveling of resistance to methyl parathion in the tobacco budworm, *Heliothis virescens*, which is a major pest of cotton as well as tobacco. In South Carolina, there is very severe, stable resistance. Although pyrethroid insecticides are very effective and there is no resistance to them in South Carolina at this time, it would be very useful to understand the genetic basis of methyl parathion resistance in case resistance to pyrethroids should arise in the future or spread eastward from Texas where it has been detected. Recent investigations with this pest will be described to illustrate certain mechanisms.

Enzymes Catalyzing Insecticide Biotransformation

Even a rather simple insecticide such as methyl parathion is transformed by insects in a complex manner. The parent insecticide is activated to methyl paraoxon, which is a more potent inhibitor of the target, acetylcholinesterase in the nerve (Figure 1). This activating desulfuration is catalyzed by monooxygenases. Both the parent and the oxon are subject to detoxication by monooxygenase and glutathione transferase, while the oxon is also more labile to hydrolysis.

Monooxygenases. These enzymes are important in the detoxication of pyrethroid, carbamate, organophosphorus and other classes of insecticides ([7]). They are microsomal, membrane associated enzymes which catalyze reactions in which one atom from molecular oxygen is inserted into the insecticide and the second oxygen atom is reduced to form water. Catalysis depends on the close association of the heme-containing cytochrome P450 terminal oxidase with NADPH cytochrome C reductase for electron transport, and it also depends on availability of NADPH and oxygen.

Monooxygenase Inhibitors as Synergists. Inhibitors of cytochrome P450 are used as preliminary diagnostic synergists ([8]) to detect resistance based on a mechanism of monooxygenase catalyzed detoxication (Figure 2). Fenvalerate resistance in diamondback moth in Taiwan and also in cotton bollworm, *Heliothis armigera*, in Australia can be synergized with piperonyl butoxide (9,10). In permethrin-resistant *Heliothis virescens*, the tobacco budworm, piperonyl butoxide was not synergistic; however, 1,2,4-trichloro-3-(2-propynyloxy)benzene (Figure 2) was synergistic with permethrin (11). In resistant Egyptian cotton leafworms, *Spodoptera littoralis* , propynyl monooxygenase synergists were 2.5-fold more active than piperonyl butoxide when used with monocrotophos (12).

Lack of synergism must be interpreted cautiously, especially with insecticides which are metabolically activated. Bioactivation of organophosphorothioate, formamidine and certain cyclodiene insecticides is catalyzed by monooxygenases. Piperonyl butoxide was synergistic with the carbamate, propoxur; however, in the same strains of house flies, it was antagonistic with the phosphorothioate, diazinon, which requires monooxygenase desulfuration to form diazoxon, the oxon inhibitor of acetylcholinesterase (13). Chordimeform was also antagonized by piperonyl butoxide in the Australian cattle tick (14). Chordimeform is bioactivated to N-dealkylation products which are much more potent octopamine mimics in the fire fly (15).

In methyl parathion resistant tobacco budworms, larvae were treated, lots of 10 were homogenized, and methyl parathion was recovered by solid phase extraction and analyzed by reversed phase high performance chromatography with ultraviolet detection. Unexpectedly, the resistant strains lost methyl parathion at a lesser rate than the susceptible Florence 1987 strain (Figure 3). Independently, T. Konno et. al. have found a slower bioactivation of [^{14}C]-methyl parathion in another resistant strain of this pest (79).

We observed that both 1,2,4-trichloro-3-(2-propynyloxy)benzene and S,S,S-tributylphosphorotrithioate (a hydrolase inhibitor) were synergists for methyl parathion in Woodrow 1983-resistant 35 mg larvae. Topical doses were applied in acetone for a 48 h exposure (Table II). This suggests both monooxygenase and hydrolase involvement in resistance. Others (Konno, T.; Dauterman, W., North Carolina State University, personal communication, 1988) have confirmed our observation of synergism using their NC-86 resistant larvae. They also found the propynyl synergist was more effective with methyl paraoxon (Table II). Both laboratories found less than 2-fold synergism by piperonyl butoxide in methyl parathion resistant strains (16).

Table II. Synergism of Methyl Parathion in *Heliothis virescens*, the Tobacco Budworm

Strain	Insecticide	Synergist	Median lethal dose, mg/kg		Ratio
			Insecticide	Insecticide + Synergist	
NC-S	m. paraoxon	TCPB[a]	5.00	4.74	1.05
NC-R	m. paraoxon	TCPB[a]	182	76.4	2.38
NC-S	m. parathion	TCPB[a]	10.8	12.7	0.85
NC-R	m.parathion	TCPB[a]	618	433	1.43
SC-R	m. parathion	none	2600		
SC-R	m. parathion	TCPB[b]	28% at 400	75% at 400	2.68
SC-R	m. parathion	TBPT[b]	28% at 400	97% at 400	3.46
SC-S	m. parathion	TBPT[c]	815	155	5.26

[a] Synergist 1,2,4-trichloro-3-(2-propynyloxy)benzene at 2857 mg/kg; data from T. Konno and W. Dauterman (see text above).
[b] Synergist 1,2,4-trichloro-3-(2-propynyloxy)benzene at 2000 mg/kg.
[c] Synergist S,S,S-tributylphosphorotrithioate (16).

Genetic Control of Monooxygenases. There are many forms of P450, even within a species, with differences in spectra, substrate specificity, electrophoretic mobilities and

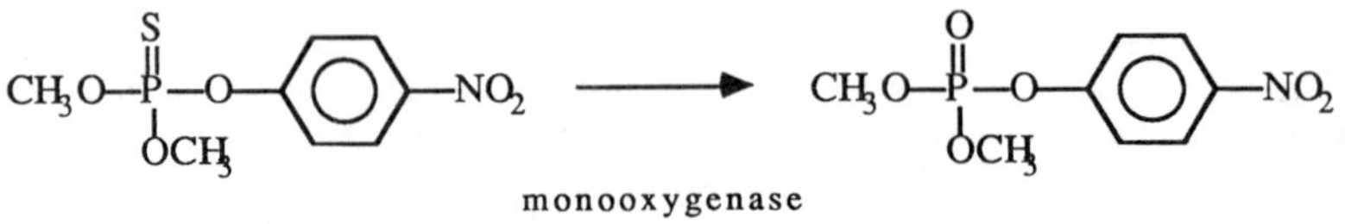

methyl parathion methyl paraoxon

Bioactivation

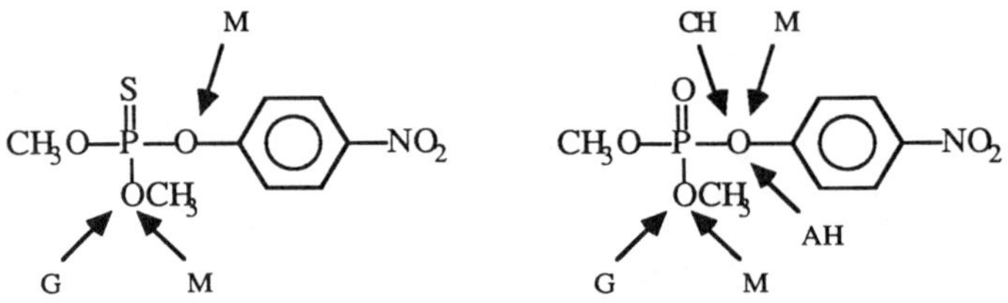

Detoxication

Figure 1. Bioactivation and detoxication of methyl parathion. AH is arylester hydrolase, CH is carboxylester hydrolase, G is glutathione transferase, and M is monooxygenase.

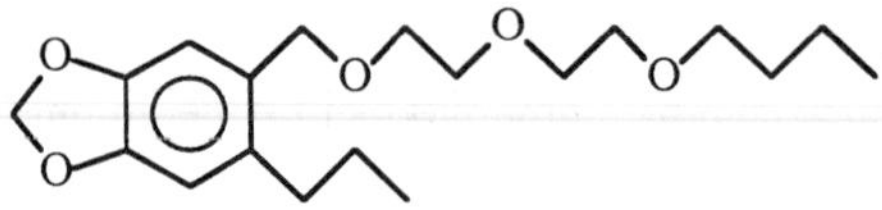

piperonyl butoxide

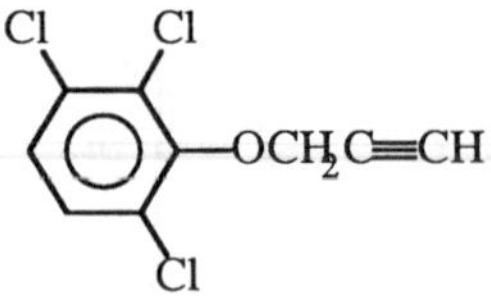

TCPB

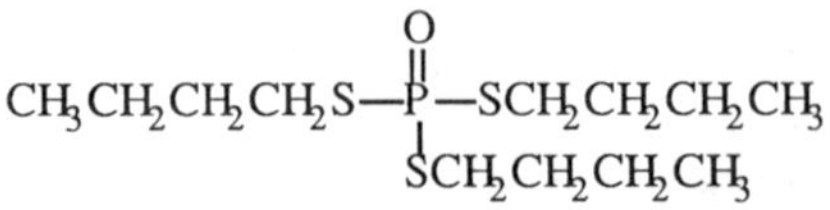

TBPT

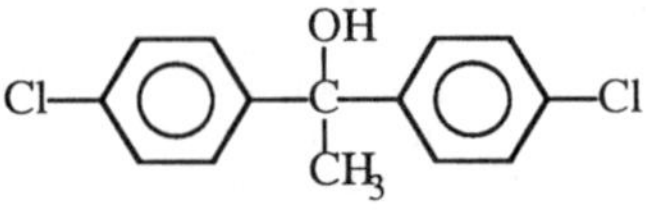

chlorfenethol

Figure 2. Diagnostic synergists for insecticide resistance described in the text.

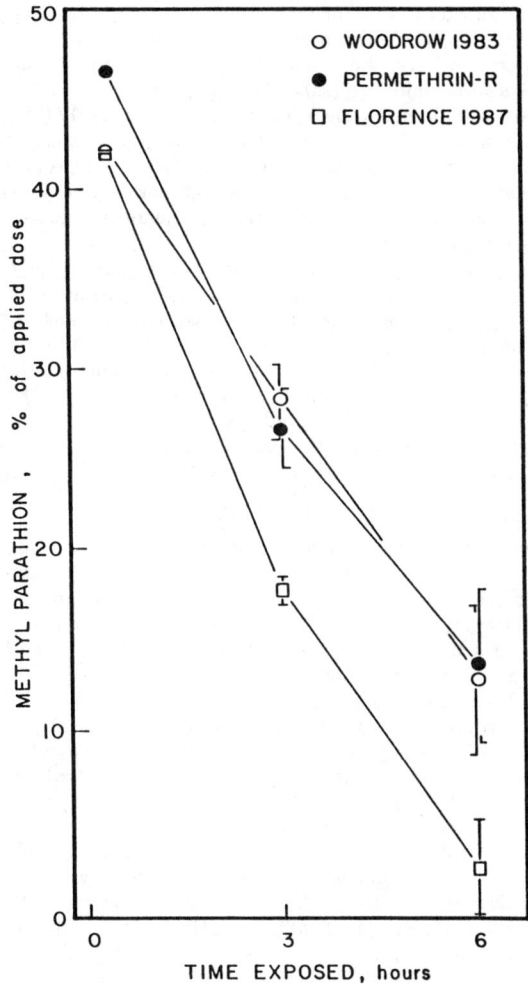

Figure 3. Methyl parathion remaining after topical application to tobacco budworm larvae *in vivo*. Florence 1987 strain was resistant; other strains were susceptible.

specificity of inducing agents. Cloning of about 70 P450 genes, primarily from mammals, has provided a systematic classification of these forms based on the amino acid sequences inferred from the nucleotide sequences. There are at least 10 gene families, each having no more than 36% similarity in amino acid sequence to the others, and families II and XI are comprised of subfamilies containing genes coding proteins with no less than 67% similarity to each other (17).

Recently, a P450 clone was prepared from phenobarbital induced, diazinon resistant, house flies (18). While the amino acid sequence near the heme-binding site is conserved compared to others, the house fly gene was assigned a new family, P450VI, because its overall amino acid sequence was insufficiently similar to any previous family. Several forms of P450 are known to exist in the house fly (19,20).

Monooxygenase activity is strongly inducible in both insects (21) and mammals (17). The quantity of the enzyme increases 10 to 100-fold upon prolonged exposure to an inducing agent, probably due to a more rapid rate of gene transcription. There is a wide chemical diversity of inducers (Figure 4) which are usually selective for the family of P450 which is induced (17).

The relationship of induction to insecticide resistance has not been established. In mice, induction of P450I results from increased transcription of the structural gene as controlled by a cytosolic receptor which is the protein product of the *Ah* locus (17). Resistance due to monooxygenases is influence by genes *ox* and *md* on chromosomes 2 and 5, respectively, in the house fly (3). Similarly, two genes confer monooxygenase-based resistance in *Drosophila,* and there is evidence that one is regulatory in nature (22). A working hypothesis is that unlinked regulatory and structural gene mutations contribute; however, multiple forms of insect P450 enzymes exist so there could be mutant structural genes on separate chromosomes.

Hydrolases. Hydrolytic mechanisms are also important in insecticide resistance, despite the apparent low activities in resistant insects when compared to mammalian enzymes (Table III). Some strains of resistant mosquitoes (23), *Tribolium* beetles (24), and Indianmeal moth (25) have specific resistance for malathion and similar carboxylester insecticides. This is due to increased catalytic hydrolysis, possibly through production of a more efficient enzyme (25,26). Californian tobacco budworms with low level permethrin resistance exhibited twice the normal activity of *trans*-permethrin carboxylester hydrolase (27).

It is clear that the insecticide of interest must be used to assay for this type of resistance since 1-naphthyl acetate, a general substrate used to stain for many hydrolases, did not detect malathion carboxylester hydrolase (assayed as the hydrolysis of [^{14}C]-malathion) in starch gel electrophoresis of *Culex tarsalis* (23). Similarly, *trans*-permethrin hydrolyzing activity of soybean looper was resolved from 1-naphthyl acetate hydrolyzing activity by polyacrylamide gel electrophoresis (28). Malathion carboxylester hydrolases were not correlated with activity toward 1-naphthyl acetate in *Drosophila, Anopheles*, and Indianmeal moth (25,29,30).

Malathion carboxylester hydrolase activity was genetically linked to chromosome 2 of the house fly as demonstrated in the Wakamatsu strain (31). Malathion resistance in this strain was nearly completely dominant, and the resistance was also linked to chromosome 2. It was found that this highly resistant strain also possessed enhanced glutathione dependent malathion detoxication, as well as acetylcholinesterase which was 76-fold less sensitive to malaoxon. These flies were cross-resistant to the non-carboxylester organophosphorus insecticides fenitrothion and trichlorfon, probably due to glutathione transferase or acetylcholinesterase factors; however, they were fully susceptible to pyrethroid insecticides containing carboxylesters, suggesting that the carboxylester hydrolase was selective. While resistance is highly dominant, malathion hydrolysis in hybrid progeny of Wakamatsu and susceptible house flies was only 60% as much as malathion hydrolysis observed in the homozygous resistant flies.

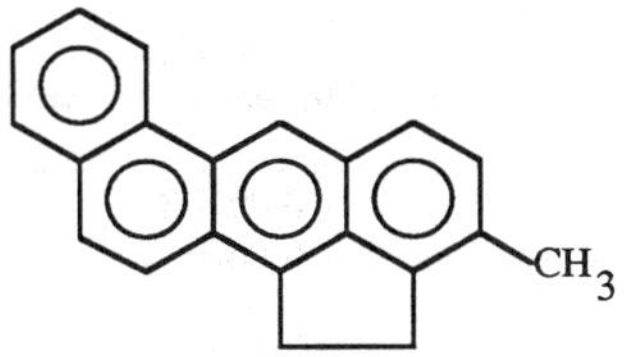

3-methylcholanthrene (P450I)

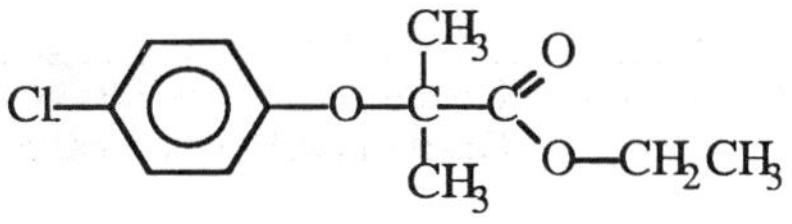

clofibrate (P450IV)

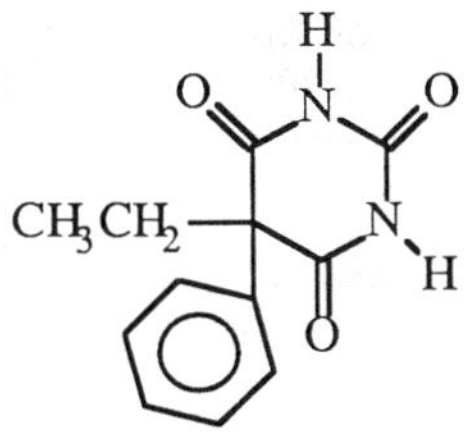

phenobarbital (P450II , insect)

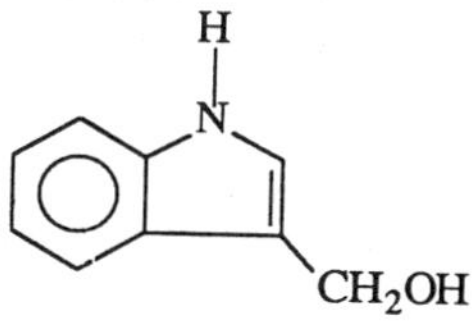

3-indolemethanol (insect)

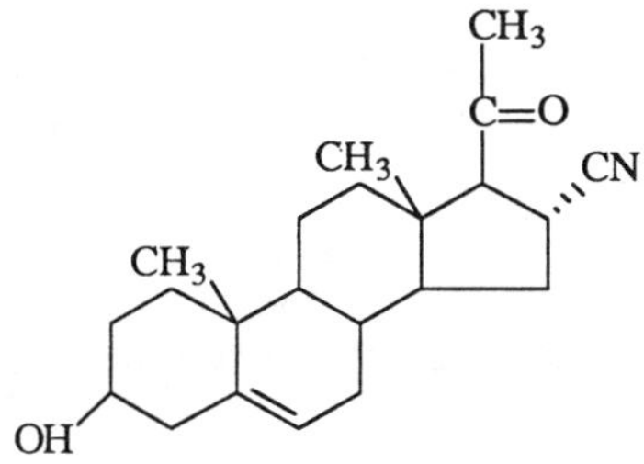

pregnenolone 16 α-carbonitrile (P450III)

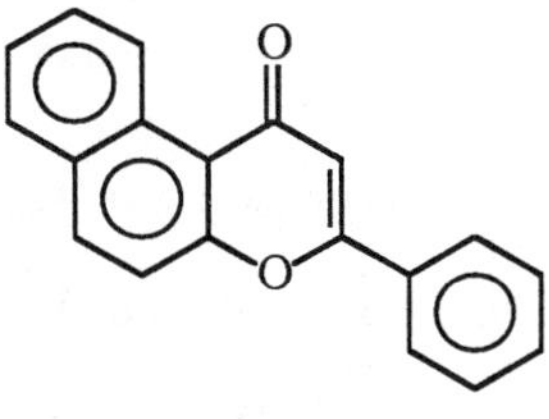

5,6-benzoflavone (insect)

Figure 4. Inducers of monooxygenases for various P450 families or for insect P450.

Malathion specific resistance was also fully dominant in Indianmeal moth, while enzyme activity was semidominant, with hybrid offspring having one-half the activity of resistant homozygotes (25). These data in house flies and moths are consistant with a mechanism of catalytic hydrolysis with one resistance allele giving sufficient enzyme activity for survival.

Table III. Insecticide Hydrolysis by Enzymes of Mammals and Resistant Arthropods

Insecticide	Enzyme or preparation	Activity (nmol/min/mg protein)	
Paraoxon	House fly carboxylester hydrolase	0.00733	(71)
	Green peach aphid carboxylester hydrolase	0.256a	(34)
	Bovine serum albumin	0.044	(72)
	Rabbit serum arylester hydrolase	642	(42)
	Sheep serum arylester hydrolase	1003	(43)
Malathion	Spider mite carboxylester hydrolase	0.0203	(26)
	House fly carboxylester hydrolase	0.937	(71)
	Indianmealmoth midgut preparation	3.94	(73)
	Rat serum	84.8a	(74)
	Rabbit liver carboxylester hydrolase	4160a	(75)
	Rat liver carboxylester hydrolase	7630; 16700a	(76)
trans-Permethrin			
	Southern armyworm cuticle preparationb	3.62a	(77)
	Soybean looper midgut preparationb	23.9a	(28)
	Rat liver carboxylester hydrolase	70.0	(76)
	Porcine liver carboxylester hydrolase	322	(28)

a V_{max}
b Insects were not reported to be resistant.

<u>Amplification of a Normal Hydrolase Results in Insecticide Sequestration</u>. Gene amplification to produce geometric increases in carboxylester hydrolase quantity provides a second type of hydrolytic mechanism in certain resistant insects. This genetic mechanism has been found using radiolabelled cDNA of the gene to estimate the quantity of the gene and its transcription in *Culex quinquefasciatus* (32) and in *Myzus persicae* (33). Resistance is conferred to many organophosphorus insecticides, including non-carboxylesters such as parathion, and to carbamates and pyrethroid in the aphids. The tremendous quantity of the enzyme is a sink for organophosphorus oxon bioactivation products which phosphorylate this enzyme more rapidly than the target.

This enzyme, purified from aphids (34), was very slow in hydrolyzing paraoxon compared to mammalian arylester hydrolases (Table III); however, it was twice as fast in recovery from paraoxon inhibition compared to a porcine carboxylester hydrolase (35) and >300 times faster than monomeric carboxylester hydrolase of rabbit liver (36). No qualitative differences were found in the enzyme, E4, isolated from resistant and susceptible aphids; E4 was one of seven electrophoretic forms of hydrolases observed (34). Recovery indicates that resistance is due to both reaction with the insecticide and a very slow turnover, or catalysis. A similar mechanism of was observed with paraoxon in resistant green rice leafhoppers (37).

Resistant aphids are identified individually in microtiter plate wells by immunological binding of E4 followed by colorimetric assay with 1-naphthyl butyrate (38). In this aphid, there appears to be no other major resistance mechanism; however, a very similar mechanism in mosquitoes can be accompanied by other mechanisms, including insensitive acetylcholinesterase (39,40).

Carboxylester hydrolase is strongly inhibited by many organophosphates such as paraoxon. This is the reason that malathion resistance is overcome by its mixture with many other organophosphorus insecticides. In fact, malaoxon is both a substrate and an inhibitor depending on the orientation of the molecule, which contains both carboxylester and phosphate ester chemistry (41). S,S,S-tributylphosphorotrithioate (Figure 2) and triphenyl phosphate, which are not insecticidal, have been used as diagnostic synergists for this mechanism. Because resistance to malathion is increasing, it may become practical to seek synergistic mixtures against resistant pests; however, such mixtures could have much greater toxicity to mammals as well.

Arylester Hydrolase and Parathion Hydrolase. Arylester hydrolase in mammalian serum (42,43) catalyzes very efficient hydrolysis of the organophosphorus oxons such as paraoxon (Table III). This enzyme does not hydrolyze the parent phosphorothioate insecticides such as methyl parathion, but only the oxon metabolites such as methyl paraoxon (44). This enzyme is has been referred to as phosphotriesterase; however, a triester is not required as seen in the rapid hydrolysis of 10 organophosphinates by rabbit serum arylester hydrolase (44). This mechanism appears only rarely in birds (45) and appears to be lacking in most insects. Recently, an arylester hydrolase requiring cobalt was isolated from a methyl parathion resistant strain of tobacco budworm (79).

A parathion hydrolase has been described to hydrolyze both phosphorothioates and phosphates and its gene, *opd*, cloned from a plasmid carried by *Psudomonas diminuta* (46). Fortunately, insects apparently lack this type of enzyme. Forty years of effective applications of organophosphorus insecticides can be attributed to some extent to the general lack of significant arylester hydrolase activities in pest insects.

Glutathione S-Transferases. These enzymes have been associated with resistance in house flies and in mites, but have not been associated with resistance in as many species as monooxygenases or hydrolases (47). This may be due in part to less testing for this mechanism. In the house fly, resistance due to glutathione S-transferase is linked to chromosome 2 and the mechanism apparently results from a qualitative change in the enzyme (48).

Cloning and analysis of glutathione S-transferase genes of mammals and plants has revealed that there are various gene families; however, this analysis has not been accomplished in insects (49). These enzymes are inducible, but generally not to the degree of the monooxygenases (48,50).

DDT-dehydrochlorinase is a glutathione-dependent enzyme which is inhibited by chlorfenethol (Figure 2). It is linked to chromosome 2 of the house fly (4).

Altered Targets

Acetylcholinesterase. Altered acetylcholinesterase less sensitive to organophosphorus and carbamate insecticides has been observed in a wide variety of insects and mites (51). Acetylcholinesterase inhibiting insecticides phosphorylate or carbamylate the serine residue in the active site of the enzyme preventing vital catalysis of acetylcholine. Resistance due to reduced sensitivity to inhibition of this target enzyme has been found in house fly, mosquitoes, green rice leafhopper, and both phytophagous and predacious species of mites.

In the house fly, at least two resistance alleles occur. They can be discriminated by relative insensitivities to azamethiphos and dichlorvos (52). A Danish strain, 49R, was less sensitive to azamethiphos. Conversely, a German strain, Weymann, was sensitive to azamethiphos, but insensitive to dichlorvos. Individual flies of these two strains, a susceptible strain and the three possible hybrids could be identified by inhibition kinetics measured in a microtitre plate reader; apparently, both resistance alleles were incompletely dominant. Previous investigations with tetrachlorvinphos resistant house flies indicated that resistance was due to a reduced affinity for binding the insecticide (53).

In resistant green rice leafhopper, an acetylcholinesterase with reduced affinity for N-methyl carbamate insecticides had increased sensitivity to inhibition by longer N-alkyl groups; however, the inverse relationship was observed against the susceptible enzyme (54). The optimal substitution appeared to be N-(n-propyl) and the use of this chemistry in a resistance breaking strategy was discussed. Unfortunately, N-(n-propyl)propoxur was inhibitory of neither susceptible nor resistant acetylcholinesterase of predatory mites, *Amblyseius potentillae* so that this type of vulnerability is not common to all resistant acetylcholinesterases (55).

In general, resistant acetylcholinesterases are less sensitive as indicated by a smaller bimolecular reaction constant, k_i, for phosphorylation of the active site. In our studies of methyl parathion resistant tobacco budworm larvae, lots of ten larval nervous systems were homogenized and k_i was determined (56). We observed 25-fold less sensitivity to inhibition to methyl paraoxon in the resistant strain (Table IV).

Table IV. Reduced Rate of Acetylcholinesterase Phosphorylation in Resistant Tobacco Budworm Larvae, *Heliothis virescens*

Strain	Median Lethal Dose, mg/kg	methyl paraoxon	k_i, M^{-1}min^{-1} ethyl paraoxon
Florence 87 (S)	< 40	93400	120000 ±35090
Woodrow 83 (R)	2600	3800 ±1430	13300 ±409

Recovery of phosphinylated acetylcholinesterase was also greatly enhanced in our studies of methyl parathion-resistant tobacco budworm larvae (Figure 5). Phosphinylated acetylcholinesterase recovery can be measured without the complication of "aging" which occurs with phosphorylated enzyme (Figure 6). Activity was totally inhibited with 4-nitrophenyl methyl(phenyl)phosphinate. Excess inhibitor was removed by solid phase extraction and recovering activity at 30°C was monitored in aliquots. We have confirmed this insensitivity with individual heads of adults from these strains and we are investigating inheritance of this trait.

There appear to be several, if not many, possible mutations which can confer various types of resistant acetylcholinesterases. Genetic approaches must be exploited to clarify this situation, including the cloning and analysis of genes for resistant acetylcholinesterases from which the critical amino acid substitutions can be discerned. To date, this gene has been cloned and sequenced from one insect, *Drosophila*, which was found to possess the identical active site sequence found in human, horse, and *Torpedo* cholinesterases (57).

Ion Channels. Voltage-gated sodium ion channel of nerve appears to be the major target for the action of pyrethroid insecticides (58). While resistant strains are known in several species, the mechanism of physiological insensitivity has not been established at the molecular level. Two recent reports presented evidence that genes *kdr* in the house fly (59) and *nap^{ts}* in *Drosophila melanogaster* (60) act through reducing the number of sodium ion channels in nerve membranes to provide pyrethroid insensitivity and resistance. Nerve insensitivity appears to be the major

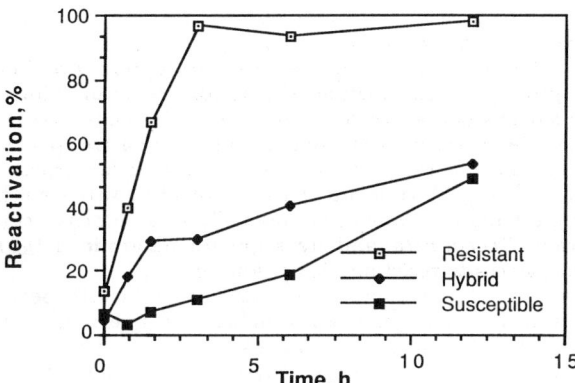

Figure 5. Spontaneous reactivation of acetylcholinesterase from tobacco budworm larvae.

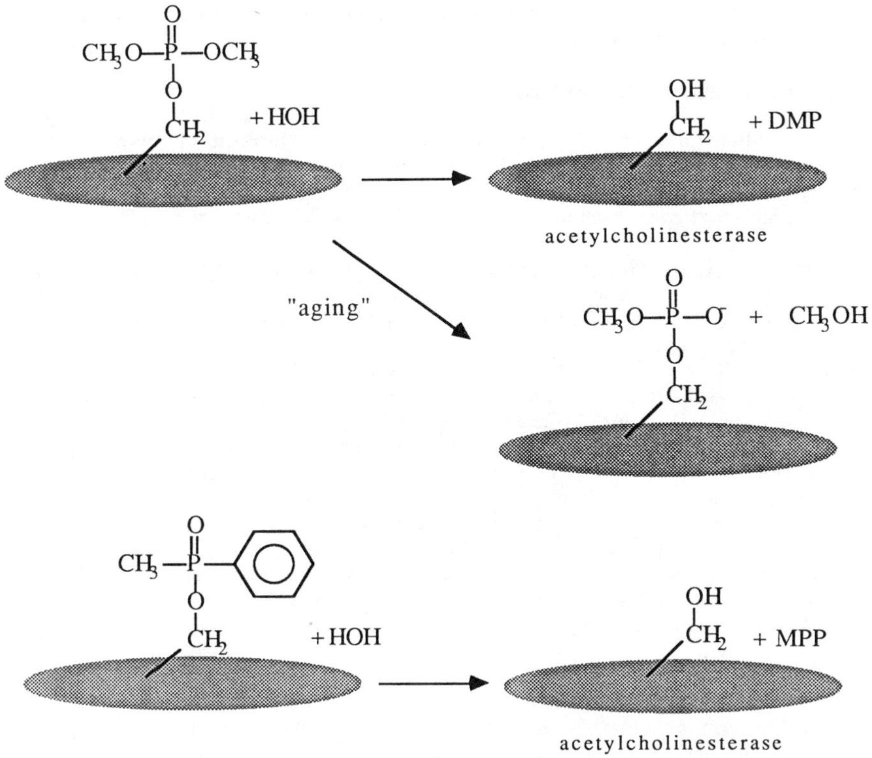

Recovery and aging

Figure 6. Comparison of phosphorylated and phosphinylated acetylcholinesterase recovery.

mechanism in a strain of the tobacco budworm in which segregation of high permethrin resistance was demonstrated (61).

Other ion channels functioning as neuroreceptors are known targets of insecticides (Figure 7). These include the GABA receptor which is a target of pyrethroids, cyclodienes and avermectin, and the acetylcholine receptor which is the target of nicotine. These receptors are similar in forming a transmembrane pore.

Only one sodium ion channel subunit is required for expression in *Xenopus* oocytes injected with mRNA from its cloned DNA (62); this gene has four similar domains, each containing six transmembrane helices. A putative sodium channel has been cloned from *Drosophila* and its sequence determined (63); however, the relationship of sequences to resistance is not known.

GABA receptor, a target for several classes of insecticides, consists of two subunits, each with four transmembrane helices as determined from cloning and sequence analysis (64). Expression in *Xenopus* oocytes requires mRNA from only one of the subunits (65). Dieldrin resistant cockroaches were found to have reduced specific binding of picrotoxinin, which acts on the GABA receptor chloride ion channel (66).

The acetylcholine receptor is also a transmembrane ion channel, and it is composed of four different subunits (67). Chimeric mRNA from recombinant clones of different species has been expressed in *Xenopus* oocytes. This is the target of nicotine, a classical agonist for discriminating among receptor subtypes. This receptor has not been evaluated as a resistance mechanism, although it is involved in the process of poisoning by many insecticides.

Prospects and Imperative Research

Genetic studies have raised many new questions. In pursuit of answers to these questions, the knowledge gained should lead to improved strategies for resistance management. In conclusion, some of these questions will be considered, and then prospects for genetic research on major pest species will be discussed.

Questions Involving Genetic Mechanisms. Gene amplification, the presence of multiple copies of the gene of interest, provides two resistant species with greatly elevated carboxylester hydrolase (see above). While gene amplification is the basis of this type of resistance in two species, is it common to other biochemical mechanisms? Other possibilities include mutations in structural genes for the enzyme or target protein of interest, or mutations in regulatory genes which could alter the rate of transcription or processing of the structural gene (68). Indeed, mutations of structural genes are already known to confer fungicide and herbicide resistance (this volume).

Is there a relationship of multiple gene families to genetic mechanisms of resistance? Multiple gene families are known for monooxygenases and glutathione transferases, two clusters of tightly linked carboxylester hydrolase genes occur on chromosome 8 of the mouse (69), and various ion channel genes have analogous construction suggesting an ancestral channel. Gene families might have arisen from duplication of an ancestral gene followed by divergence of structure through evolution. Might a similar mechanism apply to the rapid evolution of resistance?

Can individual alleles for resistance be detected in the field? This will be necessary in order to develop sound resistance management strategies. An example of the complexity of the problem is analysis of methyl parathion resistance, in which we have preliminary evidence for six possible mechanisms in tobacco budworm (Table V). Many important pests have multiple mechanisms toward each group of insecticides. Some are accumulating multiple resistance at an alarming rate.

How does multiple resistance arise genetically and why is it stable in the field? Are there any similarities to multiple drug resistant bacterial strains, in which transposable elements carrying several resistance genes can be transmitted among strains, and perhaps across species? Can we observe the dynamics of several

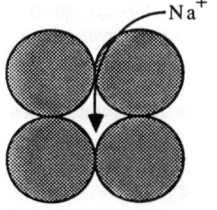 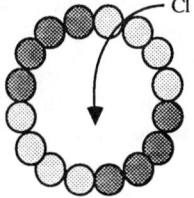 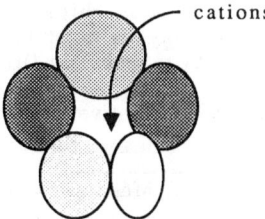

Sodium ion channel   GABA receptor   Acetylcholine receptor

DDT   avermectin
pyrethrin   cyclodienes   nicotine
pyrethroids   lindane
                            pyrethroids

Figure 7   Comparison of protein ion channels. Like-shaded areas within each channel are coded by one gene. See text for details.

individual genes in field populations? Recent progress in this area based on biochemical techniques must continue while genetic techniques are added for improved detection of specific resistance alleles (70).

Genetic Studies of Important Pest Species. Linkage mapping of insecticide resistance genes in important pests is needed for better understanding of this problem. The new power of molecular genetics provides the means to study mechanisms directly in pests of greatest interest to agriculture and public health. Linkage maps can be constructed for even those pests with many chromosomes and no prior linkage information. Extrapolation of results from only a few model species could become unnecessary.

The process can be accelerated and expanded to new species by employing restriction fragment length polymorphism (rflp) mapping. We are constructing a linkage map of the tobacco budworm using both enzyme polymorphisms, which now mark fourteen of 30 autosomal linkage groups, and adding rflp markers. We have observed linkage of acetylcholinesterase resistance to another marker enzyme and are attempting to map other resistance genes. (Heckel, D. G.; Bryson, P. K.; Brown, T. M., Clemson University, personal communication, 1988). One application of the map will be to isolate mechanisms to study their potency. If structural gene mutations confer resistance, this can be confirmed, since the gene of interest should map to the same locus as the resistance trait. Genes for resistance can be isolated from the pest of interest by recombinant DNA techniques.

Table V. Putative Mechanisms of Methyl Parathion Resistance in Several Strains of the Tobacco Budworm, *Heliothis virescens*

Mechanism	Laboratory (Reference)
Monooxygenase	
Enhanced detoxication (TCPB synergism)	This paper
Less activation to oxon	Konno et al. (78)
Arylester hydrolase	Konno et al. (79)
Carboxylester hydrolase (TBPT, EPN synergism)	Brown (16)
Altered acetylcholinesterase	
Slower inhibition reaction	This paper
Enhanced recovery from phosphinate	This paper

Acknowledgments

Technical contribution No. 2922 of the South Carolina Agricultural Experiment Station. I thank P. K. Bryson for technical assistance, and T. Konno and W. C. Dauterman of North Carolina State University for collaboration and sharing data.

Literature cited

1. Sawicki, R. M. Pestic. Sci. 1973, 4, 501-12.
2. Hoyer, R. F.; Plapp, F.W., Jr. J. Econ. Entomol. 1968, 61, 1269-76.
3. Plapp, F. W. Jr. In Pesticide Resistance Strategies and Tactics for Management; National Academy of Sciences: Washington, DC, 1986; p 74.
4. Tsukamoto, M. In Pest Resistance to Pesticides; Georghiou, G. P.; Saito, T., Eds.; Plenum Press: New York, 1983; pp 71-98.

5. Brown, A. W. A. In Pesticides in the Environment; White-Stevens, R., Ed.; Marcel Dekker: New York, 1971; Vol. 1, p 457.
6. Georghiou, G. P. Exp. Parasitol. 1969, 26, 224-55.
7. Hodgson, E.; Kulkarni, A. P. In Pest Resistance to Pesticides; Georghiou, G. P.; Saito,T., Eds.; Plenum Press: New York, 1983; pp 207-47.
8. Raffa, K. F.; Priester, T. M. J. Agr. Entomol. 1985, 2, 27-45.
9. Liu, M. Y.; Chen, J. S. ; Sun, C. N. J. Econ. Entomol. 1984, 77, 851-56.
10. Daly, J. C. Pestic. Sci. 1988, 23, 165-76.
11. Payne, G. T. Dissertation, Clemson Univ., 1987.
12. Dittrich, V.; Luetkemeier, N.; Voss; G. J. Econ. Entomol. 1979, 72, 380-84.
13. Payne, G. T. Thesis, Clemson Univ., 1982.
14. Knowles, C. O.; Roulston, W. J. J. Aust. Entomol. Soc. 1972, 11, 349-50.
15. Hollingworth, R. M.; Murdock, L. L. Science 1980, 208, 74-76.
16. Payne, G.T.; Brown, T.M. J. Econ. Entomol. 1984, 77, 294-97.
17. Nebert, D. W.; Gonzalez, F. J. Ann. Rev. Biochem. 1987, 56, 945-93.
18. Feyereisen, R.; Koener, J.; Farnsworth, D.; Nebert, D. PNAS. 1989, 86, 1465-69.
19. Agosin, M. In Cytochrome P-450, Biochemistry, Biophysics, and Environmental Implications; Hietanen, E.; Laitinen, M.; Hanniner, O., Eds.; Elsevier: Amsterdam, 1982; pp 661-69.
20. Ronis, M. J. J.; Hodgson, E.; Dauterman, W. C. Pestic. Biochem. Physiol. 1988, 32, 74-90.
21. Yu, S. J. Pestic. Biochem. Physiol. 1983, 19, 330-36.
22. Waters, L. C.; Nix, C. E. Pestic. Biochem. Physiol. 1988, 30, 214-27.
23. Zeigler, R.; Whyard, S.; Downe, A. E. R.; Wyatt, G. R.; Walker, V. K. Pestic. Biochem. Physiol. 1988, 28, 279-85.
24. Beeman, R. W. J. Econ. Entomol. 1983, 76, 737-40.
25. Beeman, R. W.; Schmidt, B. A. J. Econ. Entomol. 1982, 75, 945-49.
26. Matsumura, F.; Voss, G. J. Insect Physiol. 1965, 11, 147-60.
27. Dowd, P. F.; Gagne, C. C.; Sparks, T. Pestic. Biochem. Physiol. 1987, 28, 9-16.
28. Dowd, P. F.; Sparks, T. Pestic. Biochem. Physiol. 1986, 25, 73-81.
29. Herath, P. R.; Hemingway, J; Weerasinghe, I. S.; Jayawardena, K. G. I. Pestic. Biochem. Physiol. 1987, 29, 157-62.
30. Ashour, M-B. A.; Harshman, L. G.; Hammock, B. D. Pestic. Biochem. Physiol. 1987, 29, 97-111.
31. Shono, T. Appl. Ent. Zool. 1983, 18, 407-15.
32. Mouches, C.; Pasteur, N.; Berge, J.B.; Hyrien, O.; Raymond, M.; de Saint Vincent, B. R.; de Silvestri, M.; Georghiou, G. P. Science 1986, 233, 778-80.
33. Field, L. M.; Devonshire, A. L.; Forde, B. G. Biochem. J. 1988, 251, 309-12.
34. Devonshire, A. L. Biochem. J. 1977, 167, 675-83.
35. Krisch, K. Biochem. Biophys. Acta 1966, 122, 265-80.
36. Bryson, P. K.; Brown, T. M. Biochem. Pharmacol. 1985, 34, 1789-94.
37. Motoyama, N.; Kao, L. R.; Lin, P. T.; Dauterman, W. C. Pestic. Biochem. Physiol. 1984, 21, 139-47.
38. Devonshire, A. L.; Moores, G. D.; Ffrench-Constant, R. H. Bull. Ent. Res. 1986, 76, 97-107.
39. Raymond, M.; Pasteur, N.; Fournier, D.; Cuany, M.; Berge, J.; Magnin, M. C. R. Acad. Sci. III. 1985, 300, 509-12.
40. Takahashi, M.; Yasutomi, K. J. Med. Entomol. 1987, 24, 595-603.
41. Main, A. R.; Dauterman, W. C. Can. J. Biochem. 1967, 45, 757-71.
42. Zimmerman, J. K.; Brown, T. M. J. Agric. Food Chem. 1986, 34, 516-20.
43. Main, A. R. Biochem. J. 1960, 74, 10-20.
44. Grothusen, J. R.; Bryson, P. K.; Zimmerman J. K.; Brown, T. M. J. Agric. Food Chem. 1986, 34, 513-15.
45. Brealey, C. J.; Walker, C. H.; Baldwin, B. C. Pestic. Sci. 1980, 11, 546-54.
46. Mulbry, W. W.; Karns, J. S.; Kearney, P. C.; Nelson, J. O.; McDaniel, C. S.; Wild, J. R. Appl. Environ. Microbiol. 1986, 51, 926-30.

47. Dauterman, W. C. In Pest Resistance to Pesticides; Georghiou, G. P.; Saito, T., Eds.; Plenum Press: New York, 1983; pp 229-47.
48. Grove, G.; Zarlengo, R. P.; Timmerman, K. P.;, Li, N.; Tam, M. F.; Tu, C-P. D. Nucleic Acids Res. 1988, 16, 425-38.
49. Ottea, J. A.; Plapp, F. W. Jr. Pestic. Biochem. Physiol. 1984, 22, 203-208.
50. Wadleigh, R. W.; Yu, S. J. Insect Biochem. 1987, 17, 759-64.
51. Hama, H. In Pest Resistance to Pesticides; Georghiou, G. P.; Saito, T., Eds.; Plenum Press: New York, 1983; pp 299-331.
52. Moores, G. D.; Devonshire, A. L.; Denholm, I. Bull. Ent. Res. 1988, 78, 537-44.
53. Tripathi, R. K. Pestic. Biochem. Physiol. 1976, 6, 30-34.
54. Yamamoto, I.; Takahashi, Y.; Kyomura, N. In Pest Resistance to Pesticides; Georghiou, G. P.; Saito, T., Eds.; Plenum Press: New York, 1983; pp 579-94.
55. Anber, H. A. I.; Overmeer, W. P. J. Pestic. Biochem. Physiol. 1988, 31, 91-98.
56. Grothusen, J. R.; Brown, T. M. Pestic. Biochem. Physiol. 1986, 26, 100-109.
57. Hall, L. M. C.; Spierer, P. EMBO J., 1986, 5, 2949-54.
58. Narahashi, T. In Pest Resistance to Pesticides; Georghiou, G. P.; Saito, T., Eds.; Plenum Press: New York, 1983; pp 333-352.
59. Rossignol, D. P. Pestic. Biochem. Physiol. 1988, 32, 146-52.
60. Kasbekar, D. P.; Hall, L. M. Pestic. Biochem. Physiol. 1988, 32, 135-45.
61. Payne, G. T.; Blenk, R. G.; Brown, T. M. J. Econ. Entomol., 1987, 81, 65-73.
62. Noda, M.; Ikeda, T.; Suzuki, H.; Takeshima, H.; Takahashi, T.; Kuno, M.; Numa, S. Nature 1986, 322, 826-28.
63. Salkoff, L.; Butler, A.; Wei, A.; Scavarda, N.; Giffen, K.; Ifune, C.; Goodman, R.; Mandel, G. Science 1987, 237, 744-49.
64. Schofield, P. R.; Darlison, M. G.; Fujita, N.; Burt, D. R.; Stephenson, F. A.; Rodriquez, H.; Rhee, L. M.; Ramachandran, J.; Reale, V.; Glencorse, T. A.; Seeburg, P. H.; Barnard, E. A. Nature 1987, 328, 221-27.
65. Blair, L. A. C.; Levitan, E. S.; Marshall, J.; Dionne, V. E.; Barnard, E. A. Science 1988, 242, 577-79.
66. Kadous, A. A.; Ghiasuddin, S. M.; Matsumura , F.; Scott, J. G.; Tanaka, K. Pestic. Biochem. Physiol. 1983, 19, 157-66.
67. Imoto, K.; Methfessel, C.; Sakmann, B.; Mishina, M.; Yori, Y.; Konno, T.; Fukuda,K.; Kurasaki, M.; Bujo, H.; Fujita, Y.; Numa, S. Nature 1986, 324, 670-74.
68. Plapp, F. W. Jr. Pestic. Biochem. Physiol. 1984, 22, 194-201.
69. Berning, W.; DeLooze, S. M.; Von Diemling, O. Comp. Biochem. Physiol. 1985, 80, 859-66.
70. Brogdon, W.G., Hobbs, J.H., St. Jean, Y., Jacques, J.R. and Charles, L.B. J. Am. Mosquito Contr. Assoc. 1988, 4, 152-157.
71. Kao, L. R.; Motoyama, N.; Dauterman, W. C. Pestic. Biochem. Physiol. 1985, 23, 228-39.
72. Sultatos, L. G.; Basker, K. M.; Shao, M.; Murphy, S. D. Mol. Pharmacol. 1984, 26, 99-104.
73. Halliday, W. R. J. Stored Prod. Res. 1988, 24, 91-99.
74. Main, A. R.; Braid, P. E. Biochem. J. 1962, 84, 255-63.
75. Lin, P. T.; Main, A. R.; Motoyama, N.; Dauterman, W. C. Pestic. Biochem. Physiol. 1984, 22, 110-16.
76. Suzuki, T.; Miyamoto, J. Pestic. Biochem. Physiol. 1978, 8, 186-98.
77. Abdel-Aal, Y. A.; Soderlund, D. M. Pestic. Biochem. Physiol. 1980, 14, 282-89.
78. Wilson, T. G. J. Econ. Entomol . 1988, 81, 22-27.
79. Konno, T.; Hodgson, E.; Dauterman, W. C. Pestic. Biochem. Physiol. 1989, 33, 189-99.

RECEIVED November 3, 1989

Chapter 5

Characterization and Exploitation of Instability of Spider Mite Resistance to Acaricides

T. J. Dennehy, J. P. Nyrop, and T. E. Martinson

Department of Entomology, New York State Agricultural Experiment Station, Cornell University, Geneva, NY 14456

Resistance of spider mites to some important acaricides has been observed to increase in response to selection and to decline in the absence of selection. Management of such resistances could be based on development of methods for monitoring resistance and characterization of the processes responsible for declining resistance frequencies in the absence of selection. Studies of the dynamics of resistance to dicofol in two ecologically dissimilar acarine pests of New York apple, *Panonychus ulmi* and *Tetranychus urticae,* are described. Methods were developed which allow rapid estimation of the frequency of resistant mites in field populations. Studies conducted in 1987 related the susceptibility of field populations to acaricide use. Laboratory studies conducted in 1987-88 characterized the rate of decline of resistant genotypes in caged populations and derived selection coefficients to estimate the relative fitness of resistant individuals. Strategies for management of dicofol resistance will be based on information from dynamics of resistance and cross-resistance, and ultimately will prescribe temporal rotations of dicofol and alternate chemicals in order to keep the frequency of dicofol resistance below levels which impair acaricide efficacy.

Spider mites have demonstrated the ability to become resistant to essentially all categories of organic acaricides. In concert with the trend toward fewer new acaricides entering the marketplace each year, the successive loss of effectiveness of acaricides to resistance has resulted in a dearth of effective acaricides for many agricultural commodities. This problem is especially acute for tree fruit production in the northeastern United States, where resistance has been documented, or is suspected to exist, to every registered acaricide except petroleum oil.
An excellent overview of the history of pesticide resistance in the Tetranychidae is provided by Cranham and Helle (1) Spider mite resistance to organophosphates has been widespread for many years (2). Organotin resistance became an issue in the early 1980's

(3, 4) and resistance to the important organosulfur acaricide, propargite, was documented soon thereafter (5-8). Resistance of spider mites to formetanate hydrochloride, a carbamate acaricide, has limited the usefulness of this material in some major agricultural systems (3, 9). Resistance to the chlorinated hydrocarbon acaricide, dicofol, was first documented nearly thirty years ago (10), and continues to be a problem in some agricultural systems (11-13). The acaricidal pyrethroids represent relatively recent products for agricultural control of acarines, yet, recent reports of field failure (T. Archer, Texas A&M Univ., personal communication), and laboratory studies (unpublished results from our laboratories) indicate that resistance problems with this class of acaricides are likely. Disconcerting reports from Australia (14) have described recent field failures of the new ovicidal acaricides, clofentizine and hexythiazox. Though these ovicides are just gaining registration in the U.S., the specter of resistance is sufficiently threatening that manufacturers are prudently formulating 'resistance management' strategies for these materials from the inception of their registrations in the U.S..

Spider Mite/Dicofol System for Studying Resistance Dynamics

It appears that spider mite *resistance*, or, more precisely, spider mites which possess heritable, substantial reductions in susceptibility, occur to essentially all of the acaricides widely used in agriculture, and possibly for some newer classes of acaricidal compounds that have not been widely employed to-date. Therefore, for most acaricides it is not a question of whether resistance exists *per se*, rather it is of greatest relevance to determine if problematic frequencies of resistant mites occur at specific locations and, through research, determine what management efforts will keep frequencies of resistant pests below those which impair field performance of acaricides.

Clearly, we cannot manage (or study) what we cannot measure reliably. Yet, the development of reliable methods for estimation of resistance frequencies in spider mite populations has proven to be a major obstacle with a number of important acaricides. Detection of resistance to the organotin compounds, cyhexatin and fenbutatinoxide, and to the organosulfur, propargite, has proven to be quite difficult because these materials are repellant, and relatively slow acting, and differences between resistant and susceptible field populations are relatively small (propargite:[15];cyhexatin:[3, 16];fenbutatinoxide:[17]). Overlap in concentration-response of resistant and susceptible populations to these acaricides essentially precludes practical use of single discriminating concentrations for monitoring frequencies of resistant individuals and otherwise greatly complicates monitoring of resistance. Nevertheless, valuable and arduous documentation of resistance to these materials has been done (e.g.,6, 18, 19).

Dicofol resistance in spider mites, though of lesser practical importance in some agricultural systems than resistance to the organotin and organosulfur acaricides, provides a system conducive to the study of resistance dynamics. Resistance is generally manifest as a >1000-fold difference in susceptibility of resistant versus susceptible individuals to dicofol residues (12, 20). Dicofol is relatively fast acting and, therefore, amenable to use in 24 h, 'rapid bioassays' (21), whereby plastic petri dishes are treated with a discriminating concentration of dicofol and used to estimate the proportion of resistant individuals in populations.

The ease and reliability of monitoring dicofol resistance makes this resistance especially conducive for experimentation that involves tracking changes in resistance gene frequency in populations.

Assembling a Management Program

The fact that resistance has become a limiting factor for many spider mite management programs throughout the country is evidenced by the current commonplace use of the term *resistance management* in pest management. Unfortunately, the use of this term is, in large part, hyperbole, since only a limited number of situations exist where the prerequisite groundwork has been developed to allow evaluation, much less implementation, of specific strategies that impact resistance development. In our own situation, up to now we have *managed* very little with regard to resistance to dicofol in spider mite populations in New York apple orchards. However, in an attempt to exploit the hypothesized instability of spider mite resistance to dicofol, we have begun compiling the basic building blocks of a resistance management program. Resistant phenotypes have been characterized, and the frequency of resistance in field populations has been documented (12). The inheritance of dicofol resistance has been described for two resistant populations from New York apple orchards (22). A practical method for rapid detection of dicofol resistance has been developed and extensively used in field trials to determine the frequency of resistance at which efficacy of dicofol treatments becomes inadequate (critical frequency). Cross-resistance studies (currently underway) will show which alternate acaricides are most suitable for use against dicofol-resistant populations.

Stability of Spider Mite Resistance to Acaricides

Stability of resistance to acaricides in spider mite populations has been a long-standing point of discussion in the literature. Here we use the term *stability of resistance* to describe changes in susceptibility of populations which occur in the absence of selection. *Stable* resistances are those in which the susceptibility of populations does not change in the absence of selection. For example, organophosphate resistance in spider mites has been observed to persist at relatively high frequencies in field populations throughout the world, and has been observed to persist for long periods of time in laboratory cultures that have not been treated with acaricides. Within the context of this paper, *instability of resistance* refers to the field phenomenon of declining resistance frequencies in the absence of selection. With *unstable* resistances, the susceptibility of populations increases in the absence of selection. Whether due to immigration of susceptible individuals from refuges or other ecological or genetic processes such as assortative mating or reduced fitness of resistant organisms, unstable resistances offer the opportunity to manage resistances by using rotations of materials which do not demonstrate positive cross-resistance to one another.

Though subjectively defined, and often not intensively investigated with the necessary field and laboratory studies, the phenomenon of instability has been noted for TEPP resistance (23), the organotin acaricides (24-26), propargite (8), dicofol (see below), and formetanate hydrochloride. Noteworthy contradictions regarding resistance stability exist. Welty et al. (27) found resistance to cyhexatin in *Panonychus ulmi* to be stable in two field populations exposed for two years to three different levels of cyhexatin selection (including an untreated group). Similarly, Hoy (17) reported that greenhouse populations of *Tetranychus pacificus* held for nine months without selection with cyhexatin retained their original resistance levels. Differences in species of spider mite tested, selection pressure in the surrounding agricultural areas, and differences in magnitude and susceptibility of immigrant mites may account for some of the conflicting results.

Dicofol has long been the subject of discussions of resistance stability. This acaricide has been used for over 25 years and continues to be effective in controlling spider mites in many commodities throughout the world. Inoue (28, 29), cited field and laboratory evidence from Japan that resistance of *Panonychus citri* to dicofol increased following selection, but decreased relatively quickly after relaxation of selection. Inoue emphasized the value of studying dynamics of dicofol resistance and cited two other field studies which detail instability of dicofol resistance in Japanese citrus groves (28). Dittrich (30) referred to the comparatively slow development of resistance to dicofol, despite its worldwide use, and cited a 1965 survey by Ghobrial et al. (31) of susceptibility of spider mites collected from Egypt in which the presence of dicofol resistance in *T. arabicus* and *T. cucurbitacearum* was found in only 2 of 10 locations. Similar monitoring conducted in 1968-70 showed greatly increased frequencies of dicofol-resistant mites (32). Monitoring of resistance in Australia by Unwin (33) demonstrated declining resistance to dicofol in *T. urticae*, and showed that resistant phenotypes were present at only 2 of 11 locations tested. Soviet workers have also reported unstable resistance to dicofol (34, 35).

Dutch workers found that resistance to dicofol, parathion, and tetradifon was still present in most glasshouse populations of *T. urticae* despite the fact these materials had not been used in the glasshouses for 7 years (36, 37). However, spider mite populations collected from wild plants at varying distances from the glasshouses were resistant to parathion and tetradifon, but not to dicofol, even though adjacent glasshouses harbored the multiply-resistant spider mites. Monitoring of spider mite susceptibility to dicofol in California cotton demonstrated that, after over 25 years of use, problems with dicofol resistance were limited to ca. 10% of San Joaquin Valley cotton fields (38). Similarly, Pree and Wagner (39) reported that dicofol resistance in *P. ulmi* was detected in only 2 of 45 Niagara Peninsula, Ontario, orchards in 1984, whereas Herne (13) previously found resistance in over 50% of 35 orchards surveyed from the same area. Dennehy et al. (12) described resistance to dicofol in *T. urticae* and *P. ulmi* collected in 1985 from New York apple orchards. Of 10 apple orchards from which *T. urticae* was collected, only 3 possessed detectable levels of resistant individuals. Frequencies of dicofol resistance in *P. ulmi* populations varied from 0 to >70% and were, on the average, much higher than observed with *T. urticae*.

In this paper we detail investigations conducted in 1987-88 of the dynamics of resistance to dicofol in populations of *T. urticae* and *P. ulmi*. First, we relate the susceptibility of field populations of both spider mite species collected from over 30 different New York apple orchards to the use of dicofol within those orchards. Next, we describe the changes in susceptibility to dicofol of caged, heterogeneous populations of *T. urticae* maintained for 10-15 generations in the absence of selection with dicofol. Selection coefficients estimating the relative fitness (under the experiment conditions) of dicofol-resistant versus -susceptible *T. urticae* are derived.

Susceptibility of Field Populations versus Treatment History

Materials and Methods. Records of acaricide use in 50 Western New York apple orchards (ca. 1-10 ha in size), from 1982-87, were obtained from the Cornell University Apple IPM Program. Totals of 0-3 applications of the acaricides petroleum oil, dicofol, cyhexatin, and propargite were applied annually to each orchard. No dicofol treatments were made in any of these orchards in either 1986 or 1987. Estimates of susceptibility to dicofol were derived for a total of 30 of these orchard for *Panonychus ulmi* and from 15 orchard populations of *Tetranychusurticae*. Spider mites to be bioassayed were collected from mid-July to mid-September, 1987. Collections were made by clipping 100 terminal ends, ca. 30 cm in length, from apple branches within a nine-tree sampling unit in each orchard. The nine-tree sampling unit was composed of a central tree and the eight trees comprising a rectangle surrounding the central tree. Apple terminals were sampled from a height of 0.5-2.0 m above the orchard floor. Twenty terminals were collected from the 360 degree circumference of the central tree, and 80 terminals were collected from the 8 trees surrounding the central tree. These 80 terminals comprised the sum of ten terminals per tree, taken from the 180 degree arc of the adjacent trees oriented closest to the central tree. The 100 terminals collected at each orchard were placed in 2-3 large plastic bags and were promptly transported to the laboratory.

Bioassays of mites were conducted within 48 hr of sampling orchards. Though populations of both *T. urticae* and *P. ulmi* were sometimes collected from a single sample of terminals from an orchard, bioassays of the two species were done separately. Mites were removed from the infested foliage with fine brushes. To ensure that mites to be bioassayed were selected from throughout the 100 teminal samples, after one mite was removed from any terminal, no additional mites were removed from that terminal until all other terminals had been inspected similarly. Approximately 120 healthy adult female mites of each species present were bioassayed from each orchard. The rapid bioassay methodology described by Dennehy et al. (7) was used to estimate the frequency of resistant animals in orchard populations. This method entailed placing spider mites in small (50 x 9 mm; Falcon 1006) petri dishes that were pretreated with a discriminating concentration of dicofol (Kelthane 4 EC [emulsifiable concentrate], Rohm and Haas Co.) in 95% ethanol and allowed to dry. Control petri dishes were treated with 95% ethanol and allowed to dry. The only significant departure from the methodology employed previously by Dennehy et al. (7) was that a discriminating concentration of 320 µg/g solution was used in New York rather than the 56.2 µg/g used in California. The higher concentration was found to be appropriate for use in rapid bioassays of dicofol-resistant *T. urticae* and *P. ulmi* from New York apple orchards (unpublished results from our laboratories). Twenty spider mites were placed in each petri dish using a fine brush. One control and four or five treated petri dishes were set up for each location and species bioassayed. Assays were held at room temperature, and mortality was assessed after 24 h. To assess mortality, petri dishes were each tapped 20 times on their edge on the counter top, opened, and the mites in the top and bottom halves of the dishes viewed under the microscope. Subjects exhibiting repetitive movement of any appendages were recorded as alive.

Mean and standard error values for corrected mortality (40) were computed for all orchard populations. Dicofol selection pressure was estimated by summing the number of dicofol treatments made at each orchard during the period 1982-85. No dicofol was used in 1986-87 at any of the orchards sampled. Additionally, a weighted sum method was evaluated for estimating dicofol selection pressure. The dicofol treatments made in 1985, 1984, 1983, and 1982 were each multiplied by the weighting factors of 1.0, 0.75, 0.50, and 0.25, respectively, and then these weighted values were summed for the period of 1982-85 for each location. Mean (corrected) mortality in rapid bioassays was regressed on the two measures of dicofol selection pressure (weighted and non-weighted values) (41).

Results and Discussion. Most *T. urticae* populations sampled were relatively susceptible to dicofol (Figure 1); the average of the mean mortality values of all 15 populations was 90%. Four populations had no detectable resistance (100% mortality), and 12 of the 15 populations had >80% mortality in bioassays. Linear regression of mortality on the number of dicofol treatments (non-weighted) was done for both the complete data set (15 orchards), and after removal of two outlier points (outliers noted as open circles on Figure 1) from the complete data set. Analyses with or without the outliers yielded slopes of the regression line that were not statistically different than zero. Use of weighted values of dicofol selection pressure also produced a slope for the regression line equal to zero.

Three important findings are illustrated by the regression analysis of dicofol selection of *T. urticae* populations. First, factors other than dicofol selection pressure override the determination of population susceptibility to dicofol. Populations that had received 3, 4, or 5 dicofol treatments during the study period were just as susceptible as some populations that had received only one treatment. Clearly, factors mediating population susceptibility are strongly countering resistance development. These factors are very likely to be immigration of susceptible mites and/or reduced fitness of resistant individuals. Secondly, populations of *T. urticae* were relatively susceptible to dicofol; most populations had less than 10% resistant spider mites. Field trials evaluating the performance of dicofol in New York apple orchards possessing different frequencies of resistant mites have shown that dicofol provides very acceptable control on populations possessing up to 20% resistant individuals (W. H. Reissig, Cornell Univ., unpublished data).

The third important point illustrated in Figure 1 is that two populations that our records indicated had experienced little or no dicofol selection had unexpectedly high frequencies of dicofol resistance. These data points are from orchards that had zero or one dicofol treatment since 1982, yet they show greater than 20% dicofol-resistant individuals (<80% mortality). It is possible that these locations represent errors in pesticide use record-keeping. Alternatively, it is possible that these data reflect either effects of other pesticides (including fungicides), that may select for dicofol resistance, or immigration of resistant spider mites from nearby resistant orchard populations. Studies of cross-resistance of dicofol-resistant mites, currently underway, address this former possibility.

Susceptibility of 30 *P. ulmi* populations varied widely (Figure 2), the most susceptible having no detectable resistant individuals and the least susceptible having an average of 73% resistant individuals (mean mortality = 0.27, standard error mean [SEM] = 0.07). Only 10 of the 30 populations assayed had ≥80% mean mortality in rapid bioassays. The overall mean mortality for all 30 populations assayed was 68%. Mortality was regressed on the number of dicofol treatments (non-weighted) using the data from all 30 orchard populations. Additionally, because of high variability and low numbers of samples in the 0, 4, and 5 dicofol treatment groups (high leverage points in the regression), the 22 data points comprising orchards treated with 1, 2, or 3 dicofol applications were regressed with mean mortality (Figure 2). For the complete data set ca. 35% of the variation in *P. ulmi* susceptibility to dicofol was accounted for by differences in dicofol selection (p = 0.001). A better fit was obtained using the restricted data set; 57% of the variation in *P. ulmi* susceptibility to dicofol was accounted for by differences in dicofol selection ($p<0.001$). Weighting dicofol treatments to account for differences in temporal patterns of selection pressure did not improve the model fit.

The data indicate that the frequency of dicofol-resistant individuals in populations of *P. ulmi* is related to dicofol treatment history, much more so than for *T. urticae*. Additionally, the results suggest that resistance levels in *P. ulmi* may not decline as rapidly, in the absence of selection, as with *T. urticae*. This finding is in agreement with our understanding of differences in life history traits of these two species; *T. urticae* produces more eggs, develops faster, and has an abundance of unsprayed hosts relative to *P. ulmi*. Commercial apple orchards are essentially the only suitable habitat for *P. ulmi* in western New York. As with *T. urticae*, the regression of *P. ulmi* susceptibility to dicofol versus dicofol treatment history revealed notable outlier points (Figure 2). Two populations that our records indicated had not been exposed to dicofol since before 1982 had <80 mean mortality (>20% resistance frequency). One population was highly susceptible (mean mortality 0.97, SEM 0.01), yet records indicated that it had been treated 4 times with dicofol since 1982.

In summary, our findings illustrate high overall susceptibility of *T. urticae* and widely differing susceptibilities of orchard populations of *P. ulmi* to an acaricide that has been used for nearly thirty years. This result is contradictory to that observed for other long-used acaricides, e.g., organophosphate acaricides or tetradifon (1, 36), for which resistances have been sustained at very high levels in spider mite populations long after use of the acaricides had ceased. It merits mention that dicofol continues to be used for mite control in numerous crops throughout the world, while the majority of acaricides, many developed long after dicofol, have gone by the wayside due to resistance.

Our objective is to understand the rate at which resistance to dicofol is lost in spider mite populations and then build this information into acaricide rotation programs tailored for specific cropping systems. We intend to do this by characterizing the process(es) resulting in changes in susceptibility to dicofol. We have established 3 non-exclusive hypotheses that could account for the observed differences in susceptibility of populations to dicofol. First, it it possible that differences in susceptibility are simply a direct reflection of differences in the amount of dicofol applied to orchards since the chemical was released in the early 1960's. Second, it is possible that movement of dicofol-susceptible individuals into orchards results in declining

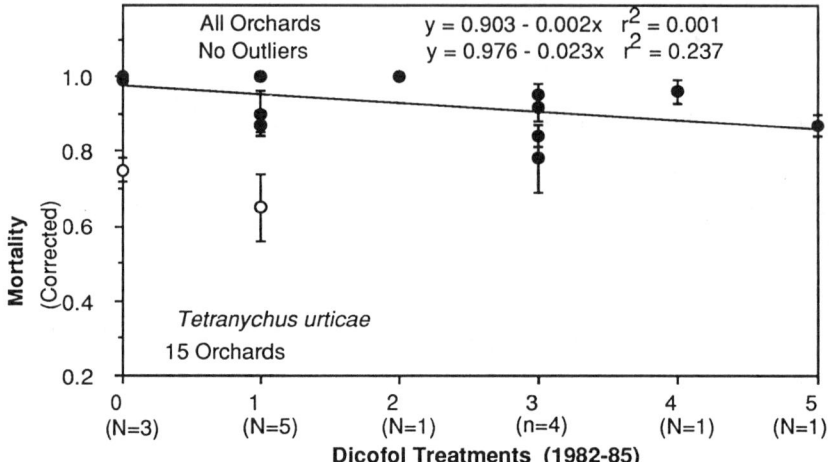

Figure 1. Regression of mean mortality observed in 1987 dicofol bioassays of 15 populations of *Tetranychus urticae* versus the number of dicofol treatments applied to orchards during 1982-85. No dicofol applications were made in 1986-87. Open circles denote outlier points. N = the number of locations sampled in each treatment category.

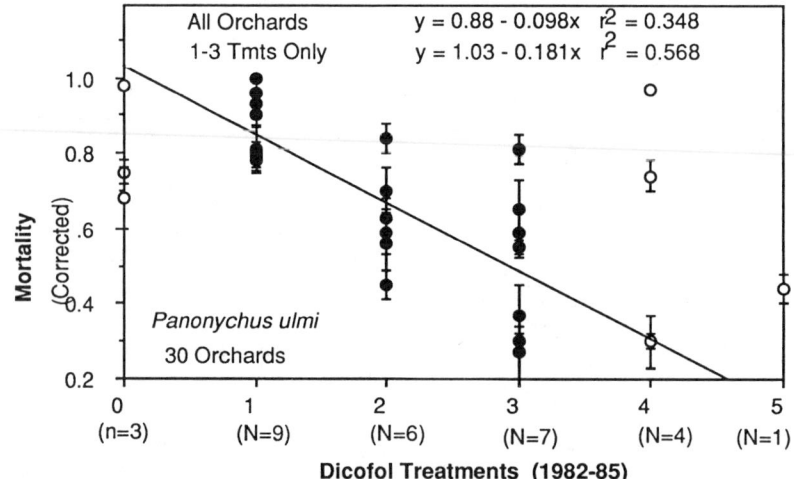

Figure 2. Regression of mean mortality observed in 1987 dicofol bioassays of 30 populations of *Panonychus ulmi* versus the number of dicofol treatments applied to orchards during 1982-85. No dicofol applications were made in 1986-87. Solid circles denote 22 orchard populations that received 1, 2, or 3 dicofol applications. N = the number of locations sampled in each treatment category.

frequencies of dicofol resistance. Finally, it is possible that dicofol-resistant spider mites increase in numbers at rates slower than dicofol-susceptible individuals (i.e., they are less fit).

Regarding this first hypothesis, we have already shown that recent history of dicofol treatments accounted for up to 57% of the variation observed in susceptibility to dicofol of *P. ulmi*, but accounted for none of the variation observed in *T. urticae*. Treatment history is important for one species but not for another. However, were no other processes involved (e.g., migration or fitness), given the nearly 30 years of use of dicofol, we would expect all populations of *P. ulmi*. and *T. urticae* to be highly resistant to dicofol; such a situation exists with the widespread and stable resistances of spider mites to organophosphate acaricides. Our second hypothesis, that migration of susceptible spider mites into orchards is responsible for differential susceptibility of populations, while very plausible for *T. urticae*, seems unlikely for *P. ulmi*, given the very limited availability of hosts for this species outside of commercial orchards in New York. If *P. ulmi* moves into orchards it is almost assured that it originated from another commercial orchard. Therefore, it appears most probable that our third hypothesis, fitness differences between resistant and susceptible spider mites, plays a critical role, in conjunction with the demonstrated importance of differences in selection pressure (number of dicofol treatments), in determining susceptibility of local populations of *P. ulmi*. It is equally possible that fitness differences are involved with *T. urticae*, though the highly dicofol-susceptible populations observed could result from immigration alone. As a result of these findings, we have focused our efforts on estimating differences in fitness of resistant and susceptible spider mites. Owing to the prohibitive difficulty and expense of rearing *P. ulmi* in the laboratory, our initial investigations of fitness costs have been done with *T. urticae*.

Change in Susceptibility of *T. urticae* in the Absence of Selection with Dicofol

The objective of this experiment was to estimate the magnitude of potential fitness differences between dicofol-resistant (RR) and susceptible genotypes (SS) of *T. urticae*. In this context, relative fitness is the ability of a particular genotype to pass alleles to the next generation, relative to other genotypes. As noted above, one cause of instability might be differences in relative fitness of resistant and susceptible animals. The general approach taken was to measure changes in the frequency of resistant female *T. urticae* in heterogeneous caged populations and to use these measurements to estimate relative fitness.

Materials and Methods. Spider mites used in these experiments were maintained in colonies on bean plants as outlined by Rizzieri et al. (22). Three colonies were used in the experiment. Two originated from field-collected populations: one (Res) which was continuously selected with dicofol and was homogeneously resistant to dicofol, and the second (Sus), which was not selected with dicofol and was homogeneously susceptible to dicofol. Each colony was started with animals collected from a single, different apple orchard. The third colony was a near-isogenic line composed of the susceptible colony genome into which the factor conferring dicofol resistance had been inserted. Six cycles of backcrosses followed by selection were conducted.

The backcrosses were conducted in the following manner. A Res male was crossed with an Sus female. The resulting female progeny were isolated as deutonymphs and allowed to lay haploid eggs. The resulting male progeny were subject to whole plant residual bioassay using 10 µg/g dicofol, a concentration previously determined to discriminate between R and S male genotypes (22). Male survivors were mated to virgin Sus females. This process was repeated six times. In the final round, the RS females were saved and the R male survivors of bioassays were mated with them. Resulting female progeny were assayed using a whole-plant residual bioassay with a discriminating concentration of 1000 µg/g dicofol. The surviving females (RR) were then used to start a colony, herein called IsoR. Susceptible (SS) spider mites used in backcrosses and for cage experiments with IsoR spider mites were from the Sus culture.

Seven experiments were conducted sequentially, each of which examined the dynamics of the resistant genotype (RR) in four separate cages, using different starting proportions of resistant and susceptible spider mites for each group of four cages. The cages measured 45 cm x 45 cm x 60 cm and were constructed of very fine-mesh screen placed over a wood frame. Each experiment was started by placing bean seedlings in four cages and then introducing the predetermined proportions of resistant (RR) and susceptible (SS) *T. urticae* into them. Two hundred adult female spider mites were released into each cage. For two of the seven experiments (groups of four cages) these mites consisted of 70% RR females from the Res colony and 30% SS females from the Sus colony. Two other experiments were started with 50% Res and 50% Sus spider mites, and 30% Res and 70% Sus spider mites, respectively. The fifth experiment served as a control and consisted of introducing 100% Res females into four cages. The sixth and seventh experiments were conducted with the near-isogenic resistant, IsoR (RR), and the susceptible, Sus (SS), colonies. For the sixth experiment, starting frequencies were 70% IsoR and 30% Sus females. In the final experiment, which served as an additional control, 100% IsoR females were released into four cages. All cages experiments were conducted in greenhouses where temperatures ranged from 20-31°C, and with illumination provided by fluorescent lights maintained on a 16 hr photophase. Bean plants in the cages were replaced each week by cutting the stems on ca. 25% of the plants in the cages, waiting for the mites to move off the cut plants, and then replacing the cut plants with fresh ones. After the first three cycles of plant replacement, the oldest plants in the cages were removed. Each of the seven groups of four cages were maintained for 100-150 days, during which time they went through ca. 10-15 generations.

Beginning 30 days after the release of mites into cages and ca. every 30 days thereafter, the frequency of RR females in each of the four cages was measured. This was done by using the rapid assay, previously described. For each cage, 10 treated dishes and 2 controls were used with ca. 20 mites assayed in each dish. Thus, 200 mites were assayed in each cage, and a mortality estimate was based on a total of ca. 800 mites. Dicofol-induced mortality was corrected for control mortality (40) and the variance for this estimate was based on the variability in mean mortality observed between cages. The grand mean of corrected mortality and 95% confidence intervals were plotted as a function of generations for each group of four cages conducted as an experiment. Each generation was assumed to be ca. 10 days.

To estimate the relative fitness of the genotypes, dicofol resistance was modelled as a monogenic, recessive trait (22) and it was assumed that any fitness cost associated with the resistance gene would also be recessive. Allelic frequencies for the resistance alleles, R and S, are denoted by q and p, respectively. In addition, the relative fitnesses of the three genotypes are denoted as w_{RR} for the RR genotype, w_{RS} for RS, and w_{SS} for SS. Assuming that the conditions for Hardy-Weinberg equilibrium hold, and that a discrete representation is a reasonable approximation for mating in the cages, then for the relative fitnesses belonging to the set {0,1}, the allele frequency for the R gene in generation t+1 can be written as in Equation 1:

$$q_{t+1} = (p_t q_t w_{RS} + q_t^2 w_{RR}) \bar{w} \qquad (1)$$

where $\bar{w} = p_t^2 w_{SS} + 2 p_t q_t w_{RS} + q_t^2 w_{RR}$ and $\bar{w}$ is known as the mean fitness. Because only the homozygous resistant animals are assumed to have a reduction in fitness, w_{SS} and w_{RS} are set to 1, and w_{RR} is written as 1-s, where s is the selection coefficient. Equation {1} can now be written as Equation 2:

$$q_{t+1} = [q_t(1-sq_t)]/[1-sq_t^2]. \qquad (2)$$

The mortality estimates (m) are estimates of the sum of the SS and RS genotype frequencies. Therefore, 1-m is an estimate of the frequency of the RR genotype and $(1-m)^{1/2}$ is an estimate for q. The method presented by Wood and Cook (42) was slightly modified and used to estimate s based on the above models and data. Wood and Cook used 1-s to represent the relative fitness of individuals susceptible to a pesticide when resistance to the pesticide was recessive. Here 1-s is the relative fitness for the RR genotype.

Results and Discussion. In all five experiments where RR and SS female *T. urticae* were introduced into the cages, the frequency of resistant animals (RR) declined greatly over the course of 10-15 generations (Figures 3a-e). Furthermore, the pattern of these dynamics among the groups of four cages used in particular experiments was remarkably consistent, as evidenced by the relatively narrow confidence intervals for grand mean mortality. In four of the five experiments involving mixtures of RR and SS mites the estimated selection coefficients (s) were very similar (0.27, 0.30, 0.31, 0.34). One estimate obtained using mixtures of Res and Sus spider mites (Figure 3d) was larger (0.46). However, this estimate may not be as accurate because the initial frequency of the R allele in the cages was relatively low, and the resulting estimate of s is strongly influenced by small changes in gene frequency. Four cages containing heterogeneous mixtures of the near-isogenic, IsoR, resistant spider mites and susceptible, SS, spider mites exhibited similar dynamics and generated a selection coefficient similar (0.34) to that obtained using the non-isogenic spider mites (Figure 3e). When only resistant spider mites from the IsoR or Res cultures were introduced into the cages, the frequency of the RR genotype remained fixed at 100% (Figures 3c and e).

In these experiments dicofol-resistant *T. urticae* showed a much lower relative fitness than dicofol-susceptible animals. Because these fitness differences were observed in colonies started from disparate field locations, as well as in near-isogenic

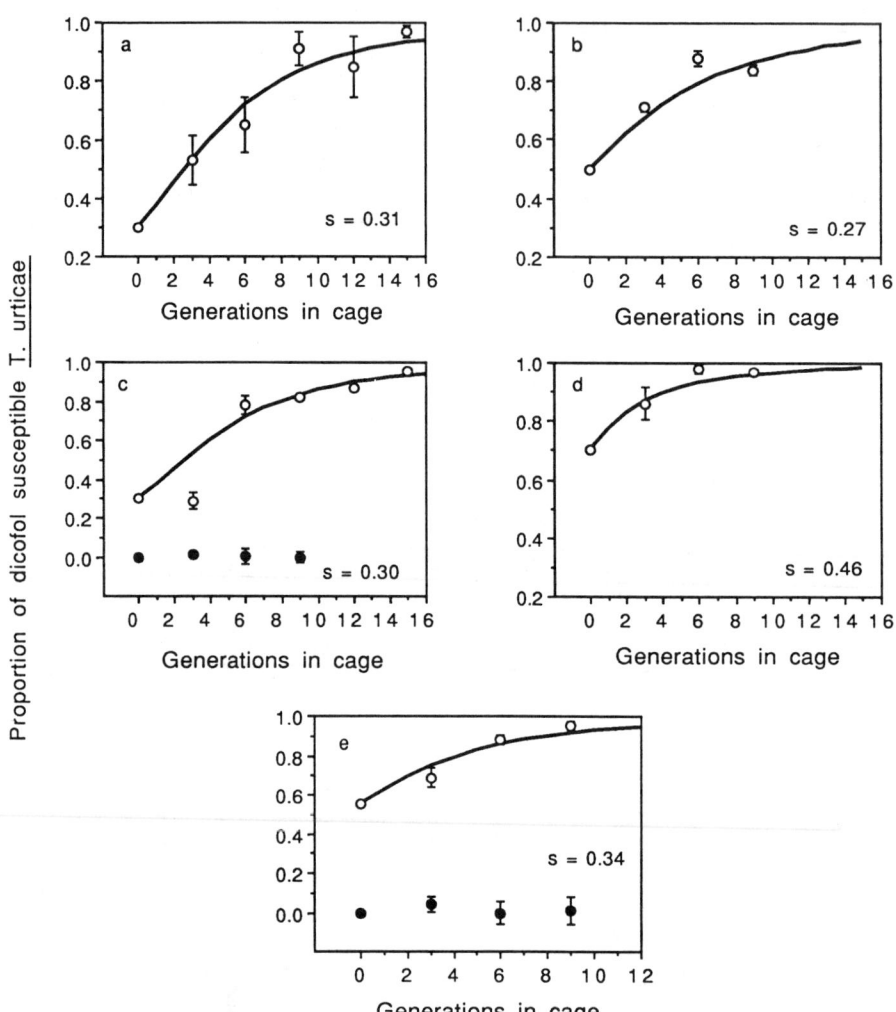

Figure 3. Proportion of dicofol-susceptible *T. urticae* in caged populations shown in relation to the number of generations populations were in cages. Grand mean mortality and 95% confidence limits are given for groups of four cages. The solid line denotes the frequency of SS plus RS genotypes predicted from the model and s is the coefficient of selection against the resistant (RR) genotype. Solid circles depict data from control groups for which 100% RR mites were introduced into cages. Figures a-d show results obtained from cages started with spider mites from the Res and Sus colonies. Figure e shows results obtained from cages started with spider mites from the near-isogenic line, IsoR, and the Sus colony.

colonies, it is highly probably that the differences observed are related directly to, or closely linked to the gene conferring dicofol resistance. If relative fitness differences of the magnitude observed in the cage studies occur in the field, then this phenomenon is a powerful force driving instability of dicofol resistance. For example, if the frequency of the RR genotype is 0.30, then, assuming a selection coefficient of 0.30, this frequency would decline to 0.22 in two generations. There are, however, unresolved questions, discussed below, that need to be addressed before we can be confident that fitness differences we have measured are manifested comparably under field conditions.

Concurrent with the cage studies, life tables were constructed for RR and SS female *T. urticae* by observing individual spider mites reared on healthy, undamaged bean seedlings. In this case the relative fitness of the RR genotype was expressed as the expected progeny from the RR females divided by the expected progeny from the SS females. The pooled results from three lifetables produced a selection coefficient, s, of 0.12. In these studies, developmental time and egg production were not significantly different in the two strains. The cause of the observed value of s was increased mortality of the resistant strain (unpublished results from our laboratory). Intrinsic growth rates for resistant spider mites were always less than those for the susceptible spider mites (SS genotype); however, the average difference was small (0.275 versus 0.282). We must conclude that the relatively large fitness differences observed in the cage studies are density dependent. It is generally accepted that fitness is a function of density, and it is certainly possible that differences in such things as viability, fecundity, and mating success vary as a function of density in different ways for RR and SS animals. Densities in the cage studies were relatively high. Therefore, the dynamics of loss of resistance observed in our cage studies may reflect an upper limit for what might be expected, under field conditions, from the process of differential fitness of dicofol-resistant and susceptible *T. urticae*.

<u>Conclusions</u>

From the standpoint of managing resistance to dicofol in spider mite pests of New York apple orchards, our findings for *T. urticae* are positive. Most populations had fewer than 10% resistant mites, and our laboratory data indicated that the resistant genotypes of *T. urticae* are substantially less fit than susceptible ones. Field studies of the critical frequency for dicofol resistance (i.e., the frequency at which resistance results in appreciable reductions in control) indicate that frequencies of resistance must be >20% to impair performance of dicofol treatments. Therefore, dicofol treatments would be expected to provide acceptable control of 9-12 of the 15 *T. urticae* populations we sampled. However, *P. ulmi* is the dominant spider mite pest of apple in western New York and, as we have demonstrated, the species most likely to exhibit resistance to dicofol.

In New York apple orchards, it appears unlikely that susceptibility to dicofol can be sustained in *P. ulmi* populations which are treated with dicofol during successive years. Mean mortality in the 9 orchards that had received only 1 dicofol treatment during 1982-85 period was 0.86 (SEM 0.30). Given a critical frequency of 20%, populations showing >80% mean mortality in rapid bioassays would be adequately controlled by dicofol. However, the mean mortality for populations that received 2 treatments (6 locations) or 3 treatments (7 locations) of dicofol was 0.63 (SEM 0.05) and 0.51 (SEM 0.08), respectively. It is very unlikely that dicofol would adequately control these populations. In New York apple orchards, it appears advisable to develop acaricide rotations that utilize dicofol no more frequently than once every 3-5 years to control *P. ulmi*. Field studies are ongoing to evaluate this preliminary conclusion.

Our findings regarding the dynamics of resistance to dicofol in *P. ulmi* and *T. urticae* illustrate the case-specific nature of the stability of resistances. Even in the same cropping system, resistance dynamics were very different for two species, *T. urticae* and *P. ulmi*. This finding should serve as warning against generalizing about resistance stability across different species and cropping systems. Though it appears that rotations of dicofol in New York apple should involve use of dicofol no more often than once every three years, major cropping systems exist throughout the world where dicofol has been used on much more intensive bases without experiencing the resistance problems documented herein for *P. ulmi*. Such examples have been already cited from Japanese citrus and California cotton. In each instance the same principles apply for exploiting instability of spider mite resistance, however, the minimal treatment interval for which dicofol treatments must be spaced in order to avoid escalating resistance frequencies is different.

Acknowledgment

We thank J. Minns, L. Fergusson-Kolmes, K. Wentworth and L. Clark for assistance with these studies. Funding was provided, in part, by the Rohm and Haas Company.

Literature Cited

1. Cranham, J. E.; Helle, W. Pesticide Resistance in Tetranychidae; Helle, W; Sabelis, M. W., Eds.; Elsevier: New York, 1985; p 405.
2. Helle, W. Resistance in the Acarina: Mites; Naegele, J. A., Ed.; Cornell University Press: New York, 1965; p 71.
3. Croft, B. A.; Miller, R. W.; Nelson, R. D.; Westigard, P. H. J. Econ. Entomol. 1984, 77, 575-8.
4. Edge, V. E.; James, D. G. J. Aust. Entomol.Soc. 1982, 21, 198.
5. Chapman, R. B.; Penman, D. R. N. Z. J. Agric. Res. 1984, 27, 103-5.
6. Keena, M. A.; Granett, J. A. J. Econ. Entomol. 1987, 78, 1212-6.
7. Dennehy, T. J.; Grafton-Cardwell, E. E.; Granett, J.; Barbour, K. J. Econ. Entomol. 1987, 80, 65-74.
8. Grafton-Cardwell, E. E.; Granett, J.; Leigh, T. F. J. Econ. Entomol. 1987, 80, 579-87.
9. Miller, R. W.; Croft, B. A.; Nelson, R. D. J. Econ. Entomol. 1985, 78, 1379-88.
10. Hoyt, S. C.; Harries, F. H. J. Econ. Entomol. 1961, 54, 12-6.
11. Dennehy, T. J.; Granett, J. J. Econ. Entomol. 1984, 77, 1381-5.
12. Dennehy, T. J.; Nyrop, J. P.; Reissig, W. H.; Weires, R. W. J. Econ. Entomol. 1988, 81, 1551-61.

13. Herne, D. H. C. Proc. 3rd Int. Cong. Acarol. Prague, 1971, p 663.
14. Edge, V. E.; Rophail, J.; James, D. G. Proc. Symp. on Mite Control in Horticultural Crops; Thwaite, W. G., Ed.; Dept. of Agric. N. S. Wales: 1987; p 87.
15. Keena, M. A; Granett, J. A. J. Econ. Entomol. 1987, 80, 560-4.
16. Welty, C.; Reissig, W. H.; Dennehy, T. J; Weires, R. W. J. Econ. Entomol. 1987, 80, 230-6.
17. Hoy, M. A.; Conley, J.; Robinson, W. J. Econ. Entomol. 1988, 57-64.
18. Flexner, J. L. Ph.D. Thesis, Oregon State University, Oregon, 1988.
19. Welty, C.; Reissig, W. H.; Dennehy, T. J.; Weires, R. W. J. Econ. Entomol. 1988, 81, 42-8.
20. Dennehy, T. J.; Granett, J.; Leigh, T. F. J. Econ. Entomol. 1983, 76, 1225-30.
21. Dennehy, T. J.; Grafton-Cardwell, E. E.; Granett, J.; Barbour, K. J. Econ. Entomol. 1987, 80, 998-1003.
22. Rizzieri, D. A.; Dennehy, T. J.; Glover, T. J. J. Econ. Entomol. 1988, 81, 1271-6.
23. Saba, F. Z. ang. Entomol. 1969, 48, 265-73.
24. Flexner, J. L.; Theiling, K. M.; Croft, B. A.; Westigard, P. H. J. Econ. Entomol. 1989, 82, 996-1002.
25. Edge, V. E.; James, D. G. J. Econ. Entomol. 1986, 79, 1477-83.
26. Pree, D. J. J. Econ. Entomol. 1987, 80, 1106-12.
27. Welty, C.; Reissig, W. H.; Dennehy, T. J.; Weires, R. W. J. Econ. Entomol. 1989 82, 692-7.
28. Inoue, K. J. Pesti. Sci. 1979, 4, 337-44.
29. Inoue, K. J. Pesti. Sci. 1980, 5, 165-75.
30. Dittrich, V. Z. ang. Ent. 1975, 78, 28-45.
31. Ghobrial, A. V.; Dittrich, V.; Hafiz, M.; Attiah, H.; Voss, G. J. Econ. Entomol. 1969, 62, 1262-8.
32. Dittrich, V.; Ghobrial, A. Z. ang. Ent. 1974, 76, 418-29.
33. Unwin, B. J. Aust. Entomol. Soc. 1973, 12, 59-67.
34. Zil'bermintz, I. V.; Fadeyev, Y. N.; Zhuravleva, L. M. Selskohozyaistwennaya Biologyia 1968, 3, 125-32.
35. Zil'bermintz, I. V.; Fadeyev, Y. N.; Zhuravleva, L. M. Selskohozyaistwennaya Biologyia 1968, 5, 96-106.
36. van Zon, A. Q.; Overmeer, W. P. Z. ang. Entomol. 1975, 79, 213-22.
37. Overmeer, W. P. J.; van Zon, A. W.; Helle, W. Ent. Exp. et. Appl. 1975, 18, 68-74.
38. Dennehy, T. J.; Granett, J. J. Econ. Entomol. 1984, 77, 1386-92.
39. Pree, D. J.; Wagner, H. W. Can. Entomol. 1987, 119, 287-90.
40. Abbott, W. S. J. Econ. Entomol. 1925, 18, 265-7.
41. Minitab 1985, Minitab Reference Manual, Release 5.1, October 1985, State College, PA.
42. Wood, R. J.; Cook, L. M. Bull. World Health Organization 1983, 61, 129-34.

RECEIVED October 16, 1989

Chapter 6

Laboratory Evaluation and Empirical Modeling of Resistance-Countering Strategies

I. Denholm, M. Rowland, A. W. Farnham, and R. M. Sawicki

AFRC Institute of Arable Crops Research, Rothamsted Experimental Station, Harpenden, Hertshire AL5 2JQ, United Kingdom

> Theoretical and practical aspects of managing resistance are studied at Rothamsted by multidisciplinary research that integrates computer modelling with laboratory experiments under simulated field conditions. For houseflies (Musca domestica), several mechanisms that differ widely in potency and cross-resistance spectra are used to establish and model how parameters such as the strength and persistence of deposits influence the expression of resistance genes. Toxicological and biochemical techniques monitor the effects of insecticides applied singly or as mixtures on selection rates. For tobacco whitefly (Bemisia tabaci), novel technology has been developed to simulate infestations on cotton and evaluate insecticide strategies for controlling resistant populations of this pest. Both programmes provide the realistic data urgently needed to test and strengthen theoretical work on this subject.

The prevention or containment of insecticide resistance requires a thorough understanding of how and why resistance genes are selected by the various control regimes likely to be implemented against pest populations. Over the last ten years, this challenge has mainly been met by developing mathematical models to describe the evolution of resistance, and to evaluate tactics such as the use of high or low doses, rotations and mixtures that may or may not delay the spread of resistance under specified conditions (eg. 1-4). A problem that is generally recognised (5,6) with this approach is that there has been far too little realistic experimental work to test the assumptions of these models or to corroborate their findings. As a result, it is very difficult to correlate field data with existing models and to translate theoretical countermeasures into practical ones that can be applied with confidence in the field (6).

To remedy this shortcoming, we have established at Rothamsted two multidisciplinary projects that in effect tackle research on resistance the other way round. Our aim is firstly to collect empirical data on the selective and suppressive effects of controllable parameters such as the rate, type and persistence of insecticide applications, then to incorporate these data into empirical computer models that in turn feed back to and support the experimentation. In this way, we hope to construct robust models capable of describing the outcome of control regimes against experimental populations.

These research programmes involve the housefly, Musca domestica, and the tobacco whitefly, Bemisia tabaci. Both are based in the laboratory in the firm belief that for these and many other pests, much of the data needed for such descriptive modelling can be collected more quickly and with greater precision under laboratory conditions than in the field. There is, however, a very important proviso to this statement, since for laboratory experiments to be of direct practical relevance they must simulate closely both the ecology and control of insects under field conditions (7). The purpose of this paper is to outline the philosophy and objectives of these two programmes and to highlight some achievements to date.

Background to the Housefly Work

Houseflies have long proved to be ideal organisms for fundamental research on resistance, and our current work on this species is directed less towards solutions to specific problems than to general principles underlying the expression and selection of resistance genes. Extensive genetic and biochemical research at Rothamsted has provided a unique collection of fly strains with well-characterised resistance mechanisms, four of which have been chosen that cover the major insecticide groups and differ considerably in phenotypic expression and cross-resistance characteristics (Table I). Kdr (knockdown resistance), the mechanism of nerve insensitivity conferring resistance to the pyrethroids and DDT (8), is represented by both the standard kdr variant and its more potent super-kdr allele. AChE-R is the mechanism of acetylcholinesterase insensitivity to organophosphorous and carbamate insecticides (9). Like kdr, this mechanism is multi-allelic (9), and CH2 and 49R are two enzyme variants that show contrasting patterns of insensitivity to this diverse group of compounds (10). Dieldrin-R confers target-site insensitivity to cyclodienes such as dieldrin and endosulfan (11), and E0.39 refers to an esterase variant implicated in moderate resistance to organophosphates such as trichlorphon and weak resistance to the pyrethroids (12). E0.39 is likely to interact with AChE-R in OP resistance and has been shown to increase two- to three-fold the level of pyrethroid resistance conferred by kdr alleles (12). All these genes have been inbred seperately into a wild-type susceptible housefly genome, and collectively they encompass genetic and toxicological phenomena such as multiple resistance, cross-resistance and allelic variation likely to occur in major resistance problems.

Table I. Resistance mechanisms isolated in houseflies

Mechanism	Autosome	Allele(s)	Resisting:
Kdr	3	kdr, super-kdr	Pyrethroids, DDT
AChE-R	2	CH2, 49R	OPs, carbamates
Dieldrin-R	4	Dld4	Cyclodienes
E0.39	2	E0.39	OPs, pyrethroids

All four of these mechanisms can be monitored readily in fly populations using discriminatory dose bioassays or, in some cases, by diagnostic biochemical techniques. Thus, E0.39 genotypes can be distinguished by polyacrylamide gel electrophoresis (12), and AChE-R is detected rapidly in individual insects by a microplate assay that identifies unambiguously all six genotypic combinations of the S (sensitive), CH2 and 49R alleles (10).

The apparatus and techniques developed to study these mechanisms (13) are simple to construct and operate, yet incorporate two important refinements when compared to most experiments selecting for resistance in the laboratory. Firstly, fly populations maintained in large cages breed continuously and are fully age-structured, rather than being reared in discrete generations. This mimics housefly ecology in the field, and is necessary to simulate realistically residual treatments that act continuously against adults of all ages and reproductive states. Secondly, applying insecticides as deposits to cage walls simulates exposure to toxins in the field and provides ample scope for expression of behaviour influencing contact and insecticide pick-up that are precluded by more routine and artificial application techniques (7).

Objectives of the Housefly Work

With this comprehensive range of resistance mechanisms and monitoring techniques to hand, the major objectives of the housefly work are as follows:

1) characterising in detail the properties of commercial formulations of several insecticides chosen to differ widely in mode of action and environmental persistence, determining in particular their efficacy at various application rates and the decline in effect following application,

2) quantifying how operational parameters such as the dose rate, frequency of application and persistence of these compounds influence the expression and selection of their respective resistance mechanisms, and examining the relative fitness of resistant and susceptible genotypes in the absence of insecticide pressure,

3) determining how insecticides and resistance mechanisms interact in proposed countermeasures, particularly when using mixtures, and the consequences for selection when the conditions that theoretically optimise the efficacy of putative tactics are not met, and

4) using these data to develop models capable of describing, and thereby predicting, the effects of different field treatments on genetic composition and population size.

Results used here to illustrate this approach refer to two insecticides widely used for housefly control, the pyrethroid permethrin and the OP trichlorphon (Figure 1), and to two mechanisms implicated in resistance to these compounds, kdr and AChE-R respectively.

Characterisation of insecticides

Efficacy of Insecticide Deposits. The overall efficacy of chosen insecticides in suppressing caged fly populations has been examined by monitoring the mortality schedules of cohorts of insecticide-susceptible adults released into cages containing deposits applied at different concentrations, and then by computing the results of continuous exposure to these treatments on adult population size. For permethrin and trichlorphon, the mortality of susceptible females varied, as expected, according to residue concentration (Figure 2). Survival was severely reduced by both insecticides applied at their recommended field rates (100 and 500 mg/m2 respectively), though somewhat less abruptly with the slower acting OP compound. Survival rates increased as concentrations were progressively reduced, and at very low rates they resembled those in untreated cages.

Computer modelling based on detailed demographic studies of laboratory populations has provided an elegant means of translating such results into figures for actual control efficacy. To achieve this, observed mortality schedules for single cohorts were input to a PASCAL program simulating our age-structuring rearing regime (13) to calculate expected changes in fly numbers before and after insecticides are applied. Predicted levels of supression, some of which have subsequently been verified by longer-term laboratory experiments, with a wide range of concentrations showed that the relationship between percentage supression and residue strength is markedly shallower for permethrin than for trichlorphon (Figure 3). At lower concentrations permethrin gives comparatively better kill (reflecting its greater intrinsic toxicity), but at higher rates there is little difference between the two compounds. Subsequent work has shown that this effect is at least partly attributable to well-documented repellancy properties of pyrethroids (eg. 14,15) causing flies to settle less readily and for shorter periods on surfaces sprayed with high concentrations of permethrin. This constrains the efficacy of the pyrethroid at high application rates.

Persistence of Deposits. While the initial strength of deposits dictates levels of kill immediately after application, their environmental persistence largely determines the duration of effectiveness. This parameter has been widely implicated in influencing the build-up of resistance (16-18). We have therefore chosen compounds with contrasting decay rates, and have quantified these rates by exposing susceptible flies to residues in population cages at various intervals post-spraying. Permethrin and trichlorphon differ considerably in this respect; starting with

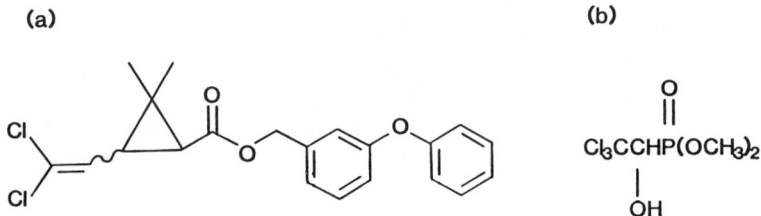

Figure 1. Structures of permethrin (a) and trichlorphon (b).

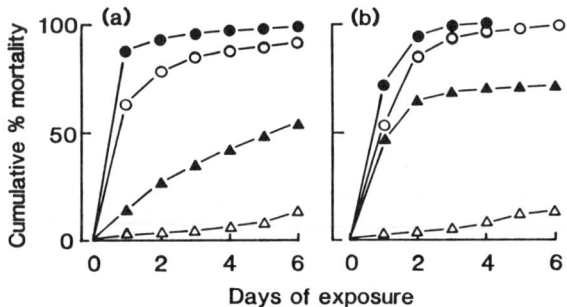

Figure 2. Cumulative mortality of susceptible female houseflies exposed to deposits of permethrin or trichlorphon applied at three rates. (a) = permethrin (● = 100, ○ = 25, ▲ = 6.25mg/m2, △ = untreated); (b) = trichlorphon (● = 500, ○ = 125, ▲ = 31mg/m2, △ = untreated). Pooled data for 2 replicates.

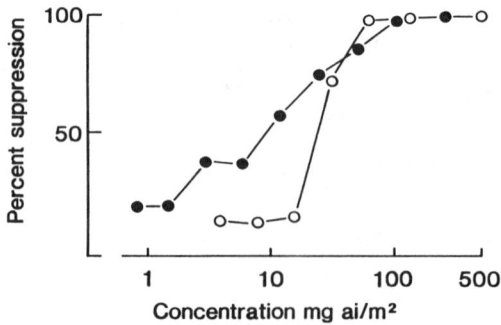

Figure 3. Predicted suppression of susceptible fly populations exposed to deposits of permethrin (●) or trichlorphon (○) applied at various rates.

fresh deposits of each insecticide at their respective field rates, trichlorphon became relatively ineffective within three weeks of application but permethrin continued to give substantial kill throughout this period (Figure 4).

These tests have been repeated for several starting concentrations of each insecticide, and we are now using these data to develop mathematical models defining 'response topographies' that relate kill at any stage after spraying to both the initial strength and age of deposits. Combining such functions with comparative data on the response of various resistance genotypes (see below) should provide a much clearer insight into the significance of pesticide persistence for selection than is available from theoretical work so far.

Appraisal of Resistance Mechanisms

Phenotypic Expression of Resistance Genes. To model realistically how genes are selected by various control regimes, it is necessary to determine how different genotypes at resistance loci respond to insecticide exposure under conditions that prevail in the field (7,19). Experiments of this kind (20-22) have clearly demonstrated that conventional laboratory bioassays involving artificial exposure methods, and which preclude behavioural interactions between insects and insecticides, can greatly misjudge the protection conferred by resistance genes under field conditions. To illustrate this further, Figure 5 shows the cumulative mortality of flies of all three genotypes at the kdr locus in cages containing permethrin deposits sprayed at two different rates. With the recomended rate (Figure 5a), kdr homozygotes enjoy substantial protection but heterozygotes respond similarly to susceptible flies. In this case kdr is behaving as an effectively recessive gene, in accordance with results of topical application tests (8). With lower concentrations, however, survival of kdr heterozygotes gradually improves, and at 6.25 mg/m2 (Figure 5b) the gene is effectively dominant in its expression. These results again emphasise that the phenotypic dominance of resistance genes is a purely dose-dependent phenomenon, and the relative superiority of heterozygotes at low concentrations may have substantially accelerated selection for kdr on farms where pyrethroids have been applied sloppily or where treatments have been allowed to decay without replenishment.

In contrast to the strong resistance conferred by kdr, the CH2 acetylcholinesterase variant alone gives little or no protection against trichlorphon applied at the field rate, and only a relatively slight advantage at lower concentrations (23, and unpublished data). This is suprising considering that this enzyme shows clearcut insensitivity to dichlorvos (the active dehydrochlorination product from trichlorphon) in the AChE microplate assay (10), and given the apparent ubiquity of OP-insensitive alleles in field populations of houseflies (9,10, and unpublished data). Such discrepancies between biochemical and toxicological assays emphasise that is impossible to extrapolate the practical significance of resistance mechanisms from biochemical tests alone. From these and other findings it appears that many of the AChE alleles in houseflies act more as modifiers or intensifiers of metabolic mechanisms (eg. E0.39) than as major resistance genes per se. We are studying populations that contain AChE-R and EO.39 both singly and in combination to test this hypothesis.

Persistence of Resistance Genes in Untreated Populations. It has been argued (eg. 24) that if resistant insects are substantially less fit than susceptible ones in the absence of insecticide pressure, any short-term build-up of resistance might be effectively countered by temporarily withdrawing the selecting agent(s) from use. This concept has been harshly criticised on theoretical grounds (19), but is nonetheless an important tenet of many existing insecticide-rotation strategies (25). Clearly, the first stage in resolving this debate is to establish whether such selective disadvantages are sufficient to cause a significant decline in resistance gene frequency in untreated populations.

Most experimental work to date (reviewed in 26) has provided little evidence that such major declines occur (but see 27 for a clearcut exception). Each of the housefly genes examined in our work has proved remarkably persistent in untreated caged populations. For example, Figure 6 shows results of monitoring biochemically at weekly intervals a population polymorphic for a sensitive (S) and two insensitive (CH2 and 49R) alleles. After some initial fluctuations, allele frequencies remained very stable over 70 days (approximately seven generations). Similar results have since been obtained for the more potent D1d4, kdr and super-kdr alleles. Under these conditions the advantage of rotating insecticides would rest solely on the immigration of susceptible insects whilst each compound is not being used, and since such migration is unlikely to be unidirectional the strategy could only give temporary respite (26).

Evaluation of Insecticide Mixtures

According to available theoretical work (3,4), insecticide mixtures can potentially delay resistance very substantially compared to the continuous or alternate use of single compounds. There are, however, several key conditions to be met for mixtures to achieve maximum effect. In particular:

- the constituents should be equally effective in controlling the target pest(s) and have similar environmental persistence,
- resistance to at least one (preferably both) compounds should still be extremely rare,
- a proportion of insects must escape exposure altogether, and
- there must be no common resistance mechanism.

Since these stringent requirements are unlikely to be met fully in practice, our main concern in appraising this tactic is to determine the consequences for selection when one or more of these conditions is violated to a greater or lesser extent.

Our initial experiments exposing different kdr genotypes to combined residues of permethrin and trichlorphon have highlighted some important important practical considerations for mixture strategies. Since trichlorphon is much less persistent then permethrin (cf. Figure 4), prolonged exposure (eg. for four weeks) of a population already containing kdr homozygotes to a standard

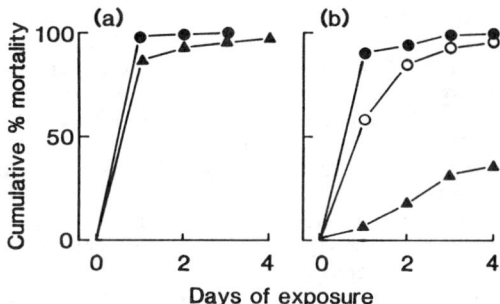

Figure 4. Cumulative mortality of susceptible male houseflies exposed to field-rate deposits of permethrin (a) or trichlorphon (b) immediately (●), 2 (○) or 3 weeks (▲) after application. Pooled data for two replicates.

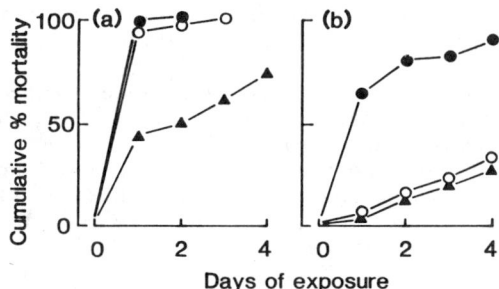

Figure 5. Cumulative mortality of male houseflies of three kdr genotypes (● = +/+, ○ = kdr/+, ▲ = kdr/kdr; + = susceptible allele) exposed to fresh deposits of permethrin applied at 100mg/m2 (a) or 6.25mg/m2 (b). Pooled data for 2 replicates.

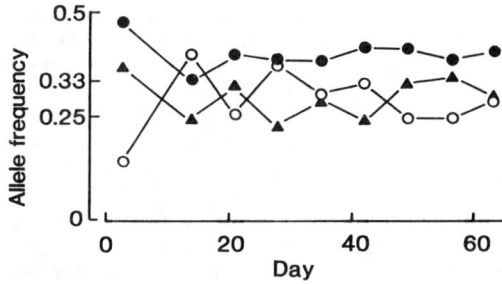

Figure 6. Frequencies of three AChE alleles monitored in an age-structured fly population unexposed to insecticides (● = S allele, ○ = CH2, ▲ = 49R).

mixture of these insecticides leads, as expected, to strong
selection for this resistance gene (Figure 7, top right). A more
surprising finding is that even with relatively uniform application
methods, fresh deposits of this mixture also discriminate
effectively between kdr and susceptible homozygotes (Figure 7, top
left). This is seemingly due to the behavioural effects of
permethrin noted earlier, causing flies to settle less on the
mixture than on a deposit of the OP alone, thereby impairing the
contribution of trichlorphon to overall control. Although these
problems of behavioural antagonism and differential persistence are
overcome to some extent by applying the trichlorphon component at
double strength (Figure 7, lower graphs), this option is probably
too costly and/or hazardous to implement under field conditions.
Longer-term selection experiments applying different mixtures at
various rates against populations containing resistance to both
constituents are now underway to appraise more widely the potential
of this approach to countering resistance.

Development and Role of Empirical Modelling

The ultimate aim of both research programmes outlined in this paper
is to integrate empirical data from the various disciplines into
rigorously-tested computer models to provide a more reliable and
predictive basis for contending with resistance. Hence throughout
the research, data on accessible ecological, genetic and
toxicological parameters are used to model the effects of control
treatments on genetically-heterogeneous populations, and the
descriptive power of these models is being tested by simulating
these control regimes. In the short-term such modelling aids the
interpretation of results and helps to identify productive areas for
additional research, but in the long-term it provides a powerful
means of formalising these results within the framework of
demographic and population genetics theory.

This modelling has, like the experimentation, progressed in
several stages. A program initially describing the ecology of
caged, age-structured populations (13) has been successively
enhanced to simulate the effects of insecticide applications against
a susceptible strain (as referred to earlier) and then to include
the observed response of genotypes at one or more resistance loci in
order to predict selection rates. In addition we have added
stochastic components to contend with sampling effects for rare
genes and small populations, and are now incorporating algorithms
that deal more realistically with the decay of pesticide residues.

To illustrate this aspect of the work, Figure 8 shows
simulations based on results described earlier to predict the effect
of two control treatments on kdr gene frequency and adult population
size in caged fly populations. In the first example (Figure 8a),
permethrin applied alone at its recommended rate leads rapidly to
control failure through selection of kdr, and this result accords
extremely well with selection experiments in the laboratory (9) and
the observed outcome of using this compound in the field (17). The
second simulation (Figure 8b) refers to a mixture of permethrin and
trichlorphon, both applied at their standard rates, and shows that
this mixture (for reasons discussed above) also selects strongly for
kdr. Predicted changes in gene frequency again reflect those

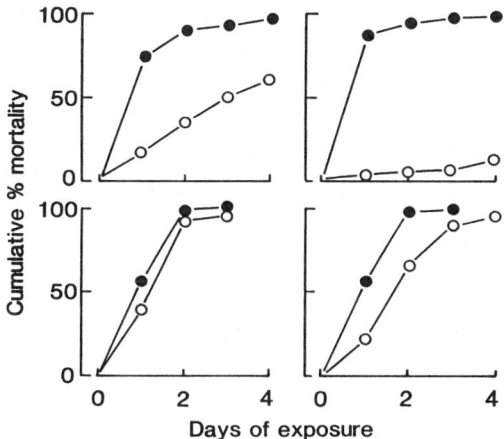

Figure 7. Cumulative mortality of female houseflies of two kdr genotypes (● = +/+, ○ = kdr/kdr) exposed to combined deposits of permethrin and trichlorphon immediately (left) or 4 weeks (right) after spraying. Trichlorphon was applied either at standard (top graphs) or double (1g/m2: bottom graphs) strength. Pooled data for 2 replicates.

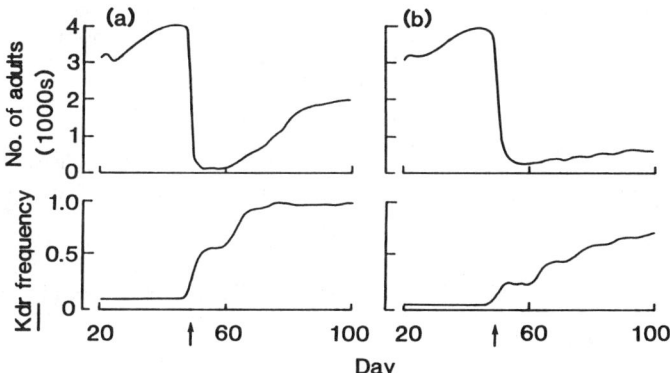

Figure 8. Predicted changes in adult numbers (top graphs) and kdr gene frequency (bottom graphs) in age-structured fly populations treated with permethrin only (a) or a mixture of permethrin and trichlorphon (b). The initial kdr gene frequency is 0.1 and treatment starts on day 50.

monitored in longer-term laboratory experiments (I. Denholm, unpublished data), but the theoretical increase in numbers is less than observed due to inadequate modelling of pesticide decay in the present program.

Work on the Tobacco Whitefly

The evolution of insecticide resistance in Bemisia tabaci has enabled this species to rise from a secondary pest to become the most important cotton pest in some areas (28). There is now an urgent need for rational control procedures to contain resistance in these countries and prevent its spread to other parts of the world. Our work on B. tabaci therefore extends the laboratory simulation approach outlined above to a crop pest of very major economic and practical significance.

Realistic laboratory simulation of the population dynamics and insecticidal control of B. tabaci obviously requires much more sophisticated technology than that employed for the housefly work. In essence, age-structured populations of this insect are reared continuously in cotton-field simulation chambers, and plants are sprayed from overhead at threshold pest densities with available or candidate insecticides applied either singly, together or sequentially at various application rates. The effects of these treatments are assessed by accurately monitoring changes in adult numbers over several generations without interfering with insects or plants (29).

Computer modelling based on empirical data is again an integral part of this research, being used to provide a quantitative description of Bemisia population biology in the chambers, and of the extent to which genetic structure and population size are modified by the control regimes applied. To illustrate this, Figure 9a shows an example of population growth observed in chambers colonised by a small number of adult insects, mimicking the start of a whitefly infestation on cotton in the field. These founders oviposit on the cotton plants, and die off according to the survivorship schedules of males and females under untreated conditions. Around day 17, the progeny of these founders start to emerge, numbers then build up rapidly, and the underlying age-structure of the population becomes progressively more complex. The emergence and growth of the next generation from day 38 or so tends to be more variable, as density-dependent regulatory factors start to take effect. These factors have not yet been examined in detail as the major aim of our work is to maintain numbers well below such high levels.

Figure 9b shows a computer prediction of these phases of population growth based on estimates of the duration of the life-cycle and of age-specific reproductive and survival rates obtained directly from the simulation chambers or from subsidiary experiments. The predictions accord well with observed results, and the model has been successfully extended to describe the outcome of insecticide treatments against laboratory populations, given experimental data on the response to these treatments of different life-stages of the pest.

Concurrent biochemical work on B. tabaci at Rothamsted has provided sensitive techniques for monitoring in treated populations the major resistance markers (esterases and AChE-R) identified so far in this species. Hence, as with houseflies, this programme now has all the components necessary to develop and appraise insecticide strategies that give effective control and minimise the selection of resistance genes.

Conclusions

Integrating experimentation and modelling in these programmes is proving an extremely productive way of applying population genetics

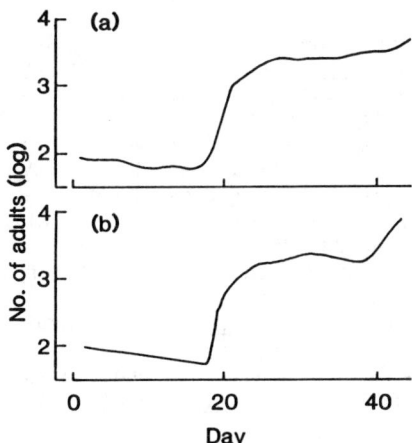

Figure 9. Observed (a) and predicted (b) growth in whitefly populations in cotton-field simulation chambers.

theory to appraise the efficacy and feasibility of resistance-countering regimes. Results presented here highlight major discrepancies in how resistance is expressed under different exposure regimes, and emphasize the importance of obtaining realistic data to underpin theoretical work. Empirical models making the minimum of hypothetical assumptions are now a key component of both studies, and existing models will undoubtedly be substantially improved in the light of further research. However, knowledge gained from fundamental studies on houseflies and the pioneering research on B. tabaci must be complemented by laboratory or field research on a wide range of insects in order to tailor tactics operable within practical constraints to different pest species. We are convinced from experience to date that there is considerable scope for extending laboratory simulation experiments to other crops and pests.

Acknowledgments

We thank K. Gorman, B. Hackett, M. Stribley and J. White for technical assistance, and CIBA-GEIGY AG for funding the work on whiteflies.

Literature Cited

1. Georghiou, G.P.; Taylor, C.E. J. Econ. Entomol. 1977, 70, 653-8.
2. Wood, R.J.; Mani, G.S. Pestic. Sci. 1981, 12, 573-81.
3. Curtis, C.F. Bull. Entomol. Res. 1985, 75, 259-65.
4. Mani, G.S. Genetics 1985, 109, 761-83.
5. Taylor, C.E. In Pest Resistance to Pesticides; Georghiou, G.P.; Saito, T., Eds.; Plenum: New York, 1983; pp 163-73.

6. Denholm, I. Proc. IVe Congres sur la Protection de la Sante Humaine et des Cultures en milieu tropical, Marseille, 1986; pp 239-46.
7. Denholm, I.; Sawicki, R.M.; Farnham, A.W. In Biological and Chemical Approaches to Combating Resistance to Xenobiotics Ford, M.G.; Hollomon, D.W.; Khambay, B.P.S.; Sawicki, R.M. Eds.; Ellis Horwood: Chichester, 1987; pp 138-49.
8. Farnham, A.W. Pestic. Sci. 1977, 8, 631-40.
9. Devonshire, A.L. In Biological and Chemical Approaches to Combating Resistance to Xenobiotics; Ford, M.G.; Hollomon D.W.; Khambay, B.P.S.; Sawicki, R.M., Eds.; Ellis Horwood Chichester, 1987; pp 239-255.
10. Moores, G.D.; Devonshire, A.L.; Denholm, I. Bull. Entomol. Res 1988, 78, 537-44.
11. Sawicki, R.M.; Farnham, A.W. Ent. exp. & appl. 1968, 11, 133-42.
12. Sawicki, R.M.; Devonshire, A.L.; Farnham, A.W.; O'Dell, K.E.; Moores, G.D.; Denholm, I. Bull. Entomol. Res. 1984, 74, 197-206.
13. Denholm, I.; Sawicki, R.M.; Farnham, A.W.; White, J.C. Bull. Entomol. Res. 1986, 76, 297-302.
14. Rust, M.R.; Reirson, D.A. J. Econ. Entomol. 1978, 71, 143-58.
15. Penman, D.R.; Chapman, R.D. Ent. exp. & appl. 1983, 33, 71-78.
16. Taylor, C.E.; Georghiou, G.P. Environ. Entomol. 1982, 11, 746-50.
17. Denholm, I.; Farnham, A.W.; O'Dell, K.E.; Sawicki, R.M. Bull. Entomol. Res. 1983, 73, 481-9.
18. Mani, G.S.; Wood, R.J. Pestic. Sci. 1984, 15, 325-36.
19. Curtis, C.F. In Biological and Chemical Approaches to Combating Resistance to Xenobiotics; Ford, M.G.; Hollomon, D.W.; Khambay B.P.S.; Sawicki, R.M., Eds.; Ellis Horwood: Chichester, 1987; pp 150-161.
20. Rawlings, P.; Davidson, G.; Sakai, R.K.; Rathor, N.R.; Alsamkhan, M.; Curtis, C.F. Bull. WHO 1981, 59, 631-40.
21. McKenzie, J.A.; Whitten, M.J. Experientia 1982, 38, 84-85.
22. Daly, J.C.; Fisk, J.H.; Forrester, N.W. J. Econ. Entomol. 1988 81, 1000-8.
23. Moores, G.D.; Denholm, I.; Byrne, F.J.; Kennedy, A.L.; Devonshire, A.L. Proc. Brighton Crop Protect. Conf. - Pests and Diseases, Brighton, 1988; pp 451-56.
24. Georghiou, G.P. In Pest Resistance to Pesticides; Georghiou, G.P.; Saito, T., Eds.; Plenum: New York; pp 769-92.
25. Sawicki, R.M.; Denholm, I. Trop. Pest Management 1987, 33, 262-72.
26. Roush, R.T.; McKenzie, J.A. Ann. Rev. Ecol. Syst. 1987, 32 361-80.
27. Rowland, M. Proc. Brighton Crop Protect. Conf. - Pests and Diseases, Brighton, 1988; pp 495-500.
28. Dittrich, V.; Hassan, S.O.; Ernst, G.H. Crop Protection 1985 4, 161-76.
29. Sawicki,R.M.; Rowland, M.W.; Byrne, F.J.; Pye, B.J.; Devonshire, A.L.; Denholm, I.; Hackett, B.S.; Stribley, M.F. Aspects Appl. Biol. 1989, 21, 121-22.

RECEIVED September 1, 1989

Chapter 7

Targeting Insecticide-Resistant Markets

New Developments in Microbial-Based Products

Wendy D. Gelernter

Mycogen Corporation, 5451 Oberlin Drive, San Diego, CA 92121

Development of microbial insecticides based on *Bacillus thuringiensis* *(Bt)* presents new alternatives in efforts to manage insects which are resistant to conventional insecticides. Growing attention to issues such as environmental pollution, food safety and pesticide resistance has focused new attention on *Bt* based products, which are non-toxic to mammals and other non-target organisms, and have been effectively used in several agricultural systems. Examples include M-One insecticide, which is based on the naturally occurring bacterium, *Bt* variety *san diego*. This product has been successfully used to control the Colorado potato beetle, an insect which has developed resistance to all major classes of synthetic chemical insecticides. In contrast, the MCap delivery system is based on a recombinant microorganism that expresses a *Bt* toxin, but has been killed via heat and chemical treatment prior to field release. Results of field trials indicate that when *Bt* toxins active against resistant insects such as the diamondback moth or the Colorado potato beetle are transferred into the MCap system, the foliar persistence and efficacy of the product is enhanced, when compared to comparable products based on naturally occurring *Bts*.

Since its introduction in the United States in 1958, the insect pathogenic bacterium *Bacillus thuringiensis* has served as the basis for the most successfully commercialized group of microbial insecticides available (1). In addition to demonstrated high levels of efficacy against key economic insect pests, benefits of *Bacillus thuringiensis (Bt)* include host range specificity, mammalian, non-target organism and environmental safety, low registration costs and economical production methods. Based on these benefits, many internationally marketed commercial products based on *Bt* have been developed for control of lepidopteran, dipteran and coleopteran pests. In recent years, interest in microbial pesticides such as *Bt* has intensified, as disadvantages associated with the use of synthetic, chemical based pesticides such as environmental contamination, broad spectrum toxicity, high registration costs and the development of pest resistance, have received increasing attention.

In this paper, the use of *Bt* -based insecticides as tools for controlling insects resistant to synthetic insecticides will be explored through a review of the development of two very different *Bt* based products: M-One Insecticide, based on the naturally occurring *Bt* variety *san diego* controls the Colorado potato beetle, a pest that has developed resistance to all major classes of synthetic insecticides in many areas of North America, Europe and Eastern Europe (2); in contrast, the MCap delivery system, based on a recombinant organism that has been engineered to produce a *Bt* toxin (3), effectively controls the diamondback moth, a pest that is rapidly developing resistance to synthetic insecticides in Asia, as well as the United States (4). This review will discuss the advantages and limitations of using *Bt* as a tool for control of resistant insects, as well as the potential of techniques in genetic engineering for improving the performance and utility of *Bt*-based insecticides.

Bacillus thuringiensis History and Biology

Biology and Taxonomy. As for other members of the genus *Bacillus*, *Bt* is a Gram positive, rod shaped bacterium that forms a terminal endospore. However, *Bt* is distinguished from other *Bacillus* species by the production, in each cell, of a crystalline protein that is toxic to insects. The toxin or toxins which make up this crystal are called delta endotoxins, and are usually produced by genes located on bacterial plasmids (5). *Bt* is grown in submerged culture in industrial scale fermentors. Towards the end of the growth cycle, the *Bt* cells break apart, or lyse, releasing spores and crystals into the culture media. It is these naked spores and crystals which serve as the active ingredient for commercial *Bt* formulations.

Each isolate of *Bt* can be characterized by a distinctive and relatively narrow insecticidal host range. The nucleotide base sequence of the *Bt* toxin plasmid, and thus the amino acid composition of the crystal's delta endotoxin, determines the unique insect host range of each *Bt* isolate. To date, literally thousands of *Bt* s have been isolated by researchers across the world. These are currently classified into 34 subspecies or varieties (1, 6) based on several characteristics, including flagellar antigens, biochemistry and insect host range (7). Of the *Bt* varieties recognized today, the majority are active against lepidopteran larvae, although several varieties are active against mosquitoes and black flies, as well as against beetle pest species (Table I). There are also several varieties with no described insecticidal activity. The general belief is that these varieties possess toxic activity, but have not yet been tested against the appropriate susceptible insect species.

Mode of Action. Despite the wide spectrum of insect host range activities displayed by *Bt* varieties, the mode of action of the insecticidal delta endotoxins is similar. When susceptible insects ingest *Bt* protein crystals, the first gross symptom observed is feeding inhibition due to paralysis of the digestive tract, including mouth parts. This usually occurs within one hour of ingestion (8).

Within the insect gut, it is believed that the *Bt* protein crystals are digested by gut proteolytic enzymes to form activated toxin molecules with molecular weights of 55 -70kda, depending on the specific *Bt* toxin. The toxin molecules then appear to pass through holes in the peritrophic membrane, and within minutes after ingestion, bind to the microvillar membrane surface of the insect's midgut epithelial cells. This is believed to be a specific binding interaction, which occurs only if the appropriate specific protein or glycoprotein receptors are present on the surface of the insect's midgut epithelium. Following the specific binding step, the epithelial cells swell, vacuoles form, and cellular organelles begin to break down. Ultimately, the microvillar membrane disintegrates, and the epithelial cells lyse, resulting in destruction of the midgut. Insect death is believed to occur as a result of the poisonous effects of changes in pH and ion balance that occur when the gut contents are mixed with the contents of the hemocoel. This, coupled with the effects of starvation caused by the initial feeding inhibition response results in death of the insect. Death usually occurs within 1 - 5 days, depending on the age of the insect (younger insects are most susceptible), the dose ingested, and the insect species and *Bt* isolate involved (9, 10).

As for many other toxicants, insects killed by *Bt* are typically stunted, darkened and appear dry and shrivelled.

Table I. *Bacillus thuringiensis* varieties and host range spectra

Serotype	Variety	Activity[+]	Serotype	Variety	Activity
1	thuringiensis	L, D	11a,11c	kyushuensis	L, D
2	finitimus	---	12	thompsoni	L
3a	alesti	L	13	pakistani	---
3a,3b	kurstaki	L	14	israelensis	D
4a,4b	sotto	L	15	dakota	---
4a,4b	dendrolimus	L	16	indiana	---
4a,4c	kenyae	L, D	---	wuhanesis	L
5a,5b	galleriae	L	17	tohokuensis	L
5a,5c	canadensis	L	18	kumamotoensis	---
6	subtoxicus	L	19	tochigiensis	---
6	entomocidus	L	20a,20b	yunnanensis	L
6a,6c	oyamensis	---	20a, 20c	pondicheriensis	L
7	aizawai	L,D	21	colmeri	---
8a,8b	morrisoni	L,D	22	shandogiensis	L
8a, 8b	tenebrionis*	C			
8a, 8b	san diego*	C			
8a,8c	ostriniae	L			
9	tolworthi	L,D			
10	darmstadiensis	L			

[+] L = Lepidoptera; D = Diptera; C = Coleoptera

* Although reported as serotype 8a, 8b, varietal names other than *morrisoni* have been utilized to reflect the unique biochemistry and insect host range of these isolates

Although current understanding of the molecular mode of action of *Bt* is still not complete, recent reports indicate that the toxin may act by forming small pores in the epithelial microvillar membrane, causing osmotic swelling and lysis and eventual destruction of the gut (11). In a contrasting but not mutually exclusive theory, it has been proposed that the toxin disrupts the potassium ion gradient which the cell normally maintains across the microvillar membrane, again leading to ionic and osmotic disruption of the cell, and lysis (12). Growing interest in the use of *Bt* in agriculture should yield a more complete view of the molecular mode of action in the near future.

Host Range Specificity. Although highly toxic to susceptible insect species, the host range of most *Bt* isolates is confined to a small number of related insect species. For example, isolates of *Bt* variety *israelensis* are highly active against *Aedes* and *Culex* mosquito larvae, but have more limited activity against *Anopheles* larvae (13). Even *Bt* isolates with the same serotype may exhibit different activity spectra. For example, three distinct isolates with the 8a, 8b serotype have demonstrated activity against lepidopteran, dipteran and coleopteran insects (14, 15) (Table I). The host range activity levels for a given *Bt* isolate are believed to be determined by a variety of factors including: structure of the delta endotoxin; internal environment of the insect midgut, which influences break down of the toxin crystals to active toxin molecules; the presence and concentration of specific receptors on the surface of insect midgut epithelial cells (10); and interactions between multiple delta endotoxins contained in one crystal, and interactions between delta endotoxin and *Bt* spores (8).

Commercial Products Based on *Bacillus thuringiensis*

Although *Bt* was first discovered in diseased silk worm larvae in 1901 by the Japanese researcher, Ishiwata, it wasn't until 1958 that a product was first developed for commercial use in the United States (1). For many years, *Bt* products were based on lepidopterous active isolates (Table II). Today, the majority of products available are still targeted towards caterpillar pests of forestry, agriculture, stored grain and field crops, and usually rely on isolates of *Bt* variety *kurstaki*. However, in 1977, Israeli researchers discovered a new isolate, *Bt* variety *israelensis*, which displayed unique activity against mosquito and black fly larvae. Today, *Bt israelensis* is the basis for several insecticides which are used to control biting flies throughout the world. Other *Bt* varieties with dipteran activity have also recently been discovered (16). Most recently, researchers in Germany and the United States have discovered *Bt* isolates with specific activity against coleopteran insects, including the Colorado potato beetle. These isolates, *Bt* variety *tenebrionis* (17) and *Bt* variety *san diego* (15), were registered for commercial use in the United States in 1988.

Table II. Commercial Products based on *Bacillus thuringiensis*

VARIETY OF *B. thuringiensis*	INSECTICIDAL ACTIVITY	PRODUCT	COMPANY
kurstaki	lepidopteran (vegetable crops, forestry, stored products)	Bactospeine Biobit Dipel Javelin Thuricide	Duphar Novo Laboratories Abbott Laboratories Sandoz, Inc. Sandoz, Inc.
aizawai	lepidopteran (greater wax moth)	Certan	Sandoz, Inc.
israelensis	dipteran (mosquitoes, black flies)	Bactimos Skeetal Teknar Vectobac	Biochem Products Novo Laboratories Sandoz, Inc. Abbott Laboratories
san diego	coleopteran (Colorado potato beetle)	M-One	Mycogen Corporation
tenebrionis	coleopteran (Colorado potato beetle)	Trident	Sandoz, Inc.

Naturally Occurring *Bt* and Control of Resistant Insects

Because of its unique mode of action targeted against the insect gut, *Bt* can be an effective control agent for insects resistant to synthetic insecticides. Unlike *Bt*, most synthetic insecticides target the insect nervous system. A case in point is that of the Colorado potato beetle, *Leptinotarsa decemlineata*. This insect is considered to be the most destructive pest of potatoes grown in many areas of the United States, Canada, Europe and the Soviet Union. The beetle's voracious appetite, (larval and adult stages are capable of consuming more than 10cm^2 of foliage per day) and high reproductive capacity are partly responsible for its pest status (18). In addition, several formerly available and effective systemic insecticides have been banned or restricted in some locations, due to fears of groundwater or environmental contamination (19). However, the most difficult issue surrounding control of the Colorado potato beetle (CPB) has been the insect's ability to rapidly develop resistance to all major classes of synthetic insecticides. In some areas of the northeastern United States where resistance is an established problem, there are currently no registered conventional insecticides available which control the CPB (2). Thus, there is a distinct need for a CPB control agent that is both effective and environmentally compatible.

Bacillus thuringiensis variety *san diego*. In 1985, scientists at Mycogen Corporation in San Diego, California, discovered a beetle-active isolate of *Bt*. The isolate, *Bt* variety *san diego* (*Btsd*) demonstrated high levels of activity against larvae of the CPB (15), but like other *Bts* was non-toxic to mammals, birds, fish and other non-target organisms. In field tests conducted in the United States, Canada and Europe during 1986 and 1987, *Btsd* proved to be a highly effective control agent for the CPB when applications were targeted against small (1st - 3rd instar) larvae, the beetle's most sensitive stages. Due to its distinctive mode of action, *Btsd* was equally effective against resistant and non-resistant populations of the CPB. In addition, *Btsd* was compatible when tank mixed with a large variety of commonly used insecticides and fungicides, and was effectively applied through ground and aerial application systems. Based on these positive results, Mycogen submitted a full registration package to the U.S. Environmental Protection Agency (EPA) for their *Btsd* product, M-One Insecticide, in September of 1987. Eight months later, Mycogen received EPA approval of their registration package and began marketing M-One in the summer of 1988.

The use of *Btsd* has given potato growers a new, effective and strongly needed tool for controlling resistant CPBs. In addition, in areas where CPB resistance to synthetic insecticides is not yet a major problem, the effective life of these insecticides is being prolonged by rotating their use with applications of *Btsd*. As a final advantage, the non-toxic characteristics of *Btsd* have helped to alleviate farmer and public concerns regarding safety and environmental pollution by insecticides.

Advantages of *Bt*-Based Insecticides

As demonstrated by the example of *Bt* variety *san diego*, *Bt*-based insecticides have several desirable traits which make them attractive to farmers, to the public, and to agricultural chemical companies in search of new development candidates. These include:
- **safety**
- **streamlined regulatory review**
- **low cost of production**
- **unique mode of action**
- **ease of use**

Safety. In toxicology tests required by the EPA for registration of microbial insecticides, *Bt* isolates have demonstrated a consistent lack of toxicity to mammals and non-target organisms including birds, fish, ducks aquatic invertebrates, beneficial insects and plants (20). In addition, the short residual activity associated with *Bt* confers the desired characteristic of biodegradability to *Bt* based products (Figures 1 and 2). As public concern for environmental and food safety has increased, the appeal of non-toxic, biodegradable pest control agents such as *Bt* has increased commensurately.

Streamlined Regulatory Review. Because *Bt* isolates have demonstrated no toxicity to non-target organisms, EPA regulatory review of new *Bt* isolates is streamlined to focus primarily on acute toxicity and infectivity to mammals and non-target organisms. Under this review process, the potential product is tested in "maximum challenge" experiments, where very high doses of the product are administered to test animals through oral, dermal, inhalation and ocular routes. If no toxic effects are demonstrated

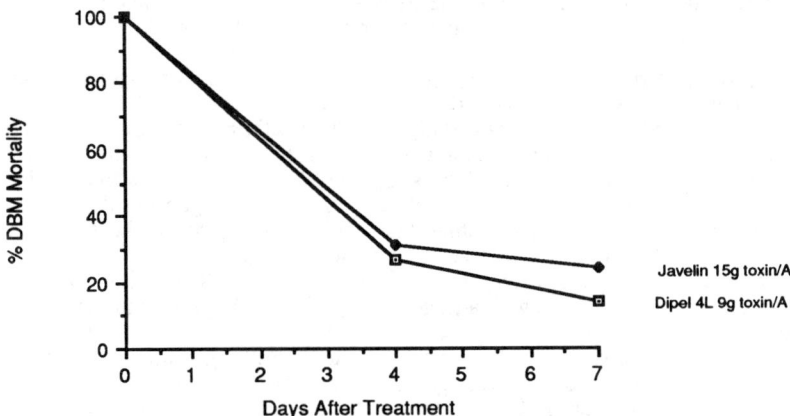

Figure 1. Foliar persistence of microbial insecticides on cabbage. Arlington, Wisconsin: 1988. Dipel and Javelin are commercial products based on *Bt* var. *kurstaki*. Persistence was measured by bioassay of field treated foliage, removed at various times post-application against 3rd instar DBM larvae. Rates of product applied are expressed in terms of grams of delta endotoxin (as measured by polyacrilamide gel electrophoresis) applied per acre.

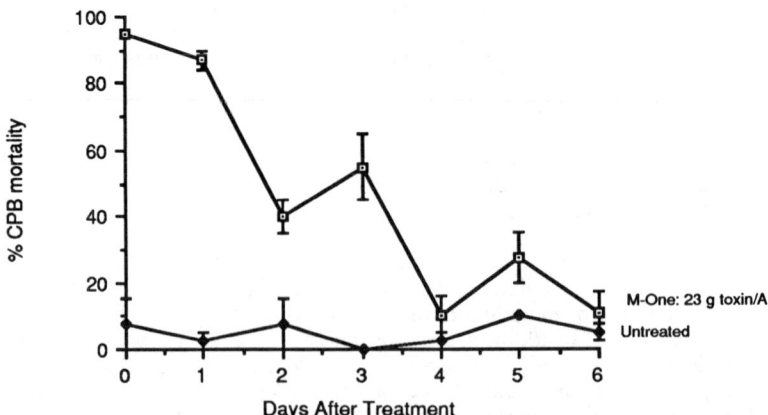

Figure 2. Foliar persistence of M-One Insecticide on potatoes. Painter, Virginia: 1988. Persistence was measured by testing field treated foliage, removed at various time post-application, in laboratory bioassays conducted against 2nd instar CPB larvae. Rates of product applied are expressed in terms of grams of delta endotoxin (as measured by polyacrilamide gel electrophoresis) applied per acre.

in this initial level of acute testing (Tier 1), then the additional testing (including residue and environmental fate, oncogenicity, teratogenicity, mutagenicity and long-term chronic studies) usually required for registration of conventional insecticides is not required (21). This tiered, streamlined approach has led to relatively short registration review periods (6 months - 1 year) and relatively inexpensive costs (up to $500,000) for insecticides based on naturally occurring isolates of *Bt*, compared to the 7-10 year and multimillion dollar costs usually associated with registration of more toxic insecticide products.

Economics of production. As for other commercially available *Bts*, *Bt* variety *san diego* is inexpensively produced in submerged liquid culture in industrial scale fermentors. While components of the culture media used remain proprietary information for each *Bt* product, nitrogen and carbon are usually supplied in the form of inexpensive agricultural or industrial by-products such as cottonseed meal, corn steep, fish meal or soybean meal (1).

Unique Mode of Action. As described above, the *Bt* delta endotoxin is a stomach poison which acts on the midgut epithelial cells of susceptible insects. This is in contrast to the typical mode of action for synthetic chemical insecticides, which act as nerve poisons. Because the *Bt* site of action is so different from that of synthetic insecticides, *Bt* has been successfully utilized for control of otherwise resistant insects such as mosquitoes, the diamondback moth and the Colorado potato beetle.

Flexibility of Use. Products based on *Bt* have been successfully incorporated into pest management programs in agriculture, forestry and vector control. This is largely due to the fact that the *Bt* delta endotoxin and spores are relatively stable when tank mixed with conventionally used fungicides, insecticides and spray adjuvants. In addition, most commercial *Bt* formulations have been designed to optimize shelf life, as well as physical compatibility with other agricultural chemicals and application systems.

Limitations of *Bt*-Based Insecticides

Over the past 30 years, insecticides based on naturally occurring isolates of *Bt* have played, and will continue to play, an increasingly important role in addressing specific pest management problems such as insecticide resistance and environmental pollution. However, there are several characteristics of *Bt* which currently limit its widespread adoption in agriculture. These include:
- **narrow insect host range**
- **lack of delivery to insects that feed inside the plant**
- **short residual activity**

Narrow Host Range A paradox associated with the use of *Bt* lies in the characteristic host range specificity of each strain to small groups of related insects. While this feature is usually regarded as an environmental benefit of the use of *Bt*, agricultural crops are seldom attacked by a single insect pest species. For example, cole crop pests such as the cabbage looper and diamondback moth are well controlled by commercial preparations of *Bt* variety *kurstaki*, but other insecticides must be used to control aphid populations. Similarly, Colorado potato beetle larval populations are highly susceptible to *Bt* varieties *san diego* and *tenebrionis*, but conventional synthetic insecticides must be applied to control occasional pests such as the potato fleabeetle and leafhoppers. While *Bt* has been successfully utilized to control the key insect pests in these and other situations, more planning and scouting time may be required of the farmer who chooses to use *Bt* when multiple insect pest populations occur simultaneously.

Delivery to Target Pests The host range of *Bt* is further limited by the fact that it is a stomach poison; it must be consumed to be effective. For this reason, commercial applications of *Bt* have been most successfully targeted against foliar feeding insects, such as the gypsy moth, the cabbage looper, or the Colorado potato beetle. In contrast, internal plant feeders such as the cotton boll weevil, or the codling moth, while susceptible to *Bt* variety *san diego* and *Bt* variety *kurstaki*, respectively, in laboratory tests, are difficult to control with foliar applications of *Bt* in the field, because *Bt* cannot be easily delivered to their internal feeding sites. In addition, piercing-sucking insects such as aphids and leafhoppers have not been investigated as potential *Bt* targets, because feeding occurs internal to the treated leaf surface, making its delivery to the insect gut difficult, if not impossible.

Residual activity. For naturally occurring *Bts,* lack of foliar persistence is perhaps the most significant factor contributing to the products' inconsistent field performance and to its concomitant limited adoption by farmers. The residual insecticidal activity of *Bt* applications is directly related to the stability of the protein delta endotoxin on the leaf surface, and to the period of time between *Bt* application and the insect's encounter with and ingestion of treated foliage. Typical degradation curves for commercial *Bt* products generated by Mycogen researchers (Figure 1 and 2) illustrate that insecticidal activity declines rapidly within the first 48 hours after field application and has nearly disappeared by 96 hours after application. Studies conducted in the laboratory and greenhouse however, demonstrate that *Bt* can remain active for up to two weeks indoors, indicating that factors relating to outdoor application are largely responsible for *Bt*'s short residual activity. Various researchers have identified ultraviolet radiation, heat, moisture, pH and enzymatic activity on the leaf surface (7, 22) as the environmental factors responsible for the decay of *Bt* activity in the field. However, there is currently little agreement on the extent to which each factor influences persistence, and to what degree these factors interact.

Modification of *Bacillus thuringiensis* Through Genetic Engineering

Application of techniques in genetic engineering to agriculture may help to reduce or completely remove the limitations on *Bt* use and applicability described above. This belief is based on the fact that the *Bt* delta endotoxin gene is usually borne on a bacterial plasmid, which can be isolated, manipulated and transferred to other organisms using standard recombinant DNA (rDNA) techniques. Thus, the insect host range might be expanded or altered through changes made in the genes that produce the toxin. Delivery to target insects that feed inside the plant could be improved through expression of the delta endotoxin protein within the crop plant tissue, while improvements in foliar persistence could be accomplished by transferring the toxin gene to naturally occurring bacteria which colonize the leaves or roots of the crop plant.

A delta endotoxin gene was first cloned from *Bt* variety *kurstaki* in 1981, by Schnepf and Whiteley (23). Since that time, several researchers have transferred *Bt* toxin genes to tobacco (24, 25) and tomato (26) in attempts to protect these plants from insect damage. Companies such as Rohm and Haas, Monsanto, Agrigenetics and Plant Genetic Systems have conducted small plot field tests with transgenic plants where a *Bt* toxin has been successfully expressed. To protect plant roots from soil dwelling insects, root colonizing bacteria have also been transformed by Monsanto researchers to produce *Bt* endotoxins (27). Crop Genetics, Inc. has demonstrated that insects feeding on the vascular system of corn plants can be targeted by the endophytic bacterium, *Clavibacter xyli*, which grows within the plant's vascular system and which has been genetically engineered to produce the *B.t. kurstaki* endotoxin.

Limitations of living rDNA products. Despite the exciting advances described above, there are still many technical obstacles that must be overcome before rDNA products can be successfully commercialized in agriculture. In addition, the regulatory position on risk and hazard assessment of living rDNA organisms is still in the process of development by government agencies (28). While there is no doubt that products based on living rDNA organisms will play an important, and possibly revolutionary role in agriculture in the future, current regulatory and technical hurdles make commercial availability within the next ten years unlikely. As will be discussed below however (see "The MCap Delivery System"), dead recombinant organisms do not present the same regulatory hurdles as their living counterparts, and may be commercialized within the next 1-2 years.

Potential for *Bt* resistance. During the relatively short gap in time between discovery and commercialization of rDNA insecticides, researchers will hopefully have an opportunity to explore an important potential limitation to the use of *Bt*-based insecticides--the development of resistance. In 1985, McGaughey reported evidence for the only documented case of *Bt* resistance in a field population of insects (29). The Indian meal moth, *Plodia interpunctella*, is a lepidopteran pest of stored grain that was successfully controlled with *Bt* variety *kurstaki* for several years. However, McGaughey's study showed that an observed decline in the efficacy of *Bt* was due to the development of resistance among treated *Plodia* populations.

As a result of McGaughey's report, researchers in industry and academia have focused greater attention on the development of methods to detect other cases of *Bt* resistance. To date, no further examples have been documented. It is possible that the case of *Plodia* is unique because, in McGaughey's words, "the stored grain environment is ideal for development of resistance, because BT is stable on stored grain and because the environment may remain undisturbed for long periods, permitting the insects to breed for successive generations in contact with the bacterial spores and toxins" (29). This is in contrast to field crop applications of *Bt*, where foliar applied product has low residual activity, thus decreasing the probability of resistance. Despite the unique conditions surrounding the development of *Plodia* resistance to *Bt*, McGaughey's data demonstrates that *Bt* resistance is possible. As it becomes technically feasible to increase the environmental persistence of the *Bt* toxin through rDNA techniques, and as the use of *Bt* -based insecticides increases, the potential for the development of resistance will increase as well. Creative approaches in integrated pest management, population genetics and molecular genetics must be combined so that the use of *Bt* as an effective and safe control agent can be maintained.

The MCap delivery system

In an attempt to improve the foliar persistence of *Bt* insecticidal activity, Mycogen Corporation has developed the MCap delivery system, based on an rDNA microorganism that expresses a *Bt* toxin, but has been killed via heat and chemical treatment prior to field release (3). Because the organisms are dead, this product has the additional advantage of relative freedom from the environmental and safety concerns associated with outdoor testing of living rDNA organisms.

Using the MCap system, researchers have transferred lepidopteran active and coleopteran active delta endotoxin genes into a non-pathogenic strain of *Pseudomonas fluorescens*. When the transformed cells are grown in submerged culture, the delta endotoxin forms a typical *Bt*-like crystal within the cell. However, unlike *Bacillus* cells, which normally burst at the end of the growth cycle releasing naked spores and crystals into the culture medium, the endotoxin-producing *Pseudomonas* cells remain intact at the completion of the fermentation cycle. While still in the fermentation tank, the cells are then chemically treated, killing the *Pseudomonas* cells and causing the bacterial wall to become more rigid through cross linking of cell wall components.

The dead bacterial cell wall now serves as a protective "microcapsule" for the enclosed delta endotoxin.

In 1985, Mycogen received EPA approval to conduct small plot field tests with a lepidopteran active toxin delivered through the MCap system (product code: MYX 7275), making it the first recombinant product approved for outdoor testing. The EPA's approval was based on the fact that extensive verification procedures were conducted by Mycogen on each batch of material to assure that all cells were dead. In addition, the EPA required that MYX 7275 be evaluated in Tier 1 toxicology tests, similar to those required for naturally occurring isolates of *Bt*. Results of these tests have demonstrated that MYX 7275 is non-hazardous to mammals and other non-target organisms.

Genetically Engineered *Bt* and Control of Resistant Insects

Due to the development of resistance to conventionally used organochlorine, organophosphate, pyrethroid and carbamate insecticides, the diamondback moth, *Plutella xylostella,* has become the key pest of cole crops grown in Asia, Central America and the southeastern United States. For this reason, commercial products based on *Bt* variety *kurstaki* have been successfully utilized for control of this resistant pest (30). To improve the consistency and efficacy of diamondback moth (DBM) control, Mycogen Corporation has developed a delta endotoxin highly potent for DBM larvae for expression in its MCap delivery system. Outdoor field tests conducted during 1988 indicated that while naturally occurring *Bts* were effective control agents for DBM larvae, insecticidal activity had dissipated by 96 hours post-application (Figure 1). In contrast, when an equivalent dose of delta endotoxin was delivered via the MCap system in MYX 7275, the product remained active for 7 days (Figure 3). This increase in persistence resulted in better insect control, and therefore higher yields in the MYX 7275 treated plots. Mycogen plans to commercialize MYX 7275 in 1990.

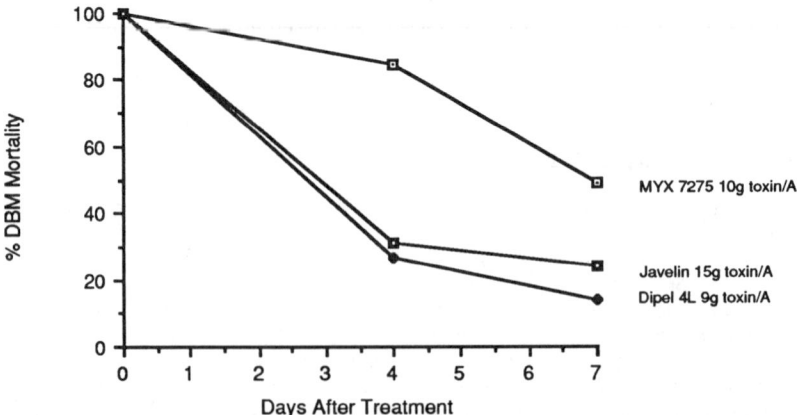

Figure 3. Foliar persistence of MYX 7275 on cabbage. Arlington, Wisconsin: 1988. Persistence was measured as described for Figure 1. Rates of product applied were based on manufacturer's recommendations and are expressed in grams of delta endotoxin applied per acre.

Similar tests have been conducted with a coleopterous-active MCap product (MYX 1806) targeted against the CPB on potatoes. In these tests, potato foliar protection and yields were compared for MYX 1806 and its naturally occurring counterpart, *Bt* variety *san diego* (M-One insecticide) (Figures 4 and 5). The results confirm that when equivalent toxin rates were applied for each product, the persistence conferred by the MCap system resulted in superior CPB control. Additionally, it was demonstrated that the protected toxin of MYX 1806 could be applied at two thirds the toxin rate, and yet still maintain higher levels of insect control than the full rate delivered by M-One.

As these examples illustrate, the MCap delivery system offers several advantages over comparable naturally occurring isolates of *Bacillus thuringiensis*. These include:

- a two fold increase in foliar persistence results in better insect control, when applied at the same rate (grams toxin/A) and same application schedule as naturally occurring *Bts*
- higher persistence levels may allow reductions in the recommended field rate, while still maintaining excellent levels of insecticidal activity
- the longevity of insect control can decrease the number of applications, and therefore the cost, required for season long insect control
- toxicity to targeted insect pests is increased by the ability to deliver the most potent delta endotoxin (or toxins) available via the MCap system. This feature will become even more attractive as the variety and number of available *Bt* toxins increases due to new discoveries of naturally occurring *Bts*, as well as from development of delta endotoxin genes engineered in the laboratory to optimize toxicity, host range and expression level
- product shelf life is enhanced by the absence of any living organism in MCap preparations.

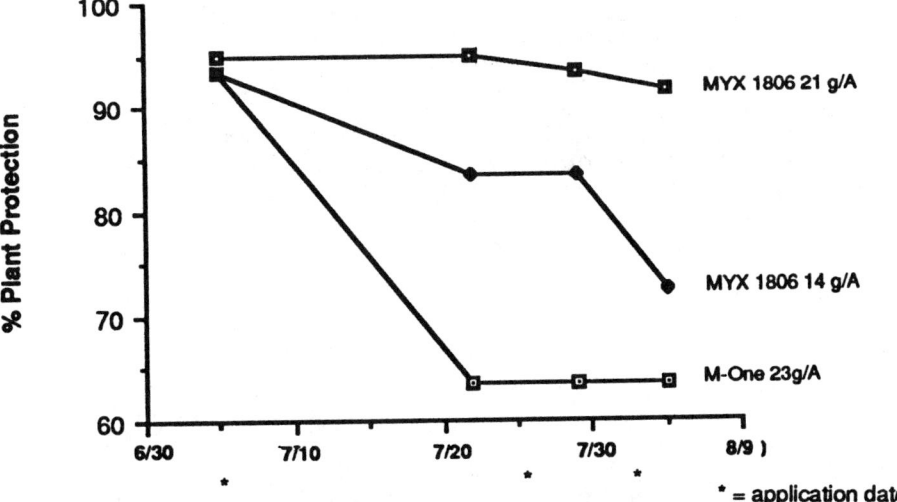

Figure 4. Effect of MYX 1806 and M-One Insecticide on potato plant protection levels. Bath, Pennsylvania 1988. Protection ratings were calculated on the basis of weekly visual estimates of percent defoliation (% plant protection = 100 - % defoliation.)

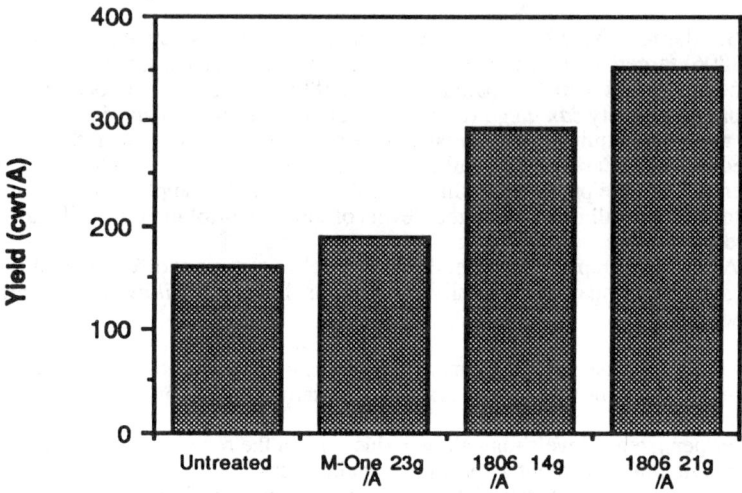

Figure 5. Effect of MYX 1806 and M-One Insecticide on potato yields. Bath, Pennsylvania, 1988. Yields are measured in hundred weight (hundreds of pounds) of potatoes per acre.

Summary

Demand for efficacious and environmentally compatible insecticides will continue to grow as issues surrounding environmental and food safety, as well as pesticide resistance receive increasing levels of attention. As regulatory review for toxic pesticides becomes more expensive and time consuming, the agrichemical industry will continue to concentrate more attention on products which are subject to streamlined regulatory review. Insect control products based on *Bacillus thuringiensis* -- from naturally occurring bacteria, to genetically engineered microorganisms, to transgenic plants -- will therefore constitute a significant portion of new products developed in coming years. The challenge now and for the future will be to apply our vast experience in pest management towards the development of strategies that will utilize the products of biotechnology in the most efficacious, environmentally conscious and judicious manner possible.

Literature Cited

1. Rowe, G.E.; Margaritis, A. CRC Critical Reviews in Biotechnology. 1987, 6, 87-127.
2. Georghiou, G. P. In Pesticide Resistance: Strategies and Tactics for Management; National Academy Press: Washington, D.C, 1986; pp 14-43.
3. Barnes, A.C; Cummings, S.E. U.S. Patent No. 4695455, 1987.
4. Tabashnik, B. E. J. Econ. Entomol. 1986, 79, 1447-51.
5. Sneath, P.H.A. In Bergey's Manual of Systematic Bacteriology; Sneath, P.H.A.; Mair, N.S.; Sharpe, M.E; Holt, J.G., Eds.; Williams and Wilkins: Baltimore, MD, 1986; Vol. 2, pp 1104-1207.
6. Ono, K.; Ohba, M.; Aizawa, K.; Iwanami, S. J. Invertebr. Pathol. 1988, 51, 296-297.

7. Dulmage, H.T.; Aizawa, K. In Microbial and Viral Pesticides; Kurstak, E., Ed.; Marcel Dekker, Inc.: New York, 1982; pp 209-237.
8. Heimpel, A. M.; Angus, T.A. J. Insect Pathol. 1959, 1, 152-170.
9. Luthy, P.; Ebersold, H.K. In Pathogenesis of Invertebrate Microbial Diseases; Davidson, E., Ed.; Allanheld, Osmun and Co.: Totowa, NJ., 1981; pp 235-267.
10. Hofmann, C. Ph.D. Thesis, Swiss Federal Institute of Technology, Zurich, Switzerland, 1988.
11. Knowles, B. H.; Ellar, D.J. Biochim. Biophys. Acta. 1987, 924, 509-518.
12. Sacchi, V. F.; Parenti, P.; Hanozet, G.; Giordana, B.; Luthy, P.; Wolfersberger, M.G. FEBS Letters, 1986, 204, 213-218.
13. deBarjac, H. In Biotechnology in Invertebrate Pathology and Cell Culture; Maramorosch, K. Ed.; Academic Press: San Diego, CA., 1987; pp. 63-73
14. Padua, L. E.; Ohba, M.; Aizawa, K. J. Invertebr. Pathol. 1984, 44, 12-17.
15. Hermstadt, C.; Soares, G.G.; Wilcox, E.R.; Edwards., D.L. Bio/Technology. 1986, 4, 305-308.
16. Ibarra, J.E.; Federici, B.A. FEMS Microbiology Letters. 1986, 34, 79-84.
17. Krieg, A.; Huger, A.M.; Langenbruch, G.A.; Schnetter, W. Z. Angew. Entomol. 1983, 96, 500-508.
18. Ferro, D.N.; Logan, J.A.; Voss, R.H.; Elkington, J.S. Environ. Entomol. 1985, 14, 343-348.
19. Jones, R. Potato Grower of Idaho. January, 1987, pp 12-17.
20. Faust, R.M.; Bulla, L. A. In Microbial and Viral Pesticides; Kurstak, E. Ed.; Marcel Dekker, Inc.: New York, 1982; pp 75-208.
21. Pesticide Assessment Guidelines. Subdivision M: Biorational Pesticides, U.S. Environmental Protection Agency, National Technical Information Service (#PB83-153965),.Springfield, VA., 1983
22. Pozsgay, M.; Fast, P.; Kaplan, H.; Carey, P.R. J. Invertebr. Pathol. 1987, 50, 246-253.
23. Schnepf, H.E.; Whiteley, H.R. PNAS USA. 1981, 78, 2893-2897.
24. Barton, K.A.; Whiteley, H.K.; Yang, N. Plant Physiol. 1987, 85, 1103-1109.
25. Vaeck, M.; Reynaerts, A.; Hofte, H.; Jansens, S.; DeBeuckeleer, M.; Dean, C.; Zabeau, M.; Van Montagu, M.; Leemans, J. Nature 1987, 328:, 33-37.
26. Fischoff, D.A.; Bowdish, K.S.; Perlak, F.J.; Marrone, P.G.; McCormick, S.M.; Niedermeyer, J.G.; Dean, D.A.; Kusano-Kretzmer, K.; Mayer, E.J.; Rochester, D.E.; Rogers, S.G.; Fraley, R.T. Bio/Technology 1987, 5, 807-813.
27. Obukowicz, M.G.; Perlak, F.J.; Kusano-Kretzmer, K.; Mayer, E.J.; Bolten, S.L.; Watrud, L.S. J. Bacteriol. 1986, 168, 982-989.
28. Conner, J.; Ebner, L.; O'Connor, C.; Volz, C.; Weinstein, K. Pesticide Regulation Handbook; McKenna, Conner and Cuneo: Washington, D.C., 1987; pp. 119-140.
29. McGaughey, W.H. Science. 1985, 229, 193-195.
30. Koshihara, T. Japan Pesticide Information. 1988, 53, 14-47.

RECEIVED October 16, 1989

Chapter 8

Strategies for Managing Resistance to Insecticides in *Heliothis* Pests of Cotton

D. L. Bull[1] and J. J. Menn[2]

[1]Veterinary Toxicology and Entomology Research Laboratory, Agricultural Research Service, U.S. Department of Agriculture, College Station, TX 77841
[2]Beltsville Agricultural Research Center, Agricultural Research Service, U.S. Department of Agriculture, Beltsville, MD 20705

Certain species of the genus Heliothis are among the most important of all phytophagous insect pests. The Heliothis complex has worldwide distribution and is responsible each year for major economic losses over a broad range of field and horticultural crops. Recent estimates indicate that, in the USA alone, Heliothis zea (Boddie) and H. virescens (F.) cause more than one billion dollars in annual losses attributable to crop damage and costs of control. Protection of crops from unacceptable damage by these pests is a continuing challenge because of the ability of some species to develop resistance to the different classes of insecticides that have served as the principal means for their control. The history and consequences of insecticide resistance in Heliothis spp. pests of cotton and the various factors that can influence development of resistance among field populations of these insects are reviewed. Different strategies for managing insecticide resistance in Heliothis are discussed, giving emphasis to the urgent need to resolve current problems and concerns relating to pyrethroid resistance among populations of Helthios virescens on cotton in the USA.

The lepidopteran genus Heliothis includes some species that are among the most important of all phytophagous insect pests. The Heliothis complex has worldwide distribution and is responsible for economic losses over a wide range of field and horticultural crops. In the USA, two species of these pests -- Heliothis zea (Boddie); otherwise known by the common names bollworm, corn earworm, tomato fruitworm, etc.; and the tobacco budworm, Heliothis virescens (F.) -- have been estimated to cause more than one billion dollars in annual losses attributable to direct crop damage and costs of control (1). A recent USDA economic study (1) indicated that in cotton alone during the period 1981-1984, these two Heliothis

This chapter not subject to U.S. copyright
Published 1990 American Chemical Society

species were responsible for average annual losses of $216 million. It is estimated that the two Heliothis species account for ca. 30% of all insect damage to U.S. cotton.

Over the past 30 years, resistance to various insecticides among populations of Heliothis spp. has been an on-going problem that usually was resolved by changing to a new type or class of compounds. Most recently this has involved a change to the pyrethroids, which are considered to be nearly the ideal class of insecticides because of their favorable toxicological characteristics with respect to mammals and other nontarget organisms, high efficacy against pests at very low rates, and generally favorable reports on environmental safety. However, if current trends in the development of resistance to the pyrethroids in the tobacco budworm are not reversed, these chemicals also could fail, and then there would be potential for substantial increases in economic losses in cotton production, as well as to chemical companies that produce and market pyrethroids. At present there is no new class of pesticide chemicals, equal to the pyrethroids, approved and available to throw into the battle with this formidable pest (2).

This paper reviews the history and the current status of insecticide resistance among field populations of Heliothis spp. in the USA and discusses certain issues and concepts related to management of the resistance problem. Since the problem of resistance among these pests is associated primarily with cotton production, that will be the focus of this discussion. We need to keep in mind, however, that the bollworm and tobacco budworm are highly mobile polyphagous pests that feed on several major crops such as corn, cotton, grain sorghum, soybeans, tomatoes, and tobacco; as well as at least 60 other species of plants both wild and cultivated (3). In addition to other factors, the development of resistance in a given species is strongly influenced by its seasonal population dynamics, which in turn is broadly influenced by the characteristics of the total agro-ecosystem throughout which the species is distributed (4).

The Target and Its Pests

Cotton grown in this country can be subjected to insect damage during much of its growth cycle. Seedling stage plants may be attacked by such pests as cutworms, thrips, and aphids. Cotton is most vulnerable during development of the fruit, beginning with the first appearance of the flower buds (or squares) and continuing through the blooming period and into the boll formation phase until the point when bolls have completed 60-70% of their development. During this period the plant can be attacked by major chewing pests such as Heliothis spp., boll weevils, pink bollworms; and to a lesser extent by armyworms, cotton leafperforators, and cabbage loopers; as well as by sucking pests such as cotton fleahoppers, Lygus bugs, spider mites, aphids, and whiteflies. Economic damage to cotton can even be incurred when bolls mature and open through contamination of lint with honeydew produced by aphids or whiteflies.

Even though this discussion centers on the Heliothis spp., management of these pests must also take into account, and quite often is compromised by, chemical control directed at other pests.

Heliothis/Plant Interaction

Although there are exceptions, Heliothis spp. usually deposit their eggs on the tender terminal growth of cotton. Newly hatched larvae feed briefly on terminal growth and then move to the fruit, where they may feed on all developmental stages of those structures -- from newly-formed flower buds to bolls that have completed about 60-70% of their development. Larvae are fairly well protected when feeding within the fruiting structures, and as a result it can be difficult to contact them with insecticide sprays. Heliothis spp. are most vulnerable to chemical insecticides during the egg stage and first larval instar. This pest has five larval instars, and insecticide susceptibility decreases as the larvae increase in size. Since adults are nocturnal, their control with present application methods relies on incidental exposure to sprays and contact with residues on foliage.

Insecticide Use on Cotton

Although natural processes and cultural methods are important components of overall crop protection strategies, the use of insecticides is, and will likely continue indefinitely to be, the primary means for suppressing Heliothis spp. on a number of important agricultural crops. This is especially true for cotton protection, which historically has been characterized by heavy use of a variety of insecticides. For example, it is estimated that almost 50% of all insecticides applied annually to crops in the USA are applied to cotton (5).
   According to recent estimates, some 35 chemical pesticides are currently used to control various pests on cotton (1). There is considerable variation across the USA cotton belt in seasonal pesticide application rates, ranging from a high average of 7.4 pounds per harvested acre in Florida to a low of 0.3 in Oklahoma. Currently, methyl parathion (21.0%), azinphosmethyl (13.0%), various pyrethroids (8.0%), chlordimeform (7.4%), propargite (6.8%), and aldicarb (6.8%) account for 63% of all insecticides applied to cotton (1).

Chronology of the Development of Resistance to Insecticides in Heliothis spp.

Synthetic organic insecticides made their debut in cotton protection shortly after World War II with the introduction of chlorinated hydrocarbons such as DDT, benzene hexachloride, toxaphene, and the cyclodienes. These chemicals were so highly effective and economical to use that they had a spectacular impact on the cotton production industry (4). Insect damage was kept under control at low cost, and yields increased dramatically. Unfortunately, this euphoric era of cotton pest management did not last long because of the development of resistance among certain major cotton pests,

especially the bollworm and tobacco budworm. Beginning with
isolated reports of control failures in the late 1950's and
continuing through the 1960's, there were progressive increases in
laboratory and field observations of resistance in both Heliothis
species to DDT and endrin as well as to the carbamate carbaryl.
Because of these resistance problems, especially with the
organochlorines, there was a major shift during the 1960's to the
use of organophosphorus ester (OP) insecticides to control Heliothis
spp. and other important cotton pests.

The OP's, notably methyl parathion, provided excellent,
cost-effective control during the first few years after their
introduction, but these successes too were short-lived. Problems of
resistance to methyl parathion among field populations of the
tobacco budworm were first observed in northern Mexico during the
latter 1960's (4) and confirmed with laboratory tests in 1969
(6,7). Although there have been a few isolated reports of
resistance to methyl parathion in the bollworm, these have not been
confirmed (8) and apparently there have never been any cases of loss
of control in the field. The problem of tobacco budworm resistance
to methyl parathion in certain areas of northern Mexico became so
severe that cotton production declined from about 700,000 acres in
early 1960 to less than 1000 acres by 1970 (4,5). During this
period there was a lack of effective, economic alternative
insecticides. Even today, when such alternative compounds are
available, there has not been a reestablishment of cotton production
to early 1960's levels in this part of Mexico. OP-resistance among
tobacco budworm populations spread to Texas in the late 1960's, and
during the 1970's was reported in most of the states where cotton is
grown. Tobacco budworm populations that are resistant to methyl
parathion are also cross-resistant to many other OP-compounds and
carbamates used in cotton protection.

Soon thereafter, it was determined that field populations of
OP-resistant tobacco budworms in the USA had varying levels of
cross-resistance to the pyrethroid insecticides, which became
available for commercial use during the late 1970's (8-13).
However, these compounds were still highly effective in the field
against Heliothis (including OP-resistant tobacco budworms) at
recommended rates of application, and they were readily adopted by
cotton producers as a major chemical control component of Heliothis
management programs.

Pyrethroid resistance among field populations of Heliothis spp.
was first reported in Australia when permethrin failed to provide
adequate control of the cotton bollworm, Heliothis armigera Hübner,
during the 1982-1983 cotton season in the Emerald region of
Queensland (14). In 1984, there were additional incidents of
pyrethroid resistance among populations of H. armigera in Thailand
(15) and Turkey (16,17). It is important to note that field
populations of H. armigera had developed high levels of resistance
to DDT and other organochlorines in Australia and Asia during the
early 1970's (16).

The first evidence of control problems in the field in the USA
was observed in two areas of western Texas in 1985 (18,19). Control
failures with the pyrethroids were encountered again in 1986 in
relatively small cotton production areas in Texas (20), Louisiana

(21), and Mississippi (22). An intensive field monitoring effort mounted by state, federal, and chemical industry researchers during the 1986 crop year revealed tolerance changes that were suggestive of potential pyrethroid resistance problems among populations of tobacco budworms in some areas of Arkansas, California, Louisiana, Mississippi and Texas (21,23); and again during the 1987 crop year in Alabama, Arkansas, Louisiana, Mississippi, and Texas (24). To date there have been no indications of increased tolerance among populations of H. zea.

Based on the results of field monitoring efforts, it appears certain that genes for pyrethroid resistance exist within tobacco budworm populations in this country. If effective wide area resistance management programs are not instituted immediately and sustained, the future for chemical control could be bleak for all segments of the cotton industry.

We now consider the main technical factors underlying attempts at resistance management: resistance mechanisms, cross tolerance/resistance, and resistance monitoring.

Resistance Mechanisms

Two factors are most commonly associated with the development of pesticide resistance in arthropods: enhanced metabolic detoxification and/or decreased target site sensitivity. Other factors that sometimes come into play are reduction in the rate at which a toxicant is absorbed into the body, or in the rate at which it is translocated to the site of action. In rare cases, there may be behavioral adaptations which allow the arthropod to minimize contact with the toxicant (25-27).

Resistance due to metabolic factors is primarily associated with elevated levels of enzyme systems that facilitate detoxification of pesticides. In most cases, reactions catalyzed by these enzymes transform pesticides into inactive polar products that are readily eliminated by the organism. Since these enzyme systems can attack different classes of insecticides, they often are important factors in the cross-resistance of certain arthropods to a variety of pesticides. In the case of the tobacco budworm, enhanced metabolic detoxification (involving microsomal oxidases, hydrolases, and glutathione-transferases and other conjugating systems) was shown to be the major factor in OP-resistance (6,28). Genetic studies with the tobacco budworm demonstrated that the inheritance of resistance to methyl parathion (29), as well as to methomyl (30), was due to a single, autosomal, incompletely dominant gene which most likely regulates synthesis of insecticide-metabolizing enzymes. Enhanced detoxification has also been shown to be a contributing factor in resistance of the tobacco budworm to DDT (31), pyrethroids (32,33), and most likely the carbamates.

Although implicated less frequently in arthropod pesticide resistance, diminished sensitivity of target sites in the nervous system is a major factor in many cases of resistance to different classes of insecticides. As more is learned about this phenomenon, we may find that it plays a more important role in resistance than is currently known. There are two major types of target site insensitivity. With insecticides such as DDT, certain other chlorinated compounds, and the pyrethroids, the development of resistance is associated with a reduced ability of pesticide molecules to inhibit certain noncholinergic receptors in the nervous system that are involved in the regulation of nerve membrane permeability to sodium ions (34). For the most part, this type of target site resistance (kdr) has been reported in insects of veterinary and public health importance, but it is also important in certain species attacking crops. With OP and carbamate insecticides, the development of resistance can be associated with a decreased sensitivity of acetylcholinesterases, enzymes which are essential to the normal transfer of electrical signals between nerve cells at specific junctions called synapses. In the tobacco budworm, diminished sensitivity of acetylcholinesterase has been shown to be a relatively minor factor in OP-resistance (35). However, decreased sensitivity of noncholinergic neural receptors has been implicated as a major factor in the resistance of the tobacco budworm to pyrethroids (32,36) and quite likely was involved in earlier incidents of organochlorine insecticide resistance.

Cross Tolerance/Resistance

Clearly the tobacco budworm is a highly adaptable species that possesses an impressive arsenal of natural physiological and biochemical defense mechanisms which can be brought to bear against a broad spectrum of xenobiotic compounds. Furthermore, recent laboratory studies have shown that the genetic traits of the pest also favor the development of significant levels of resistance to pyrethroids (36,37) when it is subjected to selection pressure. Payne et al. (36) demonstrated that a stable, ca. 540-fold level of resistance to permethrin could be established through appropriate selection regimens. The resistance was inherited via a single, major, incompletely recessive, autosomal factor that appeared to confer insensitivity of the neural target site as the major resistance mechanism. This observation, along with evidence for the existence of other minor resistance factors plus an apparent lack of reproductive disadvantage among resistant individuals, led Payne et al. (36) to hypothesize that there was potential for development of a stable pyrethroid resistance among field populations of tobacco budworms, and that this could pose a threat to resistance management. It would be very significant if such a stable resistance occurred in the field because, in many cases, insects that possess the nerve insensitivity mechanism associated with DDT and pyrethroid resistance are less fit than susceptible individuals; i.e., when selection pressure is removed the numbers of resistant genotypes within a population soon decline to insignificant levels.

It is well established that the modes of action of DDT and pyrethroids are somewhat alike, and that genes for resistance to DDT also confer cross-resistance/tolerance to pyrethroids in a number of arthropod species (25). This probably is true for the tobacco budworm as well. With its past history of DDT resistance and the likelihood that genes for resistance to DDT still exist within the population, the tobacco budworm would be expected to have a head start in developing resistance to pyrethroids. It is interesting that H. zea, which also demonstrated high levels of DDT resistance in the past, has not become resistant to the pyrethroids even though it is generally codistributed with the tobacco budworm in most of cotton production areas subjected to pyrethroid treatments. One possible explanation is that the host range for H. zea is substantially more diverse, and annually its populations expand over a much larger geographical area than the tobacco budworm. The tobacco budworm's primary cultivated hosts are cotton and tobacco, and usually its annual population distribution does not extend much beyond the geographical area comprised of cotton belt states. Thus, selection pressure for the tobacco budworm would be much more intensive owing to heavy usage of pesticides on cotton and to the restricted availability of untreated refugia. Apparently there have been no laboratory studies to determine if H. zea would develop resistance under selection pressure with pyrethroids.

Resistance Monitoring

One key to effective management of pesticide resistance is anticipation of the phenomenon before it actually occurs. Based upon knowledge of the history of pesticides used against a pest species and upon knowledge of biological/ecological factors, genetic attributes, and defense mechanisms against different classes of pesticides, it may be possible to predict the occurrence of resistance as a result of known use patterns of present pesticides. Our capability for such predictions is often inadequate, however, owing to the general lack of an appropriate data base for major pest species, especially those of agricultural importance.

Fortunately, the potential for resistance to pyrethroids in the tobacco budworm was recognized early on, and extensive susceptibility monitoring programs have been in place almost since the compounds were introduced into the U.S. cotton protection market. In 1979, in cooperation with several state and federal scientists, FMC Corporation initiated an annual program for monitoring the pyrethroid susceptibility of field populations of the tobacco budworm; this program eventually was expanded to a number of different states across the cotton belt (38). A similar monitoring effort was established by ICI Americas Inc. in 1980. Because of serious concerns with increasing reports of decreased efficacy of pyrethroids among tobacco budworm populations across the cotton belt, U.S. companies involved in the marketing of pyrethroids formed a branch of the Pyrethroid Efficacy Group (PEG) in 1987. The purpose of this move was to provide a united front that would facilitate increased, coordinated monitoring activities as a means of gaining a broader picture of potential on-going changes in the susceptibility of tobacco budworms to pyrethroids, as well as to establish an organized structure to facilitate evaluation of different monitoring techniques. This group, called PEG-US and comprised of researchers from DuPont, FMC, Hoechst-Roussel Agri-Vet, ICI, and Mobay, mounted an extensive monitoring effort across the cotton belt in 1987 whereby they began comparing three bioassay methods: (1) exposure of male adults to pyrethroid residues in glass vials, (2) topical treatment of 3rd instar larvae (20-25 mg/each), and (3) exposure of 1st instar larvae to residues on cotton leaves

(39,41). The problem is that these particular monitoring techniques must have large numbers of individuals to allow detection of resistance alleles at frequencies low enough (i.e., ca. 1%) to facilitate implementation of resistance prevention tactics. These are labor intensive, expensive monitoring methods that usually can only detect resistance after it has become established within the population. The so-called adult vial method (23) involves the use of male tobacco budworms captured in sex pheromone-baited traps, and thus might have good potential for use in testing large numbers of insects that could be collected in all areas of the cotton belt where the pest is distributed (19).

The magnitude of the present annual resistance monitoring effort on the part of the PEG-US group, as well as state and federal researchers, is unprecedented as far as field crop pests are concerned and reflects the serious concern within the cotton industry that pyrethroid resistance could get out of control. These monitoring efforts have shown a general trend in some areas to increased tolerance of tobacco budworms to pyrethroids, as well as an expansion of the geographical area within which such increased tolerance has been detected. However, except for the aforementioned incidents there have been no documented failures in field control.

Resistance Management

A number of theoretical and practical approaches to the management of pesticide resistance in arthropods have been proposed (42,43). Strategies involving such tactics as genetic manipulation, behavioral modification, directed biological controls, etc. are promising but not feasible for practical use against *Heliothis* spp. at this time. However, in the near term, strategies involving the judicious use of insecticides in conjunction with other control methods consistent with basic IPM concepts have good potential for effective resistance management (26).

Efforts are underway in the USA to establish organized programs for preventing and/or managing pyrethroid resistance among tobacco budworm populations before the problem gets out of control (44). These now rudimentary programs, which have been initiated recently in a tri-state effort (45) comprising Arkansas, Mississippi, and Louisiana, and in Texas (23), are largely based on insect control lessons learned the hard way, and draw upon the experiences and successes of the Australians in their battle with pyrethroid resistance in *Heliothis armigera*.

The key features of the Australian strategy (46) are: (1) pyrethroids are not applied during the period from planting to the early stages of the squaring cycle, (2) beginning at the peak of the squaring cycle until boll production, pyrethroids are permitted but are limited to a maximum of three applications at the full recommended rates, and (3) no pyrethroids or endosulfan are applied during the balance of the season. Primary goals of this strategy are to contain pyrethroid resistance and to prevent development of resistance to endosulfan, which still has good efficacy against these pests (17). There have been no field failures since this strategy was implemented, but annual monitoring data have revealed an ominous upward trend in the frequency of resistant genotypes among H. armigera populations -- from ca. 10-15% in 1983-1984 to 30-40% in 1986-1987 (17) -- a finding which could portend future control failures on a larger scale.

Pyrethroid resistance management programs in the USA are still in the formative stages, but likely will evolve and mature rapidly in response to apparent increases in the relative frequencies of resistant genotypes and in the extent of their geographical distribution. The current strategy has the following elements (17,45):

1. Plant within the recommended dates and try to mature the crop within 120-140 days.
2. Do not use pyrethroids during early season before cotton enters the flowering phase of fruit development.
3. Use pyrethroids at full recommended rates, alone or in mixtures, when and where necessary after flowering and boll development begins.
4. Monitor the crop carefully and time pyrethroid applications against eggs and early larval instars.

Why Earliness is Important

The importance of planting early and maturing the crop as soon as possible cannot be overemphasized. Most pyrethroid control failures reported in the USA were associated with cotton that was planted late. Lateness results in exposure of the most vulnerable developmental stages of the plant to tobacco budworm populations that are generally larger in numbers and that might have increased levels of tolerance due to selection in adjacent fields planted earlier at recommended dates. Field monitoring studies in Texas (47) have shown there are definite increases in the frequency of resistant genotypes among tobacco budworm populations in pyrethroid-treated fields as the season progresses. Similar results have been reported for H. armigera in Australia (48).

Figure 1 shows a comparison of certain phenological characteristics of a typical early-maturing cotton (TAMCOT), and Stoneville 213, a representative full-season cultivar. This computer-generated simulation using TEXCIM (49), a combination cotton-growth/insect damage model, shows how the fruiting curves for squares, green bolls, and open bolls differ between these cultivars under average central-Texas production/environmental conditions. Clearly the vulnerable square and green boll fruiting stages are compressed with TAMCOT, thus reducing the period of exposure to pests. A summary of this information (Table I) shows that the date when 80% of the bolls were open was ca. two weeks earlier with TAMCOT.

There are usually about five generations of Heliothis on cotton in central Texas, beginning with very low numbers in May and cycling at ca. 30-day intervals. Figure 2 shows a TEXCIM computer-generated simulation that describes the predicted magnitude and timing of populations of Heliothis eggs and small or large larvae on central Texas cotton. This simulation assumes that field surveys detected 10 eggs/50 plants on June 6, that populations of predators were light season-long, and that no other controls were applied. Under

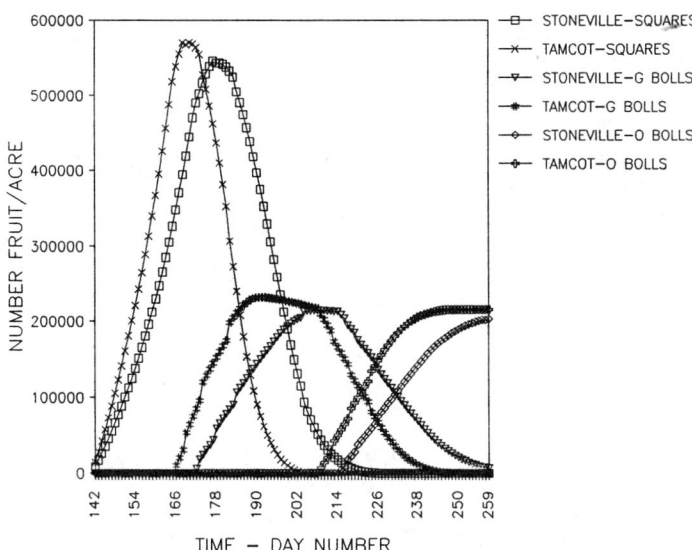

Figure 1. TEXCIM simulation of phenological information for TAMCOT and Stoneville cultivars under central Texas conditions; G = green bolls, O = open bolls.

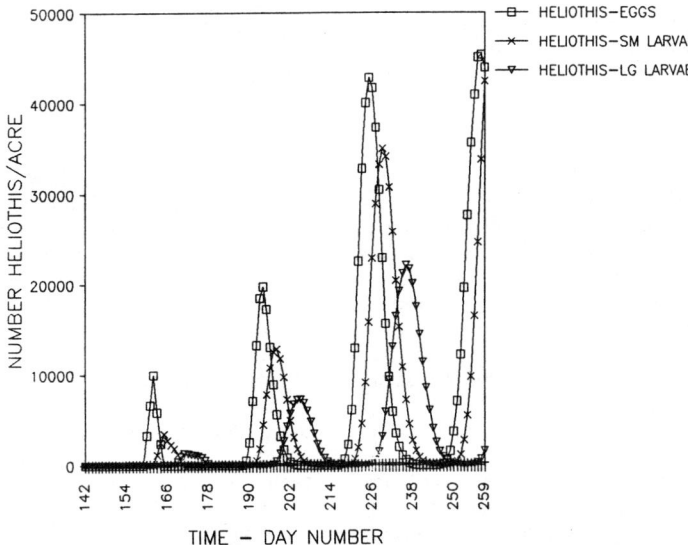

Figure 2. TEXCIM simulation of predicted magnitude and timing of Heliothis spp. generations under central Texas cotton production conditions. (Model assumes 10 eggs/50 plants on June 6 and light populations of predators.)

such conditions Heliothis populations are predicted to increase progressively with each generation.

Table I. TEXCIM Simulation of Phenological Information for Representative TAMCOT and Stoneville Cultivars Under Central Texas Conditions

	TAMCOT	Stoneville
Date cotton planted	4/10/88	4/10/88
Date of first square	5/21/88	5/21/88
Date of first bloom	6/16/88	6/22/88
Date of first open boll	7/27/88	8/02/88
Date when 80% bolls open	8/17/88	8/31/88
%Open bolls on 9/15	100.00	96.79

When this specific Heliothis population condition is factored into a cotton growth and yield simulation, the TEXCIM model predicts an average insect damage loss of $196/acre for the Stoneville variety and $115/acre for TAMCOT -- a sizeable difference of $81/acre. This of course is an artificial comparison since it ignores other pests and pesticide applications. However, the simulation provides an idea of the potential importance of earliness in the fruiting of cotton exposed to Heliothis, and it is in general agreement with results obtained in controlled field tests of the response of TAMCOT CAMD-E and Stoneville 213 cultivars to tobacco budworm infestations (50). It is especially encouraging that there are some new early-maturing cultivars of cotton being developed that incorporate high levels of tolerance to Heliothis spp. plus other highly desirable agronomic qualities. These are expected to be ready for commercial use in the near future.

Early Season Tactics

The management strategy calling for elimination of pyrethroid use during the prebloom stage of fruit development has the objective of avoiding or minimizing selection pressure on the first two field generations during May and June. This recommendation is generally supported now by the pyrethroid industry and pest control consultants, which represents a significant reversal of previous support in some circles for automatic early season applications of pyrethroids for "yield enhancement" purposes, whether or not there was a real pest problem.

If Heliothis spp. were the only pests to contend with on cotton during the critical early season period it is possible that in many cases the relatively low-level populations would be adequately suppressed by naturally-occurring beneficial species.
Unfortunately, much of the cotton belt from Texas and parts of Arizona to the eastern seaboard is also infested by the boll weevil, which can cause major problems during early season. The boll weevil is an introduced pest unfettered by any significant natural enemies in this country, and it usually must be treated with chemical

insecticides when it first appears or it will rapidly get out of control. In some areas of the cotton belt there are comparable problems with pink bollworms and Lygus spp. Thus the selection of insecticidal chemicals for early season pest control presents a challenge -- that is, how to suppress pests effectively and at the same time conserve as much of the pool of beneficial species as possible to assist in managing Heliothis during the most critical stages of fruit development from mid-June to late July. Natural enemies are also important in suppressing pests such as spider mites, whiteflies, etc., which are usually of minor importance but owing to pesticide resistance/tolerance can become significant problems in the absence of natural controls.

In a well managed area-wide program for controlling cotton pests, which is highly desirable in preventing or minimizing pyrethroid resistance problems, the outcome of early-season pest management practices should set the stage for dealing with pest problems during the even more critical mid-season stage of cotton production when bloom-stage fruit and developing young bolls must be protected.

As mentioned, conservation and maximum use of natural enemies must be given high priority in the early season Heliothis management strategy. It is essential to monitor the crop carefully and apply insecticides only if absolutely necessary. Insecticide applications should be timed so that they are directed primarily at the more vulnerable egg and early larval instar stages. Assuming that Heliothis is the predominant pest that requires control, then there could be some opportunity for using selective nonpyrethroid insecticides that would not severely impact populations of beneficial species. For example Microplitis croceipes (Cresson), an important hymenopteran parasite of Heliothis larvae (51), is highly tolerant (52), and the egg parasite Trichogramma pretiosum Riley is moderately tolerant (53), to thiodicarb and chlordimeform, which are quite effective against Heliothis larvae and eggs, respectively. Microplitis croceipes and Campoletis sonorensis (Carlson), another important parasite of Heliothis larvae, both show reasonable tolerance to certain phosphate-type OP insecticides (52,54-57), as do important predators of cotton pests such as the green lacewing, Chrysopa carnea Stephens (58-60) and the spined soldier bug Podisus maculiventris (Say) (61).

The aforementioned parasites and predators are all somewhat tolerant of the pyrethroids, an attribute that gives some hope for at least a modicum of natural-enemy conservation when these compounds are used in mid-season. It seems reasonable that careful evaluations of comparative toxicity data for Heliothis and some key major natural enemies could lead to identification of nonpyrethroid insecticides that if needed could be used with at least some degree of selectivity against the first two generations of the pest. However, such data are scarce, and more work is needed to provide an adequate data base that would allow more flexibility in choosing an appropriate chemical. Given the history of resistance in the tobacco budworm to various classes of insecticides, it would be advisable to alternate classes of insecticides early in the season to avoid selecting populations of the pest for multiple resistance.

Mid-Season Tactics

Current resistance management tactics for mid-season cotton protection, a period of ca. 4-5 weeks from first bloom until bolls are about two-thirds developed, calls for the use of pyrethroids against Heliothis spp. if they are needed. The initiation of applications should be based on results of careful monitoring; they should not be applied automatically as has often been common practice. Once the spray program begins pyrethroids should be applied at recommended rates and intervals. Combinations of pyrethroids with other nonpyrethroid insecticides have proven very effective against the tobacco budworm with suspected pyrethroid tolerance/resistance, but should be used with caution and certainly not for prolonged periods because of the risk of inducing resistance to both types of chemicals. Chlordimeform in particular has been a common additive in pyrethroid sprays during recent years. This compound enhances the activity of pyrethroids against pyrethroid-resistant tobacco budworms (18,47), through ovicidal activity (62), by synergistic effects due to increased movement of larvae (63), and perhaps by interactions at the neural receptor sites (64). Chlordimeform is scheduled for removal from the market after the 1988 season, but thiodicarb and methomyl (at ovicidal rates) are being evaluated as replacements.

If resistance monitoring during mid-season reveals a definite trend of increased tolerance to pyrethroids among tobacco budworm populations, conventional wisdom would suggest cessation of their use or at least alternating applications with nonpyrethroids. Suitable alternates include OP compounds such as sulprofos, profenofos, and acephate, which are effective against Heliothis spp. and have been shown to be almost equally effective against methyl parathion-susceptible and -resistant tobacco budworms (65). These OP-compounds have in common an asymmetric phosphorus atom and a P-S-alkyl linkage in the molecule, structural characteristics which may contribute to circumvention of metabolic resistance mechanisms.

Late Season Tactics

During late season, the use of pyrethroids should be avoided and the crop should be matured as soon as possible. For example, in the mid-South (44) the goal is to set most of the bolls by the end of July. At that time the plant and fruit will have reached a state of maturity that will minimize vulnerability to attack by Heliothis. Late irrigation should be avoided, and chemical desiccants and defoliants should be used for early crop termination.

Conclusion

There is no question that genes for pyrethroid resistance exist within populations of the tobacco budworm that infest cotton in this country, and there also is reasonable evidence from resistance monitoring efforts that the frequency of resistant genotypes is tending to increase in some areas. Thus the seeds for potential pyrethroid failures have been sown, and the future for long-term continued use of these insecticides holds scant promise unless there is a strong commitment to the establishment of organized, area-wide resistance management programs. These programs must have broad support by all parties that have a vested interest in the use of pyrethroids to protect cotton from resistance-prone tobacco budworms.

The adoption of the mid-South resistance management program by a large percentage of cotton producers in Louisiana apparently resulted in a reduction in pyrethroid-resistant genotypes during the 1987 season (66). However, the threat of significant insect damage to high yield cotton can in some cases weaken the dedication of producers and pest control consultants to the IPM concept (67). In Texas during 1987 there were cases where as many as 15-20 insecticide applications (including pyrethroids) were made in irrigated fields in the same general area where tobacco budworm control failures with pyrethroids were observed.

There is a continued need for development of better resistance monitoring methods, as well as for an expansion of such monitoring efforts to all areas of the cotton belt where the tobacco budworm is distributed. It is essential to continue emphasis on the strategic use of the pyrethroids; that is, to time their application for maximum impact on pest populations while minimizing the potential for resistance selection by limiting the number of applications during a season. Cultural methods must be used to full advantage. The use of early maturing cultivars planted early and on time is a management tactic that should be strongly encouraged.

If voluntary efforts to manage pyrethroid resistance falter then it may be desirable to establish some sort of program of structured controls to gain uniform compliance with recommended management tactics. This could include restrictions on planting and stalk destruction dates, on the number of pyrethroid applications in a season, and on the periods during crop development when pyrethroids might be applied. However, in some cases governmental regulation is considered an anathema in a free enterprise system such as that of the cotton production industry. One option might include local referenda to create pest management zones, such as the one that was recently established in the Rio Grande Valley of Texas for the purpose of enforcing compliance with cotton stalk destruction requirements as a means of reducing numbers of over wintering boll weevils.

Literature Cited

1. Suguiyama, L.; Osteen, C. The Economic Importance of Cotton Insects and Mites; U.S. Department of Agriculture, Econ. Res. Serv., AER-599, 1988.
2. Valiulis, D. Agrichem. Age 1985, 29, 20B.
3. Johnson, S. J.; King, E. G.; Bradley, J. R., Jr. Theory and Tactics of Heliothis Population Management: I. Cultural and Biological Control; South. Coop. Ser. Bull. 316, 1986; 161 pp.
4. Adkisson, P. L. Conn. Agric. Stn. Bull. 1969, 708, 155.
5. Adkisson, P. L.; Niles, G. A.; Walker, J. K.; Bird, L. S.; Scott, H. B. Science 1982, 216, 19.
6. Whitten, C. J.; Bull, D. L. J. Econ. Entomol. 1970, 60, 1492.
7. Wolfenbarger, D. A.; McGarr, R. L. J. Econ. Entomol. 1970, 63, 1762.
8. Sparks, T. C. Bull. Entomol. Soc. Am. 1981, 27, 186.
9. Crowder, L. A.; Tollefson, M. S.; Watson, T. F. J. Econ. Entomol. 1979, 72, 1.

10. Twine, P. H.; Reynolds, H. T. *J. Econ. Entomol.* 1980, **73**, 239.
11. Martinez-Carrillo, J. L; Reynolds, H. T. *J. Econ. Entomol.* 1983, **76**, 983.
12. Brown, T. M.; Bryson, K.; Payne, G. T. *J. Econ. Entomol.* 1982, **75**, 301.
13. Payne, G. T.; Disney, B. J.; Brown, T. M. *J. Agric. Entomol.* 1985, **2**, 85.
14. Gunning, R. V., Easton, C. S.; Greenup, L. R.; Edge, V. E. *J. Econ. Entomol.* 1984, **77**, 1283.
15. Collins, M. D. In *1986 British Crop Protection Conference, Pests and Diseases*; 1986; p 583.
16. Ahmad, M.; McCaffery, A. R. *J. Econ. Entomol.* 1988, **81**, 45.
17. Jackson, G. J. *Proc. Beltwide Cotton Prod. Conf.*, 1988, p 42.
18. Plapp, F. W., Jr.; Campanhola, C. *Proc. Beltwide Cotton Prod. Res. Conf.*, 1986, p 167.
19. Riley, S. L. *Proc. Beltwide Cotton Prod. Res. Conf.*, 1988, p 228.
20. Clower, D. F. *Proc. Beltwide Cotton Prod. Res. Conf.*, 1987, p 214.
21. Leonard, B. R.; Graves, J. B.; Sparks, T. C.; Pavloff, A. M. *Proc. Beltwide Cotton Prod. Res. Conf.*, 1987, p 320.
22. Luttrell, R. G.; Roush, R. T.; Ali A.; Mink, J. S.; Reid, M. R.; Snodgrass, G. L. *J. Econ. Entomol.* 1987, **80**, 985.
23. Plapp, F. W., Jr.; McWhorter, G. M.; Vance, W. H. *Proc. Beltwide Cotton Prod. Res. Conf.*, 1987, p 324.
24. King, E. G.; Phillips, J. R.; Head, R. B. *Proc. Beltwide Cotton Prod. Res. Conf.*, 1988, p 188.
25. Oppenoorth, F. J. In *Comprehensive Insect Physiology, Biochemistry, and Pharmacology, Vol. 12 - Insect Control*; Kerkut, G. A.; Gilbert, L. I., Eds.; Pergamon Press: New York, 1985; p 731.
26. Brattsten, L. B.; Holyoke, C. W., Jr.; Leeper, J. R.; Raffa, F. K. *Science* 1986, **231**, 1255.
27. Lockwood, J. A.; Story, R. N.; Byford, R. L.; Sparks, T. C.; Quisenberry, S. S. *Environ. Entomol.* 1985, **14**, 873.
28. Bull, D. L.; Whitten, C. J. *J. Agric. Food Chem.* 1972, **20**, 561.
29. Whitten, C. J. *J. Econ. Entomol.* 1978, **71**, 971.
30. Roush, R. T.; Wolfenbarger, D. A. *J. Econ. Entomol.* 1985, **78**, 1020.
31. Vinson, S. B.; Brazzel, J. R. *J. Econ. Entomol.* 1966, **59**, 600.
32. Nicholson, R. A.; Miller, T. A. *Pestic. Sci* 1985, **16**, 561.
33. Dowd, P. F.; Gagne, C. C.; Sparks, T. C. *Pestic. Biochem. Physiol* 1987, **28**, 9.
34. Lund, A. E. *Pestic. Biochem. Physiol.* 1984, **22**, 161.
35. Bull, D. L. *Bull. Entomol. Soc. Am.* 1981, **17**, 193.
36. Payne, G. T.; Blenk, R. G.; Brown, T. M. *J. Econ. Entomol.* 1988, **81**, 65.
37. Jensen, M. P.; Crowder, L. A.; Watson, T. F. *J. Econ. Entomol.* 1984, **77**, 1409.
38. Staetz, C. A. *J. Econ. Entomol.* 1985, **78**, 505.
39. Staetz, C. A.; Rivera, M. A.; Blenk, R. *Proc. Beltwide Cotton Prod. Res. Conf.*, 1988, p 339.
40. Collins, M. D.; Blenk, R.; Gouger, R. J.; Staetz, C. A. *Proc. Beltwide Cotton Prod. Res. Conf.*, 1988, p 336.

41. Simonet, D. L.; Riley, S. L.; Watkinson, I. A.; Whitehead, J. R. Proc. Beltwide Cotton Prod. Res. Conf., 1988, p 334.
42. Georghiou, G. P. Residue Rev. 1980, 76, 131.
43. National Academy of Science. Pesticide Resistance: Strategies and Tactics for Management. National Academy Press: Washington, DC, 1986; 471 pp.
44. Riley, S. L. Proc. Beltwide Cotton Prod. Conf., 1988, p 45.
45. Anonymous. MAFES Res. Highlights 1986, 49, 8.
46. Forrester, N. W.; Cahill, M. IV Congres sur la Protection de la Sante Humaine et des Cultures en Milieu Tropical, 1986; p 248.
47. Campanhola, C.; Plapp, F. W., Jr. J. Econ. Entomol. 1989, 82, 22.
48. Sawicki, R. M.; Denholm, I. Trop. Pest Manage. 1987, 33, 262.
49. Hartstack, A. W.; Sterling, W. L. The Texas Cotton-Insect Model TEXCIM: Users Guide, Version 2.3., 1988; TAES Computer Software Doc. Ser. MP-1646; 38 pp.
50. McCarty, J. C., Jr.; Jenkins, J. N.; Parrott, W. L. Crop Sci. 1986, 26, 136.
51. King, E. G.; Powell, J. E.; Coleman, R. J. Entomophaga 1985, 30, 419.
52. Powell, J. E.; King, E. G., Jr.; Jany, C. S. J. Econ. Entomol. 1986, 79, 1343.
53. Bull, D. L.; Coleman, R. J. Southwest. Entomol. 1985, Suppl. No. 8, 156.
54. Bull, D. L.; Pryor, N. W.; King, E. G., Jr. J. Econ. Entomol. 1987, 80, 739.
55. Bull, D. L.; King, E. G.; Powell, J. E. Southwest. Entomol. 1989, Suppl. No. 12, 59.
56. Powell, J. E.; Scott, W. P. Fla. Entomol. 1985, 68, 692.
57. Plapp, F. W., Jr.; Vinson, S. B. Environ. Entomol. 1977, 6, 381.
58. Lingren, P. D.; Ridgway, R. L. J. Econ. Entomol. 1967, 60, 1639.
59. Plapp, F. W., Jr.; Bull, D. L. Environ. Entomol. 1978, 7, 431.
60. Pree, D. J.; Archibald, D. E.; Morrison, R. K. J. Econ. Entomol. 1989, 82, 29.
61. Yu, S. J. J. Econ. Entomol. 1988, 81, 119.
62. Phillips, J. R. Ark. Farm Res. 1971, 4, 9.
63. Treacy, M. F.; Benedict, J. H.; Schmidt, K. M.; Anderson, R. M.; Wagner, T. L. Proc. Beltwide Cotton Prod. Res. Conf., 1987, p 318.
64. Chang, C. P.; Plapp, F. W., Jr. J. Econ. Entomol. 1983, 76, 1206.
65. Bull, D. L.; Plapp, F. W., Jr.; Sparks, T. C. In Theory and Tactics of Heliothis Population Management; Schneider, J. C.; Hammond, A. M.; Jackson, D. M.; Mitchell, E. R.; Roush, R. T., Eds.; Southern Coop. Ser. Bull. 329, 1987; p 37.
66. Graves, J. B.; Leonard, B. R.; Pavloff, A. M.; Burris, G.; Ratchford, K.; Micinski, S. J. Agric. Entomol. 1988, 5, 109.
67. Kepple, D. D. Agrichem. Age 1988, 32, 16D.

RECEIVED October 19, 1989

Chapter 9

Pyrethroid Resistance in *Heliothis* spp.

Current Monitoring and Management Programs

S. L. Riley

Crop Research Laboratory, E. I. du Pont de Nemours and Company, P.O. Box 30, Newark, DE 19714

> Since their introduction in 1978, the synthetic
> pyrethroids have become the most cost-effective
> and environmentally compatible insecticides for
> *Heliothis* spp. control in cotton. However,
> since 1983, their effectiveness has been
> threatened by reports of resistance in major
> cotton producing areas of the world. The
> increasing costs of discovering, developing and
> marketing new insect control technologies
> underscores the seriousness of this situation.
> In response, a number of programs involving
> collaboration between government, academia and
> industry have been initiated to monitor for the
> spread of pyrethroid resistance in *Heliothis*
> spp. and to manage the use of the pyrethroids.
> These programs will be reviewed along with
> recommendations for the future.

Since the 1950's, field resistance in *Heliothis* spp. has been reported for nearly every chemical used to protect cotton ([1-9]). The two *Heliothis* species that have most commonly developed resistance are *Heliothis* virescens and *H. armigera*. The loss of the organophosphorous insecticides (OP's) to resistance in the 1970's literally pushed the synthetic pyrethroids into the marketplace. Since their introduction in 1978, it was clear that these highly cost effective and environmentally compatible insecticides would be heavily relied upon to replace the OP's. The synthetic pyrethroids have become the most widely used chemicals for the control of insect pests on cotton, representing about 48% of all the insecticides applied worldwide ([10]). Most applications are directed toward controlling *Heliothis* spp.

However, the attributes that made them attractive also created concern about the evolution of resistance to these compounds. Their effectiveness has resulted in overuse in many areas of the world. In Thailand for example, farmers have been known to apply as many as 20 pyrethroid sprays per season ([11]). In the U.S., a beltwide survey in 1984 revealed that annual pyrethroid use varied from 0.1 applications per season in the Rolling Plains of Texas to 8.4 in the Central Texas River Bottom. With the exception of the San Joaquin Valley, synthetic pyrethroid use against Heliothis spp. on cotton accounted for about 70% of the total treatments directed against this pest (12).

The Costs of Resistance

The cost of insecticide resistance can be enormous. Many farmers and entomologists in the U.S. cotton belt can clearly remember the demise of the OP's (particularly methyl parathion) during the 1960's and 1970's, a situation that seriously threatened the profitable production of cotton in many areas of the cotton belt. During that period, a common response to the declining effectiveness of the insecticides was to increase the dosage and/or shorten the interval between applications. Not only was this response expensive, but it proved to be counter productive to combating resistance, accentuating the problem by increasing selection for a higher level of resistance.

In hindsight, it is now commonly accepted that no insecticide (or other control agent) is immune to the evolution of resistance in the target pest, and that the most effective way to combat resistance is to manage the use of the selection agents by practicing many of the principles of integrated pest management (IPM) (13-14). Pesticides can be an integral part of any pest management program, if they are used judiciously so as to preserve two valuable agricultural resources: the cost effective insecticides, and susceptible genotypes in a pest population.

Pyrethroid Resistance. The cost effectiveness of pyrethroids and their environmental compatibility, underscore the critical need to prolong their use. What is the potential effect of resistance to the pyrethroids on the cotton industry? First, for many growers it would mean increasing reliance on more expensive, less effective products, resulting in lower quality yield for higher costs. Recent cost analyses (Watkinson, I. A., Abbott Laboratories, personal communication, 1989) indicate that the development of pyrethroid resistance in H. virescens would approximately double control costs to mid-south cotton growers from an average of $32.00/acre to $61.50/acre and reduce yield by 11% (Graves, J. B., Louisiana State University, personal communication, 1989) This cost analysis assumes a 50% replacement of the pyrethroids by more expensive and less effective alternatives. Second, the discovery and development of replacement chemicals with novel modes of action does not come cheaply. Recent cost estimates are about $50 million and 8-10 years for the development of a compound. An additional $40-100 million can be added if a manufacturing plant is required. Third, yet another consequence is an increase in the amount of active ingredient put out into the environment. Pyrethroid application rates range from about 0.02 to 0.2 lb. ai/acre, whereas many alternative products have effective use rates of up to 1.5 lb./acre.

Early Concern About Pyrethroid Resistance

Early concern about resistance to the pyrethroids resulted in the initiation of monitoring programs. As early as 1975, studies by university researchers were initiated to determine H. virescens' susceptibility to the pyrethroids (15-17). The results of these studies generally indicated localized shifts in tolerances in field collected strains of H. virescens. However, few of these studies related "resistance" levels detected in laboratory tests of field strains to insect control in the field. Martinez-Carrillo and Reynolds (16) pointed out that even though they detected high levels of resistance, no field control failures had yet been reported. They did warn, however, that failures were possible if application trends continued. Subsequently, FMC and ICI, two of the first manufacturers of pyrethroids, established monitoring programs in the early 1980's (18).

Organization of Industry Resistance Groups

Concern about the effective life of the pyrethroids resulted in more than monitoring programs. A number of industry organizations have been formed to provide advice on the issues of insecticide resistance and to facilitate interactions between industry, local governments, scientists, consultants and growers (Figure 1). The history of these developments is outlined below.

PEG. In 1979, the pyrethroid manufacturers formed a worldwide technical committee called the Pyrethroids Efficacy Group (PEG). The principle goals of this committee have been to (1) clarify the causes of field failures; (2) provide technical advice to researchers, growers and governments on pyrethroid resistance problems, and facilitate interactions among these groups; and (3) sponsor research on pyrethroid resistance. To meet these goals, PEG has been involved in resistance studies in Thailand and Turkey, has kept a close watch on the situation in Australia, and is funding research on the mechanisms of resistance (19-20). Recently, PEG has collaborated with agencies in Colombia to implement a resistance management strategy.

IRAC. In 1984, an organization called the Insecticide Resistance Action Committee (IRAC) was formed to serve as a consulting group to GIFAP (the International Group of National Associations of Manufacturers of Agrochemical Products) on all technical and scientific matters relating to insecticide and acaricide resistance, and to develop research relationships with non-industry institutions. A sister committee to IRAC, the Fungicide Resistance Action Committee (FRAC) is discussed in this volume by Delp and Wade. To accomplish its goals, IRAC has established a number of working groups to deal with resistance issues in different crops/markets (Figure 1). Currently there are seven market workgroups (cotton, stored products, rice, animal health, field crops/vegetables, fruit and vector control). In addition, two product-based subcommittees have been added to IRAC. The existing PEG committee was added as a subcommittee in 1985 and an Insect Growth Regulator Group (IGREG) was newly formed in 1987. These group are described in some detail by Jackson (19) and Voss (21).

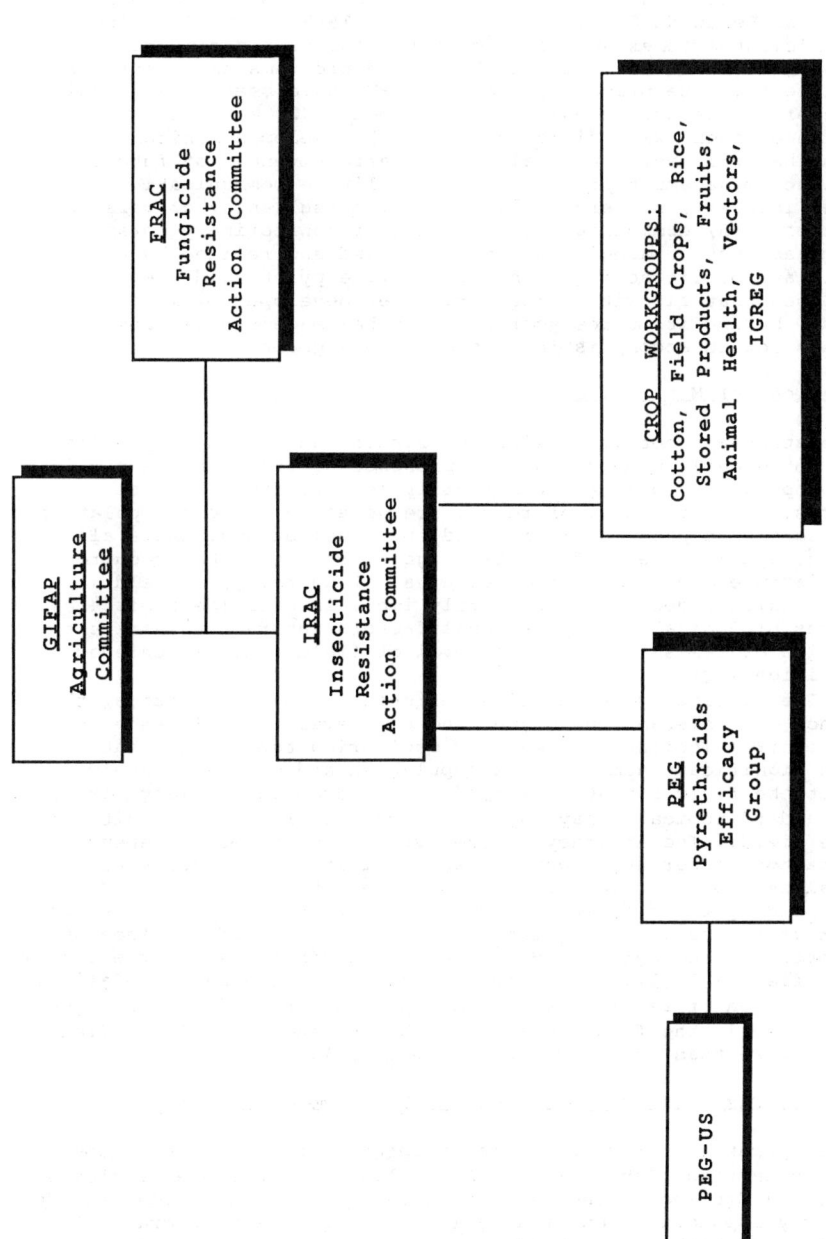

Figure 1. Organizations within the agrochemical industry concerned with insecticide resistance issues.

PEG-US. In 1986, growing evidence that pockets of pyrethroid resistance were developing in Heliothis virescens on U.S. cotton, led to the formation of a U.S. branch of PEG (PEG-US) which first met at the Beltwide Cotton Conferences in 1987. PEG-US designed and implemented an extensive program to monitor and compare monitoring techniques during 1987. These programs are described in detail below. The primary goals of PEG-US have been to (1) monitor levels of resistance across the cotton belt, (2) attempt to relate resistance levels with field control, (3) develop a monitoring system that will serve to evaluate the effectiveness of future resistance management programs, (4) facilitate communication between industry, university scientists, consultants, growers and government, (5) communicate the results of monitoring programs so that management strategies can be developed and refined by both private and public sectors, and (6) provide pyrethroid-use guidelines that fit within the strategies developed by local state groups. During its three years, PEG-US has made significant progress toward accomplishing most of these goals.

Limitations of Monitoring

The relationship between monitoring results and field control is still not clearly understood. The presence of resistance genes in a field population does not necessarily lead to field control failures. The frequency of resistance genes in a field population, the size of the pest population and the level of resistance all interact and must be sufficiently high to result in the occurrence of resistance-induced control failures. Conversely, a field control failure does not necessarily imply insecticide resistance. Numerous biological and operational factors interact to affect the evolution of resistance and the level of field control achieved by a pesticide (22).

The difficulty in relating resistance gene frequencies, the presence of resistant phenotypes and the level of resistance to field control indicates a need for monitoring techniques that sample significant numbers of a population and more accurately reflect the field situation (23-24). For example, in many cases, the standard topical assay may not adequately reflect the situation in the field. The efficacy of the pyrethroids seems to depend on many factors other than just contact activity. In cases where topical assays indicate high levels of resistance but no corresponding field failures, contact activity may not be the only factor affecting susceptibility. The repellency and antifeedant attributes of the pyrethroids may also contribute to their efficacy in the field (25-31). These behavioral attributes may also affect the evolution of resistance in field populations (32). The type of exposure in the field to the pyrethroids may also affect field control more than previously considered (33).

Worldwide Cases of SP Resistance and Management Programs

So far, pyrethroid resistance in Heliothis spp. has been documented in six countries (Table I). In Australia there has been a slow but relentless decline in the effectiveness of the pyrethroids despite the early implementation of a resistance management program. In Thailand and Turkey resistance seems to be relatively widespread and has affected cotton production in most areas. In Colombia it seems too early to tell how widespread it is. In the United States, it still seems to be confined to localized areas under intense cotton production and the most serious control failures have occurred against late-season populations. In India, resistance has so far been restricted to Andhra Pradesh, a major cotton producing state.

Table I. Documented Cases of Synthetic Pyrethroid Resistance on Heliothis spp.

Country	Year Documented	Species
Australia	1982-1983	Heliothis armigera
Thailand	1984	Heliothis armigera
Turkey	1984	Heliothis armigera
United States	1985	Heliothis virescens
Colombia	1987	Heliothis virescens
India	1987	Heliothis armigera

Pyrethroid Resistance Overseas

Australia. Pyrethroid resistance in Heliothis spp. was first documented in Australia during the 1982-1983 production season, when field control failures occurred against Heliothis armigera (3). Daly and Murray (34), in a thorough evaluation of hypotheses proposed on the evolution of resistance in Australia, concluded that an increase in the frequency of resistance genes in the Emerald area was a result of interactions between high selection pressure, population density, crop phenology and weather, but the evolution of resistance elsewhere was independent of the situation in Emerald.

A large-scale management strategy was implemented the following season throughout all inland crop areas in eastern Australia. This strategy emphasizes the rotation of insecticides and the restricted use of the pyrethroids to one Heliothis generation per year (35-36). Although voluntary in New South Wales, growers in Queensland must comply with the strategy as specified on the labels of the products sold in that state (34). The goals of this management strategy have been to (1) limit the frequency of resistance in H. armigera, (2) minimize the risk of recurrence of endosulfan resistance, (3) retard the evolution of resistance to other insecticides, and (4) maintain the use of the pyrethroids on as many crops as possible (36). This strategy is also accompanied by a large-scale monitoring program to evaluate the effectiveness of the strategy (3, 35).

The monitoring results demonstrate a cyclical change in resistance frequency, increasing during and after pyrethroid use (stages 2 & 3) and declining during the off-season (20, 36). A disconcerting trend has appeared from the monitoring results, indicating that some of the goals are not being realized. Levels of resistance in the third stage have been rising each year. Recent information (37) indicates that resistance levels during stage 3 of the 1988/89 season have been the highest ever recorded in the areas of Namoi/Gwydir (>60%) and Emerald (>40%), and probably of more concern, that levels in the unsprayed refugia areas have also increased significantly (>40% during stage 3 of 1988/89). The maintenance of susceptible H. armigera in the refugia is a major component of the resistance management program. Fortunately, endosulfan resistance appears to be relatively low and stable; however, as pointed out by Forrester (37), increased use of endosulfan in cotton or other crops could tip the balance.

To confront this situation, a number of changes in the strategy have been proposed (37). One important component is the accurate timing of pyrethroid sprays on young larvae. Daly et al. (38) report that resistant larvae less than 4 days old are susceptible to the pyrethroids. This evidence is supportive of U.S. cotton IPM programs, where the accurate and timely application of insecticides on early instar larvae has been a major component since the late 1970's. The continued increase in the level of resistance in Australia does not mean that the strategy has been a failure (imagine what might have happened if no program had been implemented), but rather indicates that a management program should be continually audited and refined if necessary. The call by Forrester (37) for changes in the strategy demonstrates a keen awareness of this fact.

Thailand. In 1984, two additional cases of Heliothis armigera resistance to the pyrethroids were added to the list. In Thailand, high numbers of applications and poor compliance with scouting procedures, recommended use rates and application timing resulted in high selection pressures on H. armigera populations. Field control with the pyrethroids has been declining since 1977 (11). During 1985, PEG funded a study that demonstrated that resistance to the pyrethroids was widespread. Cross-resistance and/or multiple resistance was also present between the pyrethroids and DDT and carbaryl. In collaboration with PEG, the Thai Department of Agriculture issued a strategy of insecticide alternations to reduce selection pressure on Heliothis spp.. However, compliance to the program has been far from ideal. Many cultural, educational and economic barriers have prevented compliance with the program (11). Cotton production in Thailand has declined steadily in recent years, from approximately 183,000 bales in 1983 to approximately 101,000 bales in 1986. Cotton production increased slightly to 138,000 bales in 1987 (39).

Turkey. The second case of pyrethroid resistance in 1984 was in Turkey. In 1985 and 1986, PEG sponsored studies which confirmed the presence of resistance in H. armigera. The Turkish form of resistance management involved the implementation of a ban on the importation of pyrethroids for their use on cotton pests (20).

Colombia. In 1984-85, Colombia began to experience control failures with the pyrethroids against H. virescens. Currently, the Colombian government, a grower cooperative and PEG are collaborating to implement a monitoring and management program in Colombia (Collins, M. D., ICI; Staetz, C. A., FMC, personal communication, 1988). The strategy proposed by this group emphasizes restricting pyrethroid use based on insect and crop phenology.

India. Recently, mismanagement of the use of the pyrethroids in India has led to resistance in H. armigera in the major cotton producing state of Andhra Pradesh (Sawicki, R. M., Rothamsted Experimental Station, personal communication, 1989). As early as 1985, some control failures were experienced in Andhra Pradesh, but no actions were taken to manage the situation. By 1987, some farmers in the state could not control outbreak populations of H. armigera, despite spray intervals of only 2-3 days (40). The use of the pyrethroids declined significantly in 1988-1989.

Pyrethroid Resistance in the U. S. Cotton Belt

In January of 1984, following the documentation of field resistance in H. armigera in Australia, seven prominent entomologists (F. L. Carter, D. F. Clower, J. R. Phillips, F. W. Plapp, H. T. Reynolds, R. T. Roush, T. C. Sparks) from major cotton producing states issued a warning about the potential for pyrethroid resistance in Heliothis spp.. It was a timely statement.

In 1985 the first reports of control problems with pyrethroids used against H. virescens in U.S. cotton came from the Wintergarden (Uvalde) and St. Lawrence (Glasscock County) areas in Texas. Testing field strains collected from these areas, Plapp and Campanhola (41) were the first to associate high LD50 values for pyrethroids in laboratory studies with reported control failures. Roush and Luttrell (42) and Luttrell et al. (20) likewise reported associations between laboratory results and control failures in Mississippi and confirmed the presence of resistant H. virescens in the Brazos River Valley of Texas. They concluded that for numerous control problems in Arkansas, Louisiana, and Mississippi, the presence of resistant phenotypes was a contributing factor. However, most of the observed problems occurred in late maturing cotton fields receiving high Heliothis pressure.

As a result of these findings Texas and the states of the mid-south (Louisiana, Mississippi and Arkansas) each implemented resistance management programs aimed at preventing the further development of resistance to the pyrethroids in Heliothis spp. by reducing the selection pressure on Heliothis spp.. In addition, in preparation for the 1987 cotton production season, PEG-US designed and implemented a beltwide monitoring program.

U.S. Cotton Belt Management Programs

So far, Texas and the mid-south are the only states in the U.S. cotton belt where formal resistance management programs have been implemented. These programs are described briefly below and in more detail by D. Bull and J. Menn in this volume. In Alabama, state researchers, private consultants and industry representatives have met to discuss the implementation of a program.

Texas. In 1986, the Texas Agricultural Extension service implemented a state-wide monitoring program, using the Adult Vial Test (AVT) developed by F. W. Plapp of Texas A&M, involving the exposure of pheromone-trapped H. virescens adult males to residues of cypermethrin in glass vials (43), and implemented a resistance management program based on the IPM principles of promoting an early crop, regular scouting and the judicious use of insecticides. Like the Australian program, the cotton production season is divided into 3 phenological periods linked to the annual Heliothis spp. population cycle. In Texas, decisions to use pyrethroids, especially late-season, are based on monitoring results. In areas where Heliothis spp. resistance levels are relatively high (Brazos Valley and Uvalde areas), it is recommended that the use of the pyrethroids be limited to the mid-season (first bloom through boll development). These areas usually are also susceptible to H. virescens attack during the late season. In areas with relatively low resistance levels, pyrethroids can be used late season, but combinations with synergists are recommended (Frisbie, R. E., Texas A&M University, personal communication, 1988).

Mid-South. State researchers, extension personnel and USDA scientists from Mississippi, Arkansas and Louisiana implemented a resistance management program during 1987, combined with a monitoring program using the AVT. This program was very similar to the Texas program. However, in the mid-south, the insecticide use periods were divided into only 2 stages (emergence to first bloom and first bloom to end of season). It was recommended that pyrethroid use be limited to the second stage. The assumption behind not restricting the late season use of pyrethroids was that most of the bolls would be set by the end of July and would be much less susceptible to attack by the August generation of H. virescens (Phillips, J. R., University of Arkansas; Graves, J. B., Louisiana State University, personal communication, 1988). Recently, suggested revisions in the mid-south program include dividing the season into 3 stages, restricting pyrethroid use to mid-season (July 1 to August 15) and more emphasis on the use of pyrethroid-ovicide mixtures.

PEG-US Monitoring Efforts

In addition to the state programs, the U.S. branch of PEG (PEG-US) implemented a beltwide monitoring program in 1987. During 1987, an extensive monitoring effort was conducted by researchers from Du Pont, FMC, Hoechst-Roussel, ICI and Mobay. Four different monitoring techniques were conducted in 16 monitoring sites from North Carolina to California. Most of the effort was directed at the AVT and a foliar spray test conducted on first instar larvae of field strains in the laboratory. Although a direct correspondence between these two tests and field control could not be attained, both tests detected patterns of susceptibility consistent with grower perceptions of product performance. Overall, including the tests conducted by university researchers, more than 40,000 adult H. virescens were tested during 1987. The overall results of these tests during the 1987 season were presented at the 1988 Beltwide Cotton Conferences (44-47). In general, eastern populations demonstrated much higher susceptibility levels than mid-south and western populations. The PEG-US data also indicates that the Brazos River Valley of Texas had the highest levels of survival of any region in the Cotton Belt during 1987. Also, based on limited tests, seasonal variation in susceptibility was detected, with LD50 levels increasing as the season progressed, peaking in mid-season and declining late-season.

A series of guidelines supporting the corrective management programs of Texas and the mid-south were also issued by PEG-US (Table II). These guidelines leave scope for a range of tactical options that fit into the needs of local programs. The strategy encompassed in the guidelines is based upon the moderate use of insecticides as a means of reducing selection pressure. Sole reliance on pyrethroid/non- pyrethroid mixtures is considered inadvisable as a corrective management tactic because of the risk of inducing multiple resistance.

Table II. PEG-US Pyrethroid Resistance Management Guidelines

---

1. Do not rely on a single chemical class.
2. Avoid treating more than 1 generation of H. virescens on cotton with pyrethroids.
3. Do not respray with a pyrethroid following a control failure. Use another chemical class.
4. Ensure the timely application of pyrethroids against early instar larvae through careful scouting.
5. Use pyrethroids at recommended rates and spray intervals.
6. Use mixtures of pyrethroids and non-pyrethroids sparingly.

<u>1988</u>. PEG-US facilitated the implementation of a beltwide monitoring program that drew on the expertise and efforts of local university personnel and private consultants. A major goal was to better document any seasonal variation across the cotton belt. The companies of PEG-US provided cooperators with all the equipment needed to conduct the AVT (vials pre-treated with cypermethrin, pheromone traps, pheromone, and data forms). More than 500 tests were conducted in 38 locations across the belt, beginning in early May. The average sample size for each test was about 20 adults per rate. In general, percent survival in the AVT increased during the season, peaking during late July and August (Figure 2) and populations east of the Mississippi delta tended to be more susceptible than those west of the delta. Once again, populations in the Brazos River Valley of Texas showed the highest levels of tolerance, with survival in the AVT as high as 61% in late September (Figure 3).

One of the most significant aspects of the 1988 program was the implementation of a system to rapidly record, analyze and distribute the test results. Data forms printed on carbonless paper and self-addressed stamped envelopes were sent to each researcher receiving treated vials. Once a test was completed, a copy of the results was sent to C. Staetz of FMC who entered the data into a central database and analyzed the results. These results were then issued in a newsletter to all cooperators in the cotton belt. This proved to be a highly effective way to collate data and communicate monitoring results over a large geographic area.

<u>1989</u>. In addition to the continuation of the monitoring program, one of the major goals of PEG-US during 1989 is to relate the monitoring results to actual control in the field. In the U.S. monitoring programs, this relationship is further complicated by the fact that the the monitoring results have been obtained on adult males while field sprays are directed at the larval and egg stages. An understanding of this relationship is important if we hope to be able to predict and confirm resistance-induced control problems and recommend specific control tactics. To fill this gap in information, PEG-US and university researchers in the cotton belt designed a field research project that will be conducted in Louisiana and Mississippi.

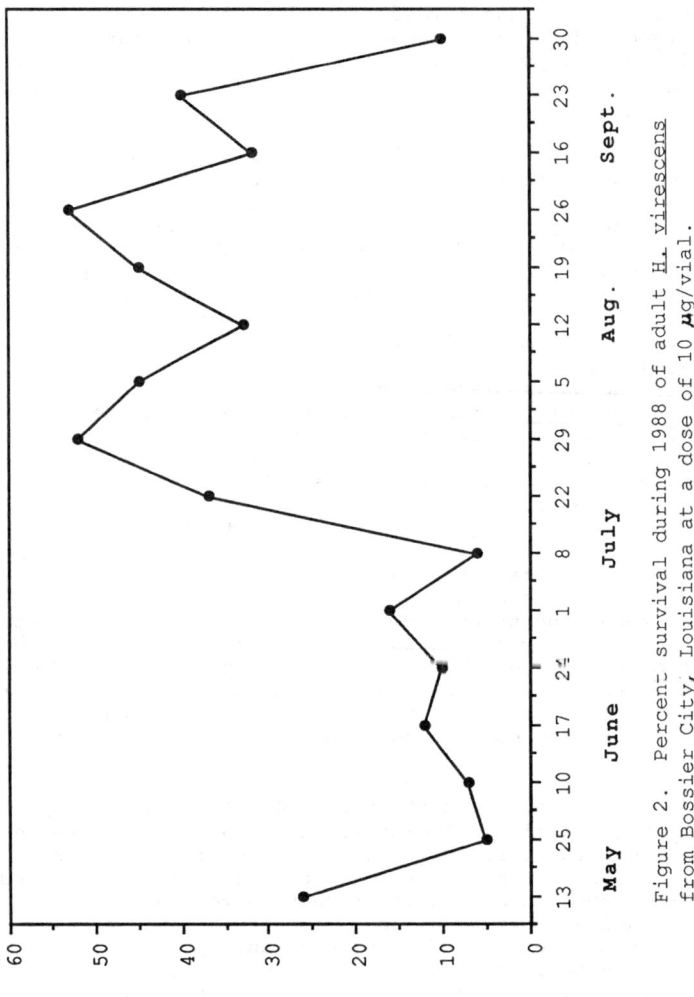

Figure 2. Percent survival during 1988 of adult H. virescens from Bossier City, Louisiana at a dose of 10 μg/vial.

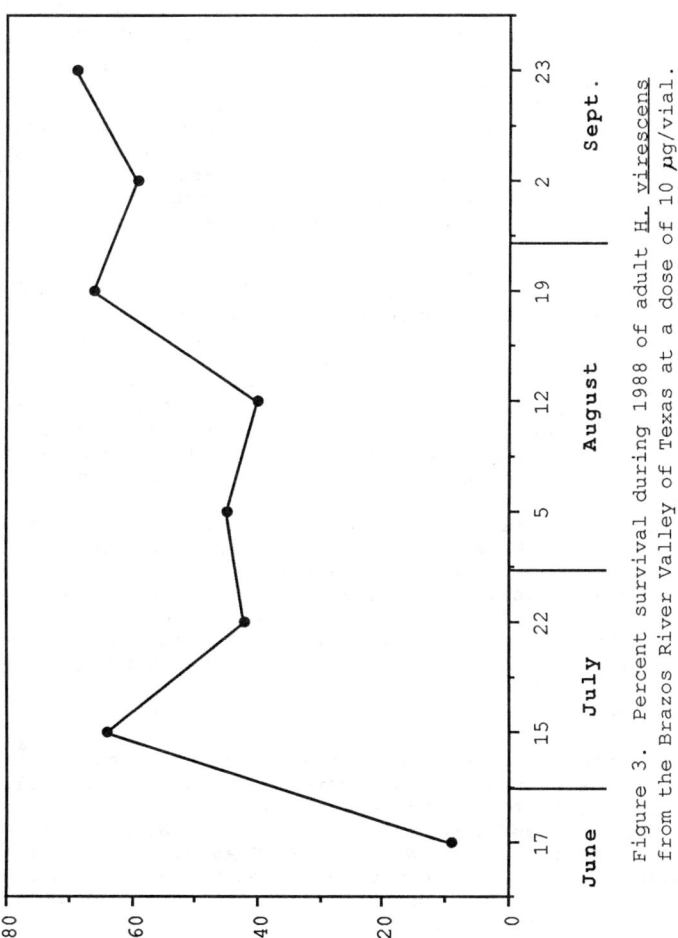

Figure 3. Percent survival during 1988 of adult H. virescens from the Brazos River Valley of Texas at a dose of 10 μg/vial.

Conclusions

Many interacting factors influence the occurrence of resistance-induced field control failures. Our ability to design successful long-term preventative and corrective management programs will depend on how well we can identify and understand these factors. Research on the mechanisms of resistance and their heritability, and the interactions between factors must be undertaken if we are to increase our confidence in proposed management strategies.

However, in the meantime, corrective strategies aimed at reducing the selection pressure for resistance must be implemented. The similar chemistry of the pyrethroids and the fact that resistance to one will most likely result in resistance to all pyrethroids is a major motivating factor.

As pointed out by Roush (48), the real challenge may not be the research, but rather the implementation of and compliance with management programs. This is a critical time for the pyrethroids as insect control agents. Our most important short-term goal should be purposeful cooperation among all involved parties. Preservation of cost effective, environmentally compatible control agents through the preservation of susceptible insects is an important global issue.

The formation of inter-company groups within the manufacturing industry (e.g. PEG-US) has been a major step forward, catalyzing cooperation between academia, industry, growers and consultants. During the demise of DDT and the OP's manufacturers did not cooperate to maintain their products, and an adversarial relationship developed between industry and academia. Hopefully, those days are gone. The seasons of 1987 and 1988 were ones of unprecedented cooperation. In the U.S., they saw the completion of probably the most geographically widespread and manpower intensive resistance monitoring effort in the history of agriculture.

Longer-term, we must remember that resistance management is a component of IPM. As we face new pressures of resistance, we can return to its basic principles: judicious use of insecticides, careful scouting, reliance on decision-making thresholds, use of alternate control tactics, and the encouragement of sound cultural practices. It is the states and countries that rely on these means of cotton production that will manage the resistance problem.

Acknowledgments

This publication represents the work of the technical representatives of the PEG-US committee (DuPont: S. L. Riley, I. A. Watkinson; FMC: C. A. Staetz; Hoechst-Roussel: J. R. Whitehead; ICI: D. Ross, H. Feese; Mobay: D. E. Simonet, W. Mullins; D. F. Clower, G. Certain) and M. Wall, J. B. Graves, J. R. Phillips and F. W. Plapp. The PEG-US committee wishes to express special appreciation to all the cooperators across the cotton belt for their dedicated hard work.

Literature Cited

1. Sparks, T. C. *Bull. Entomol. Soc. Amer.* 1981, **27**, 186-192.
2. Wolfenbarger, R. V.; Bodegas, P. R.; Flores, R. *Bull. Entomol. Soc. Amer.* 1981, **27**, 181-185.

3. Gunning, R. V.; Easton, C. S.; Greenings, E. R.; Edge, V. E. J. Econ. Entomol. 1984, 77, 1283-1287.
4. Goodyer, G. J.; Wilson, A. G. L.; Attia, F. I.; Clift, A. D. J. Austral. Entomol. Soc. 1975, 14, 171-173.
5. Goodyer, G. J.; Greenup, L. R.; General and Appl.Entomol. 1980, 12, 37-39.
6. Kay, I. R. J. Austral. Entomol. Soc. 1977, 16, 43-45.
7. Ivy, E. E.; Scales, A. L. J. Econ. Entomol. 1954, 47, 981-984.
8. Graves, J. B.; Roussel, J. S.; Phillips, J. R. J. Econ. Entomol. 1963, 56, 442-444.
9. Graves, J. B.; Clower, D. F.; Bagent, J. L.; Bradley, J.R. J. Econ. Entomol. 1967, 60, 887-888.
10. Agrochemical Monitor, Wood McKenzie & Co., Ltd., London, 1987, No. 54.
11. Collins, M. D. Proc. Beltwide Cotton Prod. and Res. Conf., 1986, p 583.
12. Bacheler, J. S. Proc. Beltwide Cotton Prod. and Res. Conf., 1985, p 120.
13. Brown, A. W. A. In Pesticides in the Environment; White-Stevens, R, Ed.; Marcel Dekker: New York, 1971; Vol. 1, part 2, p 457.
14. Georghiou, G. P. In Pest Resistance to Pesticides; Georghiou, G., P.; Saito, T, Eds.; Plenum: New York, 1983: p 769.
15. Twine, P. H.; Reynolds, H. T. J. Econ. Entomol. 1980, 73, 239-242.
16. Martinez-Carrillo, J. T.; Reynolds, H. T. J. Econ. Entomol. 1983, 76, 983-986.
17. Crowder, L. A.; Jensen, M. P.; Watson, T. F. Proc. Beltwide Cotton Prod. and Res. Conf. 1984, p 229.
18. Staetz, C. A. J. Econ. Entomol. 1985, 78, 505-510.
19. Jackson, G. J. Proc. British Crop Prot. Conf., 1986, p 943.
20. Jackson, G. J. Proc. Beltwide Cotton Prod. and Res. Conf., 1988, p 42.
21. Voss, G. Pestic. Sci. 1988, 23, 149-156.
22. Georghiou, G. P.; Taylor, C. E. In Pesticide Resistance:Strategies and Tactics for Management; National Academy Press: Washington, D.C., 1986; p 157.
23. Roush, R. T.; Miller, G. L. J. Econ. Entomol. 1986, 79, 293-298.
24. Luttrell, R. G.; Roush, R. T.; Ali, A. T.; Mink, J. S.; Reid, M. R.; Snoddgrass, G. L. J. Econ. Entomol. 1987, 80, 985-989.
25. Armstrong, K. F., Bonner, A. B. Pestic. Sci. 1985, 16, 641-650.
26. Dobrin, G. C.; Hammond, R. B. J. Kansas Entomol. Soc. 1985, 58, 422-427.
27. Gist, G. L., Pless, C. D. Florida Entomol. 1985, 68, 456-461.
28. Gist, G. L., Pless, C. D. Florida Entomol. 1985, 68, 462-466.
29. Haynes, K. F.; Baker, T. C. Archives Insect Bioch. Physiol. 1985, 2, 283-293.
30. Ho, S. H. Proc. British Crop Prot. Conf., 1984, p 553.
31. Robb, K. L.; Parrella, M. P. J. Econ. Entomol. 1985, 78, 709-713.
32. Gould, F. Bull. Entomol. Soc. Amer. 1984, 30, 34-40.

33. Scott, J. G.; Ramaswamy, S. B.; Matsumura, F.; Tanaka, K. *J. Econ. Entomol.* 1986, *79*, 571-575.
34. Daly, J. C.; Murray, D. A. H. *J. Econ. Entomol.* 1988, *81*, 984-988.
35. Forrester, N. W. *Austral. Cotton Grower* 1985, *6*, 5-7.
36. Daly, J. C.; McKenzie, J. A. *Proc. Beltwide Cotton Prod. and Res. Conf.*, 1986, p 951.
37. Forrester, N. W. *Austral. Cotton Grower* 1989, *August-October*, 62-69.
38. Daly, J. C.; Fisk, J. H.; Forrester, N. W. *J. Econ. Entomol.* 1988, *81*, 1000-1007.
39. *Cotton International*, Meister Publishing, 1988, 55th ed.
40. Sawicki, R. M. *Pestic. Sci.* 1989, *26*, 401-410.
41. Plapp, F. W., Jr.; Campanhola, C. *Proc. Beltwide Cotton Prod. and Res. Conf.*, 1986., p 167.
42. Roush, R. T.; Luttrell, R. G. *Proc. Beltwide Cotton Prod. and Res. Conf.*, 1987, p 220.
43. Plapp, F. W., Jr.; McWharter, G. M.; Vance, W. H. *Proc. Beltwide Cotton Prod. and Res. Conf.*, 1987, p 234.
44. Riley, S. L. *Proc. Beltwide Cotton Prod. and Res. Conf.*, 1988, p 228.
45. Simonet, D. E.; Riley, S. L.; Watkinson, I. A.; Whitehead, J. R. *Proc. Beltwide Cotton Prod. and Res. Conf.*, 1988, p 334.
46. Collins, M. D.; Blenk, R.; Gouger, R. J.; Staetz, C. A. *Proc. Beltwide Cotton Prod. and Res. Conf.*, 1988, p 336.
47. Staetz, C. A.; Rivera, M. A.; Blenk, R. *Proc. Beltwide Cotton Prod. and Res. Conf.*, 1988, p 339.
48. Roush, R. T. *Pestic. Sci.* 1989, *26*, 423-441.

RECEIVED October 27, 1989

Chapter 10

Management of Pesticide Resistance in Arthropod Pests

Research and Policy Issues

B. A. Croft

Department of Entomology, Oregon State University, Corvallis, OR 97331

Pesticide resistance management requires both technical progress in research and enlightened policies governing pesticide development, regulation, marketing and use. Improved monitoring systems, tactics of resistance management, models of resistance evolution, and experimentation on factors influencing resistance evolution have added to our ability to limit resistance in field populations of target pest insects. Greater emphasis on research in population genetics and ecology is needed to further extend this technology. Examples of successful resistance management with houseflies in Denmark, cotton bollworms in Australia, pear psylla in North America and a complex of pests and natural enemies on tree fruits in the United States demonstrate that policy-related factors can be a key to successful resistance management. Conversely, the lack of success in other pest systems may be due to antiquated policies. A recent case of resistance management in the spider mite *Tetranychus urticae* illustrates new policies by industry regarding research and pesticide labelling which will improve possibilities for resistance management and will extend the life of newly registered acaricides.

Pesticide resistance management is a strategy of resistance containment or suppression using a variety of tactics, including pesticides and other nonchemical control measures (e.g. biological control). Beyond research, resistance management requires enlightened policies governing pesticide development, registration, marketing, regulation, and education.
    In this paper, progress in resistance management for arthropod pests is discussed, focusing on research, successful case histories of resistance management, policy limitations to resistance management, and future needs.

0097–6156/90/0421–0149$06.00/0
© 1990 American Chemical Society

Progress in Resistance Management Research

Research on the evolution of resistance to pesticides among arthropod pests began soon after this phenomenon was widely discovered in the late 1940's and 1950's (1-2; see also Georghiou, this volume). Initially, research focused on the genetics, biochemistry, and toxicology (along with limited monitoring research) of resistance, as scientists tried to better document the basic mechanisms involved (3). The hope was to circumvent or diminish the impact of resistance, but because these initial investigations led to so little practical success, an "inevitability syndrome" set in. Given enough time and intensity of pesticide use, it was thought that certain troublesome species would eventually develop resistance and there was little one could do about it. Resistance was often considered irreversible. Once lost, a compound could not be used again and substitute materials had to be found as soon as possible. Resistance research other than that directed at documenting resistance episodes became somewhat unfashionable and neglected.

Treating pesticides as short-term tools for suppression of pests caused no serious problems as long as new materials were readily available. In fact, the obsolescence of many compounds would have occurred anyway because of the rapid progress being made in the development of safer, more effective pesticides. However, in the early 1970's, as economic and regulatory constraints on development of new pesticides increased, the number of new chemicals being introduced annually, declined worldwide (4). Coupled with a rapid increase in the number of resistance cases, the decline in the availability of pesticides only intensified the need for new approaches to pest control and resistance problems.

An important major response to resistance problems was the emergence and spread of integrated pest management (IPM) (5). Although the roots of IPM trace back to the 1930's, widespread adoption did not begin until the 1970's. IPM emphasized a more thorough study of the ecology of pests, the greater use of biological control agents, a wider array of other nonchemical tactics of pest control, and better pesticide management. IPM also rekindled interest in finding alternative ways to delay or avoid resistance in pests.

Using modeling techniques (e.g., 6-11), scientists identified a diversity of factors influencing resistance in the late 1970's-early 1980's. With this more complete and integrated theoretical framework, more extensive research on resistance development in the laboratory and field was pursued. Eventually, a more complete range of physiological, genetic, ecological and operational (i.e., those under control of the manager) factors influencing resistance were identified (6-7) and investigated (12).

By the 1980's, applied research on new tactics, monitoring methods, and integrated systems of resistance management began to be more widely developed (13-14). In addition, some of the institutional, industry-related, and regulatory constraints which previously had limited resistance management programs were beginning to be articulated (e.g., 14). With these advances, scientists began to consider resistance as a more manageable phenomenon.

The renaissance in resistance and resistance management research in the United States was summarized in a National Academy of Science/National Research Council study entitled Pesticide Resistance: Strategies and Tactics for Management (12). In this forum, scientists met to discuss the status of resistance research in a broad context. Those present represented a broad range of disciplines from agriculture and human health.

This paper focuses on progress made in research and implementation of resistance management since the NAS report. It generalizes from the reports given earlier in this symposium. While resistance management of arthropod species is specifically addressed, these principles may be applicable to other pests as well.

Research on Management of Pesticide Resistance

Important technical issues limiting the development of resistance management systems can be subdivided into basic and applied categories. While one may recognize considerable overlap in these categories, the distinction is still useful. Also, research emphasis can be organized by level of resolution ranging from molecular and cellular biology to population and community ecology (Figure 1). As noted earlier, early research efforts were conducted mostly at the suborganismal and organismal levels. However, efforts ideally must include the full spectrum of disciplines shown in Figure 1. As discussed below, there are some major gaps in the research effort presently underway in resistance management.

Basic Science Aspects. The underlying mechanisms of resistance at the cellular or organismal level, including the genetics, biochemistry, and toxicology of resistance, continues to be of extreme interest to scientists studying pesticide resistance (12-13; see Brattsten, in this volume). This type of research is fundamental to reach an understanding of resistance, including means to overcome it. For example, knowing the genetic basis of resistance is a critical to understanding the population genetics of pesticide resistance (Denholm et al., this volume). Basic aspects of resistance are also important in the design of new pesticides with novel modes of action that do not show cross-resistance to earlier pesticides (15; Brattsten, this volume); in choosing appropriate synergists to overcome resistance mechanisms (16); in identifying pesticides that select less intensively (12,17); and in developing monitoring techniques for pesticide resistance (18).

More recently, basic research has included use of molecular techniques to identify resistance genes and gene products (enzymes and receptors) (e.g., 19-20; Brown, this volume). This information adds to our understanding of pesticide target sites, pesticide pharmacodynamics and genetic mechanisms, such as changes in structural genes, amplification of structural genes and alteration in rates of gene regulation (12).

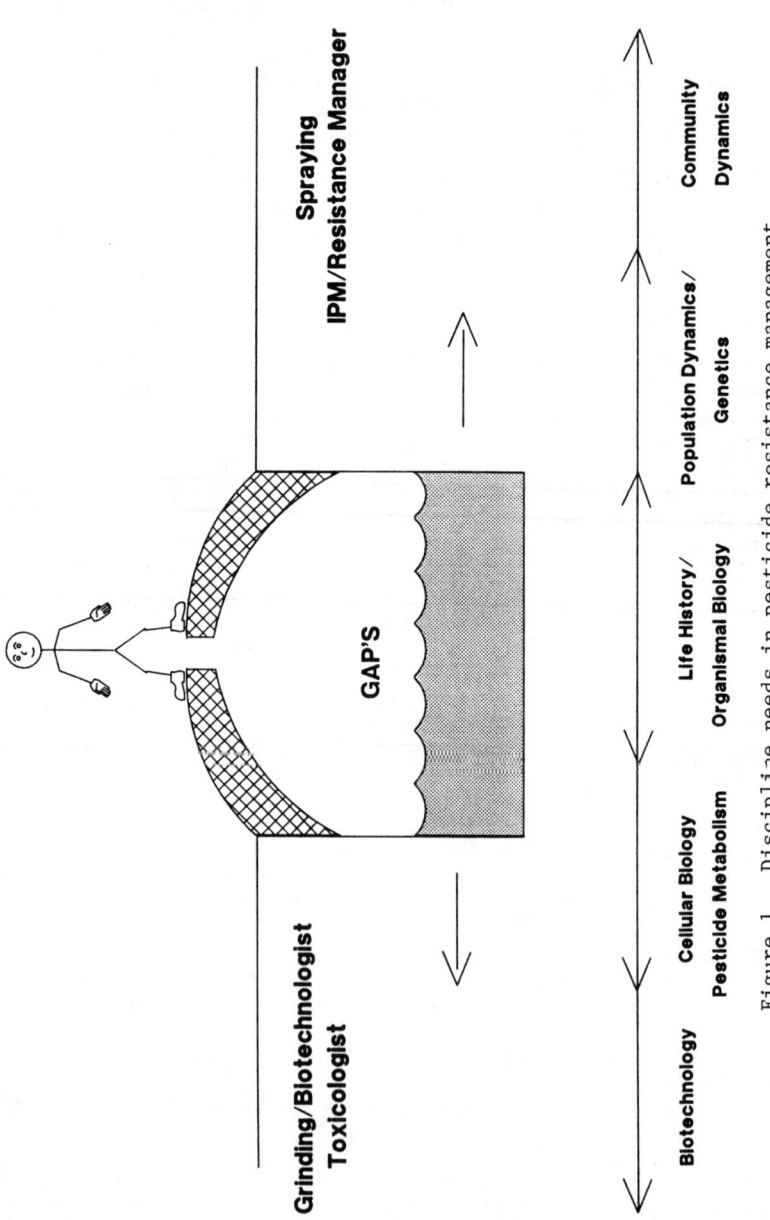

Figure 1. Discipline needs in pesticide resistance management.

Another level of factors influencing resistance development is the <u>life history level.</u> For example, it has long been known that organisms with rapid reproductive rates develop resistance more rapidly and more frequently than do pests with lower reproductive rates (6,21-22). Other life history factors influencing resistance development include fecundity, reproduction, feeding habits (e.g., monophagy/polyphagy), occurrence in refugia, and migration characteristics. For example, the relationship between the number of generations per season and the rate of resistance development has been documented for many agricultural pests exposed to pesticides over long time periods in a particular habitat (e.g., 11,21).

More recently, studies of how the life history attributes of a species influence resistance development have been investigated using models, and by comparing their output with resistance patterns observed in field populations. Denholm et al. (in this volume), illustrate this type of research for the common housefly, *Musca domestica*. In a complex of species associated with apple orchards (including both pests and natural enemies), Tabashnik and Croft (11,23) reported that variations in attributes among species of life history differentially affect resistance development. Furthermore, the influence of these factors often depends on ecological conditions present in the environment surrounding the treated habitat. For example, the level of resistance dilution by immigrating susceptible individuals greatly influences the effect of some operational factors such as dose and frequency of spraying on resistance development. In general, they reported that resistance is least common in species that have low residency levels of resistant individuals in relation to high levels of immigration of susceptible individuals into the treated environment. Furthermore, they concluded that under most ecological conditions, a low dose and low application frequency approach is the most preferable strategy of resistance management.

These studies of the individual and integrated effects of life history have identified sensitive factors which govern rates of resistance development. Generally, resistance patterns can best be explained at the population level by integrating the information obtained from the suborganismal and organismal levels discussed above (e.g., genetics, life history, etc). When this information is synthesized into a model of population dynamics, the influence of any single factor can be more completely evaluated for its influence on resistance development (see next section).

Another level of factors influencing resistance which requires much greater effort in basic research is the <u>ecological or population genetics level.</u> As compared to the two areas discussed earlier, this level has received much less emphasis (3,24; see Denholm et al. in this volume). Most models of population genetics for pesticide resistance are theoretical, with limited validation. These models often are based on simple genetic assumptions describing a very generalized pest (25-28), and usually focus on populations of a single species. Very few have explored resistance development in species complexes of pests and beneficial arthropods or communities of species (e.g., 11,23).

Another limitation to progress made in population genetics of pesticide resistance is the assumption that resistance is almost always due to single-gene phenomena (27,29). While this level of investigation may be adequate for many cases of resistance (e.g. for the housefly; 30), the incidence of polygenic resistance is becoming increasingly well documented. Very likely, study of polygenic resistance will become more important as our understanding of the population genetics of field populations improves. More complex genetic systems of resistance must be vigorously investigated particularly for factors such as the fitnesses associated with resistance both in the presence and absence of insecticides, and linkage of resistance genes to other resistance genes (3,24,27-28). In the population dynamics area, new methods for monitoring movement of highly mobile species are beginning to be developed (31), but little research to link migration phenomena to rates of gene flow in resistant species has been reported. Research remains at the elementary level of trying to develop appropriate markers (either genetic or morphological) and sampling methods that will indicate changes in the frequency of resistance genes in populations as selection (both for resistance in the presence of pesticides and against resistance in their absence), immigration and other population processes occur over time.

Contributing to the lack of study of the genetics of pesticide resistance at the population level is that many applied scientists are inadequately trained to work in this area. Furthermore, it has been very difficult to convince basic scientists with population genetics expertise to use resistance as a model for studying evolutionary processes. Even when such scientists are attracted to resistance research, they are often surprised and discouraged by the relative difficulty of obtaining funds for resistance research as compared to some more basic aspects of evolutionary biology. There is some evidence that this gap is being bridged, in part, by the involvement of basic researchers in symposia (e.g., 25,27-28; also Dehholm et al., this volume) and by the increasing use of examples of resistance management in teaching evolutionary biology, genetics, and applied sciences as well. Interactions between basic and applied scientist may occur more frequently as we view arthropod resistance as an example of biocide resistance as manifested in such diverse areas as antibiotic resistance in human and plant pathogens, heavy metal resistance in plants.

Applied Science Aspects. Progress has been made in applying basic research in the field largely because of the need to solve the resistance problem (12-14). Detection methods, monitoring programs, new tactics, and integration of these elements into resistance management programs are beginning to be used in several major cropping systems (e.g., Gelernter, Dennehy and Nyrop, Bull and Menn, Riley in this volume).

Detection and monitoring of resistance is producing promising results using classical bioassay, biochemical enzyme tests, immunological techniques (e.g., ELISA) and biotechnological probes with appropriate DNA or RNA segments (12,18; Dennehy and Nyrop, Bull and Menn and Riley, this volume). In addition, statistical methods used for detecting pesticide resistance are being better addressed (32). These tools will be useful in calculating the costs for resistance monitoring and the feasibility of resistance detection. Large databases containing base line susceptibility information for pests when first exposed to pesticides and similar resistance survey data are being maintained by WHO, FAO, EPPO and others (12). As these databases become more widely available through computer telecommunication links, better decisions about resistance management should be possible.

A variety of new tactics to manage resistance to pesticides have been proposed in recent years, with most focusing on new pesticides and their application (12). Suggestions uses of rotations, alternations and mixtures of pesticides are coming from modeling studies of resistance development (e.g., 6-8,11), but, only limited experimental studies evaluating these methods have been reported. Recently, attention has focused on use of genetically improved pesticides and crops, such those based on as modified strains of *Bacillus thuringiensis* or crops containing the endotoxin of this insect pathogen (e.g., Gelernter, this volume). Thus far, many of these measures have not been tested in the field. Problematic in testing resistance management tactics is the difficulty of maintaining experimental areas to measure impacts on regional populations of pests and the impracticality of maintaining untreated check plots for comparison (12).

Integration of tactics as part of an overall strategy of resistance management involves monitoring and modeling as well as a plan of implementation (Figure 2). Environmental monitoring and regular evaluation of changing trends in economic factors and policy must be considered. As illustrated at this conference, prototype resistance management systems for species such as spider mites on apple (Dennehy and Nyrop, this vol.; see discussion later in this paper) and *Heliothis* spp. on cotton (Bull and Menn, Riley, this vol.) are beginning to be deployed in the field. These programs illustrate that such systems vary greatly in their temporal and spatial scales of application. In the case of spider mites, resistance can be managed in small blocks of trees, whereas with *Heliothis* spp. whole regions must be considered because of the highly dispersive nature of the insect.

At present, our understanding of how a pesticide resistance management system should be organized is rudimentary (Figure 2). However, many of the principles used in the design of IPM systems are also appropriate to pesticide resistance management design. As in IPM systems, resistance monitoring, modeling and prediction, and implementation must be developed and linked in a timely manner (Figure 2). Spatial scales for monitoring and managing a pest or pest complex must be optimally determined and information often must be transmitted rapidly for the program to operate effectively. As has been pointed out by many authors (33-34; Bull and Menn, Riley, this volume), resistance management should be considered a subset of IPM and should operate within the confines of such a system. As IPM improves, so should resistance management and vice versa.

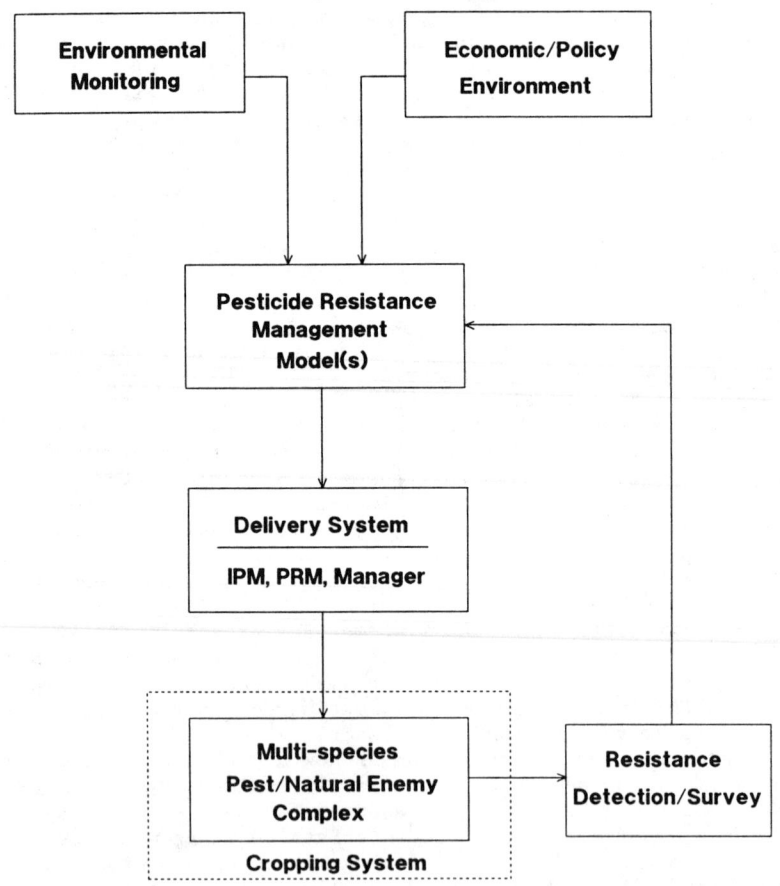

Figure 2. Components of a pesticide resistance management (PRM) system.

Case Histories of Resistance Management Implementation

Several cases histories in which resistance management has been successfully applied are reviewed in this section along with other less successful examples (Table I). These case histories highlight some of the research needs and policy limitations of resistance management. In some instances, the same pest species or a closely related species are the subject of successful resistance management as well as failure. This illustrates that success can be site specific, varying from area to area due to the "resistance environment" (i.e., organim biology, ecology and insitutional policy). In making general comparisons, key factors are identified that may have determined the success of a given resistance management attempt. As might be expected, no single factor is always responsible, but some trends are apparent. Determining why resistance management does or does not work is a challenge, given the many factors involved. Also, it is probably inappropriate to characterize certain case histories as failures, just because progress has been slow. In fact, we are learning much about what does and does not work from these examples.

Successful Resistance Management. In Australia, resistance to pyrethroid (SP) insecticides was confirmed in laboratory tests of the cotton bollworm *Heliothis armigera* after field control failures in cotton and soybean occurred in the Emerald Irrigation District of Queensland in 1983. Within a few months, a strategy to manage the resistance problem was proposed by state authorities after consultation with federal CSIRO research authorities, pesticide companies, and growers. The program involved restricting SP use on all crops in the area to a 42-day period during the middle of the growing season (35). Adherence to the program is voluntary in New South Wales, but is potentially enforceable in Queensland. Even though these restrictions in pesticide use are a financial burden in some cases, there has been no effort to enforce use of the program in either state, and compliance appears to be almost compete (Daly, J.C., pers. comm., 1988). The program has been judged a success inasmuch as there have not been further field control failures due to resistance in any of the areas where the program has been attempted, although some increase in resistance has continued in areas using the Australian program (Bull and Menn, this vol.). The key to success in this case of resistance management has been the unusually high level of voluntary cooperation.

Houseflies on Danish farms have become resistant to almost every new insecticide introduced for their control since the 1950's (36). In the mid 1970's, scientists showed that resistance to DDT and the new pyrethroid insecticides is due to a common resistance factor (the *kdr* and super-*kdr* genes), and they predicted that rapid resistance development to the more persistent SP compounds would rapidly evolve in the field. In 1978-79, surveys for SP resistance indicated that resistance to long-residual compounds was beginning to develop in the field. Overall, the survey data indicated that resistance would soon be widespread. Several steps were immediately taken to evaluate the resistance potential of nonpersistant SP's, and additional monitoring and regulatory actions were implemented. A decision was made in collaboration with scientists, industry and regulatory personnel against registering persistent compounds. Strict regulation and monitoring of SP resistance was employed. So far, SP resistance in the country has been manageable. Through these actions, continued use of these compounds has been sustained in a very difficult resistance management environment. The keys to this case history are the effectiveness of the resistance monitoring effort and the critical regulatory actions taken to insure that certain pesticidal products are not used.

Successful resistance management in the pear psylla (*Psylla pyricola*) has involved conservative pesticide programs and cooperation among industry and growers in attempting to solve a serious resistance problem. In the past, this species has developed resistance to virtually all pesticides registered for its control, often over large areas (37-38). In the late 1970's, experiments demonstrated conclusively that resistance to the SP's would occur if these products were used unilaterally prebloom and during mid-season for summer control of this pest (37). A program of SP use in prebloom on overwintering psylla and the carbamate, amitraz, in summer was recommended. Growers readily complied to SP use restricted in this manner. Furthermore, SP chemical producers did not put summer use on the label, thus making this use of SP's less likely. In some pest management districts, regulatory officials, with grower support, made summer use of SP's an illegal practice, thus going one step further in insuring that misuse of these products did not occur. To date, SP resistance has not developed in psylla populations in areas of the western United States, where these strict limits have been promoted. In contrast, in areas where SP's are use in the summer, resistance in *P. pyricola* and a related species (*Psylla pyri*) has developed rapidly (see below). The key to success in this case has been the far-sighted resistance risk assessment research that was initially done and the excellent regulatory and industry cooperation that helped maintain the program.

Another success in resistance management involves not just a single key pest species, but a whole complex of phytophagous arthropod pests of apple in the United States (39). It also focuses on resistance management of a number of beneficial arthropod predators and parasitoids that provide significant biological control of a variety of secondary pests on this crop. In this case, a rather serendipitous pattern of resistance has developed among species to certain organophosphate (OP) insecticides. This resistance or lack thereof in some species has been exploited, providing an example of IPM, as well as resistance management.

To summarize the resistance situation in broad terms, no key pest such as the codling moth, apple maggot, plum curculio has developed resistance to the OP azinphosmethyl, whereas a variety of secondary pests such as mites, aphid, leafhoppers, leafminers and their natural enemies have developed resistant strains; thus azinphosmethyl has become more selective. The example cited later of management of cyhexatin resistance management in the spider mite

T. urticae using biological control by insecticide-resistant predatory mites is just a subsystem of this larger resistance management program.

Essential management components of the program include: 1) careful monitoring for OP resistance in key pests such as the codling moth, apple maggot, plum curculio, and several leafroller species; 2) seeking to minimize further resistance development in secondary pests such as mites, aphids, leafminers, leafhoppers and scales through minimum use of selective pesticides and maximum exploitation of biological control agents; and 3) seeking to exploit resistance in key natural enemy species by monitoring for natural development of resistance or by genetic improvement of resistant strains using hybridization or artificial selection techniques.

The success of the resistance management program for the complex of pests and natural enemies of apple in the United States, while generally good, has been mixed. In some areas of the eastern United States where OP resistance has developed in key leafroller species, increased use of nonselective pesticides such as the pyrethroids has occurred. However, in the western United States, OP pesticides still provide adequate control of key pests, while not upsetting biological control of certain secondary pests by their resistant predators and parasites (39). In those areas where the program is successful, the keys to success are monitoring for resistance in the species complex (e.g., among leafrollers, 40), rapid response to early signs of resistance (e.g., 41), maximum use of IPM and alternative control tactics other than pesticides (42), and the lack of resistance development in key pests such as the codling moth.

Less Successful Resistance Management. Pyrethroid resistance in North American *Heliothis* spp. has been of great concern to scientists since these pesticides were first introduced in the late 1970's (Bull and Menn, Riley, this vol.). However, many cotton growers initially were unconcerned, apparently out of a faith that pesticide manufacturers would continue to develop replacement compounds. In spite of the efforts of cotton entomologists to promote judicious use of the SP's in the 1970's, there was a tendency to use these chemicals intensively. Not surprisingly, and with striking parallels to the Australian situation, resistance as manifested by control failures appeared in *Heliothis virescens* in certain areas of western Texas in 1985 (43). Resistance in Texas was confirmed by Roush & Littrell (44), who also documented the independent evolution of resistance in Mississippi in 1986. More recently, SP resistance in *H. virescens* has appeared in a number of cotton production areas throughout the southern United States (Bull and Menn, Riley; this vol.) An initial resistance management program was developed by entomologists for Mississippi, Louisiana, and Arkansas growers, but initially it was only *recommended* to growers. In the absence of some enforcement or unless the resistance problems becomes more severe, it is projected that the program will not be widely adopted. A larger consensus of industry, public and private groups is needed to join in supporting such a program. Either a voluntary policy of compliance or a mandated one similar to that used for stalk destruction of plant residues is needed to limit further resistance development in these key cotton pests.

Another case history of limited progress in spite of extensive research on resistance management involves the cattle tick *Boophilus microplus* in Australia. Particularly in southeastern Queensland, this species has been a critical threat to the cattle industry primarily because of the inability to control it due to resistance (45). Considerable effort has been devoted to resistance studies, and Australian tick resistance authorities now believe they have some very clear ideas on how resistance can be delayed (J. Nolan, remarks at a CSIRO/DSIR Workshop). Many of these concepts were previously described and analyzed in modeling studies of Sutherst & Comins (45). However, Australian cattle producers are a less cohesive group than Australian cotton growers. While the Australian cotton industry is only about 20 years old, cattle production has been a major enterprise for more than 100 years, and many production practices are strongly influenced by traditional approaches to management. Where the cotton growers, perhaps rather uniquely in Australian agriculture, tend to band together, cattle producers tend to be very independent, and they are slow to adopt new resistance management methods.

The diamondback moth (*Plutella xylostella*) is a major pest of cruciferous vegetables in more than 80 countries, worldwide. It has developed resistance to all major classes of insecticides (46). Three factors promote its rapid evolution of resistance to pesticides: 1) its biology, 2) the relatively low level of acceptable damage, and 3) the intensive use of insecticides necessary for its control (47). In tropical regions this pest can complete more than 15 generations yearly, which accelerates resistance development (11). Because of the direct use of the crop for human consumption, the economic threshold for control is only 1-2 larvae per plant. Insecticides have been used heavily for diamondback moth control. For example, many farmers in Taiwan spray weekly for control. Consequently, more than 2000-fold resistance to the pyrethroid insecticides has developed less than four years after initial use of these products (47). Simulation studies suggest that under tropical conditions, insecticide use must be reduced to two or fewer sprays per crop cycle to substantially slow resistance development (48). Although this moth can disperse great distances, some local variation in resistance development among individual populations does occur. This suggests that individual growers could retard resistance development in their own fields by reducing insecticide use. Integration of insecticides with biological, cultural, and microbial methods for diamondback moth control is the most promising way to retard resistance development. However, implementation of these methods is complex and difficult to achieve with growers (Tabashnik, B.E., pers. corresp., 1988).

A final example of where little progress has been made in achieving resistance management is with the pear psylla in parts of Europe (Table I). As noted earlier, in areas where both pre-bloom and summer applications of pyrethroid insecticides have been used unilaterally, resistance to the SP has developed widely.

Policy and Pesticide Resistance Management

Several generalizations can be made from the case histories cited above. They apply in many cases to the basic and applied research gaps discussed earlier. In other cases, it is antiquated policies that limit greater implementation of pesticide resistance management.

Policy is a recommended course of action followed by a person, group or institution that is selected from alternatives in light of given conditions to guide and determine present and future decisions. A variety of policy issues and constraints have been raised by academic, government, industry and public groups interested in facilitating greater implementation of resistance management in the field (12,14,49). A comparative list of policy factors identified by two study groups who considered these issues as they operate in the United States is shown in Table II. Many common issues were identified by both groups.

Table II. Issues Areas of Resistance Management Policy

Policy Issue Area/Topic[2]	Policy Study Group/Importance Rating[1]	
	Dover & Croft (1984)	NAS (1986)
Information Storage, Retrieval, & Dissemination	X X X	X X X
International Coordination and Policy	X	X X X
Regulatory Reform/ Federal	X X X	X X
Regulatory Reform/ State & Local	X X X	X X
Industry Self-Regulation	X X X	X X X
Industry Marketing	X X X	X X
Antitrust Limitations	X	X X
Implementation Infrastructure	X X	X X X

[1] XXX = great importance, XX = moderate importance, X = little importance.
[2] See Dover & Croft (14) and NAS (12) for explanation of these titles.

Policy for resistance management involves aspects of pesticide regulation, marketing, antitrust, international considerations and education (Table II). Due to time and space limitations, each of these topics can not discussed here, but several were referred to in Table I. For a more complete treatment of policy and resistance management, see (14) and (12).

Policy is usually involved in most attempts to implement resistance management programs. In fact, policy is more often a constraint to resistance management than research limitations. Many policies must be updated if the environment for resistance management approaches is to be improved.

New policies of resistance management are beginning to be developed with regard to acaricide resistance problems on deciduous tree fruits (42). Changes in pesticide industry policies have enhanced opportunities to implement resistance management in the field. Resistance management with the spider mite *Tetranychus urticae* to organotin (OT) and several new acaricides involves several physiologically selective compounds which allow for increased biological control by insecticide-resistant predatory mites (42).

In studies made in the western United States and Australia, researchers demonstrated that cyhexatin resistance rapidly reverts toward susceptibility in the absence of intense selection by the organotins or other cross-resistant acaricides (50-51). Also, a number of tactics are available to limit OT resistance including: formulations that effectively enhance toxicity to resistant mites (50-51), critical timing of organotin applications to control more susceptible immature mite stages (42,50-51), alternation of organotins with noncross-resistant acaricides (42,51), and increased use of predators in biological control (42). Using a combination of these measures allows for continued use of the organotins in areas where previous use was limited due to the presence of highly resistant mites. The key to success in this case has been the instability of OT resistance in spider mites, the wide variety of alternative tools available to combat resistance development, and the extensive research effort made to integrate these methods.

While progress in preserving the organotins was made after resistance developed, significant steps have also been taken to prevent resistance from developing to several new selective acaricides, before their registration for field use (42). The stimulus for new policy has come from an indication of high resistance risk to the ovicides, hexythiazox and clofentezine. In experimental studies in greenhouses in Australia, citrus plots in Japan and apple orchards in Spain, resistance to hexythiazox and in some cases, cross-resistance to clofentezine was noted after only 15-25 selections. At the initiative of industry, cooperative resistance management research trials were undertaken at several universities in the United States to develop monitoring methods and to evaluate the genetics of resistance and operational alternatives to the use of these acaricides. Furthermore, use of these compounds was recommended only once per season on labels, and both parent companies agreed that use of the either company's product would make subsequent use of either product undesirable. Thus a cautionary statement indicating this cross-resistance potential was stated on the label of both compounds.

These far-sighted polices by industry are examples of the types of actions that are needed to prevent or stem the tide of increasing resistance to pesticides among arthropod pests. Such actions show that the longer term perspectives of product conservation and economic performance are more widely recognized by research, management and marketing personnel from industry. At least for acaricides, these actions are essential for conserving future products considering the high costs required to bring a new pesticide to market and the short time they may last due to resistance problems. Similar actions may be appropriate for other resistance-prone pesticides and pest species.

Summary

Greater effort is needed to build management systems that help limit development of resistance to new pesticides as well as to conserve other valuable tactics of pest control, such as pest resistant host plants and even some biological control agents (Croft et al. 1988). Such effort will be required for genetically improved forms of these pest control agents, as well. Effective management systems should be put in place before these new tactics of pest control are deployed rather than after problems of pest adaptation to them are detected in the field.

In developing pesticide resistance management systems, more expertise, training and methodology are needed in the ecology and population genetics of pesticide resistance. This is a critical gap research needs. Many avenues could be pursued to solve the problem of greater expertise in these areas, ranging from better training of applied scientists, to better cooperation between applied and basic scientists to increased use of resistance models for evolutionary biology studies. Funding for research across multiple disciplines is a key element in bridging this gap.

New institutional policies of cooperation and change are needed to better facilitate the implementation of resistance management tactics across broad boundaries of societal groups including producers, industry personnel, regulators, and pesticide users. For example, resistance management districts could be organized to operate much like mosquito abatement districts (or similar units of management) that are established, at least temporarily, to solve a persistent problem. IPM districts operate in many regions where producers see a common benefit from organizing themselves. Most cases of resistance management are unique in terms of what an appropriate organizational structure might be. Each resistance episode requires a very specific, tailored response. Little effort has been made to study how to improve the implementation of resistance management in the field. Research, extension, private groups and other interested personnel must unite to develop innovative ideas on how these problems can be dealt with in the future.

Finally, a utopian goal for resistance management would be to stem the increasing tide of resistance to most chemical pest control products. Instead of experiencing an effective life of 5-15 years until a compound is rendered ineffective, we might see these products last for longer periods--up to 50 years or more--under a better resistance management scenario. This would allow time to develop other, more safe and effective products. Too often in the past the obituary of a particular product has not been written by us, but by our competitors, the pests, as they have continued to find new ways to evolve resistant strains and circumvent our best efforts to contain them.

Literature Cited

1. Brown, A. W. A. In *Pesticides in the Environment*; R. White-Stevens, Ed.; Marcel Dekker: New York, 1971; Vol. 1, Part II, p. 457-552.
2. Georghiou, G. P. *Residue Rev.* 1980, 76, 131-145.
3. Roush, R. T.; McKenzie, J. *Ann. Rev. Entomol.* 1986, 32, 361-380.
4. Patton, S.; Craig, I. A.; Conway, G. R. In *Pesticide Resistance and World Food Production*; G. R. Conway, Ed.; Imperial College Centre Environ. Tox. U.K., 1982; p. 61-76.
5. van den Bosch, R.; Stern, V. M. *Ann. Rev. Entomol.* 1962, 7, 367-386.
6. Georghiou, G. P.; Taylor, C. E. *J. Econ. Entomol.* 1977a, 70, 319-323.
7. Georghiou, G. P.; Taylor, C. E. *J. Econ. Entomol.* 1977b, 70, 653-658.
8. Comins, H. N. *J. Theor. Biol.* 1977, 64, 177-197.
9. Plapp, F. W. Jr.; Browning, C. R.; Sharp, P. J. *Environ. Entomol.* 1979, 8, 69-80.
10. Wood, R. J. *Parasitology* 1981, 82, 69-80.
11. Tabashnik, B. E.; Croft, B. A. *Environ. Entomol.* 1982, 11, 1137-1144.
12. *Pesticide Resistance: Strategies and Tactics for Management*; National Academy Press: Washington DC, 1986; 471 pp.
13. Georghiou, G. P.; Saito, T. Eds. *Pest Resistance to Pesticides*; Plenum Press: New York, 1983; 809 pp.
14. Dover, M. J.; Croft, B. A. *World Res. Inst. Policy Paper*, Nov. 1984; 80 pp.
15. Hammock, B. D.; Soderlund, D. M. In *Pesticide Resistance: Strategies and Tactics for Management*; National Academy Press: Washington, DC, 1986; p. 111-129.
16. Ozaki K. In *Pesticide Resistance to Pesticides*; G. P. Georghiou & T. Saito, Eds.; Plenum Press: New York, 1983; p. 595-614.
17. Mullin C. A.; Croft, B. A. In *Biological Control of Agricultural Integrated Pest Management Systems*; M. A. Hoy & D. C. Herzog, Eds.; Academic: New York, 1985; p. 123-150.
18. Brown, T. M.; Brogdon, N. *Ann. Rev. Entomol.* 1987, 32, 145-162.
19. Mouches, C. D.; Fournier, D.; Raymond, M.; Magnin, M.; Berge, J. B.; Pasteur, N.; Georghiou, G. P. *Comptes Rendu Acad. Sci. Paris, Ser. III*, 1985, 301, 695-700.
20. Mouches, C.; Pasteur, N.; Berge, J. B.; Hyrien, O.; Raymond, M.; de Saint Vincent, B. R.; de Silvestri, M.; Georghiou, G. P. *Science* 1986, 233, 778-780.
21. Georghiou, G. P. *Ann. Rev. Ecol. Sys.* 1972, 3, 133-168.

22. Georghiou, G. P.; Taylor, C. E. In *Pesticide Resistance: Strategies and Tactics for Management*; National Academy Press: Washington, DC, 1986; p. 143-156.
23. Tabashnik, B. E.; Croft, B. A. *Entomophaga* 1985, 30, 37-49.
24. Roush, R. T.; Croft, B. A. In *Pesticide Resistance: Strategies and Tactics for Management*; National Academy Press: Washington, DC, 1986; p. 257-270.
25. May R. M.; Dobson, A. P. In *Pesticide Resistance: Strategies and Tactics for Management*; National Academy Press: Washington, DC, 1986; p. 170-193.
26. Tabashnik, B. E. In *Pesticide Resistance: Strategies and Tactics for Management*; National Academy Press: Washington, DC, 1986; p. 194-206.
27. Uyenoyama, M. K. In *Pesticide Resistance: Strategies and Tactics for Management*; National Academy Press: Washington, DC, 1986; p. 207-221.
28. Via, S. In *Pesticide Resistance: Strategies and Tactics for Management*; National Academy Press: Washington, DC, 1986; p. 222-235.
29. Taylor, C. E. 1983. In *Pest Resistance to Pesticides*; G. P. Georghiou & T. Saito, Eds.; Plenum Press: New York, 1983; p. 163-173.
30. Plapp, F.W. Jr. In *Pesticide Resistance: Strategies and Tactics for Management*; National Academy Press: Washington, DC, 1986; p. 74-86.
31. Rabb, R. L.; Kennedy, G. G. *Movement of Highly Mobile Insects: concepts and Methodology in Research*; N. Carolina. St. Univ. Press: Raleigh, NC, 1979; 456 p.
32. Roush, R. T.; Miller, G. L. *J. Econ. Entomol.* 1986, 79, 293-298.
33. Croft, B. A. (Leader). 1986. *Pesticide Resistance: Strategies and Tactics for Management*; National Academy Press: Washington, DC, 1986; p. 271-278.
34. Croft, B. A. *Proc. CIPM IPM Project*; R. F. Frisbie & P. L. Adkisson, Eds; Texas Agric. Exper. Sta. Press: College Station, TX, 1986.
35. Daly, J. C.; McKenzie, J. A. *Proc. British Crop Prot. Meeting*, Bristol, UK, 1987.
36. Keiding, J. In *Pesticide Resistance: Strategies and Tactics for Management*; National Academy Press: Washington, DC, 1986; p. 279-297.
37. Riedl, H.; Westigard, P. H.; Bethell, R. S.; DeTar, J. E. *Calif. Agric.* 1981, 35, 7-9.
38. Follett, P. A.; Croft, B. A.; Westigard, P. H. *Can. Entomol.* 1985, 117, 565-573.
39. Croft, B. A. *Entomol. Exper. & Appl.* 1982, 31, 88-110.
40. Croft, B. A.; Hull, L. A. In *Tortricoid Pests*; L.P.S. van der Geest and H. H. Evenhius, Eds.; Elsevier: Amsterdam, The Netherlands, 1988; (in press).
41. Croft, B. A.; Miller, R. W.; Nelson, R. D.; Westigard, P. H. *J. Econ. Entomol.* 1984, 77, 574-578.
42. Croft, B. A.; Hoyt; S. C.; Westigard, P. H. *J. Econ. Entomol.* 1987, 80, 304-311.

43. Plapp, F. W. Jr.; Campanhola, C. *Proc. 1986 Beltwide Cotton Res. Conf.*, 1986, p. 167-169.
44. Roush, R. T.; Littrell, R. G. *Proc. 1987 Beltwide Cotton Res. Conf.*, 1987.
45. Sutherst, R. W.; Comins, H. N. *Bull. Entomol. Res.* 1979, *69*, 519-537.
46. Georghiou, G. P. FAO, Rome, Italy, 1981.
47. Liu, M. Y.; Tzeng, Y. J.; Sun, C. N. *J. Econ. Entomol.* 1981, *74*, 393-396.
48. Tabashnik, B. E. *J. Econ. Entomol.* 1986, *79*, 1447-1451.
49. Dover M. J.; Croft, B. A. *Bioscience* 1986, *36*, 78-85.
50. Edge, V. E.; James, D. G. *J. Econ. Entomol.* 1986, *79*, 1477-1483.
51. Flexner, J. L. Ph.D Thesis, Oregon State University, Corvallis, OR, 1988

RECEIVED October 20, 1989

FUNGICIDES

Chapter 11

Antiresistance Strategies

Design and Implementation in Practice

F. J. Schwinn[1] and H. V. Morton[2]

[1]Ciba-Geigy Corporation, CH—4002, Basel, Switzerland
[2]Ciba-Geigy Corporation, 410 Swing Road, Greensboro, NC 27419

>   Recent experience with fungicide resistance underlines
>   the need for the early development and implementation of
>   effective, realistic, and enforceable anti-resistance
>   strategies. These strategies should be developed by the
>   agrochemical industry in collaboration with extramural
>   partners. This partnership should also address the
>   manifold problems that remain to be solved to protect
>   the powerful and highly needed modern fungicides from
>   becoming obsolete due to broad resistance development.

Resistance of pathogen populations to fungicides is not only a
concern but also a threat to:
- the effective use and lifespan of modern fungicides;
- the economic interests of the farmer; and
- the image and reliability of chemical disease control.
   It is of the utmost importance to investigate at an early stage
in the development of a new product its inherent vulnerability to
fungal resistance. Once this is determined, strategies should be
designed to prevent or at least delay the development of resistant
strains in the field. The terms "anti-resistance strategies," or
"management of resistance" have come into usage for these endeavors.
These strategies may raise unrealistic expectations about the
solidity of the scientific evidence on which they are built. In
analyzing the short history of fungicide resistance, it becomes
evident that due to the fact that the phenomenon of resistance
developed so unexpectedly and quickly, short-term solutions were
urgently needed rather than having time to develop long-term strategic concepts. However, since the early 1980's, such concepts are
slowly emerging, with the corresponding endeavors clearly showing
the need for more scientific data.
   A brief review of the situation in the major groups of fungicides prone to resistance illustrates the difficulties.
   In the case of <u>pyrimidines</u> and <u>benzimidazoles</u>, the first groups
of fungicides against which resistance occurred quickly after market
introduction, to our knowledge the risk of resistance development
was not evaluated, nor was the phenomenon expected to show up. In

0097–6156/90/0421–0170$06.00/0
© 1990 American Chemical Society

retrospect, it can be said that the basic risk could have been predicted, based on in vitro/in vivo model studies. However, at the time of their introduction, the phenomenon of fungicide resistance was of no significance, and the technical advantages offered by the new products were so impressive that research focused primarily on performance and use recommendations. While resistance against pyrimidines developed quickly in all target pathogens, resistance against benzimidazole fungicides occurred to a widely varying degree in the different pathosystems and did not appear in others (1).

The history of phenylamide resistance (2) demonstrates to what extent the risk prediction depends on the experimental methods applied. Whereas the classical selection experiments did not indicate a risk (3), the use of chemical mutagens showed the contrary (4). In practice, problems emerged unexpectedly quickly, but - similar to the benzimidazoles - at different levels of resistance in the various pathosystems and regions (5). However, phenylamides remain valuable fungicides against many Oomycete diseases.

In the dicarboximides the situation is again different. In vitro and in vivo experiments easily yield resistant strains (6), but in practice some populations of the major pathogen, Botrytis cinerea, showed extreme fluctuations in sensitivity from year to year. Only recently has field resistance reached critical levels in some regions of intensive use (7).

To further complicate the picture, a brief look at the ergosterol biosynthesis inhibitors, particularly the demethylation inhibitors (DMI), shows their case is also different (8). Resistance is of a polygenic nature and, in general, it does not develop as quickly in the field as in the previously mentioned cases, although in model studies in the laboratory it was easily detected. For example, in the field, powdery mildew of small grain cereals is only very slowly shifting towards reduced sensitivity. However, strategic measures have already been introduced by way of mixtures in order to prevent further shifting. In contrast, powdery mildew of cucurbits has relatively quickly become insensitive to the DMI's.

Design of Anti-resistance Strategies

As explained above, in the earlier cases of fungicide resistance, decisions on strategies were made under time pressure, because problems requiring immediate actions emerged earlier than anticipated.

Since new types of fungicides will be developed in the future, a more structured approach should be used to allow a) earlier risk evaluation and b) the development of strategies on a more solid experimental basis. Chemical companies and researchers can learn some lessons from previous experience.

Risk Evaluation. The main factors influencing the resistance risk are shown in Table I. Characteristics and relative importance of the various different elements have been described by several authors (9, 10, 11, 12). Whereas management risk factors can be influenced, this is not true for the product- and pathogen-inherent factors. However, thorough and reliable evaluation of both product- and pathogen-related factors is needed to establish the true basis for the design of any realistic strategy. Needless to say, this is a difficult task.

Table I. Factors Influencing Resistance Risks After Staub and Sozzi (12)

Inherent Factors (Fungus Biology, Fungicide Chemistry)	Management Factors Fungicide Usage
• Biochemical mode of action • Fitness and population dynamics of resistant strains • Reproduction rate of target fungus and spore mobility • Duration of high disease pressure (climate)	• Duration of exposure (in generations) • Presence of other controlling factors (effective mixture partners, host resistance) • Size of target population, escape, overkill, (protective vs. curative use) • Proportion of crop area treated

Laboratory experiments on nutrient media, the use of mutagenic agents, the search for naturally occurring R-strains, studies on cross-resistance of R-strains to known fungicides, greenhouse testing on pathogenicity, and fitness and survival of R-strains selected in vitro or from natural sources are all tests which may indicate the probability of resistant strains occurring and their potential behavior in competition to wild-type sensitive strains. However, all such studies suffer from the fact that they have to be done with relatively small populations and numbers of isolates, and thus little genetic diversity. Further, they cannot include the impact of natural mixing with wild-type populations, or crop-related and weather-related factors. Therefore, the results of such studies need cautious interpretation. Only in cases where none of the above methods lead to the selection of a resistant strain can a clear conclusion be drawn, i.e., that the risk for resistance development is very low. In all probability this will be a rare case; on the other hand the fact that resistance does develop in such studies does not justify the exclusion of such a candidate fungicide from further development. This we can learn from all the aforementioned commercially successful systemic fungicide groups.

It has been stated repeatedly that field experiments are indispensable in order to define the risk under realistic conditions covering crucial elements like size of treated population, continuous mixes in the pathogen population by influx and efflux of spores, overwinter refugia, and impact of overkill, partial kill, and escapes on the population (11, 12). The problem here is with the risks derived from releasing resistant laboratory strains into the field, and/or the risk of selecting in such field experiments resistant strains, the dissemination of which cannot be excluded. Industry is definitely reluctant to run such risks.

Even if this risk factor could be managed and field experiments carried out, they may not yield typical results because the size and dynamics of the treated population are atypically small. This means that despite favorable results of such studies, resistance may still develop after broad commercial introduction of a new fungicide. Such situations arose many years after commercial introduction in the cases of Dodine for apple scab control, Kitazin P for rice blast control, and benomyl for eyespot control in cereals. However, negative findings in glasshouse and field experiments at least indicate that a build-up of resistance is unlikely to happen quickly, so that there is more time left for development and implementation of antiresistance strategies.

If early experiments show that a risk exists, the immediate development of anti-resistance strategies is of the utmost importance.

In conclusion, early risk evaluation is an absolutely essential feature or element in the development of new products. The range and sequence of tests to be done at this stage are subject to discussion and various approaches can be chosen. A first attempt to establish a structured risk evaluation scheme was made by Gisi and Staehle-Czech (13). This is an interesting and valuable approach towards a more quantitative risk estimation.

Design of Strategies. We have stressed the importance of early investigations of the inherent resistance risk factors because their results have a far-reaching impact on the strategy's design. As shown in Table I, the managerial factors are of similar importance. After evaluation of the risk factors, the design of strategies can be initiated. In this phase the following key elements have to be considered (Table II):

Table II. Key Elements of Anti-Resistance Strategies

---

1. Technical elements
   - Early evaluation of inherent risk during product development
   - Establishment of sensitivity baselines for each pathosystem and development of monitoring methods
   - Design to be based on product-, crop-, and pathogen-specific parameters
   - Detection and monitoring programs under conditions of practical product use

2. Managerial elements
   - Use recommendations (dosage, number of applications/season, duration of pathogen exposure, proportion of area treated)
   - Integration with other disease-suppressing methods
   - Enforceability in practice
   - Acceptance by companies, extension service, and users
   - Coordination with manufacturers of products with identical resistance pattern
   - Early implementation before resistance becomes a problem

---

Among the key technical elements is the establishment of sensitivity baselines, since they form the basis for all monitoring studies on sensitivity shifts. In past cases of rapid development of resistance, the lack of such figures has caused problems.

Monitoring techniques have been developed for a broad range of pathogens (14). They are useful tools to:
- check the occurrence of resistant subpopulations in the field before they cause damage to the crop;
- evaluate survival of resistant strains;
- track the progress of resistance with time; and
- eventually, check the value of resistance management strategies.

For further details, see Brent (15); Sozzi and Staub (16).

However, it is necessary to recognize the limitations of the methods. The methods developed so far, which are mainly based on sampling and in vivo sensitivity screening under laboratory conditions, are not very sensitive; they will only allow resistance detection when resistance has already reached a level of more than 1% in the population. Thus, in the case of a very dynamic pathogen, monitoring results may be too late for taking actions (17). Here the development of monoclonal antibodies or DNA probes may lead to much more sensitive and faster methods of detection.

It also has to be stressed that monitoring is only an auxiliary element of a strategy; it does not by itself prevent product failures and crop losses, but is only an indicator of shifts which helps to judge the validity of a chosen strategy. Last but not least, monitoring is an important tool in analyzing product failures and resistance rumors.

Integration of Technical Parameters and Management Elements. In contrast to the product- and pathogen-inherent factors which are beyond our control, there is a range of operational factors, as shown earlier in Table I. Their role and importance have been described (9, 12, 10, 11). Therefore they will not be addressed here in detail.

Instead we will elaborate on the specific difficulties in combining technical parameters and management elements into a solid and realistic strategy.

Whereas technical or inherent elements can be studied by experiments which allow, at least to some extent, reasonable extrapolations, it is difficult to scientifically evaluate the role and impact of management factors and also the value of product use strategies. Since the publication of the first mathematical model studies by Kable and Jeffery (18), several computer-based models for resistance management have been developed (9, 19, 20, 21, 22, 23, 24, 25). These models vary in their emphasis between theoretical and applied, and pathogen- and product-oriented aspects, and vary from the general to the specific. These are all useful, but their model character has to be underlined. The experimental validation of such models has only recently been addressed by such researchers as Staub and Sozzi (12); Sanders, et al. (26); Lalancette, et al. (27); and Milgroom and Frye (23).

To illustrate why largely pragmatic approaches had to be taken to date to develop a timely strategy, only a few factors which are subject to variation are used as a basis for the strategies, viz:

Reducing selection pressure:
- No excessive use rates
- Limit the number of applications per growing cycle

- Timing of applications to be aimed at a critical part of pathogenesis/epidemiology
- Mode of application should not favor extended exposure (e.g., soil application vs. foliar pathogen)

These strategies translate basically into two main tactics: (1) The exclusive use of <u>mixtures</u> of the product at risk with a strong, chemically unrelated companion product, and (2) The <u>alternating</u> use of straight products that are at risk and others of different mode of action.

There are pros and cons for the two tactics, as shown in Table III. In general the mixture strategy looks superior, particularly for products with high inherent risk. However, the judgment also varies with the inherent characteristics of the product at risk, the type of pathogen, growing pattern of the crop, climatic conditions, etc.

Table III. Comparative Merits of the Two Basic Anti-Resistance Use Strategies

Merit	Mixtures vs. Alternations
• Reduction of selection pressure	=
• Reduction of crop loss risk in case of resistance	>
• Potential for synergistic effects	>
• User compliance with strategy	>>
• Competitor compliance with strategy	>>
• Overall disease control	>

Adapted from Staub and Sozzi (<u>12</u>).

This can be illustrated by the <u>phenylamide</u> use strategy. It splits into two separate strategies for foliar and soil/seed-borne pathogens as showin in Table IV. Since introduction in 1981 these strategies have clearly helped to slow down development of resistance when they were introduced prior to its occurrence, and thus have maintained the high level of product performance and confidence of the user in the products' usefulness. Whereas the mixture strategy of FRAC (<u>5</u>) recommended full rates of phenylamides in mixtures, Sanders, <u>et al</u>. (<u>26</u>) obtained better results in terms of resistance-delaying efficacy at half rates.

Table IV. Elements of Phenylamide Anti-Resistance Strategy
After Urech and Staub (5)

A. Against foliar pathogens
- Prepacked mixtures with fungicides of different mode of action at high rate
- Application intervals no longer than 14 days in case of residual mixture partner
- Only two to four consecutive sprays, preferably early in the season
- Preventive but no curative use of mixtures
- No soil use against airborne diseases
- Integrated approaches

B. Against seed and soil-borne pathogens
- Against systemic pathogens use straight product in view of lack of suitable systemic mixture partners
- On perennial crops alternate with unrelated fungicide
- For seed-dressing against non-systemic pathogens use mixtures with conventional fungicides

Looking at the established anti-resistance strategies for the other major groups of systemic fungicides, most of them are also based on the use of prepack mixtures (Table V). However, they each take into account the differences in inherent risk levels of the various product groups. In the case of the phenylamides, with a proven high risk potential, the strategies need to be much more stringent and enforceable than in the case of the DMI's, where resistance has a multigenic base and apparently is developing slowly.

Table V. Anti-Resistance Strategies for Major Groups of Systemic Fungicides

Class	Mixtures	Alternat.	Limited Use	Other(s)
Benzimid-azoles	XX	X	Single treatment post-harvest	
Dicarb-oximides	XX	-	2 (-3) (for straight product)	
DMI Fungicides	XX	X	For single product	Alternation singly or in blocks Example: Bananas

Last but not least, we should like to address the importance of another element in the design of anti-resistance strategies, i.e., integration with other disease-controlling factors under the concept of Integrated Crop Production. Here careful consideration of crop management and - again - product use factors is crucial. Details are shown in Table VI (28). Much stronger efforts than in the past need to be made towards this goal.

Table VI. Elements of Anti-Resistance Strategies Utilizing Integrated Crop Production Practices

Crop Management		
• Resistant cultivars	-->	Lower disease levels
• Sites with low disease occurrence	-->	Reduce selection pressure
• Lower N fertilization	-->	Lower disease levels
• Sound sanitation	-->	Less inoculum

Fungicide Use		
• Based on forecasting	-->	Reduce selection pressure
• Protective use	-->	Hit small pathogen populations
• Mixtures	-->	Reduce risk of resistance selection
• Alternating use	-->	Reduce risk of resistance selection

Implementation of Anti-Resistance Strategies

The successful implementation of anti-resistance strategies and tactics requires:
- acceptance by in-house marketing and salesmen;
- acceptance by extension service, regulatory authorities, and users;
- coordination with manufacturers of products with identical resistance pattern; and
- continuous open communication among all parties involved.

As a first step, company researchers have to reach agreement with their product management and sales colleagues on the principles and elements of the strategies to be chosen for a given product at risk. It is easily conceivable that in cases like metalaxyl, due to its outstanding performance, strategy elements like limited number of applications or exclusive use of prepack mixtures against foliar pathogens were not easily accepted, particularly when resistance had not appeared in a given crop or region. It must be stressed that it is of crucial importance that a uniform, consistent strategy is applied in multinational companies.

Early communication with the extension service is necessary in order to achieve their support for the chosen strategy. The extension service can help in the early development of use instructions for the farmer and in education towards understanding the threat of resistance and the need for the strategic measures. Early coordination with competitors is another important element. Implementation and surveillance of anti-resistance strategies require constant and intensive communication within a complicated network, as shown in Table VII and as explained in detail by Urech (29). The difficulties in coordinating strategies among several competing companies should not be overlooked. Even though industry started early on to establish coordination groups under the guidance of the Fungicide Resistance Action Committee (FRAC) of GIFAP (30), it was often difficult to reach consensus, not only at the table but also at the sales front. It also needs to be mentioned that consensus between academic research/extension service personnel and commercial companies is not always easily reached. In certain situations, the views of the two parties may clearly differ, especially if researchers base their conclusions on theoretical or model studies with no clear proof under conditions of agricultural reality, and if industry researchers have to respect marketing considerations.

As indicated above, industry is not the only partner in the resistance management puzzle. Extension consultants, regulatory agencies, and pesticide users are also largely involved in designing and implementing strategies, as described by Urech (29), Johnson (31), Hawkins (32), and Frisbie, et al. (33). The roles of the various groups involved are shown in Table VIII. Industry sees an important role of regulatory agencies in assuring the availability of conventional multisite-inhibiting fungicides as companions, and in supporting the mixture strategy by approving adequate mixtures for use in their country (34, 35). The role of extension/consultants focuses on the education and and guidance of farmers so that they are motivated and supportive in following anti-resistance strategies.

Conclusions

In conclusion, the industry faces problems in connection with anti-resistance strategies as outlined in Table IX.

Beyond industry's problems in designing and implementing realistic anti-resistance strategies, there are clear research deficiencies in this field, as indicated earlier. They are summarized in Table X. We should like to stress the need for basic research on these topics, the results of which will be of great help for coping with future fungicide resistance problems.

In a more long-term perspective, strategies for lowering the risks of resistance and enlarging the spectrum of tools to cope with

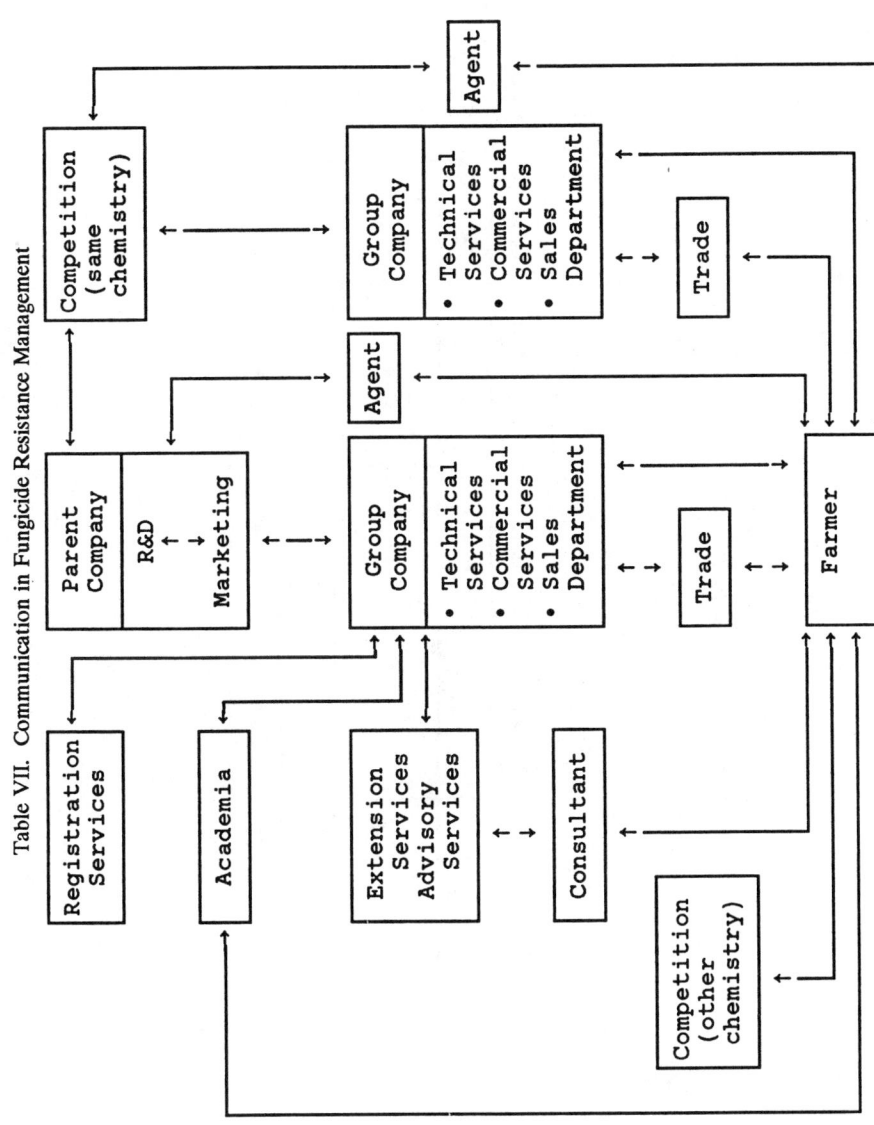

Table VII. Communication in Fungicide Resistance Management

Table VIII. Management of Fungicide Resistance in Practice

Active Group	Action				
	Determination of Inherent Resistance Risk	Risk Evaluation and Design of Resistance Strategy	Coordination Between Manufacturers	Implementation: Communication, Education	Monitoring	Adaption, New Design of Strategy
Industry	•••	•••	•••	•••	•••	•••
Academia Official Services	•	•		•••	•••	•
Farmers				•••	•	
	└─ Development Stage ─┘			└─ Marketing Stage ─┘		

SOURCE: Reprinted with permission from ref. 29. Copyright 1985 EPPO Bulletin.

it should include a range of additional factors, as shown in Table XI. Most of them point to areas of research which have received little attention so far and yet are of great importance. We hope that this meeting contributes to initiating work toward these objectives.

Table IX. Problems with Anti-Resistance Strategies: Problems from an Industry Viewpoint

- In view of low success rate, need to develop new products despite inherent resistance risks

- Risk of selecting resistant strains in field experiments or of releasing resistant strains produced in the laboratory during risk evaluation

- Technical requirements of strategies vs. marketing interests/priorities

- Technical and commercial availability of suitable companion fungicides

- Delay in knowledge about mode of action (single vs. multisite, single gene vs. multistep resistance mechanism)

- Need to establish a strategy before all key factors are known and prior to experimental validation of model studies

- Coordination of strategies with competitors, academia, and extension services.

Table X. Anti-Resistance Strategies: Problems and Prospects

- Present strategies have limited scientific basis
- Research needed on:
    - Dynamics of resistant field populations for major pathogens
    - Genetic basis and stability of resistance
    - Comparative field evaluation of alternative strategies
    - Dose-response in product mixtures
    - Quick, quantitative monitoring methods
    - Role of refugia
    - Validation of computer models under field conditions

Table XI. Anti-Resistance Strategies: Long-Term Aspects

- Search for systemic compounds with low inherent risk or negative cross-resistance including biorational approaches
- Systematically explore synergistic effects in mixtures
- Maintain availability of multisite-inhibiting conventional fungicides as companion products
- Investigate possibilities of safening agents against resistance
- Validate resistance models by practical experiments
- Develop and introduce integrated disease control programs in collaboration with plant breeders

Literature Cited

1. Smith, C. M. In *Fungicide Resistance, Research and Development Goals and Their Implementation in North America*; Delp, C. J., Ed.; American Phytopathology Society: St. Paul, Minnesota, 1988, pp. 23-24.
2. Schwinn, F. J., and Urech, P. A. In *Fungicide Chemistry. Advances and Practical Applications*; Green, M. B., and Spilker, D. A., Eds.; ACS Symposium Series 304, 89-106.
3. Staub, T. Dahmen, H., Urech, P., and Schwinn, F. *Plant Dis. Rep.* 1979, 63, 385-89.
4. Davidse, L. C. *Acta Bot. Neerl.* 1980, 29, 216 (Abstract).
5. Urech, P. A., and Staub, T. Bulletin OEPP/EPPO 1985, 15, 539-543.
6. Lorenz, G. In *Fungicide Resistance, Research and Development Goals and Their Implementation in North America*; Delp, C. J., Ed.; American Phytopathology Society: St. Paul, Minnesota, 1988, pp. 45-51.
7. Löcher, F. J., Lorenz, G., and Beetz, K. J. *Crop Prot.* 1987, 6, 139-147.
8. Köller, W., and Scheinpflug, H. *Plant Disease* 1987, 71, 1066-1074.
9. Delp, C. J. *Plant Disease* 1980, 64, 652-657.
10. Dekker, J. In *Fungicide Resistance in Crop Protection*; Dekker, J., and Georgopoulos, S. G., Eds.; Pudoc: Wageningen, 1982, pp. 177-186.
11. Dekker, J. In *Pesticide Resistance. Strategies and Tactics for Management*; National Research Council, Board of Agriculture, Ed.; National Academy Press: Washington, D.C., 1986, pp. 347-354.
12. Staub, T., and Sozzi, D. *Plant Disease* 1984, 68, 1026-1031.

13. Gisi, U., and Staehle-Czech, U. In Fungicide Resistance, Research and Development Goals and Their Implementation in North America; Delp, C. J., Ed.; American Phytopathology Society: St. Paul, Minnesota, 1988, pp. 101-106.
14. Georgopoulos, S. G. FAO Bulletin 1982, 30, 2, 36-77.
15. Brent, K. J. In Pesticide Resistance. Strategies and Tactics for Management; National Research Council, Board of Agriculture, Ed.; National Academy Press: Washington, D.C., 1986, pp. 298-312.
16. Sozzi, D., and Staub, T. Plant Disease 1981, 71, 422-425.
17. Staub, T., and Sozzi, D. Proc. 10th Internat. Congr. Pl. Prot., Brighton, England, 1983, pp. 591-598.
18. Kable, P. F., and Jeffery, H. Phytopathology 1980, 70, 8-12.
19. Josopovits, G., and Dobrovolszky, A. Pesticide Sci. 1985, 16, 17-22.
20. Levy, Y., Levy, R., and Cohen, Y. Phytopathology 1983, 73, 1475-1480.
21. Skylakakis, G. Phytopathology 1981, 71, 1119-1121.
22. Skylakakis, G. Crop Protection 1982, 1, 249-262.
23. Milgroom, M. G., and Fry, W. E. Phytopathology 1987, 77, 565-570.
24. Chin, K. M. Phytopathology 1987, 77, 666-669.
25. Arneson, P. A., Ticknow, B. E., and Sandlau, K. P. In Fungicide Resistance, Research and Development Goals and Their Implementation in North America; Delp, C. J., Ed.; American Phytopathology Society: St. Paul, Minnesota, 1988, pp. 107-109.
26. Sanders, P. L., Houser, W. J., Parish, P. J., and Cole, H. Jr. Plant Disease 1985, 69, 939-943.
27. Lalancette, N., Hickey, K. D., and Cole, H. Jr. Phytopathology 1987, 77, 86-91.
28. Staub, T. Parasitis '86 Conference Abstracts, Ergesa S.A. Geneve, Switzerland.
29. Urech, P. A. Bulletin OEPP/EPPO 1985, 15, 571-575.
30. Wade, M., and Delp, C. J. Bulletin OEPP/EPPO 1985, 15, 577-583.
31. Johnson, E. L. In Pesticide Resistance. Strategies and Tactics for Management; National Research Council, Board of Agriculture, Ed.; National Academy Press: Washington, D.C., 1986, pp. 393-402.
32. Hawkins, L. S. In Pesticide Resistance. Strategies and Tactics for Management; National Research Council, Board of Agriculture, Ed.; National Academy Press: Washington, D.C., 1986, pp. 403-409.
33. Frisbie, R. E., Weddle, P., and Dennehy, T. J. In Pesticide Resistance. Strategies and Tactics for Management; National Research Council, Board of Agriculture, Ed.; National Academy Press: Washington, D.C., 1986, pp. 410-421.
34. Schwinn, F. J., and Staub, T. In Modern Systemic Fungicides; Lyr, H., Ed.; Fischer: Jena, 1987; pp. 259-273.
35. Urech, P. A. In Fungicide Resistance, Research and Development Goals and Their Implementation in North America; Delp, C. J., Ed.; American Phytopathology Society: St. Paul, Minnesota, 1988, pp. 74-75.

RECEIVED October 24, 1989

Chapter 12

Sterol Biosynthesis Inhibitors

Model Studies with Respect to Modes of Action and Resistance

D. Berg, K.-H. Büchel, G. Holmwood, W. Krämer, and R. Pontzen

Bayer AG, Pflanzenschutz-Zentrum Monheim, D–5090 Leverkusen-Bayerwerk, Federal Republic of Germany

So far all mode of action studies on sterol biosynthesis inhibitors have been performed in model systems using yeasts or non-obligate fungal pathogens. We report on sterol biosynthesis in powdery mildew and compare these data to those from more artificial systems. With respect to molecular studies on fungal resistance to azoles we again have to rely on studies with non-obligate pathogens. Using triadimenol-resistant isolates of Ustilago avenae and Saccharomycopsis lipolytica, we looked for possible changes in uptake, detoxification and binding capacity of cytochrome P-450 as the target protein as compared to the wild types. No significant changes with respect to these parameters were observed. Major differences between resistant strain and wild type have been found with respect to fatty acid composition. These differences as well as possible consequences for the mechanism of resistance are discussed.

Almost two decades of research on the so-called "azoles" has yielded extensive knowledge of the chemistry, biochemistry and biological properties of this family of antifungal compounds ([1]). Originally the term "azoles" from chemical definition was restricted to imidazoles and 1,2,4-triazoles ([2]). In the meantime there has been a dramatic increase in knowledge of mechanisms of action, and, from a mechanistic viewpoint, fungicides from other chemical classes such as the pyridines and pyrimidines really have to be considered with the azoles ([3], [4]). Together they form the group of sterol biosynthesis inhibitors (SBI's), which interfere with C-14-demethylation by inhibiting the mono-oxygenase system involved ([5]). In the literature they are sometimes referred to as DMIs, an abbreviation for demethylation inhibitors.

Even though the mechanism of action has been intensively studied there is still cause for dissatisfaction among the biochemists working in this field. The studies mainly had to be performed with model microorganisms such as yeasts; the target organisms - mildews, rusts and scabs, for example - could not easily be studied directly. There have been several attempts to grow obligate pathogens in axenic culture - a certain degree of success has been reported in some cases of rusts (6) - but systematic studies have not been performed owing to the extremely long incubation periods and associated sterility problems.

Mode of action studies in model systems

One way out of this situation was to study model systems such as yeasts or non-obligate pathogens, particularly those that, at least in principle, could be combatted with azoles in any case (7). Such pathogens include Pyricularia oryzae, Botrytis cinerea and various Fusarium species. Botrytis is particularly significant in this respect (8); several azoles such as etaconazol, hexaconazole, flutriafol and tebuconazole show remarkable efficacy against Botrytis in vines and are at least under serious consideration as candidates for management of grey mould (9).

In principle the same is valid for rice blast, Pyricularia oryzae as shown in Figure 1. The efficacy of azoles against this fungus can be demonstrated in vitro, and hence rice blast can also serve as a model system. The data obtained with triadimenol and terbuconazole are typical. The formation of ergosterol is suppressed and there is a concomitant accumulation of sterol precursors. A trimethylsterol, 24-methylenedihydrolanosterol is the main precursor to accumulate; sometimes other 14-α-methylsterols such as obtusifoliol are also detected, depending on test organism and culture conditions. The data for triadimenol and tebuconazole show this accumulation, but tebuconazole, like other hydroxyethylazoles, produces an additional accumulation of Δ^5-sterols. Δ^5-Ergostenol, Δ^5-stigmastenol and $\Delta^{5,22}$-stigmastadienol, which are normally regarded as typical plant sterols, are also found (8).

With respect to their mechanism of action which is shown in Figure 2, both azoles inhibit ergosterol biosynthesis at the C-14-demethylation stage, as indicated by the accumulation of the trimethylsterol precursor. The cytochrome P-450 system involved in the oxidative demethylation of the 14-methyl group has been studied extensively, inter alia with isolated enzyme systems (10-14). Interaction of an azole with this system is quite well understood; there is a direct binding of the azole to the prosthetic group, i.e. to the Fe-porphyrin system. Accumulation of Δ^5-sterols is only observed in filamentous fungi, not in budding cells of yeasts (8). One possible explanation as shown in figure 2 is an additional interference with sterol synthesis by blocking of Δ^7 double bond formation (15). On the other hand a more unspecific, direct interaction of the azole with the fungal membrane lipids is also not unlikely and might also explain the differences between yeast budding cells and filamentous fungi. The administration of unspecific membrane effectors such as thiram (TMTD) also leads to an accumulation of these particular Δ^5-sterols (16), and may be regarded as support for the interpretation as an unspecific membrane effect in the case of tebuconazole.

sterol [%]	Botrytis cinerea			Pyricularia oryzae		
	control	triadimenol 10 ppm	HWG 1608 10 ppm	control	triadimenol 10 ppm	HWG 1608 10 ppm
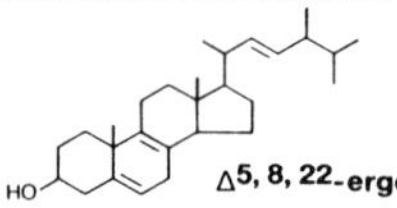 $\Delta^{5,8,22}$-ergostatrienol	5,7	–	–	5,2	–	–
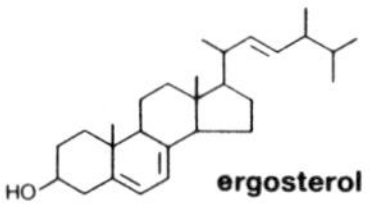 ergosterol	83,4	55,7	22,0	78,2	57,0	–
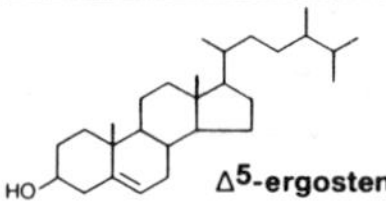 Δ^5-ergostenol	1,5	4,4	8,1	3,4	3,2	15,0
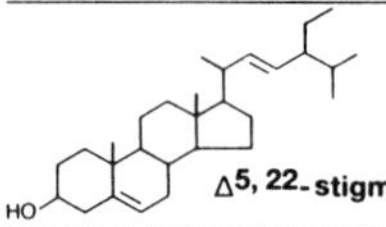 $\Delta^{5,22}$-stigmastadienol	2,8	2,5	6,9	4,2	8,2	11,2
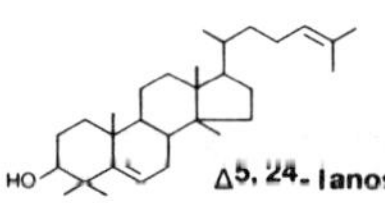 $\Delta^{5,24}$-lanostadienol	–	10,1	9,3	–	–	–
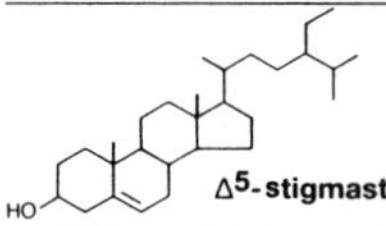 Δ^5-stigmastenol	4,5	6,3	25,8	8,9	12,2	45,7
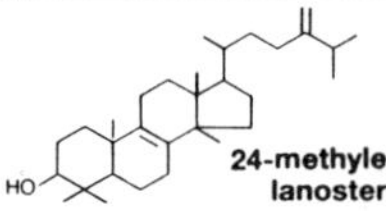 24-methylenedihydro-lanosterol	2,2	20,9	28,0	–	19,4	28,0

Figure 1. Effect of triadimenol and HWG 1608 (terbuconazole) on sterol biosynthesis in Botrytis cinerea and Pyricularia oryzae.

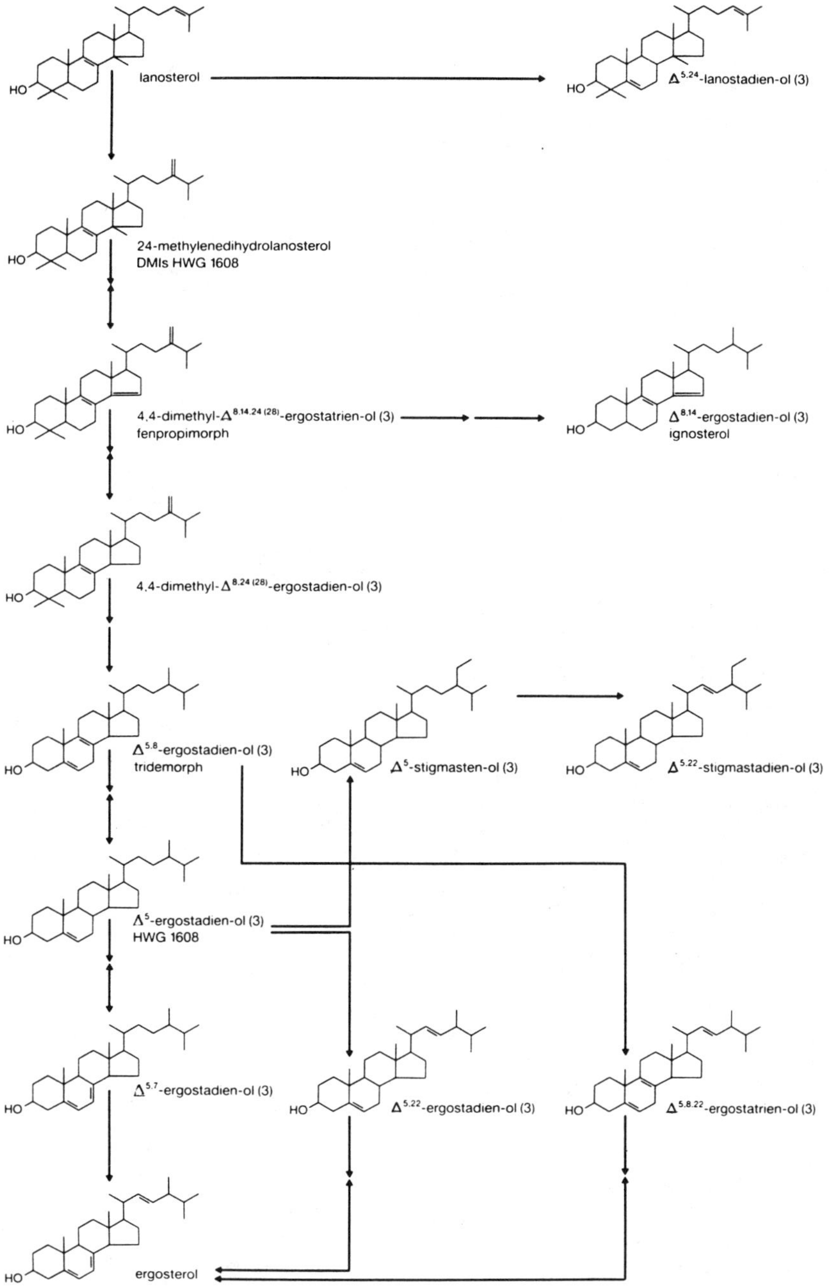

Figure 2. Mode of action of DMIs. (Reprinted with permission from ref. 8. Copyright 1987 Pflanzenschutz-Nachrichten Bayer)

Mode of action studies with powdery mildew

The unsatisfactory situation that the mechanism of action cannot easily be investigated directly on mildews, rusts or scabs has led to attempts to isolate mildew sterols, for instance, from the plant matrix. One of the first reports on mildew sterols came from Loeffler and Hollomon (17), who isolated $\Delta^{5,24(28)}$-ergostadienol from conidia of barley, apple and cucurbit powdery mildews as the end-product of sterol synthesis.

Actually there was some overlap in the work here as we had also isolated sterols from barley powdery mildew treated with an azole and from untreated control mildew, and confirmed that $\Delta^{5,24(28)}$-ergostadienol is the functional sterol, and not ergosterol. However, $\Delta^{5,24(28)}$-ergostadienol must also be regarded as a quasi-planar sterol, able to assume membrane functions just like ergosterol and the mammalian sterol, cholesterol. The main difference between the two experimental approaches was that Loeffler and Hollomon isolated sterols from conidia that were tapped from the leaves; we wanted to isolate the haustoria as well and therefore stripped off the plant epidermis, complete with fungus. The reasoning was that changes in the haustoria sterols might be more relevant for disease control than those in the conidia sterols. The disadvantage was, of course, that the epidermal matrix itself contains sterols, the typical plant sterols that have already been subtracted in figure 3. This naturally prevented us from looking for any possible accumulation of Δ^5-sterols following tebuconazole treatment.

As a result of azole treatment as shown in figure 3, two 14-α-methylsterols accumulated, namely the dimethylsterol, obtusifoliol, and the trimethylsterol 24-methylenedihydrolanosterol, with a concomitant reduction in $\Delta^{5,24(28)}$-ergostadienol.

In considering the mechanism of action of azoles on sterol biosynthesis in powdery mildew (see figure 4), we should discuss a variation in the biosynthetic sequence to the end-product. The occurrence of obtusifoliol and 24-methylenedihydrolanosterol following azole treatment confirms that the oxidative removal of the C-14-methyl group by the cytochrome P-450 system is also inhibited. The fact that side chain alkylation is performed prior to this step in powdery mildew seems worthy of note, reflecting as it does a principal difference to cholesterol biosynthesis in mammals.

A comparison of sterol formation in untreated barley powdery mildew with that in rice blast exposes as the major difference a variation in the structure of the functional membrane sterol. In the mildew sterol the Δ^7 double bond of ergosterol is absent and isomerisation of the $\Delta^{24(28)}$ double bond into the 22-position of the sidechain has not occurred. Nevertheless, the mildew sterol is a quasi-planar functional membrane component.

The mechanism of action of azoles in both organisms is identical in that C-14-demethylation is inhibited, as indicated by the accumulation of 14-α-methyl sterols.

Whether the observed accumulation of Δ^5-sterols in filamentous fungi following tebuconazole treatment has a parallel in mildew remains unanswered because interference from the plant matrix prevents analysis of these sterols. In any case, the studies on azoles in yeast and non-obligate fungi do reflect the in vivo situation, at least in barley powdery mildew.

Sterol (%)	control	KGW 0519 triadimenol		HWG 1608 terbuconazol	
		25 ppm	50 ppm	2 ppm	5 ppm
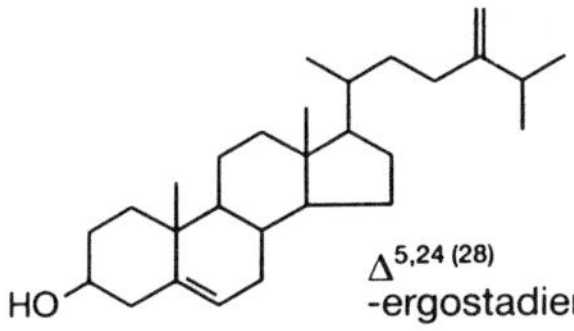 $\Delta^{5,24(28)}$-ergostadienol (3)	100	60.3	35.6	72.5	48.0
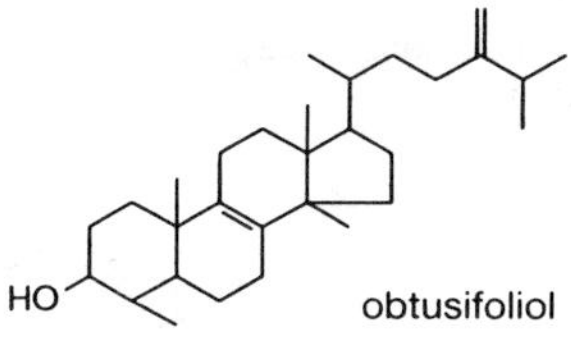 obtusifoliol	n.d.	18.1	28.7	5.5	13.7
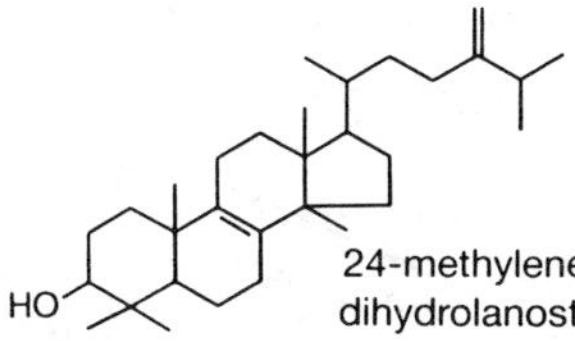 24-methylene-dihydrolanosterol	n.d.	21.6	35.7	22.0	38.3

Figure 3. Sterol distribution in barley powdery mildew (E. graminis f. sp. hordei) after azole application.

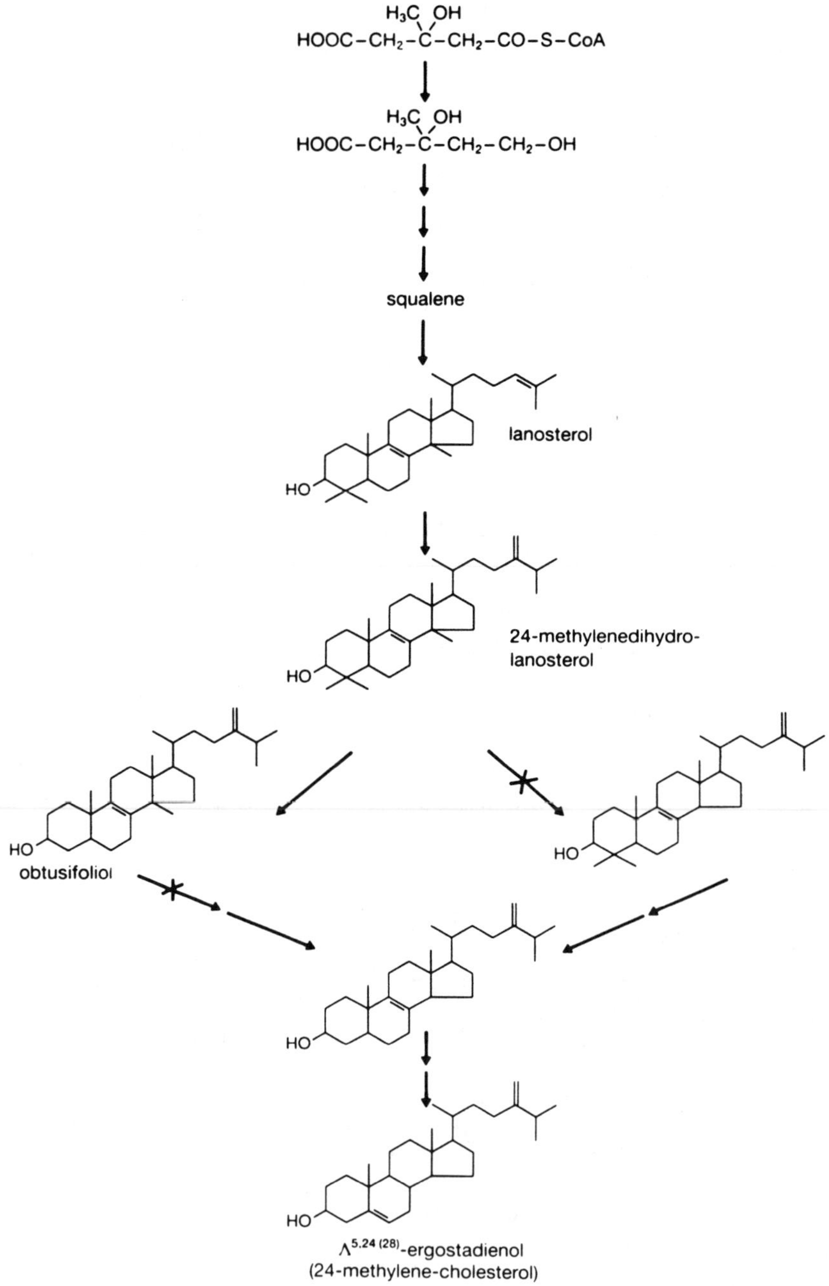

Figure 4. Mode of action of azoles on sterol biosynthesis in barley powdery mildew (E. graminis f. sp. hordei).

Even though the situation in powdery mildew is not completely identical with that in our model systems, results from non-obligate fungi may still be extrapolated, at least with respect to mechanism of action studies.

Model studies on resistance

The occurrence of the first field resistance phenomena of mildew towards azoles provoked the question of the molecular origin of this resistance. Again we had to trust in the reliability of our model systems, biochemical studies at the molecular level with resistant mildew strains being practically impossible.

Resistance and the rate at which it develops are obviously of great concern to fungicide users and producers. Skylakakis (18) has developed models for the rate of resistance development, based on two different concepts; these are shown in figure 5. He distinguishes between "disruptive selection" and "directional selection", each being characterised by typical Gaussian distribution curves for the ED-50 values of a pathotype. In the case of "disruptive selection" resistant pathotypes already exist from the start as an isolated minor population in the field prior to the use of a particular fungicide. The suppression of the sensitive strains leaves the way clear for the resistant population, which gains the upper hand after only a few treatments. In the case of "directional selection" there is a slow build-up of resistance; one observes a shift of the whole Gaussian distribution curve towards less sensitive strains.

These differences in the way in which fungal resistance develops towards a particular fungicide are thus also of crucial importance for decisions concerning chemical synthesis (19). If indications of "disruptive selection" are found during biological testing then the commercial risk involved in developing that compound must be regarded as much greater. In the case of azoles and powdery mildew resistance interpretation of the field monitoring data is that selection is occurring in a directional manner (20). The main biochemical origins of resistance (21) have to be studied to understand this phenomenon.

Mutation of the system transporting the fungicide into the target cell has been observed in the case of the polyoxins (22). This is an unusual case as the polyoxins normally undergo active transport into the fungal cell via a dipeptide transporter system. Mutation of this system confers resistance. Enhanced detoxification of the fungicide has been demonstrated in Kitazin-resistant mutants of rice blast (23); oxidation of the thiophosphate sulphur occurs. Incidentally, detoxification is the most common mode of resistance towards insecticides. A third origin of resistance, target mutation, has also been observed. A good example is resistance towards the benzimidazoles, which are specific inhibitors of tubulin polymerisation (24). Resistance due to target mutation is considered to be a particularly high risk with the so-called "single-site" effectors (25).

We thus had to investigate whether azole resistance can be explained by one of these mechanisms. Resistance to azoles can be produced quite easily in vitro, and we selected a triadimenol-resistant yeast strain by UV-irradiation on a nutrient solution containing triadimenol (8). The resistant isolate grew rapidly in the presence of a 100ppm suspension of a mixture of triadimenol diastereomers. The MIC value was shifted from about 5ppm for the susceptible isolate to above 1000ppm for the resistant one.

Uptake kinetics, detoxification, target mutation

The uptake kinetics of ^{14}C-labelled triadimenol, shown in figure 6 were examined in both the susceptible and triadimenol-resistant strains of the yeast Saccharomycopsis lipolytica. To minimize the effect of unspecific absorption on the cell surface the incubated cells were separated by suction filtration and washed twice with unlabelled triadimenol. The passive uptake curve shows the typical characteristics of saturation by diffusion, equilibrium being reached after about 20 minutes. There are no real differences between the susceptible and resistant isolates, and no indication of transport mutation or induction of efflux transport can be derived from these data.

The next possible mechanism to be examined was enhanced metabolic degradation of triadimenol by the resistant yeast isolate. The experimental approach is quite simple; the growing fungus was incubated with sublethal doses of radiolabelled triadimenol for various exposure times. Figure 7 shows the situation after incubation for 24 hours. Under these conditions both the susceptible and the resistant isolates degraded the triadimenol to a certain degree. The highly polar metabolite at the origin was itself degraded by glucosidase, yielding triadimenol again, so this metabolite should be a glucoside of triadimenol. It is important to note that there are no differences between the two isolates in the degree of conjugation nor is the diastereomer ratio significantly changed. There is no evidence here for different metabolic degradation rates as an explanation for resistance - at least in the case of S. lipolytica.

The third possible resistance mechanism was target mutation. Amino-acid sequencing or sequencing of the corresponding gene is not yet possible, and so we had to rely on estimates of affinity towards the prosthetic group in the microsomal cytochrome P-450 enzymes from the susceptible and resistant yeast isolates. Microsomal fractions with identical protein concentrations were obtained from the two isolates. The differential binding spectra of the P-450, see figure 8, with triadimenol showed no marked variations between the two sources. In addition, earlier studies (19) had shown that both isolates possess ergosterol as end-product, im somewhat decreased amount in the resistant strain. The sterol synthesis, however, in this case could not be inhibited by triadimenol.

To summarise, the typical resistance mechanisms one would expect, namely transport mutation, enhanced detoxification and target mutation, have been investigated in the case of Saccharomycopsis lipolytica without positive results.

Lipid composition

Various reports in the literature have shown that, following azole treatment, secondary changes in lipid composition have been observed relative to controls, especially with respect to fatty acids (26-28). Interestingly, free fatty acids have also been reported to

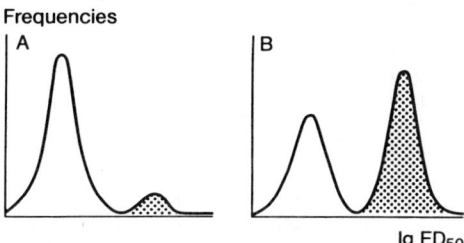

"Disruptive Selection"

A = distribution before application of the fungicide
B = distribution after prolonged selection

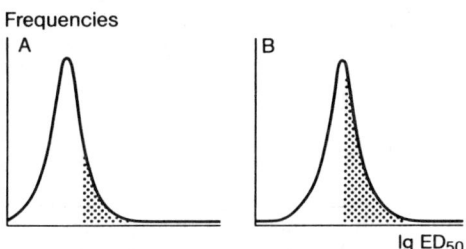

"Directional Selection"

A = distribution before application of the fungicide
B = "shifting" of the distribution of ED_{50}-values after fungicide application

Figure 5. Models for development of resistance.

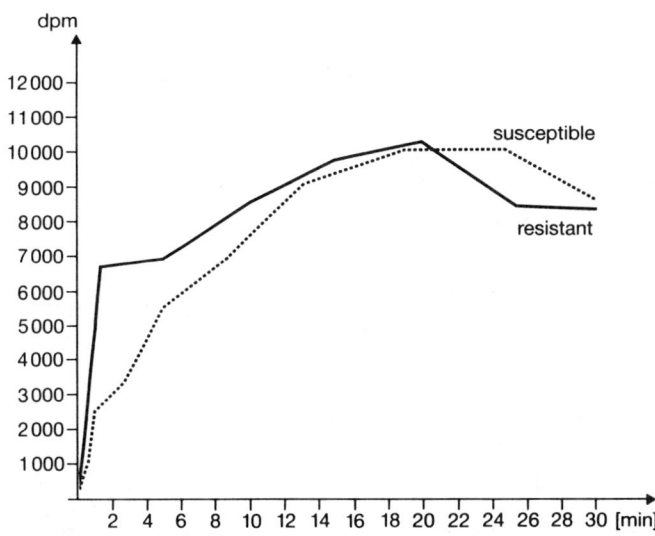

Figure 6. Uptake kinetics of (^{14}C)-triadimenol by susceptible and triadimenol-resistant Saccharomycopsis lipolytica.

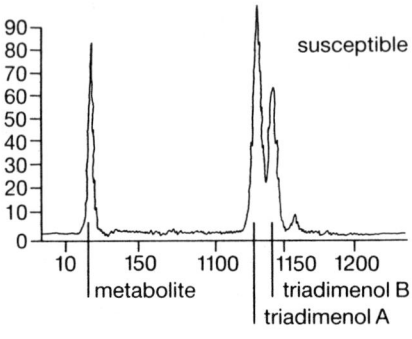
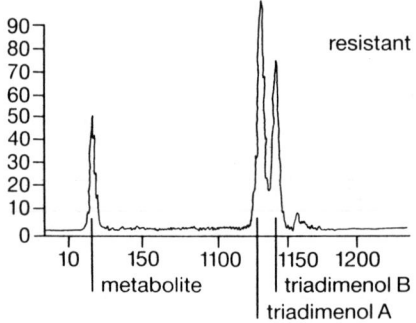

Figure 7. TLC-analysis of (^{14}C)-triadimenol after exposure to susceptible and resistant S. lipolytica.

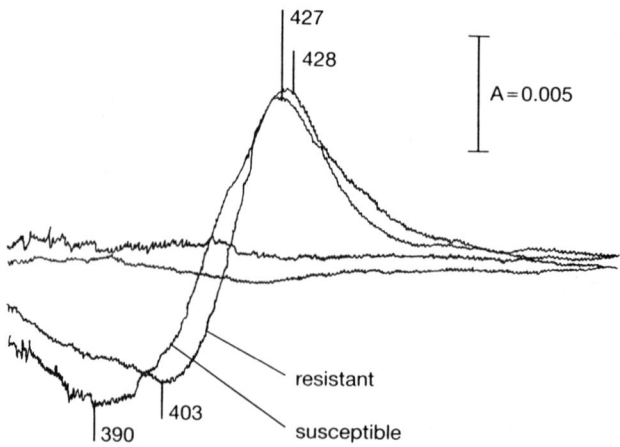

Figure 8. Triadimenol binding spectra of microsomal cytochrome P-450 from susceptible and resistant S. lipolytica.

influence azole efficacy in vitro (29,30). We therefore compared the fatty acid distribution in the susceptible and resistant Saccharomycopsis isolates and were surprised to find a major difference. The resistant isolate contained an increased amount of 18:1 oleic acid; a 25% increase in relative terms (see Table I).

Table I. Fatty Acid Analysis of susceptible and Triadimenol-resistant S. lipolytica

FA (%)	S. lipolytica	
	susceptible	resistant
16:1	15.0	9.2
16:0	8.9	9.8
18:2	33.0	27.5
18:1	43.1	53.5

This result caused us to repeat an old experiment that had already been published in 1976/77 by various authors (29,30), namely to study the reversion of azole efficacy by fatty acids in general and oleic acid in particular (Table II).

Table II. Influence of Oleic Acid on Efficacy of Triadimenol towards S. lipolytica (susceptible)

	diameter inhibition zone (mm)							MIC (ppm)
	1000	500	250	125	62.5	31.3	15.6	
oleic acid	-*	-	-	-	-	-	-	>1000
triadimenol	36	34	32	28	25	20	17	3.2
triadimenol + 1000ppm oleic acid	34	30	26	22	16	-	-	21.4

* no inhibition zone could be detected

In fact oleic acid is able to antagonize the efficacy of triadimenol against Saccharomycopsis. Among a larger set of fatty acids, oleic acid was also the most effective antagonist. We had the impression that the changes in fatty acid content could perhaps be more than a mere indicator to differentiate between susceptible and resistant isolates. If the increase in the 18:1 fatty acid content is **causally** related to resistance then inhibitors of fatty acid biosynthesis should restore azole efficacy (Table III).

Table III. Cooperative Effect of Cerulenin and Triadimenol towards S. lipolytica (susceptible)

	diameter inhibition zone (mm) (ppm)						MIC (ppm)
	62.5	31.3	15.6	7.8	3.9	1.9	
triadimenol	23	18	14	-	-	-	8.2
cerulenin	24	19	13	-	-	-	10.2
triadimenol + 7.8ppm cerulenin	24	20	17	14	11	-	3.4

As a standard inhibitor we chose cerulenin, a secondary metabolite of <u>Cephalosporium caeruleus</u>, which is itself fungicidally active and is known to interfere with fatty acid biosynthesis (31) as well as HMG-CoA-synthase (32). Mevalonate formation on the level of HMG-CoA-synthase and HMG-CoA-reduction, however, to our knowledge is a relatively poor fungicidal target (33). Nevertheless the potency of cerulenin to interfere with early steps of terpenoid synthesis complicates interpretation.

Cerulenin:

[structure of cerulenin]

A more than additive effect was observed when cerulenin and triadimenol were tested together against the <u>Saccharomycopsis</u> isolates. The results are shown in table III.
We have to regard this as the starting point for a more detailed understanding of resistance towards azoles. Possibly the increase in 18:1 fatty acid levels leads to a direct interaction between oleic acid and triadimenol. Lipid binding of the triadimenol might cause a corresponding decrease of azole at the target enzyme.

Conclusions
In conclusion it may be stated that experiments with barley powdery mildew have confirmed the validity of mechanism of action studies with azoles on yeasts and non-obligate fungi. Even though there is a structural difference in the end-product of sterol biosynthesis, this does not affect the concept of C-14-demethylation inhibition.
Resistance towards azoles has been observed in the field particularly in the case of powdery mildews. Field monitoring has shown that the development of resistance proceeds by directional rather than disruptive selection.
Studies have been performed on an isolate of the yeast <u>S. lipolytica</u> with artificially induced resistance as a model system. No evidence has been found that uptake kinetics, enhanced metabolic degradation or target mutation are responsible for resistance, but changes in fatty acid composition have been observed.
The data presented show that oleic acid has an antagonistic effect upon triadimenol but that cerulenin has a synergistic effect.
For the immediate future, experiments have to be aimed at answering the question whether variations in fatty acid composition do cause changes in intracellular azole distribution, resulting in decreased availability at the target site. And again the situation in powdery mildew has to be studied in parallel.
In principle inhibitors of fatty acid biosynthesis could enhance or restore azole efficacy, and if binding of triadimenol to fatty acids is the reason for resistance development then even compounds with an identical mechanism of action should not necessarily show cross-resistance provided that their physio-chemical properties such as lipophilicity are sufficiently different.

Literature Cited

1. Vanden Bossche, H., In *Sterol Biosynthesis Inhibitors: Pharmaceutical and Agrochemical Aspects*; Berg, D., Plempel, M., Eds.; Ellis Horwood, VCH-Verlagsgesellschaft: London, 1988; p 79-119.
2. Worthington, P., *ibid.*, 19-55.
3. Ragsdale, N.N.; Sisler, H.D. *Pestic. Biochem. Physiol.* 1973, 3, 20-29.
4. Buchenauer, H. *Z. Pflanzenkrankheiten Pflanzenschutz* 1977, 84, 286-299.
5. Wiggins, T.E.; Baldwin, B.C. *Pestic. Sci.* 1984, 15, 206-9.
6. Kuck, K.H.; Reisener, H.J. *Physiol. Plant Pathol.* 1985, 27, 259-268.
7. Berg, D., In *Fungicide Chemistry - Advances and Practical Applications*; Green, M.B., Spilker, D.A., Eds.; ACS-Symposium Series, 1986; 304, 25-51.
8. Berg, D.; Born, L.; Büchel, K.-H.; Holmwood G.; Kaulen, J. *Pflanzenschutz-Nachrichten Bayer* 1987, 40 (58), 111-132.
9. Kaspers, H.; Brandes, W.; Scheinpflug, H. *Pflanzenschutz-Nachrichten Bayer* 1987, 40 (58), 81-110.
10. Yoshida, T.; Aoyama, Y. In *In vitro and in vivo Evaluation of Antifungal Agents*; Iwata, K.; Vanden Bossche, H., Eds.; Elsevier Science Publishers: Amsterdam, 1986, p 123-134.
11. Sanglard, D.; Käppeli, O.; Fiechter, A. *Archiv. Biochem. Biophys.* 1986, 251, 276-286.
12. Vanden Bossche, H. In *Recent Trends in the Discovery, Development and Evaluation of Antifungal Agents*; Fromtling, R.A., Ed.; J.R. Prous, Science Publishers: Barcelona, 1987, p 207-221.
13. Wiggins, T.E.; Baldwin, B.C. *Biochem. Soc. Trans.* 1983, 11, 712.
14. Yoshida, Y.; Aoyama, Y.; Kamaoka, H.; Kubota, S. *Biochem. Biophys. Res. Commun.* 1977, 78, 1005-1010.
15. Berg, D.; Büchel, K.-H.; Plempel, M.; Zywietz, A. *Mykosen* 1986, 29 (5), 221-229.
16. Berg, D.; Plempel, M.; Büchel, K.-H.; Holmwood, G.; Stroech, K. *Ann. N.Y. Acad. Sci* 1988, 544, 338-347.
17. Loeffler, R.S.T.; Butters, J.A.; Hollomon, D.W. *Brit. Crop Protect. Conf. Proceeding* 1984, Vol. 3, 911-916.
18. Skylakakis, G. *Crop Protection* 1987, 1, 249-262.
19. Krämer, W.; Berg, D.; Köller, W. In *Combating Resistance to Xenobiotics. Biological and Chemical Approaches*; Ford, M.G.; Hollomon, D.W.; Klambag, B.C.S; Sawicki, R.M., Eds.; Ellis Horwood Ltd.: Chichester, 1987, p 291-305.
20. Köller, W.; Scheinpflug, H. *Plant Disease* 1987, 71 (2), 1066-1074.
21. Dekker, J. In *Progress in Pesticide Biochemistry and Toxicology*; Hutson,D.A.; Roberts, T.R., Eds.; 1985, Vol. 4, 165-218.
22. Misato, T.; Kakiki, K.; Hori, M. *Neth. J. Plant Pathol.* 1977, 83, Suppl. 1, 253-260.
23. Uesugi, Y.; Katagiri, M. In *Pesticide Human Welfare and the Environment*; Miyamoto, Y.; Kearney, P.C., Eds.; Proc. 5th IUPAC Congr. Pestic. Chem.: Pergamon, Oxford, 1983; Vol. 3, 165-170.
24. Davidse, L.C. In *Fungicide Resistance in Crop Protection*; Dekker, J.; Georgopoulos, S.G., Eds.; Pudoc: Wageningen 1982; 60-70.

25. Dekker, J.; ibid.; 128-138
26. Sisler, H.D.; Ragsdale, N.N. Neth. J. Plant Pathol. 1977, 83, Suppl. 1, 81-91.
27. Leroux, P.; Gredt, M.; Fritz, R. Phytiatric.-Phytopharm. 1976, 25, 317-334.
28. Weete, J.D.; Sancholle, M.S.; Montant, C. Biochem. Biophys. Acta 1983, 752, 19-29.
29. Yamaguchi, H. Antimicrob. Agents Chemother. 1977, 12, 16-25.
30. Georgopapadakou, N.H.; Dix, B.A.; Smith, S.A.; Freudenberger, J.; Funke, P.T. Antimicrob. Agents Chemother. 1987, 31, 46-51.
31. Vance, D. Biochem. Biophys. Res. Commun. 1972, 48, 649.
32. Omura, S. J. Biochem. 1972, 71, 783-796
33. Berg, D.; Plempel, M. Biotech-Forum, in press

RECEIVED September 1, 1989

Chapter 13

Resistance to Sterol Biosynthesis-Inhibiting Fungicides

Current Status and Biochemical Basis

D. W. Hollomon, J. A. Butters, and J. A. Hargreaves

Department of Agricultural Sciences, University of Bristol, AFRC Institute of Arable Crops Research, Long Ashton Research Station, Long Ashton, Bristol BS18 9AF, United Kingdom

Inhibition of sterol biosynthesis has proved a fertile area for discovery of broad spectrum fungicides (SBI's), many of which are active against foliar diseases of small grain cereals, and especially against powdery mildews. Two groups of SBI's are important agriculturally. The morpholine group inhibit Δ^8-Δ^7 isomerase or $\Delta^{14(15)}$ reductase, and have not encountered resistance problems despite twenty years intensive use. The second group of diverse chemicals all inhibit sterol 14 α demethylase (DMI's), and although resistant mutants have been generated to DMI fungicides in the laboratory in many different fungi, resistance has only occurred in practice in five cases, mainly involving powdery mildews. Genetic analysis of both laboratory and field resistant mutants suggests that DMI resistance is controlled by many independent genes, and the biochemical data reflect this, pointing to several posssible mechanisms of resistance. Failure to accumulate sufficient fungicide, tolerance of abnormal sterols, or a lack of sterol 14 α demethylase, are just a few of the resistance mechanisms considered. Uncertainty surrounds the possibility that changes in sterol 14 α demethylase may alter fungicide or substrate binding, but definitive data on this may be obtained through cloning and sequencing of the demethylase. Sufficient homology exists between a sterol 14 α demethylase probe from <u>Saccharomyces cerevisiae</u> and <u>Erysiphe graminis</u> DNA, and we have identified a 3.0 KB fragment of <u>Eyrsiphe graminis</u> f.sp. <u>hordei</u> DNA, which appears to contain the demethylase sequence.

Sterols are important membrane lipids in Eukaryotes. Various functions have been identified for sterols within the membrane (1), but in general a planar molecule with a 3-hydroxyl group is required to modulate movement of phospholipids, to ensure correct membrane fluidity, and to control both permeability and the activity of at least some membrane bound enzymes. In addition, sterols may have a hormonal (so-called "sparking") function, which can only be satisfied by Δ^5 sterols (2). The biosynthesis of sterols from lanosterol involves a matrix of reactions (3), and although the chemical changes are limited (demethylation, desaturation, isomerization, reduction), different enzymes generally catalyse each step in this matrix. Consequently, there is ample scope for specifity in this region of the pathway where most sterol biosynthesis inhibiting (SBI) fungicides act; for example, inhibition of oxidative demethylation of the C-14 methyl group of lanosterol in fungi does not generally interfere with the corresponding step in plant sterol biosynthesis, where obtusifoliol is the substrate for this demethylation.

The last twenty years has witnessed the successful development of a number of fungicides that inhibit different steps in the sterol biosynthesis pathway (Table I).

Table I. Sites of Action of SBI Fungicides

Fungicides	Target site	Reference
Allylamines	Squalene epoxidase	8
Triazoles, imidazoles, pyrimidines, pyridines, piperazines,	Lanosterol C-14 demethylase	10
Triazoles, pyrimidines	C_7-22 desaturase	4
Ethylazoles, pyrimidines	Δ^7_8 reduction	5
Tridemorph, fenpropimorph	$\Delta^8\text{-}\Delta^7_{14(15)}$ isomerase	7, 36
Fenpropimorph, fenpropidin	$\Delta^{14(15)}$ reductase	7, 36

Twenty-six different SBI fungicides are currently used in agriculture and horticulture, and these involve 150 different products. By far the major target are diseases of small grain cereals (30m ha treated annually; Williams, R.J., Ciba-Geigy Basle, personal communication, 1988) followed by vine diseases (2.2m ha treated) and many other crops. More than 75% SBI fungicides are used in Europe, although as a result of new registrations, use is expanding in other areas such as North America. Inhibition of sterol C-14 α demethylase (14DM) is the target site for a diverse, but widely used, group of DeMethylation Inhibiting (DMI) fungicides which generally are systemic, and have a broad spectrum of activity. Some of these fungicides may inhibit other steps in the pathway, such as C-22 desaturase (4) or Δ^7 reductase (5,6), although more detailed work is needed in some instances to confirm these observations. DMI fungicides account for at least 75% of all SBI's used. A second group of SBI fungicides, the morpholines and 3-phenylpropylamines, are especially active against cereal powdery mildews, but seem to block more than one step in the pathway. They may inhibit $\Delta^8-\Delta^7$ isomerase, $\Delta^{14(15)}$ reductase, or both depending on the target fungus (7). A third SBI group, the allylamines, inhibit squalene epoxidase (8) but although these reactive compounds have been successfully used as medical antimycotics, none are yet used in agriculture.

Despite a good understanding of the site of action of SBI fungicides, no clear reason has emerged as to why fungal growth is stopped. Methylated sterols are apparently not toxic since mutants lacking 14DM activity accumulate these sterols, but can grow normally under certain conditions (9). Mucor rouxii accumulates methylated sterols after propiconazole or etaconazole treatment, but is apparently unaffected by these fungicides (10). In many cases SBI fungicides reduce the total amount of sterols, and this may restrict membrane synthesis. Changes in membrane sterols will affect fluidity and alter the integrity of membranes which may become leaky. Activity of membrane bound enzymes, such as chitin synthase, may be altered (11), and it is perhaps these effects that limit growth.

Current Status of Practical Resistance to SBI Fungicides

Some variation in sensitivity to SBI fungicides seems inevitable in all target pathogens. The extent of this variation, and whether monitoring, mutation or recombination can expose it, depends on the pathogen, and the sensitivity of the assay system available. It is perhaps not too surprising that resistance to all SBI fungicides has been identified in at least some fungi, either in laboratory or field studies. The practical significance of any variation will depend on many factors, including its association with relative fitness, which are discussed elsewhere in this volume.

Despite early indications to the contrary (13), there are now five plant pathogens against which DMI fungicides have failed to sustain adequate control (Table II), because of resistance. In three other diseases, available information suggests that significant shifts towards less sensitive populations has occurred following repeated use of DMI fungicides (14). As yet these shifts can not be correlated with decreased performance, but similar situations were encountered when the first signs of DMI resistance emerged in cereal powdery mildews. Field resistance to DMI's was first detected in Erysiphe graminis f.sp. hordei in the UK in 1981 (15), some three years, or 90 mildew generations, after the first use of

triadimefon. Since then, gradual evolution towards higher
levels of resistance, and their greater frequency within
populations, has eroded the effectiveness of all current DMI
fungicides in regions where their use remains high. Elsewhere
mildew populations have remained more sensitive, and effective
disease control is still possible. A similar pattern of change
has occurred in wheat powdery mildew in the Netherlands (16).

Table II. Resistance to DMI fungicides

Pathogen	Year resistance first reported	Crop	Reference
Resistance in practice			
Erysiphe graminis f. sp. hordei	1982	barley	15
Sphaerotheca fuliginea	1982	cucurbits	21
Pyrenophora teres	1985	barley	52
Erysiphe graminis f. sp. tritici	1986	wheat	16
Penicillium digitatum	1987	citrus	53
Field resistance Significant shifts in sensitivity not yet leading to loss of control			
Rhynchosporium secalis	1986	barley	14
Uncinula necator	1986	apple	14
Venturia inaequalis	1987	grape	14

Cross resistance generally extends to all DMI fungicides,
although "resistance factors" may not be identical for all DMI's
in all fungal strains. No cross resistance was observed to
morpholine fungicides, and to the non-SBI fungicide, ethirimol,
evidence of negatively correlated cross resistance was reported
in several studies (17, 18, 19). These observations on cross
resistance patterns form the basis of strategies to combat
further spread of DMI resistance, which use mixtures of a DMI
fungicide with either a morpholine or a hydroxypyrimidine
fungicide.

Resistance to DMI fungicides has also emerged gradually in
cucurbit powdery mildew (Sphaerotheca fuliginea), and has been
identified as the cause of poor performance, especially on
indoor crops (20, 21). Any lack of fitness associated with
resistance is less significant in glasshouses and polythene
tunnels, since the enclosed environment largely excludes fitter,
more sensitive strains. Similarly, in citrus packing sheds
Penicillium digitatum resistant to imazalil has caused control
difficulties (53, Eckert, J.N., University of California,
Riverside, personal communication), despite the lower relative

fitness of these strains (van Gestel, Janssen Pharmaceuticals, personal communication, 1988).

Although one report identified field isolates of E. graminis f.sp. hordei resistant to tridemorph (22) no practical case of resistance to morpholine fungicides has emerged in any disease. This is remarkable given the specificity in their mode of action, and their continuous and increasing use against cereal mildews since tridemorph was introduced in 1970.

Genetics of Resistance to DMI Fungicides

Once stable variation in sensitivity to SBI fungicides was identified, either in field strains or laboratory induced mutants, genetic analysis was possible. Only a few studies, however, have attempted to unravel the genetic basis for control of DMI resistance, partly because of the difficulties of genetic analysis with obligate powdery mildews. No attempt seems to have been made to analyse genetically morpholine resistant mutants of Ustilago maydis that have been generated in laboratory studies. From a study of 21 u/v induced mutants of Aspergillus nidulans selected for resistance to imazalil, resistance was shown to be multigenic (23). Eight different genes were identified, one having 11 different alleles, giving an eight to ten-fold increase in resistance. Two additional genes were identified in other work (cited in 23), so that 10 unlinked genes were recognised as contributing to DMI resistance. The effects of two of these genes, ima A and ima B, together with a modifier gene of ima A, were additive. Pleiotropic effects were associated with some of these genes, which also conferred increased sensitivity to cycloheximide, indicating that they were possibly permeability mutants. Similar low levels of resistance to DMI fungicides in Venturia inaequalis were found to be controlled by a single gene (24). It is not yet known how many genes are involved in the higher levels of DMI resistance now present in field isolates of the apple scab pathogen.

In obligate powdery mildews, fungicide assays are more variable, and multigenic resistance is likely to be blurred into continuous variation, and equated with polygenes. The distinction between multigenic and polygenic is, in any case, a poor one (25), but DMI resistance in barley powdery mildew is not simply controlled by one or two genes. Crosses between two DMI sensitive barley mildew strains yielded continuous variation in the progeny (26), with some progeny distinctly less sensitive than either parent, indicating that they apparently did not have the same genetic architecture. Other crosses between DMI resistant and sensitive strains confirmed the polygenic nature of resistance.

All available genetic evidence points, therefore, to more than one mechanism of resistance operating against DMI fungicides. Whilst it is possible to construct A. nidulans strains with readily identified single gene resistance, in order to study the biochemical mechanism of resistance, similar constuction of strains with a single identifiable resistance gene is not yet possible in powdery mildews.

Mechanisms of Resistance to DMI Fungicides

Decreased Fungicide Uptake. Genetic analysis of DMI resistant mutants of A. nidulans paved the way for studies in the biochemical mechanisms involved (27), and these have been extended to include resistant Penicillium italicum mutants (28). Less fungicide accumulated within resistant cells of both fungi than in sensitive cells, and consequently insufficient fungicide reached the site of action to exert a fungistatic effect. Dissection of the accumulation process into passive influx, and an energy dependent efflux, revealed that efflux was constitutative in resistant mutants, but needed to be induced in wild-type strains. Efflux was induced by many, seemingly unrelated toxicants including some SBI fungicides. The process was inhibited by ATP-ase inhibitors, which synergised the action of DMI's against resistant strains. It was suggested that efflux of DMI's is mediated by an ATP-dependent proton gradient across the fungal membrane, and this is more efficient in resistant strains. In other experiments, accumulation of imazalil sulphate was similar in both resistant and sensitive strains (29), although this may be because imazalil is likely to be protonated at physiological pH (29,30). Reduced uptake was also implicated as the mechanism of resistance to ketoconazole in clinical isolates of Candida albicans (31), although whether "active efflux" was involved was not established.

Fungicide Transfer to Barley Powdery Mildew. Similar experiments with powdery mildews are not possible, but we have attempted to measure the transfer of C^{14} triadimenol from barley to mildew strains differing in their sensitivity to DMI fungicides. A total of 1.75 µCi of $[C^{14}_3]$ triadimenol (sp. act. 6.5 µCi/µmol) and 70 µCi of L-[methylH^3] methionine (sp. act. 75 curies/mM) were taken up by 40 heavily infected 6.0 cm long leaves for 4 h. at 20°C in daylight. Radio-labelled triadimenol was kindly prepared by Dr. K. Chamberlain (Rothamsted Experimental Station) and it was a 50 : 50 mixture of the two diastereomers. Mycelium and conidia were removed from the apical 4.0 cm of each leaf with nail varnish, which was then dissolved in 60 : 40 (v/v) ether : ethanol. Insoluble material was washed twice with ethanol : acetic acid and the supernatants combined with the ether : ethanol fraction. Insoluble material was dried, digested overnight in N-Chlorosuccinimide at 40°C, and bleached with benzoyl peroxide in toluene (51), and radioactivity counted by liquid scintillation spectrometry. Treatment of this insoluble residue with proteinase rendered all radioactivity soluble in 10% trichloroacetic acid, indicating that methionine was incorporated into protein.

The ether : ethanol fraction was reduced in volume to
0.5 ml, and ether (4.0 ml) : water (3.0 ml, pH 4.0) added to
precipitate the nail varnish. The ether supernatant was removed
and washed twice with water. This water fraction contained no
significant amounts of [C^{14}] radioactivity. After determining
the [C^{14}] radioactivity in the ether fraction, the remainder was
dried down, chromatographed on Silica Gel 60 F254 zone
concentrating plates (Merck 13794) using hexane, ethylacetate,
acetic acid (9 : 4 : 1) and autoradiographed to detect possible
triadimenol metabolites.

To allow for inevitable differences in infection levels
between experiments, transfer was expressed relative to the
incorporation of H^3 methionine into trichloroacetic acid
insoluble material (= protein), which provided a measure of
mildew growth. This technique only accounted for radioactivity
in conidia and surface mycelium, not haustoria, but significant
differences in the transfer of triadimenol were apparent between
strains (Table III). These differences were not, however,
correlated with triadimenol sensitivity, so that if reduced
accumulation is involved in DMI resistance in barley powdery
mildew, it must not operate in all resistant strains. In the
same experiments we looked at the possibility that
detoxification of triadimenol might account for resistance, but
were unable to find any evidence of triadimenol metabolites in
the six strains examined.

Table III. Transfer of C^{14} triadimenol from barley to
Erysiphe graminis f.sp. hordei strains differing in
sensitivity to DMI fungicides

Strain	Triadimenol Sensitivity (ED50 µg/ml)	C^{14} Triadimenol accumulation*
DH49	0.002 ± 0.0002	0.66 ± 0.01
DH14	0.004 ± 0.0006	1.52 ± 0.54
23D5	0.061 ± 0.008	1.37 ± 0.60
JB115	0.391 ± 0.042	0.18 ± 0.10
JB152	1.034 ± 0.108	1.03 ± 0.26
JB214	1.380 ± 0.140	0.44 ± 0.28

* expressed as total incorporation of C_3^{14} triadimenol into
mildew relative to incorporation of H^3 methionine into mildew
protein. Values are derived from at least two separate
determinants for each strain.

Sterol Changes and Resistance. DMI-resistant mutants of
U. maydis lacking 14DM have been isolated (32). These mutants,
which were initially selected for polyene resistance, have no
detectable demethyl sterols and grow more slowly than wild-type

strains. Attempts to select similar mutants in filamentous
fungi have failed (26, 27, 33), and so it seems unlikely that a
similar mechanism would operate in many plant diseases. Indeed,
other DMI resistant mutants of U. maydis have unaltered sterol
composition (34, 35), as do triadimenol resistant strains of
powdery mildews (36), Rhynchosporium secalis (37) and
Cladosporium cucumerinum (38) (Figure 1). In a recent study of
nine ketoconazole-resistant mutants of S. cerevisiae, all were
found to have a lesion in the Δ^5-Δ^6 desaturase suggesting that
this somehow compensated for the 14-methyl sterols that
otherwise accumulated in the presence of ketoconazole (39).

Changes in sterol 14 α demethylase. Changes in 14DM may also
account for resistance to DMI fungicides. A DMI resistant
clinical isolate of C. albicans also appeared to have an altered
14DM from both spectral and kinetic analysis of crude cell
homogenates (40). Binding of both the pyridine DMI fungicide
buthiobate, and the substrate lanosterol, to isolated 14DM was
abolished in one DMI resistant mutant of S. cerevisiae, although
the imidazole ketoconazole still bound normally to this altered
enzyme (41). Another mutant was obtained in which ketoconazole
did not bind to the 14 α demethylase. The primary structures of
14DM from a ketoconazole resistant mutant (SG1), and its wild-
type parent, were deduced from the DNA sequences of their
structural genes. A single amino acid substitution of aspartic
acid to glycine so altered the protein, that a neighbouring
histidine residue gained access to the heme iron as the sixth
ligand, blocking entry of molecular oxygen, and presumably
ketoconazole (42). Changes influencing the sterol substrate
binding site are less likely to have such a dramatic impact,
since considerable chemical diversity contained in the non-polar
part of DMI fungicides, suggests that the configuration of the
sterol binding region is not too critical. Changes of this type
may account for the low levels of resistance often observed to
DMI fungicides in many plant pathogens.

In U. avenae, the CO difference spectra of the target
cytochrome P450 was almost the same in sensitive and triadimenol
resistant strains, indicating that fungicide binding was not
impaired in the resistant strain (43). Although this may
indicate that resistance is not caused by a mutation in the
target demethylase, until binding studies can be done with
purified enzymes, this conclusion is perhaps premature.

A summary of possible mechanisms of resistance to SBI
fungicides is given in Table IV.

Detoxification. A ten minute incubation of both triadimenol
resistant and sensitive U. avenae strains in C^{14} acetate,
labelled 4,4 dimethyl sterols equally in both strains, when
triadimenol was present. After further incubation in the
absence of radio-isotope, methyl sterols were no longer labelled
in the resistant strain, suggesting that rapid turnover

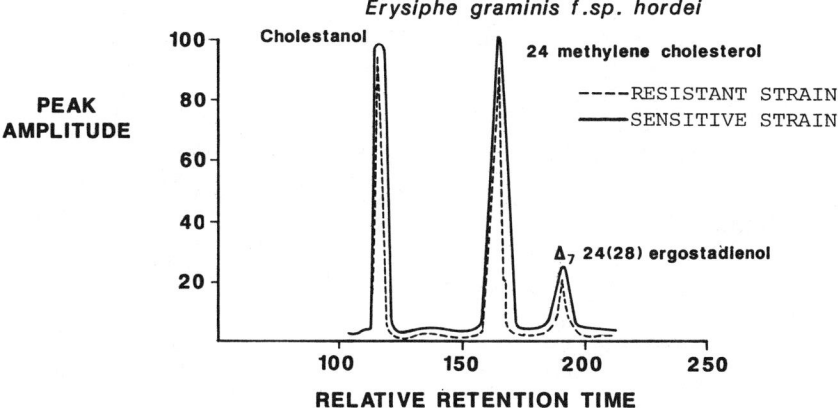
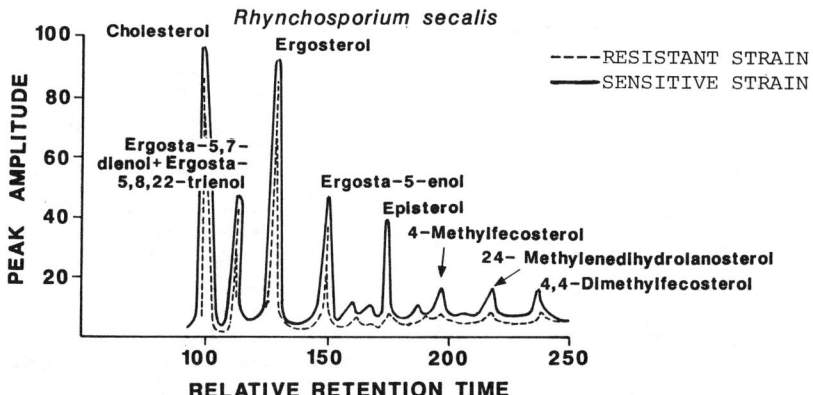
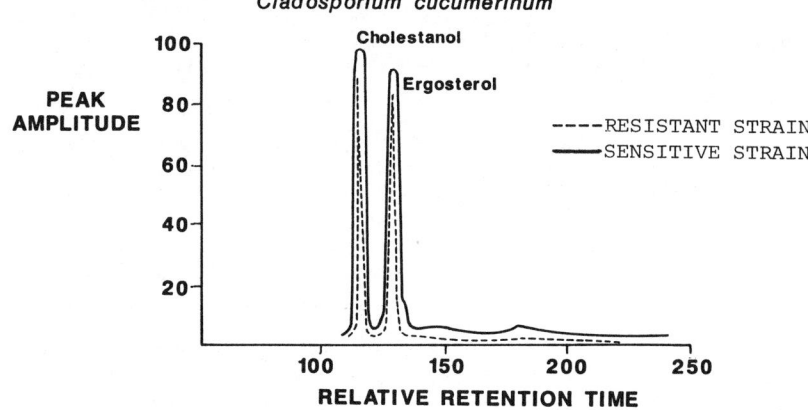

Figure 1. Sterol composition of DMI resistant and sensitive strains of three fungal pathogens. Details of the methods used for the extraction and analysis of sterols are given in the relevant source papers (36-38). Either cholesterol or cholestanol were used as internal standards.

(detoxification) of methylated sterols might account for resistance (Koeller, W., Cornell University, personal communication, 1988).

Table IV. Possible mechanisms of resistance to SBI fungicides

Mechanism	Reference
1. Deficiency in 14DM	32
2. Tolerance of toxic sterols	2, 9
3. Detoxification of sterols	43
4. Reduced affinity for 14DM	41
5. Decreased accumulation of fungicide	27, 28, 29
6. Detoxification of fungicide	
7. Failure to activate fungicide	54, 55
8. Deposition of fungicide in lipid droplets	56
9. Change in pH leading to protonation of fungicide	28

Molecular Analysis of Sterol 14 α Demethylase Gene from Barley Powdery Mildew

One of our most triadimenol resistant barley powdery mildew strains has a normal sterol composition in the conidia collected from fungicide treated plants (36), although in sensitive strains this, and lower doses of triadimenol, cause the appearance of 24-methylene dihydrolanosterol. Since uptake of triadimenol by this resistant strain was apparently normal, and it did not appear to metabolise the fungicide in any way, this suggests that the 14DM may be altered, so that normal sterol biosynthesis can continue.

Evidence that changes in the 14DM are responsible for DMI resistance might be obtained through studies in cell free systems (42) and purified enzymes. We have described an <u>in vitro</u> system from barley powdery mildew conidia which converts mevalonate to sterols, which was intended as a first step towards answering this question (45). However, purification of this membrane bound enzyme from a filamentous fungus has not yet been achieved, although lanosterol 14 demethylase has been purified to homogeneity from yeast (46). In addition, the technical difficulties of obtaining sufficient material for biochemical work, makes this an arduous approach for powdery mildews, and some other plant pathogens.

Cytochrome P450's form a family of NADPH dependent terminal oxidases, of which 14DM is one. There is considerable structural homology in P450's including the 14DM's from different organisms, not only in the heme binding domain, but elsewhere in the protein (47). Recently, lanosterol 14DM was cloned by transformation of a S. cerevisiae host using a yeast genomic DNA library in a multicopy plasmid vector (pVK2). Transformants with the amplified 14DM sequence were identified by selection for reduced ketoconazole sensitivity (48). The gene has now been sequenced (49), together with some of the flanking sequences, and shows common features with P450's from other eukaryotes.

We have used this gene (kindly supplied by Professor J.C. Loper, Cincinatti College of Medicine, USA), together with fragments containing either the 5' or 3' regions of the gene, to probe restriction enzyme digests of E. graminis f.sp. hordei DNA using Southern blot analysis (50, Figure 2). Even at moderate stringency, sufficient homology exists between the probes and mildew DNA to identify hybridization bands. Analysis of DNA from several strains is now under way to establish if there is polymorphism within this gene, and if so, whether the difference can be correlated with differences in DMI sensitivity, or simply reflect the general level DNA polymorphism between mildew strains.

DNA isolation from Erysiphe graminis conidia. Washed, freeze-dried conidia (100 - 200 mgs) were ground with an equal volume of glass beads (150 - 212 μm) and suspended in 4 mls 200 mM Tris-HCl, pH 8.0, 250 mM NaCl, 25 mM EDTA and 14 mM β-mercaptoethanol. 200 μl of 10% SDS was added gently, and heated for 1 h at $65^{\circ}C$. When cool, 1 ml 4 M K acetate (pH 5.5) was added and mixed gently. After 2h on ice insoluble material was removed by centrifugation at 12,000 xg for 1h at $4^{\circ}C$. DNA in the supernatant was precipitated by addition of 5 mls redistilled ethanol, and cooling at $-70^{\circ}C$ for 1h. The supernatant was discarded after centrifugation, and the pellet air-dried and resuspended in 4 ml TE (10 mM Tris, 1 mM EDTA, pH 8.0). DNA was further purified by adding 4 mls phenol (equilibrated with TE), mixed, followed by 4 mls chloroform : pentanol (24 : 1) and mixing for 30 min. After centrifugation, the aqueous phase was transferred to a new tube, and DNA precipitated by adding one tenth volume 4 M Na acetate pH 6.0 and one volume redistilled ethanol. After storage at $-70^{\circ}C$ for 1 h, DNA was pelleted, air-dried, and resuspended in 600 μl TE. Ethanol precipitation was repeated, and the pellet resuspended in 50 - 100 μl TE. DNA was digested with RNA-ase (125 μg/ml, Sigma) for 2h at $37^{\circ}C$, precipitated with 4 M NaAc and ethanol as before, dried, and finally resuspended in TE. 100 mg freeze-dried conidia typically yield 6 - 9 μg DNA.

Hybridization. Genomic DNA (c. 3 μg)of E. graminis f. sp. hordei was digested with the restriction endonucleases Hind III, Sst 1, Pst 1, Xho 1, Sal 1, Eco Rl and Bam Hl in accordance with the manufacturer's instructions (BRL), separated by 0.7% agarose gel electrophoresis, and then Southern blotted (45) onto nylon membrane (Hybond-N Amersham International). Prehybridization and hybridization reactions were done at $42^{\circ}C$ using 40% deionised formamide, and 5 x SSC and 5 x Denhards solution.

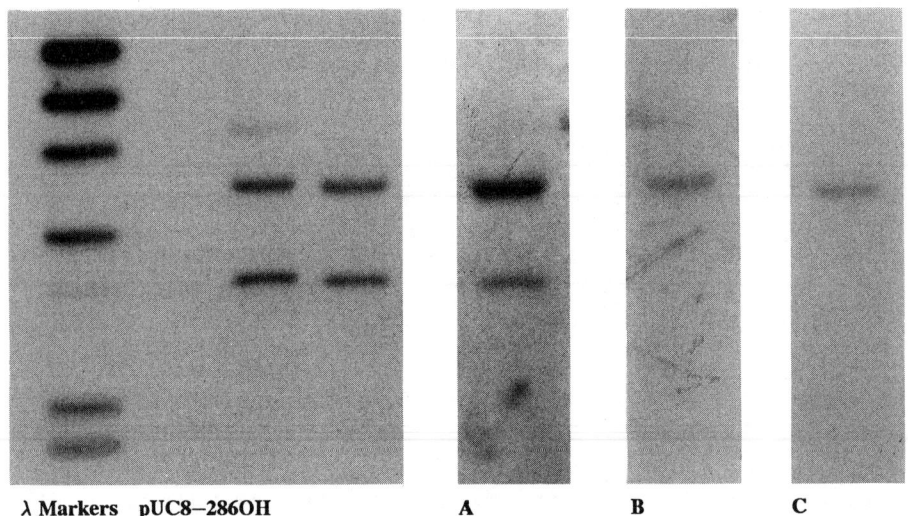

Figure 2. Southern analysis of mildew DNA restriction digests probed with plasmid pUC8 containing the sterol 14 α demethylase sequence from yeast, (pUC8-286OH), the whole coding sequence (A), 3' end (B) and 5' end (C).

Sna Bl/Hinc 11 and Hinc 11/Bgl 11 fragments of the 14 DM gene, as well as the whole gene, were labelled with P^{32} deoxycytidine and used as probes. After hybridization, membranes were washed twice in 2 x SSC and 0.5% SDS for 15 mins, followed by 2 x 2h washes in 0.5% SSC and 0.5% SDS, and autoradiographed.

Conclusions

Inhibitors of sterol biosynthesis provide many, useful, broad spectrum, fungicides, and it is essential that their effectiveness is retained as long as possible. Of the two major groups of SBI fungicides, only the morpholines, which inhibit $\Delta^{14(15)}$ reductase and/or $\Delta^8-\Delta^7$ isomerase depending on their chemistry, have not yet encountered practical field resistance. Unfortunately, the reasons for this stability, despite their apparent site specific mode of action, are not clear. It is speculated that the flexibility in the molecule arising from the "floppy" side chain, might make it more difficult to exclude morpholines from target sites by small changes in protein confirmation.

Inhibition of 14DM is the site of action for an important, but chemically diverse, group of fungicides (DMI's) used in both agriculture and medicine, and here cases of practical resistance have developed in important pathogens. Although something is known about the mechanisms of resistance in "model" fungi, we are largely ignorant of mechanisms operating in field resistant strains, and especially in powdery mildews. This in part is due to the fact that many different mechanisms probably contribute to resistance, and it is difficult to genetically construct strains likely to have only one mechanism.

Our work suggests, however, that uptake and metabolism do not play a major part in the resistance of field strains of cereal mildews, although they may do in some laboratory induced mutants. The extent to which changes in the target 14DM occur is also unclear, although application of molecular techniques should provide answers to questions that have hitherto been difficult to resolve directly through conventional enzyme purification and analysis. Cytochrome P450's represent a family of proteins and there may well be isozyme variants of 14DM which differ in their sensitivity to DMI fungicides, and which could be increased in DMI resistant strains. Amplification of genes leading to increased production of target enzymes also provides a mechanism for the gradual increase in resistance, which has been charted in cases leading to practical resistance. Analysis of the target enzyme at the DNA level should provide answers to these possibilities.

Acknowledgments

We thank Dr. B.C. Baldwin (ICI Agrochemicals) for helpful discussions and encouragement during some aspects of this work.

Literature Cited

1. Rodriguez, R. J.; Taylor, R. F.; Parks, L. W. Biochem. Biophys. Res. Comm. 1982, 106, 435-41.
2. Parks, L. W.; Rodiguez, R. J.; Low, C. Lipids, 1986, 21, 89-91.
3. Pierce, A. M.; Mueller, R. B.; Umrau, A. M.; Oehlschlager, A. C. Can. J. Biochem. 1978, 56, 794-800.
4. Ragsdale, N. N. Biochem. Biophys. Acta, 1975, 380, 81-96.
5. Berg, D.; Born, L.; Buchel, K. H.; Holmwood, G.; Kaulen, J. Pflanz. Nach. Bayer 1987, 40, 111-32.
6. Berg, D. In Fungicide Chemistry Advances Practical Applications; Green, M.B.; Spilker, D.A., Eds.; ACS: Washington D.C. 1986; 25-51.
7. Mercer, E. I. In Sterol Biosynthesis Inhibitors: Pharmaceutical and Agrochemical Aspects; Berg, D., Ed.; Ellis Horwood: Chichester UK, 1988; 120-500.
8. Ryder, N. S. In Sterol Biosynthesis Inhibitors: Pharmaceutical and Agrochemical Aspects; Berg, D., Ed.; Ellis Horwood: Chichester UK, 1988; 151-67.
9. Trocha, P.J.; Jasne, S.J.; Sprinson, D.B., Biochemistry, 1977, 16, 4721-6.
10. Weete, J.D.; Wise, M.L. Expt. Mycol. 1987, 11, 214-22.
11. Barug, D.; Kerkenaar, A. Pestic. Sci. 1984, 15, 78-84.
12. Koeller, W. In Fungicide Resistance; Delp, C. J., Ed.; Am. Phtyopath. Soc.: St. Paul, 1988;
13. Fuchs, A.; Drandarevski, C. A. Neth. J. Pl. Path., 1976, 82, 85-7.
14. Brent, K. J.; Hollomon, D. W. In Sterol Biosynthesis Inhibitors: Pharmaceutical and Agrochemical Aspects; Berg, D., Ed.; Ellis Horwood: Chichester UK, 1988; 332-46.
15. Fletcher, J. S.; Wolfe, M. S. Proc. 1981 Br. Crop Prot. Conf. - Pests and Diseases, 1981, pp 633-40.
16. De Waard, M. A.; Kipp, E. M. C.; Horn, N. M.; van Nistelrooy, J. G. M. Neth. J. Pl. Path., 1986, 92, 21-32.
17. Hollomon, D. W. Phytopath. Z., 1982, 105, 279-87.
18. Hunter, T.; Brent, K. J.; Carter, G. A. Proc. 1984 Br. Crop Prot. Conf. - Pests and Diseases, 1984, pp 471-6.
19. Buchenauer, H.; Budde, K.; Hellwald, K. H.; Traube, E.; Kirchner, R. Proc. 1984 Br. Crop Prot. Conf. - Pests and Diseases, 1984, pp 483-8.
20. Huggenberger, F.; Collins, M. A.; Skylakakis, G. Crop Prot. 1984, 3, 137-49.
21. Schepers, H. T. A. M. Neth. J. Pl. Path. 1985, 91, 105-18.
22. Walmsley-Woodward, D. J.; Laws, F. A., Whittington, W. J. Ann. Appl. Biol., 1979, 92, 211-9.
23. van Tuyl, J. M. Meded. Landbouw. Wageningen, 1977, 77-2, 1-136.
24. Stanis, V. F.; Jones, A. L. Phytopathology, 1985, 75, 1098-1101.

25. Grindle, M. In <u>Combatting Resistance to Xenobiotics: Biological and Chemical Approaches</u>; Ford, M. G.; Hollomon, D. W.; Khambay, B. P. S.; Sawicki, R. M., Eds.; Ellis Horwood: Chichester UK, 1987; pp 74-93.
26. Hollomon, D. W.; Butters, J.; Clark, J. <u>Proc. 1984 Br. Crop Prot. Conf. - Pests and Diseases</u>, 1984, pp 477-82.
27. De Waard, M. A.; van Nistelrooy, J. G. M. <u>Neth. J. Pl. Path.</u>, 1982, <u>88</u>, 96-112.
28. De Waard, M. A.; van Nistelrooy, J. G. M. <u>Pestic. Sci.</u> 1988, <u>22</u>, 371-82.
29. Siegel, M. R.; Solel, Z. <u>Pestic. Biochem. Physiol.</u> 1981, <u>15</u>, 222-33.
30. Grimmett, M.R. In <u>Advances in Heterocyclic Chemistry</u>; Katritzky, A.R. and Boulton, A.J., Eds.; Academic Press: London, 1970; Vol. 12, ppd 140-142.
31. Ryley, J. F.; Wilson, R. G., Barrat-Bee, K. J. <u>J. Med. Vet. Mycol.</u> 1984, <u>22</u>, 53-63.
32. Walsh, R. C.; Sisler, H. D. <u>Pestic. Biochem. Physiol.</u> 1982, <u>18</u>, 122-31.
33. Ziogas, B. N.; Sisler, H. D.; Lusby, W. R. <u>Pestic. Biochem. Physiol.</u> 1983, <u>20</u>, 320-9.
34. Hippe, S.; Koeller, W. <u>Pestic. Biochem. Physiol.</u> 1986, <u>26</u>, 209-19.
35. Leroux, P.; Gredt, M. <u>Pestic. Sci.</u> 1984, <u>15</u>, 85-9.
36. Loeffler, R. S. T.; Butters, J. A.; Hollomon, D. W. <u>Proc. 1984 Br. Crop Prot. Conf. - Pests and Diseases</u>, 1984, pp 911-6.
37. Girling, I. J.; Hollomon, D. W.; Kendall, S. J.; Loeffler, R. S. T.; Senior, I. J. <u>Proc. 1988 Br. Crop Prot. Conf. - Pests and Diseases</u>, 1988, 385-90.
38. Carter, G. A.; Kendall, S. J.; Burden, R. S.; James, C. S.; Clark, T. <u>Pestic. Sci.</u> 1989, <u>26</u>, 181-92.
39. Watson, P. F.; Rose, M. E.; Kelly, S. L. <u>J. Med. Vet. Mycol.</u>, 1988, <u>26</u>, 153-62.
40. Vanden Bossche, H.; Marichal, P.; Gorrens, J.; Bellens, D.; Verhoeven, H.; Coene, M. C.; Lauwers, W.; Janssen, A. J. <u>Pestic. Sci.</u> 1987, <u>21</u>, 289-306.
41. Yoshida, Y.; Aoyama, Y.; Nishimo, T.; Katsuki, H.; Maitra, U. S.; Mohan, V. P.; Sprinson, D. B. In <u>Cytochrome P450, Biochemistry, Biophysics and Induction</u>; Vereczkey, L.; Magyar, K., Eds.; Elsevier: Amsterdam; 1985; pp 439-43.
42. Ishida, N.; Aoyama, Y.; Hatanaka, R.; Oyama, Y.; Imajo, S.; Ishiguro, M.; Oshima, T.; Nakazato, H.; Noguchi, T.; Maitra, U. S.; Mohan, V. P.; Sprinson, D. B.; Yoshida, Y. <u>BBRC</u>, <u>155</u>, 317-23.
43. Koeller, W.; Scheinpflug, H. <u>Plant Dis.</u> 1987, <u>71</u>, 1066-74.
44. Gadher, P.; Mercer, E. I.; Baldwin, B. C.; Wiggins, T. E. <u>Pestic. Biochem. Physiol.</u> 1983, <u>19</u>, 1-10.
45. Hollomon, D. W.; Loeffler, R. S. T.; Butters, J. A. <u>Tagungs. Akad. Landwirts. DDR.</u> 1987, <u>253 S</u>, 137-42.
46. Aoyama, Y.; Yoshida, Y.; Smada, Y.; Sato, Y. <u>J. Biol. Chem.</u> 1987, <u>262</u>, 1239-43.

47. Nelson, D. R.; Strobel, H. W. J. Biol. Chem. 1988, 263, 6038-50.
48. Kalb, V. F.; Loper, J. C.; Dey, C. R.; Woods, C. W.; Sutter, T. R. Gene, 1986, 45, 237-45.
49. Kalb, V. F.; Woods, C. W.; Turi, T. G.; Dey, C. R.; Sutter, T. R.; Loper, J. C.; DNA, 1987, 6, 529-37.
50. Southern, E. M. J. Mol. Biol. 1975, 98, 503-17.
51. Burrell, N. M.; Brunt, P. Ann. Bot. 1981, 48, 395-7.
52. Sheridan, J. E.; Grbavac, N.; Sheridan, M. H. Trans. Br. Mycol. Soc. 1985, 85, 338-41.
53. Eckert, J. W. Phtyopathology, 1987, Abstr. 328.
54. Gasztonyi, M. Pestic. Sci. 1981, 12, 433-438.
55. Deas, A. H. B.; Clifford, D. R. Pestic. Biochem. Physiol. 1984, 22, 276-84.
56. Hippe, S. Histochemistry, 1987, XX, 309-15.

RECEIVED September 1, 1989

Chapter 14

Biochemical Basis of Resistance to Phenylamide Fungicides

L. C. Davidse

Department of Phytopathology, Wageningen Agricultural University, P.O. Box 8025, 6700 EE Wageningen, Netherlands

> Phenylamide fungicides selectively inhibit ribosomal RNA synthesis of sensitive Oomycete fungi by interference with the activity of the RNA polymerase I-template complex.
> Phenylamides bind to an as yet unidentified component of this complex. The phenylamide binding site in resistant strains has a very low affinity for metalaxyl, whereas values of the dissociation constant are in the order of 0.1-0.2 μM for phenylamide-sensitive strains. Apparently a change at the phenylamide receptor is responsible for resistance.
> Identification of the phenylamide receptor is hampered by loss of its ability to bind phenylamides during purification. To overcome this a photoaffinity probe for the phenylamide receptor has been synthesized. The biological properties of this probe allow its use to tag the components of the enzyme complex involved in binding.

Commercially available phenylamide fungicides include the acylalanines (metalaxyl, furalaxyl, and benalaxyl), the butyrolactones (ofurace and cyprofuram), and one oxazolidinone (oxadixyl), the structures of which are shown in Figure 1. These systemic fungicides show a high and selective activity against fungi of the order Peronosporales (1).

Phenylamide fungicides have met with severe resistance problems in the field. Major target organisms such as *Bremia lactucae* of lettuce, *Peronospora tabacina* of tobacco, *Phytophthora infestans* of potato, and *Pseudoperonospora cubensis* of cucumber have the ability to develop a high level of resistance (2). It requires the implementation of resistance management strategies involving limited use of phenylamides and use against foliar pathogens only in mixtures with residual fungicides to prevent or delay the build-up of resistant populations (3).

The mechanism of action of the phenylamides as yet is only understood in general terms (4). Especially information regarding molecular interactions involved in binding of the fungicides at the receptor site is lacking. This knowledge and that on structural changes of the receptor that result in resistance are essential to redesign phenylamide structures that might be less prone to development of resistance or to design compounds that selectively inhibit strains with resistance to the existing phenylamides. Practical experiences on combating benzimidazole-resistant strains of several pathogens with N-phenylcarbamates or N-phenylformamidoximes which selectively act against these strains (5,6), indicate that this approach can be succesfull. However, understanding this phenomenon at the molecular level is necessary to evaluate the full potential of this approach. In this contribution recent progress in research on the molecular basis of antifungal activity of phenylamides is discussed.

Mechanism of Action and Biochemical Basis of Resistance

Effect on Biosynthetic Processes. Among the biosynthetic processes studied in various fungi, incorporation of uridine into RNA proved to be the most sensitive one to the phenylamides (Table I, 7-12).

Table I. Effect of Metalaxyl (0.1 µg/ml) on Incorporation of Labeled Precursors by Mycelium of *Phytophthora megasperma* f.sp. *medicaginis*

Precursor	Incorporation as % of Control
[methyl-^{3}H]thymidine	84
[5,6-^{3}H]uridine	26
L[U-^{14}C]phenylalanine	81

Incorporation of precursors into DNA, proteins, and lipids is affected less rapidly or to a lesser extent. Respiration is not inhibited. Since the phenylamides at low concentrations do not affect the uptake of precursors or the conversion of nucleosides into nucleotides, inhibition of RNA synthesis must be responsible for the observed effects on uridine incorporation. However, at phenylamide concentrations that are fully inhibitory to growth, a complete inhibition of uridine incorporation does not occur. Depending on the fungal species used incorporation is reduced to 20-60 % of the control value (12). This indicates that only part of the cellular RNA synthesis is sensitive to phenylamides.

Analysis of radiolabeled RNA formed during incubation with metalaxyl showed that in mycelia of various *Phytophthora* species, synthesis of polyA(denylated) RNA that represents most of the m(essenger) RNA appeared to be less sensitive than that of total RNA, the majority of which is r(ibosomal) RNA (10,11,13). Results of the fractionation of total RNA by agarose gel electrophoresisis are shown in Figure 2. Labeling of both high molecular weight rRNA species is almost absent in metalaxyl-treated mycelia, whereas in control cultures these species are distinctively labeled. Inhibition

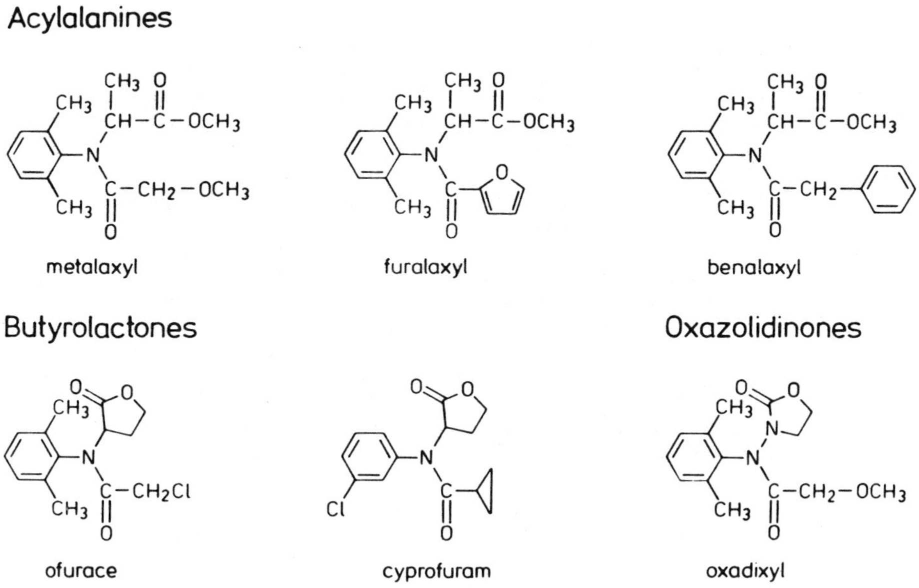

Acylalanines
metalaxyl
furalaxyl
benalaxyl
Butyrolactones
Oxazolidinones
ofurace
cyprofuram
oxadixyl

Figure 1. Structures of Phenylamide Fungicides.

Radioactivity (dpm×10⁻³)
7.5
5.0
2.5
5
10
15
20
Slice
control
metalaxyl

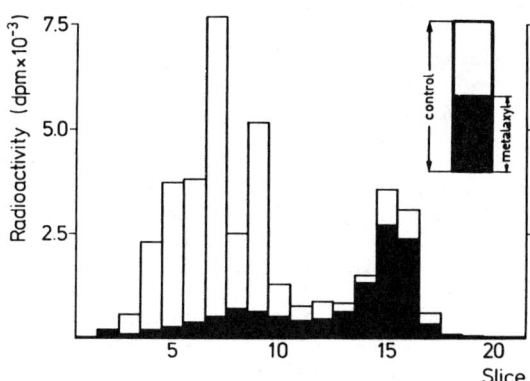

Figure 2. Electrophoretic Analysis of Total RNA isolated from Mycelium of a Phenylamide-Sensitive Strain of *P. megasperma* f. sp. *medicaginis* after Incubation with [5,6-^{3}H]Uridine in the Presence or Absence of Metalaxyl (1 μg/ml).

of rRNA synthesis, therefore, can be considered to be the primary mode of action of the phenylamides.

In eukaryotes RNA is synthesized by three different RNA polymerases, each of which mediates the synthesis of a distinct product. RNA polymerase I (or A) synthesizes 25S and 18S rRNA. RNA polymerase II (or B) produces mRNA, whereas RNA polymerase III (or C produces t(ransfer)RNA and 5S rRNA. Each enzyme has its own characteristics that can be used to follow the purification of the individual enzymes. Sensitivity to α-amanitin is a useful property to characterize the polymerases. RNA polymerase II is highly sensitive to this toxin, whereas polymerase I is insensitive. Polymerase III varies in sensitivity from species to species but is in general less sensitive than polymerase II.

Metalaxyl does not inhibit the activity of partially purified polymerase I from P. megasperma f. sp. medicaginis (10) or Phytophthora nicotianae (11). Endogenous RNA polymerase activity of nuclei isolated from these fungi, however, is sensitive to metalaxyl, indicating that metalaxyl only interferes with the activity of the intact polymerase template complex. About 40 % of the endogenous RNA polymerase activity of isolated nuclei of P. megasperma f. sp. medicaginis is metalaxyl sensitive and about 30 % is α-amanitin sensitive (Table II). The effects of metalaxyl and

Table II. Effect of Metalaxyl (10 µg/ml) and α-Amanitin (5 µg/ml) on Endogenous RNA Polymerase Activity of Nuclei isolated from P. megasperma f. sp. medicaginis

Inhibitor added	Activity as % of Control
none	100
metalaxyl	64
α-amanitin	72
metalaxyl and α-amanitin	38

α-amanitin are additive, indicating interference of metalaxyl with an RNA polymerase activity different from the α-amanitin sensitive polymerase II. The lack of any effect of metalaxyl on the synthesis of mRNA in isolated nuclei confirms the inability of metalaxyl to interfere with the synthesis of poly(A) RNA by RNA polymerase II in the intact organism.

In phenylamide-resistant strains, phenylamides are not able to inhibit ribosomal RNA synthesis, as is indicated by the inability of metalaxyl to interfere with uridine incorporation in a phenylamide-resistant strain of P. megasperma f. sp. megasperma obtained in the laboratory after mutagenic treatment (14) and a phenylamide-resistant strain of Phytophthora infestans (Table III) originating from a potato field (15).

Likewise, endogenous polymerase activity of isolated nuclei from these strains is significantly less sensitive to phenylamides than that of nuclei from sensitive strains (Table IV). This indicates that in both species a change in the target site of metalaxyl is responsible for resistance. Apparently both mutagen induced resistance and resistance that develops in the field have a similar basis. This once more proves the validity of laboratory studies aimed at evaluating the potential of a fungus to develop resistance to a fungicide.

Table III. Effect of Metalaxyl (1 μg/ml) on Incorporation of Uridine by Mycelium of a Phenylamide-sensitive (S) and a Phenylamide-resistant (R) Strain of P. megasperma f. sp. medicaginis and P. infestans

Species	Incorporation as % of Control	
	S	R
Phytophthora megasperma	18	103
Phytophthora infestans	61	93

Table IV. Effect of Phenylamides (1 μg/ml) on Endogenous RNA Polymerase Activity of Nuclei isolated from a Phenylamide-Sensitive (S) and a Phenylamide-Resistant (R) Strain of P. megasperma f. sp. medicaginis and P. infestans

	Activity as % of Control			
	P. megasperma		P. infestans	
Inhibitor added	S	R	S	R
metalaxyl	56	99	60	95
cyprofuram	80	96	74	86
benalaxyl	56	100	54	83
oxadixyl	75	100	69	86

<u>Binding at the Phenylamide Receptor.</u> To substantiate the idea that a target site change is responsible for phenylamide resistance, binding studies were performed with [^{3}H]metalaxyl and cell-free mycelial extracts of sensitive and resistant strains. Binding was determined with a Sephadex G25 gel filtration assay. Specific binding occurred only in extracts of sensitive strains, as can be seen in the Scatchard plots shown in Figures 3 and 4. The apparent dissociaton constants (K_d) as determined from these plots were 0.17 and 0.07 μM for P. megasperma f. sp. medicaginis S and P. infestans S, respectively. Cyprofuram, benalaxyl and oxadixyl competed for the binding site (Table V).

This indicates that the apparent dissociation constants for benalaxyl and metalaxyl are similar. The binding site has a lower affinity for cyprofuram and oxadixyl, which is in agreement with the lower activity of these compounds against RNA synthesis measured as [^{3}H]uridine incorporation by mycelium (12).
These experiments clearly show that phenylamide resistance in both a laboratory mutant of P. megasperma f. sp. medicaginis and a field isolate of P. infestans is due to a change at the target site.

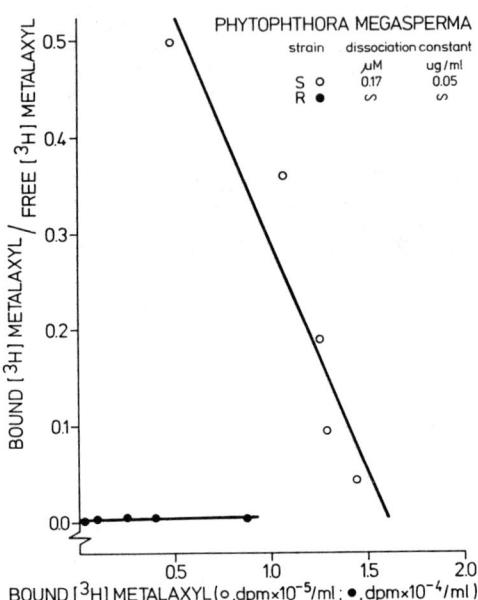

Figure 3. Scatchard Plot of Binding Data for [^{3}H]Metalaxyl by Cell-Free Mycelial Extracts of a Phenylamide-Sensitive (S) and a Phenylamide-Resistant (R) Strain of *P. megasperma* f. sp. *medicaginis*.

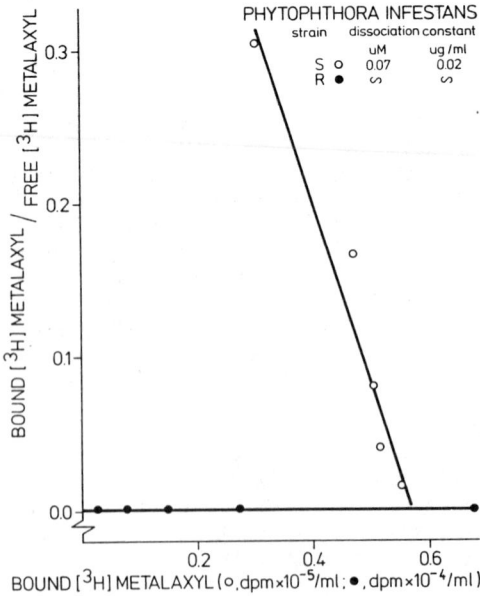

Figure 4. Scatchard Plot of Binding Data for [^{3}H]Metalaxyl by Cell-Free Mycelial Extracts of a Phenylamide-Sensitive (S) and a Phenylamide-Resistant (R) Strain of *P. infestans*.

Table V. Effect of Phenylamides on the Binding of [^{3}H]Metalaxyl (0.14 µg/ml) by a Cell Free Mycelial Extract of a Phenylamide-Sensitive Strain of P. megasperma f. sp. medicaginis

Test Compound	Binding of [^{3}H]Metalaxyl as % of Control in the Presence of the Test Compound at the Indicated Concentration (µg/ml)			
	0.01	0.1	1	10
metalaxyl	94	53	13	3
cyprofuram	110	109	78	24
benalaxyl	94	49	18	4
oxadixyl	93	86	51	14
azidometalaxyl	88	86	55	24

Biological Properties of a Photoaffinity Probe for the Phenylamide Receptor

Although binding activity could be easily demonstrated in crude cell free mycelial extracts of phenylamide-sensitive Phytophthora strains, binding activity decreased considerably when the extracts were subjected to Heparin Sepharose affinity chromatography, a first step in the purification of RNA polymerases (16). Purification of the phenylamide receptor, therefore, cannot be followed using the binding assay. An alternative approach would be to covalently label the receptor with a photoaffinity probe. Purification of the labeled complex can then be followed either by a radioassay or an immunoassay, which respectively require a radiolabeled photoaffinity probe or antiserum that selectively recognizes covalently bound phenylamide residues.

A suitable photoaffinity probe would be 3-azidometalaxyl (Figure 5), since substituents like amino, nitro and chloro at the 3- and 5-position do not cause a drastic change in biological activity of the compound (17). This compound, therefore, was synthezised and its biological properties were determined. Azidometalaxyl indeed inhibited mycelial growth of the phenylamide-sensitive strain of P. megasperma f. sp. medicaginis (EC_{50}: 4 µg/ml), whereas the compound was almost inactive against growth of the phenylamide-resistant strain. Like metalaxyl, azidometalaxyl inhibited endogenous RNA polymerase activity from isolated nuclei, although at least a tenfold higher concentration was required to achieve a similar level of inhibition (Table VI). Azidometalaxyl also bound to the phenylamide receptor, as indicated by its ability to compete with [^{3}H]metalaxyl (Table V). These data indicate that azidometalaxyl is a suitable photoaffinity probe for the phenylamide receptor.

Preliminary studies involving a model system in which antibodies to metalaxyl were photoaffinity labeled with azidometalaxyl and detection of the phenylamide-antibody complex in a Western blot with metalaxyl antiserum, raised again a formaldehyde-conjugated 3-aminometalaxyl-bovine serum albumin complex, indicated that an immunoassay to detect phenylamide residues covalently bound to the receptor in mycelial extracts might be suitable as an assay in the purification procedure of the labeled receptor.

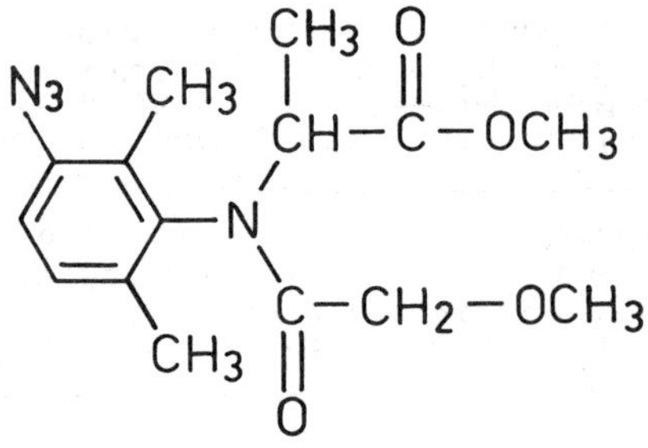

Figure 5. Structure of Azidometalaxyl.

Table VI. Effect of Azidometalaxyl and Metalaxyl on Endogenous RNA Polymerase Activity of Nuclei isolated from a Phenylamide-Sensitive strain of *P. megasperma* f. sp. *medicaginis*

Inhibitor Concentration (μg/ml)	Activity as % of Control	
	metalaxyl	azidometalaxyl
0.1	63	92
1.0	53	67
10.0	52	57

Conclusions

Studies on the mechanism of action of phenylamides have identified these compounds as highly specific inhibitors of rRNA synthesis. This property makes the phenylamides excellent tools to study details of this complex process. The research described in this contribution hopefully can lead to characterization of the phenylamide receptor and its role in rRNA synthesis. When the receptor, presumably a protein, is characterized, isolation of its structural gene would be feasible. Once this has been accomplished, an extensive research area is accessible to be explored. A search for homologous genes in phenylamide-resistant strains and fungi that are naturally resistant would be possible, thereby providing information on structural differences of the receptor and changes which influence the binding of phenylamides.

Knowledge on the structure of the receptor and the molecular interactions involved in binding of phenylamides is a prerequisite to understand their fungicidal action at the molecular level. Once a three-dimensional model of the receptor can be constructed, it may be possible to redesign phenylamide structures or even design completely new structures that fit into the binding site.

It is along these lines that research on the mode of action of existing fungicides should make progress. It will help us to develop the tools with which to design new compounds that interact with prechosen target sites. It is my opinion that both in academia and industry a firm commitment to this type of research continues to be necessary in order to meet the future demand for new and highly active chemicals for the control of plant diseases.

Acknowledgments

The author wishes to thank Ciba-Geigy, Basel, Switzerland for providing a number of phenylamide compounds.

Literature Cited

1. Schwinn, F.J.; Staub, T. In Modern Selective Fungicides; Lyr, H., Ed.; Gustav Fischer: Jena, 1987; pp 259-273.
2. Morton, H.V.; Urech, P.A. In Fungicide Resistance in North America; Delp, C.J., Ed.; APS Press: St Paul, 1988; pp 59-60.
3. Urech, P.A. In Fungicide Resistance in North America; Delp, C.J., Ed.; APS Press: St Paul, 1988; pp 74-75.
4. Davidse, L.C. In Modern Selective Fungicides; Lyr, H., Ed.; Gustav Fischer: Jena, 1987; pp 274-282.
5. Kato, T. In Fungicide Resistance in North America; Delp, C.J., Ed.; APS Press: St Paul, 1988; p 40.
6. Nakata, A.; Sano, S.; Hashimoto, S.; Hayakawa, K.; Nishikawa, H.; Yasuda. Y. Ann. Phytopathol. Soc. Jpn. 1987, 53, 659-62.
7. Arp. U.; Buchenauer, H. Mitt. Biol. Bundesanstal. 1981, 236-7.
8. Fisher, D.J.; Hayes, A.L. Pesticide Sci. 1982, 13, 330-9.
9. Kerkenaar, A. Pest. Biochem. Physiol. 1981, 16, 1-13.
10. Davidse, L.C.; Hofman, A.E.; Velthuis, G.C.M. Exp. Mycol. 1983, 7, 344-61.
11. Wollgiehn, R; Braütigam, E,; Schuhman, B.; Erge, D. Z. Allg. Mikrobiol. 1984, 24, 269-79.
12. Davidse, L.C.; Gerritsma, O.C.M.; Ideler, J.; Pie, K.; Velthuis, G.C.M. Crop Protection 1988, 7, 347-55.
13. Kang, K.Y.; Eckert, J.W. Tag.-Ber., Akad. Landwirtsch-Wiss. DDR, Berlin 1987, 253, 109-14.
14. Davidse, L.C. Neth. J. Pl. Path. 1981, 87, 11-24.
15. Davidse, L.C.; Looijen, D.; Turkensteen L.J.; Van der Wal, D. Neth. J. Pl. Path. 1981, 87, 65-8.
16. Hammond, C.I.; Holland, M.J. J. Biol. Chem. 1983, 258, 3230-41.
17. Hubele, A; Kunz, W.; Eckhardt, W.; Sturm, E. In Pesticide chemistry, Human Welfare and the Environment; Vol.1 Synthesis and Structure-Activity Relationships; Doyle, P; Tujita T., Eds.; Pergamon: Oxford, 1983; pp 233-42.

RECEIVED September 1, 1989

Chapter 15

Mechanism of Action of N-Phenylcarbamates in Benzimidazole-Resistant *Neurospora* Strains

Makoto Fujimura[1], Kenji Oeda[1], Hirokazu Inoue[2], and Toshiro Kato[1]

[1]Takarazuka Research Center, Sumitomo Chemical Company, Ltd., Takatsukasa Takarazuka, Hyogo 665, Japan
[2]Laboratory of Genetics, Saitama University, Shimo-ohkubo, Urawa, Saitama 338, Japan

Mutant strain F914 of Neurospora crassa had two contrary phenotypes to chemicals; hypersensitivity to N-phenylcarbamates and resistance to benzimidazoles. These phenotypes were caused by a single mutation of the beta-tubulin gene. After UV-mutagenesis of the F914 strain, 5 revertants resistant to N-phenylcarbamates were isolated and characterized. Two strains showed normal sensitivity to benzimidazoles, but the remaining strains were still resistant to benzimidazoles. Genetic analysis indicated that second mutations of all these revertants were located on the beta-tubulin gene. These data suggested that N-phenylcarbamates had binding activity to beta-tubulin, as observed with benzimidazoles. To determine the molecular change of the beta-tubulin gene in the F914 strain, gene cloning and sequencing were carried out. The results showed that the amino acid at position 198 of beta-tubulin was glutamic acid in the wild type strain, but was glycine in the F914 strain. Thus a single change in the beta-tubulin protein governs sensitivity to benzimidazoles and N-phenylcarbamates so dramatically that negatively correlated cross resistance results. A model to explain this phenomenon is described, and it is proposed that in future such genetic studies on fungal strains resistant to currently used chemicals may aid in the design of new fungicides effective on the resistant population.

Benzimidazole fungicides (Figure 1) are in wide use, as they are highly effective against various diseases present in commercially important crops. However, the intense, continuous use of benzimidazoles can result in the emergence of resistant plant pathogens. For example, a prominent pathogen of grapes, Botrytis

0097–6156/90/0421–0224$06.00/0
© 1990 American Chemical Society

cinerea, has gained resistance to benomyl, following a long period of use (5).

In principle, one of the most effective means to cope with resistance of plant pathogens is the use of fungicides to which resistant strains show negatively correlated cross resistance. However, until now this strategy has seen limited use in practice, principally for lack of suitably effective compounds. Herein we discuss one of the first practically useful examples of negatively correlated cross resistance, and report on its genetic and molecular basis.

Leroux and Gredt reported that benzimidazole-resistant strains of Botrytis cinerea and Penicillium expansum exhibited negatively correlated cross resistance to herbicidal N-phenylcarbamates such as barban, chlorpropham, and chlorbufam (10, 11, 12). Based on their observation, Sumitomo scientists evaluated many examples of N-phenylcarbamates to search for compounds with potent fungitoxicity and no phytotoxicity. This effort was eventually successful, leading to new fungicides such as MDPC (9, 21) and diethofencarb (14, 23, 24) (Figure 2).

Diethofencarb is highly toxic to benzimidazole-resistant isolates of B. cinerea, whereas its toxicity to the benzimidazole-sensitive isolates is weak. In contrast, MBC inhibited the growth of wild type isolates at a low concentration but had low toxicity to resistant isolates. In pot tests, diethofencarb was effective for gray mold caused by resistant isolates of Botrytis cinerea, but was not effective for wild type isolates. Diethofencarb has systematic activity in plants and both preventive and curative activity in controlling gray mold.

Davidse has shown that benomyl blocks mitosis in the fungus Aspergillus nidulans, in a manner similar to that seen with colchicine in animal and plant cells (3). Mutants resistant to benomyl have tubulin showing lower affinity to it than do the wild type strain(4). Mutants of Aspergillus nidulans with resistance to benomyl were sometimes found to have a tubulin protein with an altered electrophoretic mobility (20). Most of the mutations conferring benomyl resistance in other fungi, such as A. nidulans (20), Neurospora crassa (18), Saccharomyces cerevisiae (16), and Schizosaccharomyces pombe (7), were also mapped in the beta-tubulin gene. These findings led to the idea that benzimidazoles probably inhibit microtubule-mediated cellular functions. However, the mechanisms of negatively correlated cross resistance commonly observed in benomyl resistant isolates of plant pathogens are not well understood.

Negatively correlated cross resistance is of great interest for workers in basic sciences, and could have significant practical benefits in resistance management. A clarification of the mechanism related to resistance should aid in overcoming resistance problems. Ultimately, such knowledge could potentially help researchers to design new classes of compounds showing negatively correlated cross resistance with benzimidazoles, and perhaps even with fungicides acting at the other target sites.

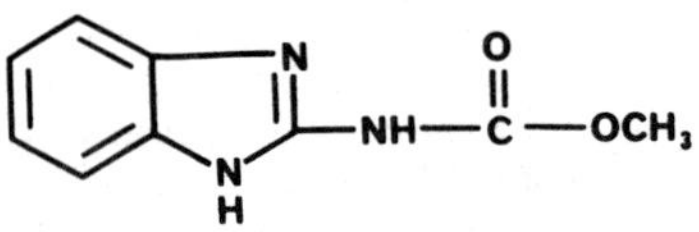

CARBENDAZIM (MBC)

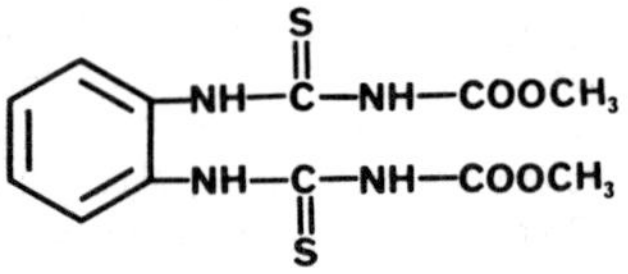

THIOPHANATE-METHYL

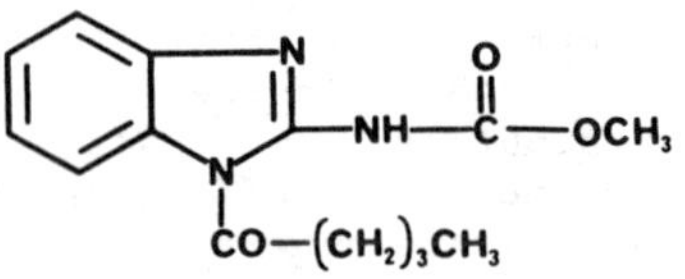

BENOMYL

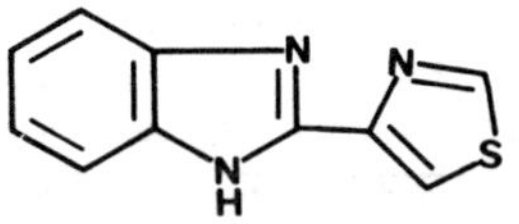

THIABENDAZOLE (TBZ)

Figure 1. Chemical structures of benzimidazoles.

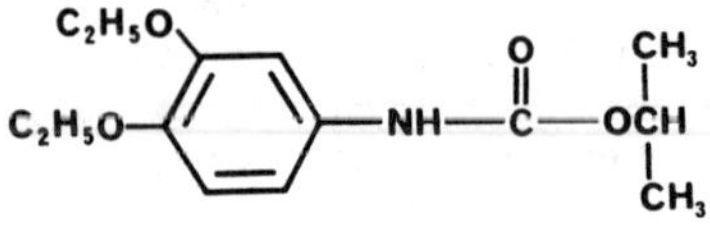

DIETHOFENCARB

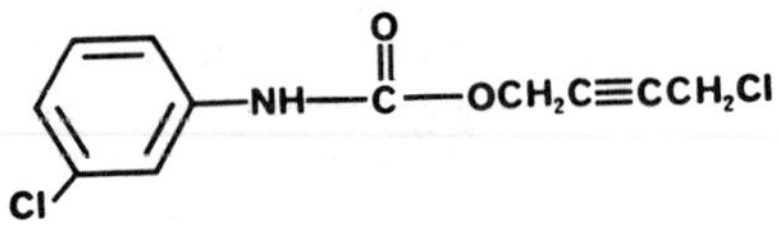

BARBAN

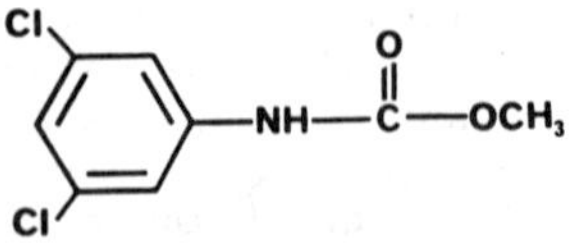

MDPC

Figure 2. Chemical structures of N-phenylcarbamates.

Isolation and characterization of diethofencarb supersensitive mutants of N.crassa

We used N. crassa as a model fungus and attempted to elucidate the mechanism of negatively correlated cross resistance between benzimidazoles and N-phenylcarbamates. MBC resistant mutants were isolated from the wild type strain of N. crassa by UV treatment. Selection of the MBC resistant mutants was carried out on medium containing 50ppm MBC. MBC (carbendazim), the degradation product of benomyl and thiophanate-methyl, appears to be the active form in fungi (2).
Nineteen independent MBC resistant mutants were isolated, and two isolates were sensitive to diethofencarb (Table I). Strains F914 and F939 showed positively correlated cross resistance to benzimidazoles, MBC, thiophanate-metyl, and TBZ, and negatively correlated cross resistance to N-phenylcarbamates, diethofencarb, MDPC, and barban. Other mutants, including the Bm1511 strain isolated by Borck and Braymer (1), showed positively correlated cross resistance to benzimidazoles but no negatively correlated cross resistance to any of the N-phenylcarbamates. Negatively correlated cross resistance of the N. crassa F914 strain to N-phenylcarbamates was similar to that seen for B. cinerea. As the sensitivities to these chemicals of strains F914 and F939 were similar, the mutants seemed to be identical.

Table I. Cross resistance in MBC resistant mutants of N. crassa

	Minimal inhibitory concentration (µg/ml)					
	Benzimidazoles			N-phenylcarbamates		
Strains	MBC	Thiophanate-methyl	TBZ	Diethofencarb	MDPC	Barban
WT	0.4	1.6	1.6	>100	100	50
F914	>100	>100	100	0.1	0.8	6.2
F939	>100	>100	100	0.1	0.8	6.2
Bm1511	50	100	>100	>100	100	50

Cytological studies showed that diethofencarb-treated F914 conidia and MBC-treated wild type conidia showed a similar abnormal morphology, wherein the germ tubes were distorted and swollen. Diethofencarb had no effects on the morphology of the wild type strain.
Diethofencarb affected the nuclear morphology of the F914 strain; scattered nuclei were apparent. This was similar to the effect of MBC on the wild type strain. Thus the mode of action of diethofencarb in the F914 strain resembles that of MBC in the wild type strain. We noted similar findings in the case of B. cinerea treated with MBC and MDPC (21).

Genetic studies of negatively correlated cross resistance between benzimidazoles and N-phenylcarbamates.

In the cross between the wild type strain and the F914 strain, half of the progenies exhibited wild type phenotype (MBC sensitive and diethofencarb insensitive) while another half of the progenies exhibited F914 phenotype (MBC resistant and diethofencarb supersensitive, see Table II). There was no progeny resistant or sensitive to both chemicals, thereby indicating that two phenotypes of the F914 strain were caused by a single mutation.

The F914 strain was crossed to strain Bm1511, the mutation of which had been already mapped on the beta-tubulin gene (18). In this cross, half of the progenies showed Bm1511 phenotypes; that is double resistance to both diethofencarb and MBC. Another half of the progenies showed F914 phenotypes. If recombination between those mutations had occurred in this cross, some progenies should show the phenotype of the wild type strain (MBC sensitive and diethofencarb insensitive). One hundred and twenty-five progenies were tested, but recombinant progenies were not obtained. Thus, the mutation of the F914 strain occurred in the beta-tubulin gene and conferred negatively correlated cross resistance. This seems to be the first description that the mutation in beta-tubulin is responsible for both MBC resistance and diethofencarb supersensitivity. The same result was obtained with the F939 mutant.

Table II. Genetic analysis of MBC resistant mutants of $\underline{N}$. crassa

Cross	Numbers of progenies			
	BsNr	BrNs	BrNr	Total
WT x				
F914	98	105	0	203
F939	78	76	0	154
Bm1511 x				
F914	0	57	68	125
F939	0	54	50	104
F914 x				
F914	0	109	0	109
F939	0	149	0	149

BsNr : Sensitive to MBC and insensitive to diethofencarb
BrNs : Resistant to MBC but sensitive to diethofencarb
BrNr : Resistant to both MBC and diethofencarb

To determine whether the mutations in F914 strain and F939 strain occurred at different sites within beta-tubulin gene, we searched for wild type progenies as the recombinants in the cross between these two mutants. However, all the progenies (among approximately 5×10^6) did not grow on medium containing 1ppm diethofencarb, thereby lending support to the idea that mutation sites of these mutants were either identical or extremely close.

Revertant analysis

To elucidate the mechanism of negatively correlated cross resistance and mode of action of diethofencarb, we isolated diethofencarb resistant mutants from mutagenized F914 strain.

We isolated 11 diethofencarb resistant mutants from the F914 strain, and 5 revertants were further examined(Table III). The strains F422 and F424 exhibited wild type phenotypes; that is, they were not sensitive to diethofencarb but were sensitive to MBC. The growth pattern of these strains was also similar to that of the wild type strain. Three other revertants (strain F513, F421 and F423) were moderately resistant to diethofencarb but retained sensitivity to diethofencarb at 50ppm, These mutants did not recover sensitivity to MBC and most exhibited abnormalities in mycelial growth.

Table III. Diethofencarb resistant mutants of N. crassa

		MBC (µg/ml)		Diethofencarb (µg/ml)	
	none	0.1	50	0.1	50
WT	+	−	−	+	+
F914	+	+	+	−	−
Bm1511	+	+	−	+	+
F422	+	−	−	+	+
F424	+	−	−	+	+
F421	+	+	+	+	−
F423	+	+	+	+	−
F513	+	+	+	+	−

+ : normal growth
− : no growth

We analyzed these mutants, genetically (Table IV). In the cross between the wild type strain and the F422 strain, or the F424 strain, all progenies exhibited wild type phenotype (sensitive to MBC and not sensitive to diethofencarb). If these strains have an additional mutation outside the beta-tubulin gene, we should be able to obtain recombinant progenies in this cross. However, we found no F914-type progenies, hence they may be true revertants. We considered these mutants to have been obtained by the back mutation.

In another cross of moderately diethofencarb-resistant mutants (strain F421, F423 and F513) to the wild type strain, we obtained no diethofencarb supersensitive progenies, such as the F914 strain. This indicates that these mutants have a second mutation within the beta-tubulin gene, in addition to the mutation of the F914 strain. These additional mutations are responsible for the moderate resistance to diethofencarb. In cold sensitive mutants of S. pombe, increased sensitivity to N-phenylcarbamates is not necessarily associated with resistance to benzimidazoles. A mutation in the nda2 gene that codes for alpha-tubulin confers supersensitivity to N-phenylcarbamates as well as benzimidazoles (25). Thus, mutation in the tubulin gene may decrease stability of the microtubule and increase the sensitivity to benzimidazoles and N-phenylcarbamates. We assumed that N-phenylcarbamates bind to beta-tubulin but not to alpha-tubulin. The observation that the reverants from the F914 strain were caused by mutation of the beta-tubulin gene supports this idea.

Table IV. Genetic analysis of revertants of N. crassa

Cross		Numbers of progenies			
		BsNr	BrNs	BrNr	Total
W x					
	F422	76	0	0	76
	F424	84	0	0	84
	F513	55	0	50	105
	F421	58	0	63	121
	F423	44	0	48	92
F914 x					
	F422	50	49	0	99
	F424	59	53	0	112
	F513	0	64	59	123
	F421	0	62	52	114
	F423	0	54	63	117

BsNr :Sensitive to MBC and insensitive to diethofencarb
BrNs :Resistant to MBC but sensitive to diethofencarb
BrNr :Resistant to both MBC and diethofencarb

Molecular biology of negatively correlated cross resistance

To obtain structural information on the beta-tubulin in our mutants, we cloned the beta-tubulin gene from the F914 strain, using the Bml511 beta-tubulin gene as a probe. This tubulin gene(Bml511), which has been cloned on plasmid pSV50 (27), was a gift from Dr S. Vollemer. Only one band, approximately 3.1kbp, was detected in HindIII fragments of the genomic DNA of the F914 strain, a finding consistent with the data of Orbach et al (18). We then constructed a gene library of HindIII fragments, using the vector pUC13.

One positive clone was isolated from approximately 20,000 colonies by colony hybridization, using the beta-tubulin gene as a probe. This plasmid, named pEF50, cross-hybridized to the probe, even under conditions of high stringency. The digestion pattern of pSV50 and pEF50 by restriction enzymes indicated that this plasmid contained the beta-tubulin gene.

To determine the region of the beta-tubulin gene responsible for MBC resistance and diethofencarb supersensitivity, we constructed chimeric beta-tubulin genes from the plasmid pSV50 and pEF50(Figure 3). For the controls, the parent plasmids pSV50 and pEF50 were introduced into wild type spheroplasts. Transformants containing the plasmid pSV50 were obtained on either MBC medium or MBC+diethofencarb medium. On the other hand, transformants containing the plasmid pEF50 were obtained on MBC medium but not on MBC+diethofencarb medium. The pSV50 was responsible only for MBC resistance but not for the diethofencarb supersensitivity. The pEF50 was responsible for MBC resistance and diethofencarb supersensitivity.

We obtained MBC resistant transformants on introduction of each plasmid (from pEF51 to pEF54) into wild type cells. This indicates that the mutations of pEF50 and pSV50 are relatively close. However, on introduction of pEF52 and pEF53, we obtained transformants on MBC+diethofencarb medium but not transformants by introduction of pEF51 and pEF54. Therefore, the mutation of pEF50 may be between the EcoRI and BamHI sites. Orbach et al. reported that the mutation point of pSV50 is at position 167 on the amino acid sequence. This mutation site of pSV50 is located in this EcoRI-BamHI region.

We determined the nucleotide sequence of this region by the dideoxy method, and detected one amino acid substitution. An A to G transition occurred in the second base of coden 198. This means that the amino acid substitution is glutamic acid to glycine at position 198. The amino acid of the position 167 of plasmid pEF50 was the same as that of the wild type strain.

Discussion

Mutations in beta-tubulin gene and fungicide binding. Table V summarizes the amino acid substitutions of beta-tubulin in benzimidazole-resistant fungi. In Bm1511 of N. crassa, the amino acid at position 167 was changed from phenylalanine to tyrosine (18). In S. cerevisiae, the amino acid at position 241 was arginine in the wild type strain but histidine in ben-R (22). In the case of the F914 strain of N. crassa, the amino acid was changed from glutamic acid to glycine at position 198. This amino acid change may be in the binding region for both MBC and diethofencarb. Alternatively, this change could be the outside of the compound binding region but may alter the conformation of beta-tubulin. A third possibility is that diethofencarb binds to some other protein, and that this complex in turn interacts with the altered beta-tubulin but not with wild type beta-tubulin. However, the mutations responsible for diethofencarb resistance were mapped in the beta-

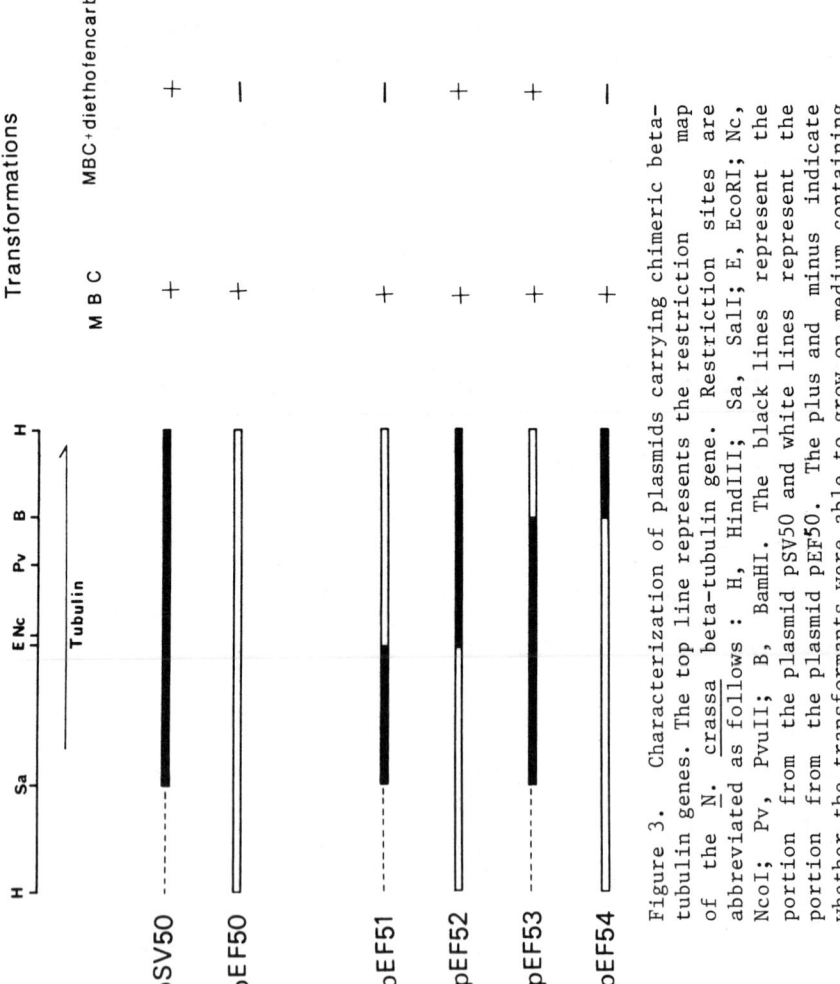

Figure 3. Characterization of plasmids carrying chimeric beta-tubulin genes. The top line represents the restriction map of the N. crassa beta-tubulin gene. Restriction sites are abbreviated as follows: H, HindIII; Sa, SalI; E, EcoRI; Nc, NcoI; Pv, PvuII; B, BamHI. The black lines represent the portion from the plasmid pSV50 and white lines represent the portion from the plasmid pEF50. The plus and minus indicate whether the transformants were able to grow on medium containing the chemicals.

tubulin gene, and the precise locations of a mutation causing benzimidazole resistance in these fungi were relatively close. Therefore, this region most likely includes the binding domain of both benzimidazoles and N-phenylcarbamates.

Table V. Amino acid substitutions of benzimidazole resistants

	Position of amino acid substitution		
	167	198	241
N. crassa WT	phe	glu	arg
N. crassa Bm1511	tyr	glu	arg
N. crassa F914	phe	gly	arg
S. cerevisiae WT	phe	glu	arg
S. cerevisiae ben-R	phe	glu	his
Chicken	phe	glu	arg

A model for explanation of negatively correlated cross resistance. Compound binding studies with extracts from the wild type strain and benzimidazole-resistant mutants in fungi such as B. cinerea, Fusarium oxysporum f. sp. lycopersici, Penicillium brevicompactum, P. corymbiferum, Venturia nashicola showed that only the sensitive strains showed positive binding activity to benzimidazoles (4, 6, 8). Thus it appears that the affinity of the target site to benzimidazoles determines whether or not benzimidazoles have antifungal activity. In addition, Ishii et al., showed that only the proteins extracted from strains exhibiting a negatively correlated cross resistance to MDPC and DCPF (15) had binding activity to DCPF.

Based on these data we describe a model of negatively correlated cross resistance between MBC and diethofencarb at the molecular level (Figure 4). However it must be kept in mind that this model is speculative, given the absence of rigorous physical and chemical analysis of the protein. For diethofencarb binding, it seems that the change of glutamic acid to glycine at position 198 is essential, and that this change reducing the affinity of MBC to beta-tubulin. We propose that the imidazole of MBC has affinity to glutamic acid, but that this residue has no affinity to diethofencarb. When this polar residue of glutamic acid is lost, diethofencarb fits the open space.

Validity of Neurospora as a model organism. When one is examining the resistance mechanisms of plant pathogens using model organisms, it is important to show that similar mechanisms works in the field isolates. In this case, benzimidazole resistant B. cinerea isolates from the field show several characteristics supporting the validity of using N. crassa as model system. First, both the F914 strain and MBC resistant B. cinerea showed clear negatively correlated cross resistance to N-phenylcarbamates. Second, the cytological effects of

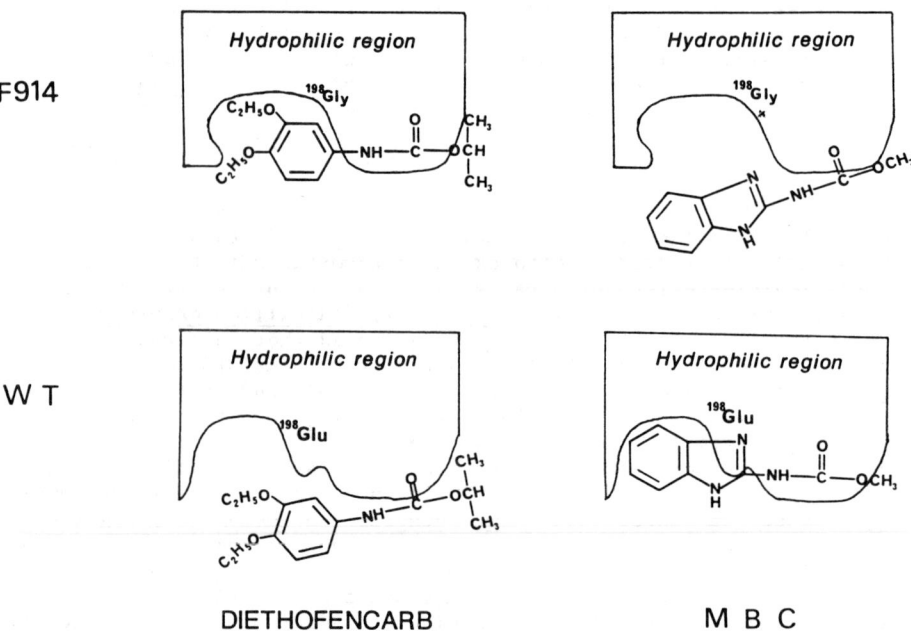

Figure 4. A putative model of binding of MBC and diethofencarb to wild type and altered beta-tubulin.

diethofencarb are similar in both N. crassa and B. cinerea. Third, both the F914 strain and B. cinerea(MBC-resistance) showed negatively correlated cross resistance to diphenylamine(19). Therefore, it will be considered that the resistance mechanism seen with N. crassa is essentially the same to that of B. cinerea.

However difference aspect was observed in resistance to MBC between N. crassa isolates in laboratory and B. cinerea isolates from the field. Laboratory-generated mutants of N. crassa showed various levels of resistance to MBC, and a specific one showed negatively correlated cross resistance to diethofencarb. Others showed double resistance to these chemicals. On the other hand, most of the benzimidazole resistants isolated from the field, especially in in the case of B. cinerea, were supersensitive to diethofencarb. This difference between laboratory and field strains may derive from a difference in fitness among the resistant strains of the plant pathogen.

We studied diethofencarb- and benzimidazole- resistant mechanisms, using the model fungus, N. crassa and determined the mutation site of the beta-tubulin gene responsible for the benzimidazole resistance and N-phenylcarbamate supersensitivity. Studies on the mode of action of benzimidazole and N-phenylcarbamates are useful because the phenomenon of negatively correlated cross resistance provides information on the mode of binding more than does the common positively correlated cross resistance. Bassed on these studies, we proposed a model for the differential binding of the two classes of fungicides with the two types of tubulin. These approaches should aid in clarifying the binding mode of N-phenylcarbamates to beta-tubulin. It might be helpful to design the new potent fungicide, which inhibits the microtubule functions, to overcome the resistant problem.

Literature Cited

1. Borck, K.; Braymer, H.D. J. Gen. Microbiol. 1974, 85, 51-56.
2. Clemons, G. P.; Sisler, W. D. Phytopathology 1969, 59, 705-706.
3. Davidse, L. C. Pest. Biochem. Physiol. 1973, 3, 317-325.
4. Davidse, L. C.; Flach, W. J. Cell. Biol. 1977, 72, 174-194.
5. Delp, C. J. Plant Diseases 1980 64, 652-657.
6. Gessler, C.; Sozzi, D.; Kern, H. Ber Schweiz. Bot. Ges. 1980, 90, 45-54.
7. Hiraoka, Y.; Toda, T.; Yanagida, M. Cell 1984, 39, 349-358.
8. Ishii, H.; Davidse, L. C. British Crop Protection Conference-Pest and Diseases 1986, 567-572.
9. Kato, T.; Suzuki, K.; Takahashi, J.; Kamoshita, K. J. Pesticide Sci. 1984, 9, 489-495.
10. Leroux, P.; Gredt, M.C.R. Acad. Sci. Paris Ser. D 1979, 289, 691-693.

11. Leroux, P.; Gredt, M. Phytiatr. Phytopharm. 1979, 28, 79-86.
12. Leroux, P.; Gredt, M. Weed Res. 1980, 20, 249-254.
13. May, G.S.; Tsang, M.L.-S.; Smith, H.; Fidel, S.; Morris, N.R. Gene 1987, 55, 231-243.
14. Nakamura, S.; Kato, T.; Noguchi, H.; Takahashi, J.; Kamoshita, K.; Kato, T. Abstracts 6th Int. Congr. Pestic. Chem. Ottawa,Canada E-01 1986.
15. Nakata, A.; Sano, S.; Hashimoto, S.; Hayakawa, K.; Nishikawa, H.; Yasuda, Y. Ann. Phytopath.Soc. Japan 1987 53, 659-662.
16. Neff, N.F.; Thomas, J.H.; Grisafi, P.; Botstein, D. Cell 1983, 33, 211-219.
17. Oakley, B. R.; Morris, N. R. Cell 1981, 24, 837-845.
18. Orbach, M.J.; Porro, E.B.; Yanofsky C. Mol. Cell, Biol. 1986 6, 2452-2461.
19. Rosenberger, D. A.; Meyer, F. W. Phytopathology 1985, 75, 74-79.
20. Sheir-Neiss, G.; Lai, M.; Morris, N. Cell 1978, 15, 639-647.
21. Suzuki, K.; Kato, T.; Takahashi, J.; Kamoshita, K. J. Pesticide Sci. 1984, 9, 497-501.
22. Thomas, J.H.; Neff, N. F.; Botstein, D. Genetics 1985, 112, 715-734.
23. Takahashi, J.; Nakamura, S.; Noguchi, H.; Kato, T.; Kamoshita, K. J. Pesticide Sci. 1988 13, 63-69.
24. Takahashi, J.; Kirino, O.; Takayama, C.; Nakamura, S.; Noguchi, H.; Kato, T.; Kamoshita, K. Pestic. Biochem. Physiol. 1988 30, 266-271.
25. Umesono, K.; Toda, T.; Hayashi, S.; Yanagida, M. J. Mol. Biol. 1983, 168, 271-284.
26. Vanlenzuela, P.; Quiroga, M.; Zaldivar, J.; Rutter, W.J.; Kirschner, M. W.; Cleaveland, D. W. Nature (London) 1981, 289, 650-655.
27. Vollemer, S. J.; Yanofsky, C. Proc. Natl. Acad. Sci. USA 1986, 83, 4869-4873.

RECEIVED September 1, 1989

Chapter 16

Binding of Cellular Protein from *Venturia nashicola* Isolates to Carbendazim

Its Relationship with Sensitivity to N-Phenylcarbamates, N-Phenylformamidoximes, and Rhizoxin

H. Ishii[1], S. Iwasaki[2], Z. Sato[3], and I. Inoue[1]

[1]Fruit Tree Research Station, Tsukuba, Ibaraki 305, Japan
[2]Institute of Applied Microbiology, University of Tokyo, Tokyo 113, Japan
[3]National Institute of Agro-Environmental Sciences, Tsukuba, Ibaraki 305, Japan

> Most of the highly benzimidazole-resistant isolates of Venturia nashicola (the scab fungus of Japanese pear) show increased sensitivity to the N-phenylcarbamate compound MDPC and the N-phenylformamidoxime compound DCPF, but the intermediately and weakly benzimidazole-resistant isolates did not. Doubly resistant isolates, i.e. highly benzimidazole-resistant isolates with the resistance to MDPC and DCPF were found. It is suggested that a genetic change of binding site(s) is involved in the change of sensitivity to the benzimidazole fungicide carbendazim. The ^{14}C-carbendazim binding in a carbendazim-sensitive isolate was exchangeable with nonradioactive carbendazim, but no inhibition of the binding was observed by pretreatment with unlabeled MDPC or DCPF. Mycelial growth of this fungus was strongly inhibited by the microtubule inhibitor rhizoxin, irrespective of benzimidazole resistance. This result suggests that different binding site(s) exist in the molecule of tubulin-like protein besides the binding site(s) to carbendazim, and new approaches to combating resistance are suggested.

The benzimidazole fungicides benomyl and thiophanate-methyl have been used extensively for the control of important plant diseases. During the last decade, however, benzimidazole resistance of phytopathogenic fungi has increased throughout the world, making it difficult to control many plant diseases with benzimidazole fungicides. In fields where the resistant strains have predominated, use of benzimidazole fungicides

0097–6156/90/0421–0237$06.00/0
© 1990 American Chemical Society

was stopped or restricted, and these fungicides were replaced by other fungicides with a different mechanism of action. However, monitoring data in the field demonstrated the persistence of benzimidazole resistance (1). More rational approaches, therefore, would be required to compete with fungicide resistance to this class of compounds.

Some years ago, two groups of compounds, i.e. N-phenylcarbamates and N-phenylformamidoximes were synthesized in Japan and their specific antifungal activity against benzimidazole-resistant strains was reported (2, 3). In France, where benzimidazole-resistant strains of the gray mold Botrytis cinerea predominate in some areas, a N-phenylcarbamate compound diethofencarb is already in use as a mixture with a benzimidazole fungicide carbendazim.

Although the mechanism of action of N-phenylcarbamates and N-phenylformamidoximes to benzimidazole-resistant strains of fungi remains unknown, recent studies indicate that the differential binding activity of cell-free mycelial extracts to a N-phenylformamidoxime compound, DCPF, might be involved in the increased sensitivity to this compound (4). Furthermore, Ishii and Davidse (5) suggested that the binding of ^{14}C-carbendazim to tubulin-like protein of benzimidazole-resistant isolates of Venturia nashicola was lower than to the protein from a sensitive isolate, and that decreased affinity of the binding site(s) for carbendazim might result in decreased sensitivity to carbendazim.

A microtubule inhibitor, rhizoxin (Fig. 1) was found in the culture filtrate of Rhizopus chinensis (6). Subsequently, Iwasaki et al. (7) determined the chemical structure of this compound, and the antitubulin activity of rhizoxin was characterized using porcine brain tubulin (8).

In the present paper, binding of carbendazim to the cell-free mycelial extracts of V. nashicola and potential interference of this binding by MDPC, DCPF or rhizoxin were examined. Antifungal activity of rhizoxin to the benzimidazole-resistant isolates of this fungus was also tested.

Materials and Methods

Isolates. Pear leaves with sporulating V. nashicola were collected from orchards in Japan, and conidia from the lesions were suspended in sterile distilled water. Drops of conidial suspensions were placed on water agar plates and incubated at 15 C for 3 to 5 days. Agar blocks containing germinated conidia were individually isolated with a steel needle after microscopic observation and transferred onto potato dextrose agar (PDA) slants. After incubation at 20 C for about two months, the

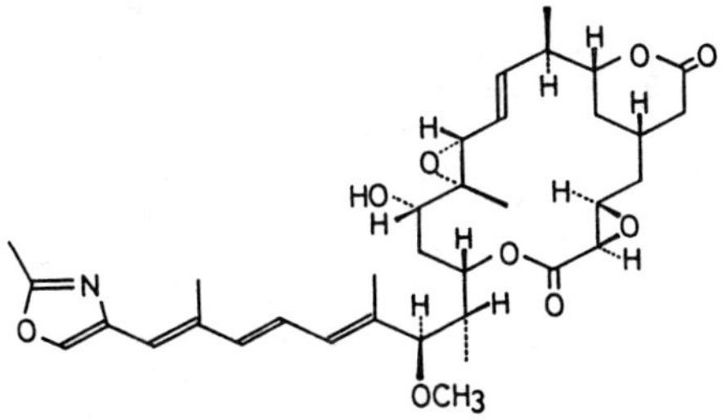

Figure 1. Structure of rhizoxin.

cultures were stored at 10 C. Twenty-one isolates in total were used. An isolate which acquired high resistance to carbendazim after repeated subculture of an intermediately resistant isolate on carbendazim-amended culture medium described earlier (9), was also used.
Chemicals. Carbendazim and ^{14}C-carbendazim (specific activity 4.5 mCi/mmole) were kindly supplied by Du Pont Japan Ltd., Tokyo and Hoechst AG, Frankfurt am Main, West Germany, respectively. N-(3,5-dichloro-4-propynyloxyphenyl)-N'-methoxyformamidine (DCPF) was a gift from Nippon Soda Co. Ltd., Tokyo. Methyl N-(3,5-dichlorophenyl)carbamate (MDPC) was synthesized by the method of Kato et al. (10). Rhizoxin was isolated from the culture broth of R. chinensis as reported earlier (7).
Sensitivity to carbendazim, MDPC, DCPF and rhizoxin. Each isolate was previously cultured on PDA plates at 20 C for 45 days. Mycelial discs, 4 mm in diameter, were cut from the margins of the colonies and transferred onto PDA plates amended with carbendazim, MDPC, DCPF or rhizoxin. Stock solution of these compounds were obtained by dissolving technical grades in acetone. Addition of the proper amount of stock solution to molten PDA after autoclaving gave a series of concentrations of each compound. The final acetone concentration in PDA was 2 %. After incubation at 20 C for 3 weeks, mycelial growth of the isolates on PDA plates was observed, and the minimum inhibitory concentrations (MIC) of carbendazim, MDPC or DCPF necessary for complete inhibition of mycelial growth of the isolates were determined. In addition, the maximum allowable concentrations (MAC) of rhizoxin which allows similar degree of mycelial growth of the isolates to that in untreated control were determined. Based on the difference in sensitivity to carbendazim, isolates were devided into four groups as follows: highly carbendazim-resistant, MIC > 100 μg/ml; intermediately carbendazim-resistant, 100 μg/ml ⩾ MIC >10 μg/ml; weakly carbendazim-resistant, 10 μg/ml ⩾ MIC > 1 μg/ml; and carbendazim-sensitive, 1 μg/ml ⩾ MIC.
Preparation of cell-free mycelial extracts. Mycelial discs of a carbendazim-sensitive isolate JS-18, cut from the cultures grown on PDA at 20 C for 45 days were transferred to a liquid medium containing 0.3 % yeast extract, 0.3 % malt extract, 0.5 % peptone, and 1 % glucose (wt/vol). After incubation at 20 C for one month, mycelia collected from the cultures were washed with sterile distilled water and homogenized aseptically. The mycelial fragments thus obtained were further used to inoculate the liquid medium described above and incubated at 20 C for one week. Fresh mycelia, harvested from cultures in the liquid medium, were washed in cold 0.05 M potassium phosphate buffer, pH 6.8 containing 0.1 M KCl and 0.005 M MgCl$_2$ (PKMg solution, ref. 11) and placed in a previously cooled (-20 C) X-Press Cell Disintegrator (AB BIOX, Sweden) with 1 ml of PKMg solution per gram wet

weight of mycelium. After freezing, the mycelia were disintegrated, and the homogenates were centrifuged at 40,000xg for 10 min. The 40,000xg supernatant was further centrifuged at 48,000xg for 30 min. The supernatant was amended with 0.1 mM guanosine triphosphate (GTP) to stabilize carbendazim-binding activity of tubulin and immediately used for binding assays. All procedures were done at 4 C.

Binding assays. Aliquots of the 48,000xg supernatant from a benzimidazole-sensitive isolate JS-18 were incubated with increasing concentrations of unlabeled carbendazim, MDPC, DCPF or rhizoxin for 1 hr and then with 2 or 4 μM ^{14}C-carbendazim for an additional hour at 4 C. Binding of ^{14}C-carbendazim was measured with a charcoal assay (11). Aliquots of the incubation mixture were placed in centrifuge tubes containing an equal volume of a charcoal suspension at 6 mg/ml in PKMg solution. The tubes were placed in a microtube-mixer, and the mixture was heavily agitated for 10 min followed by the centrifugation at 10,000xg for 2 min at 4 C. Aliquots of the supernatant were taken and the radioactivity present in each was measured by liquid scintillation spectrometry. To determine nonspecific interaction of carbendazim with proteins, blanks which contained 10 mg/ml bovine serum albumin in PKMg solution were incubated with ^{14}C-carbendazim and handled in the same way mentioned above. The difference in amount of radioactivity bound in supernatant aliquots of sample and blank was considered to represent bound carbendazim.

Results

Sensitivity to carbendazim, MDPC, DCPF and rhizoxin. Sensitivity of 21 isolates to carbendazim, MDPC, DCPF and rhizoxin in vitro is listed in Tables I. and II. Out of 6 highly carbendazim-resistant isolates, five isolates showed a large increase of sensitivity to MDPC and DCPF as compared with five carbendazim-sensitive isolates (Table I). One highly carbendazim-resistant isolate (a laboratory mutant) JS-40M did not show such an increased sensitivity to the N-phenylcarbamate and the N-phenylformamidoxime compounds. The increased sensitivity to MDPC and to DCPF was not observed in five intermediately carbendazim-resistant isolates nor in five weakly carbendazim-resistant isolates. On the other hand, the mycelial growth of all the isolates tested were strongly suppressed on the culture medium treated with rhizoxin, irrespective of benzimidazole resistance (Table II).

Binding assays with ^{14}C-carbendazim. If MDPC, DCPF or rhizoxin binds to the fungal cellular protein at the carbendazim-binding site(s), then a decrease of ^{14}C-carbendazim binding to tubulin-like protein would be expected in the presence of these compounds. In order to

Table I. Sensitivity of Venturia nashicola isolates to carbendazim, MDPC and DCPF

Isolate	Reaction to carbendazim[a]	Minimum inhibitory concentration (µg/ml) of:		
		carbendazim	MDPC	DCPF
JS- 18	S	≤0.19	25	>800
JS- 19	S	≤0.19	50	>800
JS- 20	S	≤0.19	25	>800
JS- 49	S	≤0.19	25	>800
JS- 75	S	≤0.19	50	>800
JS- 53	WR	1.56	50	>800
JS- 54	WR	1.56	50	>800
JS- 55	WR	1.56	25	>800
JS- 56	WR	1.56	50	>800
JS- 57	WR	3.12	50	>800
JS- 39	IR	25	50	>800
JS- 40	IR	50	25	>800
JS- 41	IR	50	50	>800
JS- 42	IR	50	50	>800
JS- 43	IR	50	50	>800
JS-134	HR	>800	3.12	0.19
JS-137	HR	>800	3.12	0.19
JS-140	HR	>800	3.12	0.19
JS-111	HR	>800	1.56	0.19
JS-114	HR*	>800	1.56	0.19
JS- 40M	HR*	>800	25	>800

[a] Abbreviations: HR, Highly resistant; HR*, Highly resistant (laboratory mutant); IR, Intermediately resistant; WR, Weakly resistant; S, Sensitive.

Table II. Sensitivity of *Venturia nashicola* isolates to rhizoxin

Isolate	Maximum allowable concentration (μg/ml) of rhizoxin
Carbendazim-sensitive	
JS- 18	<0.19
JS- 19	<0.19
JS- 20	<0.19
JS- 49	<0.19
JS- 75	<0.19
Weakly carbendazim-resistant	
JS- 53	<0.19
JS- 54	<0.19
JS- 55	<0.19
JS- 56	0.19
JS- 57	<0.19
Intermediately carbendazim-resistant	
JS- 39	<0.19
JS- 40	<0.19
JS- 41	<0.19
JS- 42	<0.19
JS- 43	0.19
Highly carbendazim-resistant	
JS-134	<0.19
JS-137	<0.19
JS-140	<0.19
JS-111	<0.19
JS-114	<0.19
JS- 40M[a]	<0.19

[a] Laboratory mutant.

check this point, cellular protein from a carbendazim-sensitive isolate was previously treated with unlabeled carbendazim, MDPC, DCPF or rhizoxin and then incubated with ^{14}C-carbendazim. The ^{14}C-carbendazim binding activity of the protein was reduced when progressively higher concentrations of unlabeled carbendazim was added to the incubation mixture. In the presence of 80 µM of unlabeled carbendazim and 4 µM of ^{14}C-carbendazim incubated, approximately 80 % inhibition of the ^{14}C-carbendazim binding was observed (Fig. 2). Similar experiments were carried out using unlabeled MDPC, DCPF or rhizoxin. However, no inhibition of the ^{14}C-carbendazim binding was observed by the pretreatment with unlabeled MDPC, DCPF or rhizoxin (Fig. 3, 4). Although the reason was unclear, the binding activity increased in the presence of excess unlabeled MDPC (Fig. 3).

Discussion

A potentially attractive strategy to cope with fungicide resistance is development of chemicals to which the resistant strains show negatively correlated cross-resistance or increased sensitivity. However, this has been difficult to achieve in practice until recently. The compounds studied in this investigation, having specific antifungal activity against the benzimidazole-resistant strains, represent one of the first cases where this concept can be tested. For example, introduction of diethofencarb in vineyards in France seemed to be a successful example of the practical use of negatively correlated cross-resistance or increased sensitivity. Leroux et al. (12) reported that diethofencarb applications alone or a mixture with carbendazim controlled benzimidazole-resistant strains of B. cinerea where such strains had predominated. In Israel also, mixtures of diethofencarb and benzimidazoles gave good results in the control of gray mold in vineyards (13).

However, to properly use such compounds in resistance management, information is needed on the molecular basis for negatively correlated cross-resistance or increased sensitivity, and on the behavior of various pathogens possessing the target site in question. In V. nashicola, for example, increased sensitivity to diethofencarb and to MDPC was observed in highly carbendazim-resistant isolates but not in isolates with intermediate or weak resistance (14). The same results were obtained for the sensitivity to a N-phenylformamidoxime compound DCPF (Table I.). Furthermore, as reported recently (14), doubly resistant strains, i.e. highly carbendazim-resistant strains with the resistance to MDPC, can be obtained from the cross of a highly carbendazim-resistant isolate (sensitive to MDPC) with an intermediately carbendazim-resistant or a carbendazim-sensitive isolate (both resistant to MDPC).

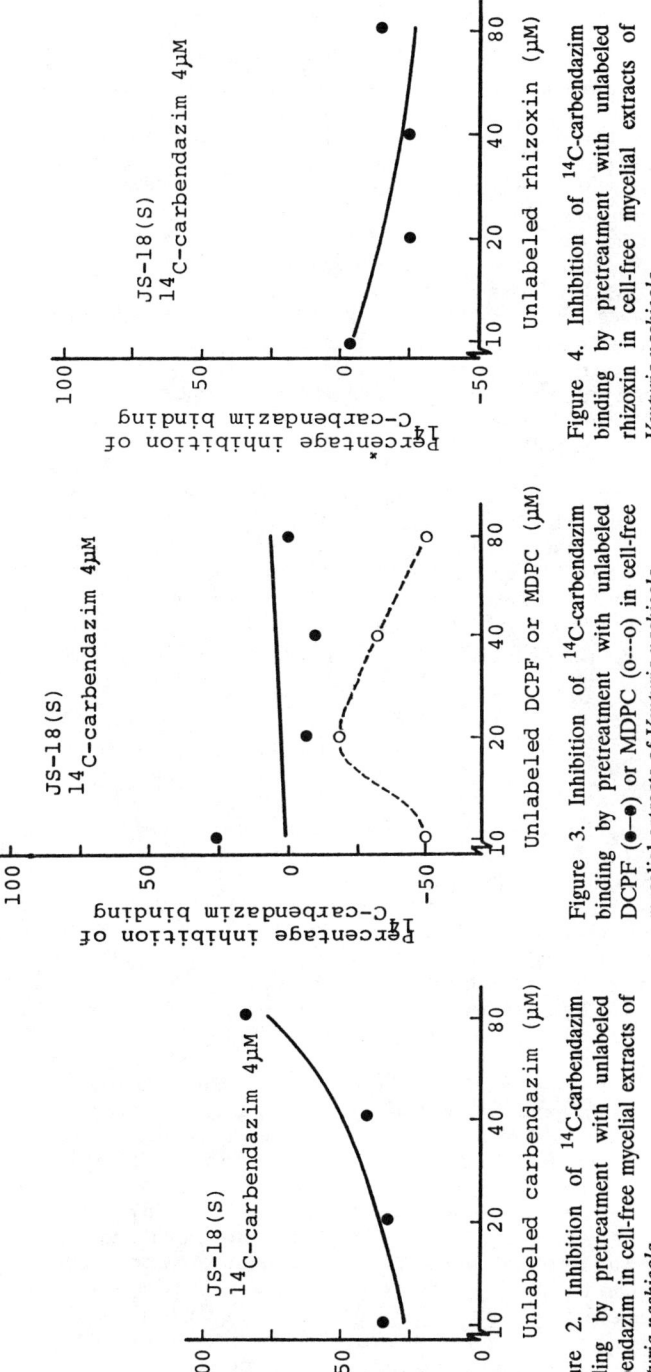

Figure 2. Inhibition of ^{14}C-carbendazim binding by pretreatment with unlabeled carbendazim in cell-free mycelial extracts of *Venturia nashicola*.

Figure 3. Inhibition of ^{14}C-carbendazim binding by pretreatment with unlabeled DCPF (●—●) or MDPC (o--o) in cell-free mycelial extracts of *Venturia nashicola*.

Figure 4. Inhibition of ^{14}C-carbendazim binding by pretreatment with unlabeled rhizoxin in cell-free mycelial extracts of *Venturia nashicola*.

The doubly resistant strains also were detected among field isolates of this fungus, even though N-phenylcarbamates have never been used in practice in Japan yet. Inheritance of the double resistance was further demonstrated experimentally. These results point out a potential limitation of the use of N-phenylcarbamates as a "resistance-breaker".

Ishii et al. (9) have already demonstrated that the occurrence of three different levels of benzimidazole resistance is due to three allelic mutations in a single gene, and that each level is controlled by one of the multiple alleles. Subsequently, it was suggested that the increased sensitivity to N-phenylcarbamates is controlled by a single chromosomal gene(14). Therefore, one important question to be clarified is whether the increased sensitivity to N-phenylcarbamates in highly benzimidazole-resistant isolates results from a pleiotropism of the same allele coding high benzimidazole-resistance.

We propose that genetic changes may have occurred to alter the characteristics of binding site(s) for benzimidazole fungicides in resistant strains. Differential affinities of these binding sites then account for the differential sensitivity observed. Although the experimental evidence has not yet been fully obtained, in V. nashicola, the binding site(s) of benzimidazoles is very likely to be β-subunit of the microtubule protein tubulin, as found in studies on Aspergillus nidulans (15). In this study, it was shown that decreased affinity of benzimidazoles for tubulin-like protein was the case of benzimidazole resistance (11). As reported previously (5), the decreased affinity of tubulin-like protein from benzimidazole-resistant isolates of V. nashicola to carbendazim is involved in the resistance. Furthermore, cellular protein from a highly benzimidazole-resistant isolate which was sensitive to DCPF exhibited higher binding activity to ^{14}C-DCPF than that from the intermediately benzimidazole-resistant, weakly resistant and sensitive isolates (4). Therefore, the difference of sensitivity to DCPF might be due to the difference of affinity level of the binding site(s) to this compound.

Moving to the molecular level, a second question is whether the binding site(s) of N-phenylcarbamates or N-phenylformamidoximes is the same as the binding site(s) of benzimidazoles or not. In the present paper, the binding of ^{14}C-carbendazim to cellular protein from a benzimidazole-sensitive isolate decreased in preparations which were pretreated with increasing concentrations of unlabeled carbendazim (Fig. 2). It is suggested that the cell-free mycelial extracts from a benzimidazole-sensitive isolate of V. nashicola contains a protein that specifically binds to carbendazim. On the other hand, no inhibition of the ^{14}C-carbendazim binding was observed by pretreatment with unlabeled MDPC or DCPF (Fig. 3). It is clear that DCPF does not interfere with the carbendazim binding to the protein from a benzimidazole-sensitive isolate at reasonable concentrations. It is possible that DCPF could bind to the carbendazim binding site(s) on the protein isolated from a benzimidazole-sensitive isolate but not alter the configuration/function of this protein. This paper does present data disproving this hypothesis. Stimulation of the carbendazim binding by MDPC may be very interesting and deserves more attention. Similar enhancement of binding was reported in the interaction of phomopsin A and related compounds with purified sheep brain tubulin (16).

An especially interesting case is the microtubule inhibitor rhizoxin, for which antifungal activity has already been reported (17). In this paper, the effect of rhizoxin on mycelial growth of benzimidazole-resistant isolates of phytopathogenic fungi has been examined for the first time. In V. nashicola, remarkable inhibition of the growth was observed in all isolates tested, irrespective of benzimidazole resistance of the isolates (Table II). Takahashi et al. (8) recently reported that rhizoxin binds to tubulin from porcine brain, and that the binding was completely independent from the binding of antitubulin colchicine. Furthermore, Davidse and Flach (11) found that the carbendazim binding was competitively inhibited by colchicine in A. nidulans. In Fig. 4, no competitive inhibition of the ^{14}C-carbendazim binding was observed by rhizoxin. Therefore, it is likely that rhizoxin does not bind to the tubulin-like protein at the carbendazim binding site(s). Rhizoxin represents an exciting new case, since it indicates that other binding site(s) in tubulin molecule can be targeted for fungal control.

Acknowledgments

The authors gratefully acknowledge Dr. L. C. Davidse, Department of Phytopathology, Agricultural University, Wageningen, The Netherlands for technical advice. Contribution No.: Fruit Tree Res. Stn., A-246.

Literature Cited

1. Ishii, H.; Udagawa, H; Yanase, H.; Yamaguchi, A. Plant Pathol. 1985, 34, 363-368.
2. Kato, T.; Suzuki, K.; Takahashi, J.; Kamoshita, K. J. Pestic. Sci. 1984, 9, 489-495.
3. Nakata, A.; Sano, S.; Hashimoto, S.; Hayakawa, K.; Nishikawa, H.; Yasuda, Y. Ann. Phytopath. Soc. Japan 1987, 53, 659-662.
4. Ishii, H.; Takeda, H. Neth. J. Pl. Path. 1989, 95 Supplement 1, in press.
5. Ishii, H.; Davidse, L. C. Proc. 1986 Br. Crop Prot. Conf. - Pests and Diseases, 1986, Vol. 2, pp 567-573.

6. Sato, Z.; Noda, T.; Goh, N.; Kobayashi, H.; Okuda, S. Ann. Phytopath. Soc. Japan 1978, 44, 349 (in Japanese).
7. Iwasaki, S; Namikoshi, M.; Kobayashi, H.; Furukawa, J.; Okuda, S.; Itai, A.; Kasuya, A.; Iitaka, Y.; Sato, Z. J. Antibiotics 1986, 39, 424-429.
8. Takahashi, M.; Iwasaki, S.; Kobayashi, H.; Okuda, S.; Murai, T.; Sato, Y. Biochimica et Biophysica Acta 1987, 926, 215-223.
9. Ishii, H.; Yanase, H.; Dekker, J. Med. Fac. Landbouww. Rijksuniv. Gent 1984, 49/2a, 163-172.
10. Kato, T.; Takahashi, J.; Kamoshita, K. Japan Kokai Tokkyo Koho (A) 57-80309 (in Japanese).
11. Davidse, L. C.; Flach, W. J. Cell Biol. 1977, 72, 174-193.
12. Leroux, P.; Gredt, M.; Massenot, F.; Kato, T. Proc. Bordeaux Mixture Centenary Meeting, 1985, Vol. 2, pp 443-446.
13. Elad, Y.; Shabi, E.; Katan, T. Plant Pathol. 1988, 37, 141-147.
14. Ishii, H.; van Raak, M. Phytopathology 1988, 78, 695-698.
15. Sheir-Neiss, G.; Lai, M. H.; Morris, N. R. Cell 1978, 15, 639-647.
16. Lacey, E.; Edgar, J. A.; Culvenor, C. C. J. Biochem. Pharmac. 1987, 36, 2133-2138.
17. Iwasaki, S.; Kobayashi, H.; Furukawa, J.; Namikoshi, M.; Okuda, S.; Sato, Z.; Matsuda, I.; Noda, T. J. Antibiotics 1984, 37, 354-362.

RECEIVED October 16, 1989

Chapter 17

Edifenphos Resistance in *Pyricularia oryzae* and *Drechslera oryzae*

In Vitro Techniques for Detection and Biochemical Studies

D. Lalithakumari and P. Annamalai

Centre for Advanced Studies in Botany, University of Madras, Guindy, Madras 600 025, India

> Resistance to edifenphos (Hinosan, EDDP) in *Pyricularia oryzae* and *Drechslera oryzae* has been noticed in the field. EDDP resistant strains of *P. oryzae* and *D. oryzae* were obtained *in vitro*, by UV irradiation, adaptation and chemical mutagenesis. The EDDP resistant mutants were found to be stable and pathogenic. Morphological variations were not observed in EDDP mutants of *P. oryzae* while the field mutants of *D. oryzae* distinctly differed in morphology and pigmentation. The Mechanism of resistance to EDDP was presumed to be alteration in the membrane integrity and not in its site of action, as phosphatidylcholine remained unaffected in mutants. Plasmid mediated resistance to EDDP was presumed, as the intensity of plasmid bands was high in resistant mutants of *P. oryzae* and *D. oryzae*. *P. oryzae* mutants showed high sensitivity to Ziram and *D. oryzae* mutants to Mancozeb in cross-resistance studies.

The National Agricultural Research System in India brings out a number of new crop varieties and hybrids but it is well known that in sub-tropical and tropical climates crop resistance to obligate parasites does not last for more than 4 or 5 years. Modern plant breeding technology and improved crop improvement practices do not solve this problem. Hence chemical control is indispensable and highly effective in controlling many major diseases. Systemic fungicides have improved tremendously. There are many lines of evidence suggesting that systemic fungicides not only control the pathogen but also alter the host physiology thereby enhancing the defence mechanism of the host plant (1-3). Problems of pathogen resistance to conventional fungicides have been reported by many workers in India (4-7). But increasing reports of fungal resistance to systemic fungicides are left unnoticed or only very few occurrences are reported so far (7,8-12). Work on molecular basis of resistance and control measures is completely lacking in India. Hence, the present investigation emphasizes the mechanism, molecular basis of fungicide resistance and an easy method to control or prevent fungicide resistance in Pyricularia oryzae and Drechslera oryzae against edifenphos (EDDP, Hinosan).

MATERIALS AND METHODS

Fungal strains: Pyricularia oryzae and Drechslera oryzae (parent), sensitive to edifenphos were maintained on potato dextrose agar (PDA). The edifenphos-resistant mutants were maintained on edifenphos-amended PDA. For the present study, uniformaly 8 and 10-day-old cultures of D. oryzae and P. oryzae, respectively, were used throughout the study.

Mutagenesis: Chemical mutagens ethyl methane sulphonate (EMS) and N-methyl-N-nitro-N-nitrosoguanidine (NTG) were used as mutagenic agents. Ungerminated conidia of P. oryzae and D. oryzae at 2.5×10^5 conidia/ml were treated with EMS, 6 mg/ml in pH 7.0 sodium phosphate buffer and NTG, 0.5mg/ml in pH 9.0 Tris maleic acid buffer (13). Treated conidia were plated on EDDP-amended medium containing EDDP 5 times the effective dose required to inhibit 50% of parent strain (ED_{50}). Colonies appeared on the amended medium were picked out and maintained on fungicide-amended medium.

The Ultraviolet light induced mutants were obtained by exposing the conidial suspension of P. oryzae and D. oryzae to UV light at wavelength of 254 nm for 30 minutes. The UV-exposed conidia were again plated on EDDP-amended medium as mentioned above and surviving mutants were isolated.

Adapted mutants: Adapted mutants were selected by the natural screening of 0.5×10^5/ml conidia of P. oryzae and D. oryzae over the EDDP-amended medium. Colonies obtained thus on five times higher concentration of the ED_{50} dose of sensitive strain were further trained to grow on higher concentrations.

Field mutants: Four EDDP-resistant mutants of D. oryzae were obtained from EDDP-treated paddy rice fields showing no disease control.

Test for stability and pathogenicity: All of the isolated EDDP-resistant mutants of both P. oryzae were grown on fungicide-free and amended medium for 10 transfers, to test their stability of resistance. Additionally, pathogenicity of all mutants was confirmed on the rice cultivar IR 50. Only stable and pathogenic mutants were chosen for further studies.

Stable and virulent mutants: For P. oryzae, one EMS (POLR-1), the adapted mutants (POLR-2 and POLR-3) and one UV mutant (POLR-4) and for D. oryzae, one EMS mutant (DOLR-1) and four field mutants (DOFR-1, DOFR-2, DOFR-3 and DOFR-4) were chosen based on their stability and pathogenicity.

Morphological Characterization of EDDP-resistant mutants and sensitive (parent) strains

All of the tested mutants were grown on fungicide-free medium (PDA) and compared with sensitive strain for growth rate, colony morphology, pigmentation and conidial production.

Biochemical Characterization of EDDP-resistant mutants and sensitive (parent) strains

Rate of efflux of electrolytes: Efflux of electrolytes from the fungal mycelium was measured using a Type CM82 T conductivity bridge with a dip electrolyte cell (1). The conductance was expressed as specific conductance in μ mhos/cm^2/h/g/dry wt of the mycelium.

Estimation of total DNA, RNA and protein: Isolation of degraded DNA fraction was carried out by the modified method of Munro and Fleck(14). DNA was estimated by Giles and Myers(15) further modified by Lalithakumari et al.(16). Total RNA was estimated by the method of Wu(17). Total protein was estimated by the method of Lowry et al.(18).

Extraction of undegraded DNA: The extraction of undegraded DNA was carried out by using Marmur's method(19). The DNA strands recovered at the end were dissolved in 10 ml of sodium saline citrate (SSC; NaCl, 0.15M; trisodium citrate, 0.15 M at pH 7.0). Aliquots of undegraded DNA in SSC solution were taken for UV spectral analysis that were carried out between 220 and 320 nm and compared with the spectrum obtained with calf thymus.

Determination of the DNA base composition: Two different methods were used for the estimation of DNA base composition, Guanine plus Cytosine (GC) content.

    (a)   UV absorption ratio OD 260 / OD 280(20.).
    (b)   Thermal denaturation method(21).

Electrophoretic pattern of proteins: Protein pattern of mutants and parent strains was analysed by polyacrylamide gel electrophoresis (PAGE) using the method of Davis(22).

Estimation of total lipids and phospholipids: The total lipid content was estimated according to Tsuda et al.(23) and purified according to Bligh and Dyer(24). Phospholipids were separated by thin layer chromatography (TLC) method(25). The phospholipids were quantified by estimating the phosphorous content(26).

Isolation of plasmid from P. oryzae and D. oryzae protoplasts

Isolation of protoplasts: Pure suspensions of protoplasts from P. oryzae and D. oryzae were obtained using the method of Hashiba and Yamada(27). Mycelia from shaken cultures were harvested aseptically by vacuum filtration and washed twice with distilled water and once with 0.6 M Sucrose mannitol osmotic stabilizer.

One gram mycelial mat was weighed and suspended in a 10 ml enzyme mixture (containing cellulase 30mg/ml; pectinase 10mg/ml; β-glucuronidase 0.1ml/ml and chitinase 25mg /ml in 0.6 M Sucrose mannitol pH 5.5) in a 100ml Erlenmeyer flask. The flask was placed on a reciprocal shaker at 75 strocks/min. at 28°C for 3 h. The culture was filtered through six layers of cheese cloth to remove mycelial fragments and the filtrate was centrifuged at 1,000 rpm for 5 min to remove traces of the enzymes. Intact protoplasts were further separated from mycelial fragments and cell debris by an aqueous two-phase system. Protoplasts were counted using a haemocytometer and expressed as number of protoplasts per gram dry or fresh weight. Isolated protoplasts were checked for regeneration on Czapek's (Dox) yeast extract agar medium.

Isolation of plasmid DNA from the protoplasts of **P. oryzae** and **D. oryzae**: For the isolation of plasmid DNA in P. oryzae, the method of Takai et al.(28) was followed and for D. oryzae Kistler and Leong's(29) method was followed.

Agarose gel electrophoresis: From the purified protoplasts of P. oryzae and D. oryzae plasmid DNA were isolated by preparative 0.7% agarose gel electrophoresis at 60 V for 4 h. The electrophoresis buffer contained 0.8M Tris, 0.4M sodium acetate, 0.04M EDTA and acetic acid 1ml (pH 8.3). Sample was mixed with 1/10 volume of 25% Ficoll (Sigma) and 0.25% bromo phenol blue dye was added before electrophoresis. Staining of the gel was performed using ethidium bromide, 5μg/ml. Plasmid DNA was viewed and photographed using RP4 polaroid camera. Curing of plasmid was done using ethidium bromide(30).

Counter measure for edifenphos resistance

Cross resistance studies: All of the edifenphos-resistant strains were tested for their cross-resistance to related and unrelated systemic and contact fungicides. The growth rate of both mutant and sensitive strains were compared and their ED_{50} values were estimated. The resistance level was expressed as Q value (ratio of ED_{50} of sensitive strain/ED_{50} of mutant strains)(31).

RESULTS

Morphology, Growth and Sporulation: There were no differences in color and colony morphology of edifenphos-resistant mutants and sensitive strain of P. oryzae. However edifenphos-resistant mutants showed a slow growth rate when compared to sensitive strain of P. oryzae. In D. oryzae, the field mutants (DOFR-1, DOFR-2, DOFR-3 and DOFR-4) differed in morphology and mycelial growth, showed puffy and lobbed growth (Fig. 1 and 2) on the fungicide free medium and color changed from greenish black to blackish brown. Sporulation was reduced in both P. oryzae and D. oryzae mutants than sensitive strain. All the tested P. oryzae and D. oryzae mutants were as infective as the sensitive strains of these pathogens.

Figure 1. Colony morphology of sensitive strain mutants of *D. oryzae*.

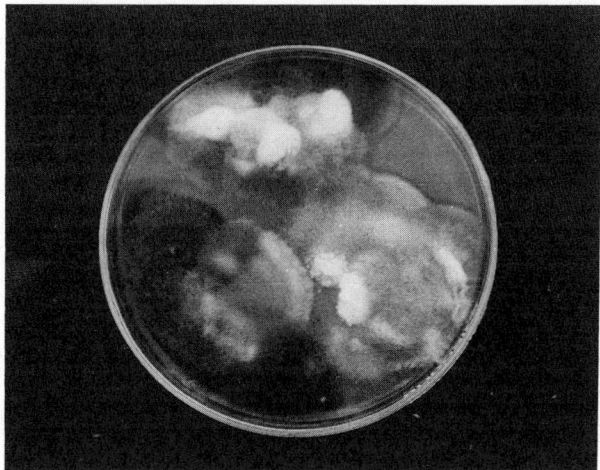

Figure 2. Colony morphology of field resistant mutants of *D. oryzae*.

Biochemical Characterization of resistant mutants and sensitive (parent) strains

Rate of efflux of electrolytes: The results in Table I clearly indicates a significant reduction in efflux of electrolytes from all the edifenphos resistant mutants of P. oryzae and D. oryzae throughout the observation period.

Table I. Rate of efflux of electrolytes of sensitive strain and resistant mutants of P. oryzae and D. oryzae

	Specific conductance μ mhos/cm^2/h/g/dry wt of mycelium				
	Incubation Time (hrs)				
	2	4	6	8	12
P oryzae					
Parent(Sensitive)	2750	4800	5206	6173	8039
POLR-1	1460	2220	3403	3611	5277
POLR-2	1450	2210	3250	3500	5200
POLR-4	860	1090	1318	1355	1864
D. oryzae					
Parent(Sensitive)	2454	3818	4363	4818	5636
DOLR-1	2370	2727	3272	4045	5063
DOFR-1	2352	2647	3529	4117	4803
DOFR-2	2100	2450	2860	3215	3800

Macromolecular synthesis: The macromolecular contents increased in all of the mutants. The Total DNA content (Table II) was observed to be higher in the edifenphos resistant mutants. UV-induced P. oryzae mutants showed only 20% increase of DNA over sensitive strain. Similarly D oryzae mutants also showed an increase in the DNA content. Among the phospholipids the increase in phosphatidylcholine was significant in the resistant mutants.

Table II. Macromolecular content of sensitive strain and resistant mutants of P. oryzae and D. oryzae

	dry weight					
	DNA mg/g	RNA mg/g	Protein mg/g	Lipid %	Phospholipids	
					PE μg	PC μg
P. oryzae						
Parent (sensitive)	3.19	2.50	11.80	15.10	0.720	0.840
POLR-1	4.92	3.25	13.60	14.00	0.830	0.930
POLR-2	4.82	3.57	15.72	14.60	0.900	0.950
POLR-3	3.82	3.20	15.65	14.90	0.890	0.930
D. oryzae						
Parent (sensitive)	7.80	4.60	15.30	10.20	0.850	1.220
DOLR-1	9.75	5.70	20.10	11.00	0.950	1.350
DOFR-1	10.00	5.75	21.10	13.00	0.975	1.385
DOFR-2	9.70	5.85	20.10	12.00	0.965	1.360

Electrophoretic protein pattern: The electrophoretic pattern of protein (Table III) revealed extra bands in two of the edifenphos-resistant mutants (POLR-1 and POLR-4) of P. oryzae. POLR-2 did not show any extra band. In D. oryzae there was no difference in the number of protein bands but the intensity was greater in all the edifenphos-resistant mutants.

Table III. Electrophoretic protein pattern of sensitive strain and resistant mutants of P. oryzae and D. oryzae

	Total No. of bands observed
P. oryzae	
Parent (Sensitive)	10
POLR-1	13
POLR-2	10
POLR-4	13
D. oryzae	
Parent (Sensitive)	9
DOLR-1	9
DOFR-1	9
DOFR-2	9

Characterization of undegraded DNA for GC% and melting temperature (Tm)

The results in Table IV show a value of 50.2 and 49.2 as GC% for the parent strain of P. oryzae by UV absorption and thermal denaturation methods respectively. Similarly, in D. oryzae parent strain the GC% was 57.2 and 58.2 by both methods. The GC% was higher in the resistant mutant than in the parent strain of both P. oryzae and D. oryzae. The melting temperature (Tm) was 87.0, 90.0, 91.0 and 96.0°C respectively for P. oryzae parent, POLR-1, POLR-2 and POLR-4. For D. oryzae it was 87, 88, 90, 90°C respectively for parent, DOLR-1, DOFR-1, DOFR-2.

Table IV. Characterization of undegraded DNA for GC% and melting temperature (Tm) of P. oryzae and D. oryzae

	GC% of undegraded DNA		
	UV absorbancy method	Thermal denaturation method	Tm value (°C)
P. oryzae			
Parent (Sensitive)	50.2	49.2	87
POLR-1	51.4	52.0	90
POLR-2	55.1	52.9	91
POLR-4	65.4	65.1	96
D. oryzae			
Parent (Sensitive)	57.2	58.2	87
DOLR-1	61.4	61.8	88
DOFR-1	61.8	61.2	90
DOFR-2	60.8	59.8	90

Isolation of Plasmid DNA from the protoplasts of **P. oryzae** and **D. oryzae**: Recovery of protoplasts on an average was 1.6×10^5 in P. oryzae and 2×10^5 in D. oryzae (Fig. 3). The protoplasts of P. oryzae were distinctly circular and D. oryzae protoplasts were bigger than P. oryzae protoplasts.

Analysis of edifenphos-resistant mutant and sensitive strains of P. oryzae and D. oryzae (Fig.4, 5a and 5b) showed a single plasmid DNA band in the agarose gel electrophoresis. Further, the intensity of Plasmid bands was more in both P. oryzae and D. oryzae resistant mutants than the sensitive strain.

Counter measure for edifenphos resistance

Cross resistance studies:

The results in Table V show that the chemical mutant of P. oryzae (POLR-1) was highly sensitive to Carbendazim, Ziram and Mercuric chloride. Adapted mutant POLR-2 was found sensitive to Bitertanol, Ziram, and Mancozeb. The UV mutant POLR-4 was sensitive to Wettable sulphur, Ziram and Mancozeb. EMS mutant of D. oryzae (DOLR-1) showed high sensitivity to Mancozeb and Bitertanol while the two resistant field strains (DOFR-1 and DOFR-2) were sensitive only to Mancozeb. The resistance level (Q value) was less than one for the above chemicals.

Table V. Cross resistance to fungicides

| Fungicide | Q value |||||||
| | P. oryzae | | | | D. oryzae | | |
	POLR-1	POLR-2	POLR-3	POLR-4	DOLR-1	DOFR-1	DOFR-2
Benomyl	1.5	1.8	2.0	1.5	1.20	1.11	1.21
Carbendazim	0.6*	2.8	1.7	1.3	1.15	1.05	1.03
Bitertanol	3.2	0.7*	1.6	1.5	0.50*	1.60	4.00
IBP	2.1	2.0	3.0	1.8	1.39	1.13	1.21
Pyroquilon	-	-	-	-	1.16	1.04	1.15
Thiophanate methyl	2.4	2.1	1.2	1.1	1.09	1.26	1.07
Mancozeb	3.8	0.8*	1.0*	0.8*	0.80*	0.80*	0.90*
Ziram	0.5*	0.8*	0.9*	0.5*	1.38	2.69	1.76
Copper oxy chloride	1.5	1.2	1.5	1.5	1.23	1.50	1.19
Wettable sulphur	1.0	1.0	0.8*	0.3*	1.40	1.44	1.23
Methoxy ethyl mercury chloride	3.5	1.0	1.0	1.2	-	-	-
Mercuric chloride	0.8*	1.5	1.7	0.9*	-	-	-

* Negative correlation

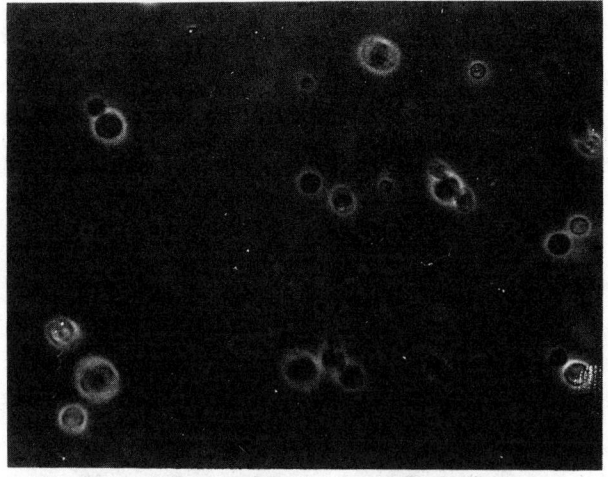

Figure 3. Protoplasts of D.oryzae (320 x).

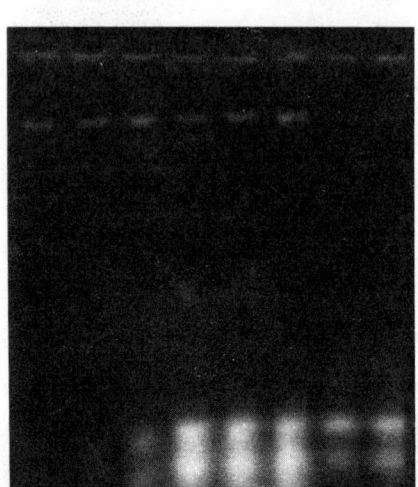

Figure 4. Agarose gel electrophoresis of Plasmid DNA of *P. oryzae*.

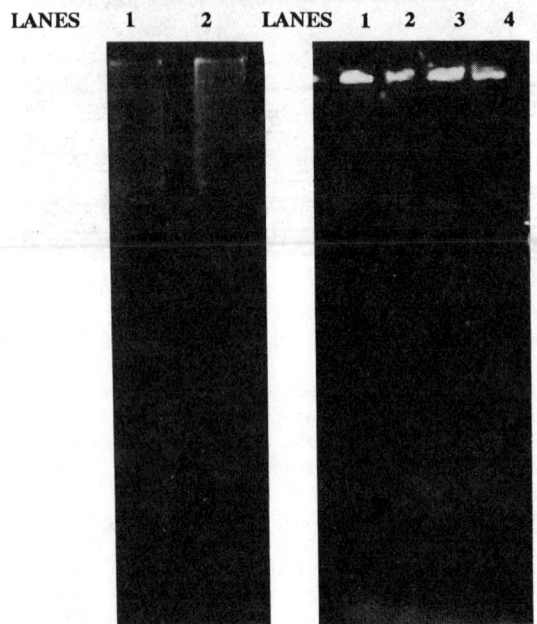

a. Lane 1–2: Sensitive strain   b. Lane 1–2: DOLR-1, Lane 3–4: DOFR-1.

Figure 5. Agarose gel electrophoresis of Plasmid DNA of *D. oryzae*.

DISCUSSION

Edifenphos (EDDP) is effectively used for the control of both P. oryzae and D. oryzae in India. Level of resistance found in the field of EDDP is usually rather low and therefore it is difficult to detect and measure the EDDP- resistance. Since fungicidal activity and fungal resistance are easier to evaluate in vitro, the present investigation employed in vitro methods to determine the probability level, mechanism and molecular basis of resistance and control measures to break resistance in P. oryzae and D. oryzae to EDDP.

Screening of 0.5×10^5/ml conidia of P. oryzae and D. oryzae on EDDP-amended media revealed the variability in sensitivity among the conidia, besides the development of two colonies of P. oryzae resistant to EDDP. Selection of mutants resistant to the Kasugamycin, Kitazin and EDDP from a large number of conidia has been reported by Katagiri and Uesugi(32). These two colonies (POLR-2 and POLR-3) were further trained to grow in higher concentration of EDDP by an adaptation technique. Many colonies of D. oryzae developed on high concentration of EDDP but they did not grow well in the absence of fungicide. No resistant mutants of D. oryzae developed by the UV irradiation technique. Georgopoulos et al.(33) also were unable to obtain resistance mutants of Ustilago maydis by UV irradiation. Selected in vitro mutants of P. oryzae and D. oryzae induced by EMS treatment, UV irridation and adaptation, served as good examples of EDDP-resistance owing to their survival in the absence of fungicide and their ability to infect on paddy rice. EDDP-resistant mutants of P. oryzae did not show any variation in colony morphology and color, but the field resistant mutants of D. oryzae showed a puffy and lobbed growth with changes in pigmentation. All of the resistant mutants of P. oryzae and D. oryzae showed slight reduction in sporulation. The reduction in sporulation by resistant mutants has been reported by many workers(34-37). Pathogenicity of all the resistant mutants of P. oryzae and D. oryzae to rice plants proved their fitness in the field to compete with the parent strain in the absence as well as in the presence of EDDP. Hence, the selected mutants in the present investigation served as very good examples of EDDP resistance.

Various cellular contents analysed, showed significant increase in all the macromolecular contents viz. DNA, RNA Protein, lipids and phospholipids of the EDDP-resistant mutants of P. oryzae and D. oryzae. The efflux of electrolytes was reduced in the resistant mutants when compared to control. In interpreting the results, it should be realized that the phospholipids, the target site of inhibition for edifenphos seem to be unaffected in the mutants as evidenced from the higher values of phosphatidylcholine compared sensitive strain. These data indicate that the mechanism of resistance to edifenphos is not related to the target site of edifenphos. De waard and Van Nistelrooy(38) also reported that the mechanism of resistance in P. oryzae to pyrazophos (PP) was not related to the change of target site of PP. It seems probable that resistance in P. oryzae and D. oryzae to edifenphos influences cell membrane permeability as evidenced from the reduction in the efflux of electrolytes from the resistant mutants. Similarly, Rank et al.(39) demonstrated that decreased membrane permeability was the mechanism of resistance in Saccharomycos cerevisiae resistant to various toxicants, in Aspergillus nidulans resistant to fenarimol (40) and dicarboximide and benzimidazole resistant mutants of Gerlachia nivalis (41). The acquired mechanism of resistance in D. oryzae by in vitro mutants is identical with the natural mechanism of resistance in field resistance mutants of D. oryzae isolated from the field.

There are many reports on the chromosomal control of resistance to agricultural fungicides (42-44). Chromosomal gene involvement in fungicide resistant has been reported so far owing to the reason that toxicants mostly act outside the mitochondria. In eucaryotic cells, mitochondrial DNA cannot be involved in resistance of the toxicants acting outside the mitochondria. Involvement of plasmids, the extrachromosomal genes in resistance in common in bacteria (45 - 48). But, plasmids have been recognised quite recently in fungi, involvement of plasmid in benzimidazole resistance in Mycospherella musicaola (49) and Neurospora crassa (50) have already been reported. In addition, Saccharomyces cerevisiae resistant to oligomycin was reported to be due to extrachromosomal DNA (51). Besides, a linear plasmids have been described in a number of Fusarium sp. (52), Cochliobolus heterostrophus (53) and Ascobolus immersus (54), however, no definite role has been explained. Plasmid DNA was isolated from sensitive strain and resistant mutants of P. oryzae and D. oryzae, but in the resistant mutants of P. oryzae and D. oryzae the intensity of plasmid DNA was more than the sensitive strain. The high intensity of plasmid DNA bands in the resistant mutants may be due to increased number of plasmid copies which might have resulted from rapid amplification of plasmids in the presence of edifenphos. This high intensity in plasmids was also observed in plantomycin-resistant mutants of Xanthomonas campestris pv. oryzae (55). Similar increase in copy number of plasmids due to resistance has been reported by Clewell (48) and Nordstrom (56). Further, Annamalai (57) has reported that curing with ethidium bromide for 48 h in darkness resulted in the loss of resistance to edifenphos in D. oryzae. Protoplasts isolated from plasmid cured protoplasts did not show plasmid DNA bands undoubtedly confirming the definite involvement of plasmid DNA in edifenphos resistance in P. oryzae and D. oryzae. There are several reports on plasmid mediated resistance to various toxicants. For instance, Escherichia coli resistant to Kasugamycin by Yoshikawa et al. (58), Xanthomonas campestris pv. vesicatoria resistant to Kanamycin and Neomycin (59), Pseudomonas aeruginosa resistance to chloramphenicol (60), Flexibacter spp. and Agrobacterium tumifaciens resistant to tetracycline (61). Although the specificity of plasmids and molecular weight of plasmids in the present investigation could not be explained, it is confirmed that plasmid coded resistance can occur in the non-site specific resistance mechanism or in membrane permeability of fungi. Cross resistance studies were carried out to select a suitable chemical for breaking the EDDP-resistance. The results on the sensitivity of selected EDDP-resistant mutants of P. oryzae and D. oryzae against unrelated fungicides were interesting. Also the pleiotropic effect observed regarding cross-resistance to unrelated compounds in the present investigation confirm the mechanism of action to decreased membrane permeability. The overall conclusion on the type of mutation in P. oryzae, and D. oryzae to edifenphos was presumed to be pleiotropic mutation. Because, such a mutation may not only cause resistance to unrelated compounds (62). Such pleiotropic mutation has been reported for resistance to triarimol in Botrytis cinerea (63) and imazalil in Aspergillus nidulans (64). All of the mutants of P. oryzae were sensitive to Ziram and D. oryzae to Mancozeb though number of chemicals to which each mutant showed varied sensitivity.

Cross resistance in an organism to two different chemicals cannot be assumed without evidence that sensitivity to both chemicals is controlled by the same gene (62), but it does influence the selection of alternate chemicals for immediate control of the pathogen in the field. For a country like India, information on alternate chemicals to overcome disease control failure is very important as the farmers cannot afford yield losses at any cost. The present observation on cross resistance to unrelated compounds have been confirmed by many workers (55, 65-70). It is, therefore, concluded that, to overcome disease control failure of P. oryzae due to EDDP-resistance, Ziram may be used as an alternate chemical. In D. oryzae Mancozeb is recommended as an alternate chemical to break edifenphos resistance. The mechanism of cross resistance is not clearly understood from these results.

ACKNOWLEDGMENTS

We are extremely grateful to our Director, Prof. A.Mahadevan, for his encouragement and to the Department of Science and Technology Govt. of India for funding the above work.

LITERATURE CITED

1. Swaminathan, K. Ph.D. Thesis, Madras University, India. 1983.
2. Nageswara Rao, M. Ph.D. Thesis, Madras University, India. 1984.
3. Lalithakumari, D; Ganesan, T & Rao, M.N. Indian Phytopath. 1984. 37, 111-114.
4. Gupta, P.K.S., Dass. S.N., Curr. Sci. 1971, 40, 168-69.
5. Rana, J.P. Sen Gupta, P.K. z. Pflkrankh Pflchutz. 1977, 84, 738-742.
6. Thind, T.S., Jhooty, J.S. Indian Phytopath. 1980, 33, 570-573.
7. Mukhopadhyay,A.N., Singh.U.S., Pesticides (India) 1985, 23-29.
8. Putto, B.L., Basuchaudhary, K.C. Ind. J. Mycol. Pl. Path. 1985, 15, 15.
9. Natarajan, M.R., Lalithakumari D. Contemporary themes in Biochemistry, ISCU Issue, Singapore. 1986, 4, 363-364.
10. Chauhan, V.B., Singh, U.P. J. Phytopathol. 1987, 120, 93-96.

11. Lalithakumari, D., Annamalai, P. ISPP News letter, U.K. 1988. 10, 26-27
12. Gangawane, L.V., Kareppa, B.M., Suman Waghnare. Abstract 5th ICCP, Kyoto, Japan, 1988.
13. Sanchez, L.E., Leary, J.V. Endo, R.M. J. Gen. Microbiol. 1975. 37, 326-332.
14. Munro, H.N., Fleck, A. In. D. Glick, Meth Biochem. Analysis. Inter Science Publishers, New York, 1966, 14, 113-176.
15. Giles, K.W., Myers, A. Nature. 1965. 206, 93.
16. Lalithakumari, D., De callone, J.R., Meyer, J.A. J. Gen. Microbiol. 1975, 88, 245-252.
17. Wu, L. Ph.D. Thesis University of Illinois, Urbana. 1959.
18. Lowry, O.H., Rosebrough, N.J., Farr, A.L., Randell, R.J. J. Biol. Chem. 1951, 193,, 265-275.
19. Marmur, J. J. Mol. Biol. 1961, 3, 208-218.
20. De Ley, J. Antonie Van Leuwenhok. 1967, 33, 203-208.
21. Marmur, J. Doty, P. J. Mol Biol. 1962, 5, 109-118.
22. Davis, B.J. Ann. Aca. Sci. 1964, 2, 409-423.
23. Tsuda, M. Ueyama, A. Nakano, M. Fujina, Y. Ann. Phytopath. Soc. Japan, 1972, 38, 60-67.
24. Bligh, E.G., Dyer, W.J., Can. J. Biochem. Physiol. 1959, 37, 911-917.
25. Marinetti, G.V. In New biochemical separations. D. Van Nostrand company Inc., Princeton, New Jersey. 1964. p.339.
26. King, E.J. Biochem. J. 1932, 26, 292-297.
27. Hashiba, T.Y., Yamada. N. Phytopath. 1981, 72, 849-853.
28. Takai, S., Lizuki, T., Richards, W.C. Phytopath. 1984, 74, 833.
29. Kistler, C.H., Leong, S.A. J. Bacteriol. 1986, 167, 587-593.
30. Samac. D.A., Leong, S.A. Plasmid, 1988, 19, 56-67.
31. Dekker, J. Fungicide resistance in crop protection practical manual. 1984. p.10-13.
32. Katagiri, M., Uesugi, Y. Ann. Phytopath. Soc. Japan. 1978, 44, 218.
33. Georgopoulos, S.G., Geerlings, J.W.G., Dekker, J. Neth. J. Pl. Path. 1975, 81, 35-41.
34. Warrd de, M.A., Sisler, H.D. Rijksuniv, Gent 1976, 41/2, 571-578.
35. Grindle, M. Trans. Br. Mycol. Soc., 1984, 82, 635-643.
36. Grindle, M., Temple, W. Trans. Br. Mycol. Soc. 1985, 84, 635-643.
37. Rose Maria., Sullia, S.B. J. Phytopathol. 1986, 116, 60-66.
38. Waard de, M.A., Van Nistelrooy, J.G.M. Neth. J. Pl. Pathol. 1980, 86, 251-258.
39. Rank, G.H., Robertson, A., Phillips, K. J. Bacteriol. 1975, 122, 593-595.
40. Waard de, M.A., Van Nistelrooy, J.G.M. Pestic Biochem physiol. 1980, 13, 255-66.
41. Ressler, H., Buchenauer, H. Z. Pflanzenkrankn Pflanzensch. 1988, 95, 156-68.
42. Georgopoulos, S.G., Proc. Amer. Phytopathol. Soc. 1976, 3, 533-60.
43. Shrivastava, S., Sinha, V. Gent. Res. Camb. 1975, 25, 29-38.

44. Van Tuyl, J.M., Med. Fac. Land. Bouww. Rijkswniv. Gent. 1975, 25, 29-38.
45. Natarajan, M.R., Lalithakumari, D. Proc. Phytobacteriology, 1987.
46. Weisblum, B., Siddhikel, C., Lai, C.J., Demohn, V.J., J. Bacteriol. 1971, 106, 835-47.
47. Courvalin, P., Weisblum, B., Davis, J. Proc. Natl. Sci. Acad. USA. 1977, 74, 999-1003.
48. Clewell, D.B., Microbiol Rev. 1981, 45 409-36.
49. Stover, R.H. Trans. Brit. Mycol. Soc. 1977, 69, 500-502.
50. Orbach, M.J., Porro, E.B., Yanofsky, Z.C. Mol. Cell. Biol. 1986, 6, 2452-61.
51. Guerineau, M., Slonimski, P.P., Avner, P.R. Biochem. Biophys. Res. Commumic. 1974, 61, 462-69.
52. Samac, A.D., Leong, S.A. Plasmid. 1988, 19, 57-67.
53. Garber, R.C., Turgeon, B.G., Yoder, O.C. Mol. Gen. Genet. 1984, 196, 301-10.
54. Francou, F.S., Randscholt, N., Decaris, B., Gregoire, A. J. Gen. Microbiol 1987, 133, 311-316.
55. Natarajan, M.R., Ph.D. Thesis, Madras University, India, 1988.
56. Nordstrom, K., Ingram, L.C., Lund Back,A. J. Bacteriol. 1972, 110, 562-69.
57. Annamalai, P. Ph.D. Thesis, Madras University, India, 1989.
58. Yoshikawa, M., Okuyama, A., Tanaka, N. J. Bacteriol. 1975, 122, 796-97.
59. Lai, M., Shaffer, S., Panopoulos, N.T., Phytopath. 1977, 67, 1527-30.
60. Gaffney, D.F., Gundliffe, E., Foster, T.J. J. Gen. Microbiol. 1981, 125, 113-21.
61. Barcak, C.J., Burchard, R.P. J. Bacteriol. 1985, 161, 810-812.
62. Georgopoulos, S.G., In Antifungal compounds. M.R. Siegel., H.D. Sisler, Eds. Vol 11, New York, 439-95.
63. Leroux, P., Gredt, M., Fritz, R., Phytiotric Phytopharmacie, 1976, 25, 317-34.
64. Van Tuyl, J.M., Ph.D. Thesis. Modedelingen. Landbouwhogeschool. Wageningen. 1977.
65. Leroux, P., Fritz, R., Gredt, M. Phytopath. Z. 1977, 89, 347-58.
66. Fuchs, A., DeRuig, S.P., Van Tuyl, J.M., Devries, F.W. Neth. J. Pl. Pathol. 1977. 83, 189-205.
67. Buchenauer, H. Pestic. Sci. 1978. 9, 509-12.
68. Barug, D. Kerkenaar, A. Rijksuniversiteit. Gent. 1979, 41, 421-27.
69. Rosenberger, D.A., Meyer, F.W. Phytopath, 1985. 75, 74-79.
70. Ishii, A., VanRaak, M. 1988. Phytopath. 1988, 78, 695-98.

RECEIVED September 1, 1989

Chapter 18

Management of Fungicide Resistance by Using Computer Simulation

Phil A. Arneson

Department of Plant Pathology, Cornell University, Ithaca, NY 14853

>Management of fungicide resistance requires some means of predicting the quantitative response of the target fungus population to various management alternatives. This can be accomplished with so-called "mechanistic" computer simulation models--models that represent the underlying biological mechanisms with sufficient fidelity that it is possible to mimic the behavior of both the fungicide-sensitive and resistant subpopulations in response to a particular spray program under a specific set of conditions. There are two major ways in which computer simulation can be used in the management of fungicide resistance--preseason planning of fungicide spray programs and day-by-day forecasting for timing of fungicide applications. Examples of two models are presented, "Resistan", a streamlined model for preseason planning, and "Sigatoka", a detailed model for day-to-day decision making. Both of these models can be used as teaching tools as well as management decision aids.

Management of fungicide resistance implies purposeful manipulation not only of the frequency of fungicide resistance but also of the total population of the target fungus in a specified area. To manage fungicide resistance, both the total population of fungal propagules and the proportion of that population that is resistant must first be monitored. There must then be a way of predicting what will happen to that population when different fungicides are applied. Methods for monitoring fungicide resistance have been worked out for many of the fungal pathogens of agricultural importance (1). We now need predictive models that, given the frequency of fungicide resistance at the start of the season, will allow the manager to simulate different spray schedules with particular fungicides and to see the likely effects on frequency of resistance, disease level, and profitability at the end of the season.

General models of the selection of fungicide resistance have given us overall management strategies (2-8). However, the fungi

we attempt to control with fungicides differ widely in their modes
and rates of reproduction and in their sensitivity to different
fungicides. Likewise, the fungicides we use to control these fungi
differ widely in their behavior on plant surfaces and in their
fungicidal modes of action. Furthermore, the physical mode of
resistance, the mode of inheritance of the resistance, and the
fitness of the fungicide-resistant biotypes vary widely with both
the fungus and fungicide. Therefore, general models are not
adequate for day-to-day decision making and may even lead to a
wrong decision in a specific situation. To predict the effects of
a particular fungicide on a given combination of fungus, crop,
site, and weather conditions requires models that include
considerable mechanistic detail. The level of detail depends upon
the objectives of the model.

This paper considers two different models to illustrate two
quite different objectives. The first model, which we call
"Resistan", is a streamlined model with limited mechanistic detail
and is intended for preseason planning of fungicide spray programs.
The second, "Sigatoka", is a highly detailed, mechanistic model
that responds to hourly weather conditions and is intended as a
research tool and for day-to-day decision making. Both of these
models are still in the early stages of their development and will
need field validation before they can be used in fungicide
resistance management.

The "Resistan" Model

One objective of a fungicide management model might be to plan
fungicide spray schedules for the coming growing season, based upon
resistance monitoring results. Monitoring might be done at the end
of the previous season, between seasons, or at the start of the
current season, depending on the fungus. The model would be used
to predict the relative efficacy of various combinations of
fungicides, rates, and frequencies of application. The model would
not necessarily have to simulate the effects of weather on the
behavior of the system, since it is not necessary (or possible) to
predict the incidence or severity of the disease at any point
during the season. It would, however, have to realistically
represent the relative effects of the different fungicides on the
resistant and sensitive fungal subpopulations. Since the manager
would likely be using a microcomputer and would want to simulate
the whole growing season or perhaps even several growing seasons in
a few minutes, the model would have to be somewhat streamlined in
order to execute with sufficient speed.

An example of this kind of model is "Resistan", a mechanistic
simulation of the process of selection of fungicide-resistant
biotypes of a clonally reproducing organism with many generations
per season (9,10). In this situation selection alone is by far the
most important factor in the development of resistant populations,
and therefore in this model no attempt is made to simulate genetic
recombination. "Resistan" was written for the IBM-PC and PC-com-
patible computers. It has a generalized structure that allows the
simulation of different fungi, crops, and fungicides by changing
the parameter sets. Currently we have a parameter set only for one
fungus, Venturia inaequalis, and four fungicides used for the

control of apple scab, captan, mancozeb, benomyl, and myclobutanil. The user's manual (9) details the methods of estimating these parameters.

The Structure of the Simulation. Development of the fungus is simulated with three life-stages (Figure 1). A population of spores, dispersed and landed on susceptible tissue, germinate and infect, giving rise to a population of latent lesions. The latent lesions develop into sporulating lesions, whose spores are then dispersed to complete the cycle. The spore population can be augmented by spores blowing in from outside the treated area. At each stage of development there are losses from the population (represented by the clouds in Figure 1), resulting from both natural mortality and the effects of the fungicide.

The flow diagram in Figure 1 actually represents only one subpopulation of the fungus in the model. The model, which simultaneously simulates the effects of two fungicides, has four such subpopulations developing in parallel, one sensitive to both fungicides, one resistant to each of the fungicides individually, and one resistant to both fungicides. The subpopulations differ only in their mortality rates, both in the presence of toxic levels of the fungicides and also in the absence of fungicides. If there is a fitness cost associated with fungicide resistance, the mortalities of the resistant subpopulation are slightly higher, so without continued suppression of the sensitive biotype by applications of the fungicide, the sensitive subpopulation will increase slightly faster than the resistant subpopulation, resulting in a gradual reversion to a low frequency of resistance in the whole population.

The effects of the fungicides in "Resistan" are simulated as daily mortalities to each of the three growth stages (spores, latent lesions, and sporulating lesions. A single probit mortality/log dose function is parameterized for each fungicide, and the mortality (proportion) that is calculated for each dose is multiplied by a weighting factor to calculate the mortality for each growth stage. In this way it is possible to approximate fairly simply the effects of fungicides with different modes of action.

Weather variables are not considered in this model. Its parameters are estimated assuming optimum environmental conditions for fungal growth and development and remain constant throughout the simulation.

"Resistan" is a generalized model that can be made to simulate different polycyclic fungi, different crops, and different fungicides simply by changing sets of parameters. The parameters needed to describe the development of the fungus in the model include:
1. The proportions of the spores, latent lesions, and sporulating lesions that would survive each day in the absence of fungicides.
2. Infection efficiency--the proportion of landed spores that successfully infect per day in the absence of fungicides.
3. Latent period--the time from infection to sporulation.
4. Spores produced per lesion per day in the absence of fungicides.
5. Proportion of the dispersed spores that actually land on susceptible tissue.

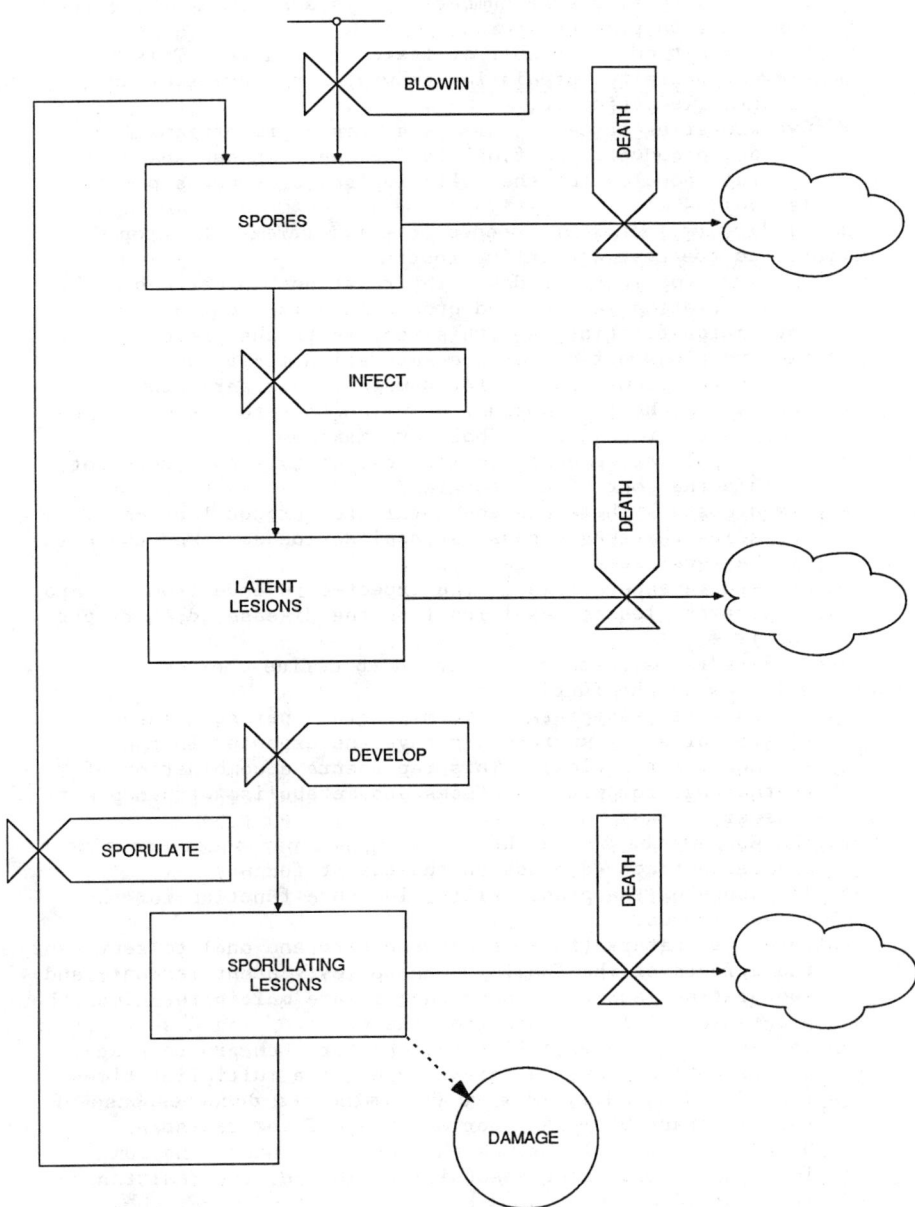

Figure 1. Diagrammatic representation of the "Resistan" simulation model. Rectangles represent measurable quantities that change with time, valves represent flow rates, and clouds represent disappearance from the system. (Reproduced with permission from Ref. 10. Copyright 1988 APS Press.)

6. Infectious period--the number of days a sporulating lesion continues to produce spores.
7. Upper limit on the number of lesions per acre. This is to prevent unlimited population growth in the event of an uncontrolled epidemic.
8. Overwintering factor. This is a simple proportionality factor to convert the final lesion count in one season to initial inoculum for the following season (spores per final lesion).

The following values are needed to parameterize the crop damage functions and the cost accounting routine:
1. Maximum crop loss per day. The relationship between number of sporulating lesions and crop damage is modeled as a saturation function, and this represents the proportion of the crop loss per day at the saturation lesion count.
2. Half-saturation constant for damage. This parameter represents the lesion count at which the proportion of the crop lost per day is one-half its maximum.
3. Spray application cost (dollars per acre). This does not include the cost of the fungicide.
4. Fixed costs. These are the total crop production costs, excluding the fungicide spray application cost and the cost of the fungicides (dollars per acre).
5. Maximum revenue. This is the expected revenue from the crop without any losses resulting from the disease (dollars per acre).

The following parameters are needed to define the characteristics of the fungicides:
1. The rate of disappearance in micrograms per square centimeter of plant surface per day (the exponent in a negative exponential function). This represents a combination of weathering from plant surfaces and metabolism within plant tissues.
2. The EC_{50} of the fungicide in micrograms per square centimeter of leaf surface (depends on the target fungus).
3. The slope of the probit kill / log dose function for the target fungus.
4. Three parameters (factors between zero and one) to represent the effects of the fungicide on spores, latent lesions, and sporulating lesions. These factors are multiplied times the proportion kill obtained from the probit / log dose function to determine the mortality factors for each growth stage.
5. The resistance level. This is used as a multiplier times the EC_{50} of the fungicide to determine the dose response of the resistant biotype. For example, if the resistant biotype required 100 times the dose to achieve the same level of control as the sensitive biotype, the resistance level would be 100.
6. The relative fitness of the resistant biotype compared with the sensitive biotype. This also is a factor between zero and one that is multiplied times the natural daily survival at each growth stage. If the resistant and sensitive biotypes are equally fit, this factor is one.
7. Finally, to help in a rough benefit-cost analysis at the end

of the simulation, the last parameter is the cost of the fungicide in dollars per pound.

Using "Resistan" as a Management Tool. To use "Resistan" one first enters the parameters appropriate for the fungus of interest and the fungicides used to control it. (Alternatively, one could simply accept the default parameters defined in an external data file.) Next one enters the level of initial inoculum of the fungus and the frequency of resistance to each of its fungicides. The desired fungicides are selected from the menu choices, and the desired spray dates and application rates are entered. ("Resistan" currently offers a choice of four different fungicides, any two of which can be applied in any combination in any given season.) The simulation is then run for the designated number of days in the growing season. As the simulation is progressing, it is possible to observe day by day the numbers of lesions and the percent resistance to each of the fungicides in response to the spray applications. At the end of the simulated season there is a cost accounting that can be used to compare different spray schedules on a cost/benefit basis. The manager could then repeat the simulation, testing different proposed spray schedules, perhaps including tank-mixes of fungicides, alternating fungicide sprays, or some program of reduced rates or fewer sprays of the at-risk fungicide. If one wishes to test a spray program over a period of several growing seasons, "Resistan" will simulate the overwintering of the inoculum and carry over the resistant populations from one year to the next.

Although the "Resistan" model parameterized for *V. inaequalis* has not been rigorously validated in the field, its overall behavior is consistent with field observations of fungicide resistance and apple scab to date. The effectiveness of control, the rates of selection of resistant fungus populations, and the persistence of the resistant populations differ quite markedly for the different fungicides in the simulation, emphasizing the importance of developing use strategies for each particular one on a case-by-case basis. Reducing the dose of the "at-risk" fungicide in the simulations or applying it in combinations with "multi-site" fungicides, either in tank-mixes or programs of alternating sprays reduces the rate of selection of resistance to the "at-risk" fungicide. The "best" spray program depends upon the simulated characteristics of the fungicide. Reversion of a resistant population to a sensitive one occurs in the simulator only if there is a fitness cost to the resistance or if there is a significant dilution of the population by sensitive spores blowing in from outside the treated area. If disease control has been very effective and the simulated levels of inoculum are very low, the selection of resistance can continue undetected for several seasons until the resistant population appears to "explode" in an uncontrollable epidemic.

The "Sigatoka" Model

Deciding on a day-by-day basis what fungicide to apply, how much to apply, and when to apply it to satisfactorily control a plant disease while at the same time manipulating the frequency of fungicide resistance requires detailed knowledge of the inoculum

available, the frequency of resistance, the concentration and
distribution of fungicide residues on the plant surfaces, and the
environmental conditions in the field on any given day. It is not
practical to obtain this information solely by monitoring, but an
appropriately designed computer simulation model might mimic
reality sufficiently well to substitute partially for monitoring in
day-to-day decision making. Using biological monitoring to
initialize the model at the beginning of the season, the model
might be driven by daily environmental observations up to the
current date. To predict where the epidemic will go in the future,
one could use canned weather data sets representing the mean or the
extreme values of the key weather variables, or one could even use
weather variables generated by Monte Carlo simulation to introduce
realistic random variability in the weather.

This kind of management decision making requires a considerably
more complex simulation model than that represented by "Resistan".
Such a model must mimic the growth and development of the host
plant, sporulation of the fungus, spore dissemination, infection,
the development of disease, and the effects of fungicides on the
fungus, all in response to appropriate environmental variables.
Since only a few days at a time would have to be simulated to make
a decision about the next spray application, a slow execution speed
is not a serious drawback. Representative of this kind of model is
"Sigatoka", a simulator of an important foliar disease of plantains
and bananas (11).

"Sigatoka" consists of 8 major modules:
1. a weather module that simulates micrometeorological variables, driven by daily observations from a standard weather station,
2. a plant module that simulates leaf development in 15 leaf positions from the unexpanded candle leaf to the first expanded leaf downward through the successively lower leaf positions to the removed leaves on the ground (See Figure 2),
3. a lesion distribution module that moves the latent infections and sporulating lesions downward as the leaves develop,
4. a sporulation module that simulates the production of conidia and ascospores (See Figure 3),
5. a dispersal module that simulates the dispersal of conidia and ascospores among the 15 leaf positions,
6. an infection module that simulates the infection by conidia and ascospores independently for each of the 15 leaf positions,
7. a fungicide weathering module that simulates the decrease in average fungicide residue and the redistribution of the residues among the different leaf positions, and
8. a fungicide effects module that simulates the effects of the fungicide residues on spore mortality, infection, lesion expansion, and sporulation for the subpopulations sensitive and resistant to each fungicide.

"Sigatoka" simulates the effects of 3 fungicides
simultaneously, and thus the developmental cycle in Figure 3 is not
only executed for each of the 15 leaf strata but also for each of
the 8 fungal subpopulations representing the possible combinations

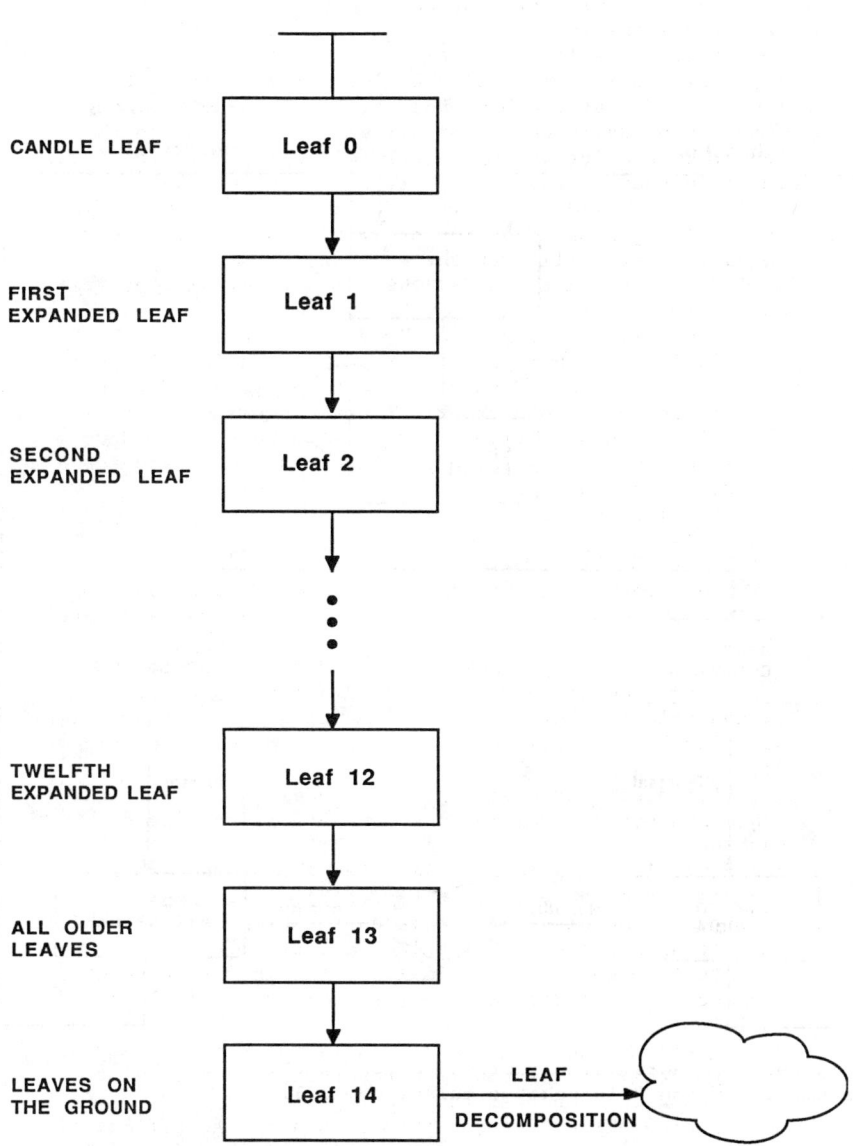

Figure 2. Diagrammatic representation of the banana leaf submodel of the "Sigatoka" simulation model.

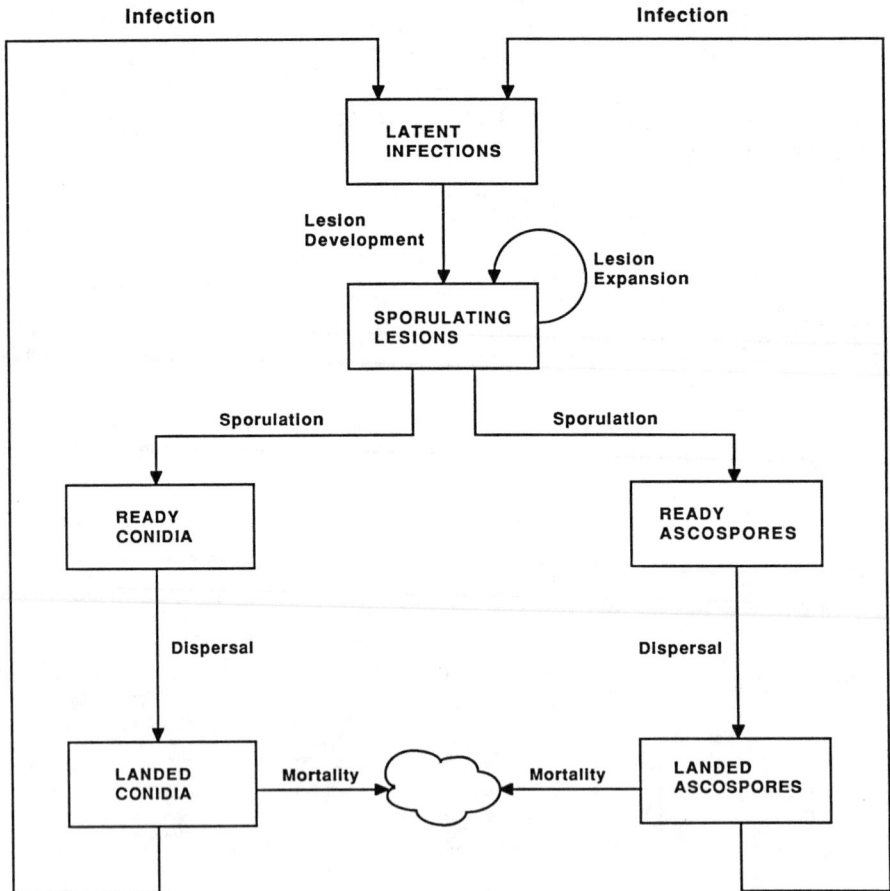

Figure 3. Diagrammatic representation of the fungus development submodel of the "Sigatoka" simulation model.

of fungicide resistances (sensitive or resistant to each of the 3 fungicides = 2^3). Furthermore, the state variables in "Sigatoka" are updated on a 2-hour time-step. This kind of bookkeeping pushes the memory available in the IBM-PC to the limit and makes execution very slow, even with a math coprocessor (more than 2 minutes per simulated day on a 8088-based system). While "Resistan" has only 20 parameters to be estimated, "Sigatoka" has 73 basic parameters, which when combined with the values of the parameters that vary with leaf position, fungicide, and fungal biotypes, make a total of 283 parameters to be estimated.

The fungicide submodels in "Sigatoka" are considerably more detailed than in "Resistan". The fungicide residues on leaf surfaces are redistributed downward in response to rainfall and to plant development. Each fungicide can independently affect spore mortality, infection rate, rate of lesion development, and rate of sporulation, making it possible to mimic with considerable fidelity the different effects of the different fungicides on each stage of development of the fungus.

Currently "Sigatoka" is nothing more than a logical structure representing our present understanding of the biological mechanisms important in the development of the pathogen and its interaction with its host. None of its parameters has yet been estimated, their current values simply based upon the "educated guesses" of its author. In this early stage of development, of course, the model has not yet been validated in the field.

Conclusions

There are two major ways in which computer simulation can be used in the management of fungicide resistance, preseason planning of fungicide spray programs and day-by-day forecasting for timing of fungicide applications. Each of these uses requires a different type of simulation model, each with a different set of inputs and a different kind of outputs. "Resistan" is an example of a streamlined, although somewhat mechanistic, model for simulating whole seasons or multiple seasons for use as a preseason management decision aid. "Sigatoka", on the other hand, is an example of a detailed mechanistic model, driven by hourly weather, that can give day-by-day predictions of disease severity, frequency of resistance, and fungicide residues. As noted above, both these models are still in the early stages of development as management tools, and considerably more work is needed before they can be used with confidence to aid in management decision making. However, both have already proven extremely useful as teaching tools to illustrate the concepts of fitness and selection to students and to those attempting to manage fungicide resistance in the field.

Acknowledgments

The author wishes to acknowledge the programming efforts of Barr E. Ticknor in the creation of the user interface for "Resistan", the source code of which exceeds the number of lines in the simulation itself by more than 2 orders of magnitude. Barr's creative thinking and knowledge of program design conventions have immensely enhanced the utility and "user-friendliness" of the program. We express appreciation to Dr. Ken P. Sandlan for programming consultation. The author gratefully acknowledges the technical assistance and logistic support in the creation of "Sigatoka" from the Centro Agronómico Tropical de Investigación y Enseñanza (CATIE) in Turrialba, Costa Rica. This research was supported through Hatch project 153425, USDA.

Literature Cited

1. Delp, C. J., Ed.; Fungicide Resistance in North America; APS Press: St. Paul, 1988; Chapter 11, 16, 21, 26.
2. Delp, C. J. Plant Disease 1980, 64, 652-7.
3. Josepovits, G.; Dobrovolszky, A. Pesticide Science 1985, 16, 17-22.
4. Kable, P. F.; Jeffery, H. Phytopathology 1980, 70, 8-12.
5. Levy, Y.; Levi, R.; Cohen, Y. Phytopathology 1983, 73, 1475-80.
6. Milgroom, M. G.; Fry, W. E. Phytopathology 1988, 78, 559-70.
7. Skylakakis, G. Phytopathology 1981, 71, 1119-21.
8. Skylakakis, G. Phytopathology 1982, 72, 272-3.
9. Arneson, P. A.; Ticknor, B. E.; Sandlan, K. P. Resistan: A Mechanistic Simulation of the Selection of Fungicide Resistance; diskette and manual, Cornell University, Ithaca, NY, 1988.
10. Arneson, P. A.; Ticknor, B. E.; Sandlan, K. P. In Fungicide Resistance: Research and Management Goals and Their Implementation in North America; Delp, C. J., Ed.; APS Press, 1988; pp 107-9.
11. Arneson, P. A.; Sigatoka; diskette and manual, Centro Agronómico Tropical de Investigación y Enseñanza: Turrialba, Costa Rica, 1988.

RECEIVED October 16, 1989

Chapter 19

Population Biology and Management of Fungicide Resistance

W. E. Fry and M. G. Milgroom

Department of Plant Pathology, Cornell University, Ithaca, NY 14853

> Both absolute population size and relative frequency of resistant forms are important criteria in evaluating management strategies. Selection is the most important force in the evolution of fungicide resistance in agriculture. The speed at which a population changes is related to population growth rates. Strategies which slow fungal population growth rates will slow the rate at which the resistant form increases in frequency. Simulation models describing the *Phytophthora infestans*: metalaxyl system identified various combinations of strategies that retard the rate at which resistant forms displace sensitive ones.

Theoretical population biologists and plant pathologists have much to offer each other. Theoretical population biologists have developed theory concerning the evolution of populations in response to various forces (mutation, selection, migration, etc.). Plant pathologists identify specific situations in which the theory can be tested and applied. The overall goal of this paper is to illustrate the benefit of this exchange.

From the plant pathologist's perspective, there are interesting and important messages that come from the application of theoretical population biology to the question of fungicide resistance management. When making this application, it is important to remember that population size may be unimportant to many theoretical population biology models, but it is critically important to managing plant pathogen populations. Theoretical population biologists frequently deal primarily with gene frequencies, regardless of population size. Plant pathologists, on the other hand, are vitally concerned with population sizes in order to minimize disease levels. These two perspectives (Table I) may result in conflicting recommendations. For example, the plant pathologist's goal of maintaining small pathogen populations has contributed directly to the fungicide resistance problem by using

Table I. Dual Goals of Fungicide Resistance Management

1. Maintain a low frequency of resistant individuals in a pathogen population.

2. Maintain small pathogen population.

---

especially effective fungicides extensively and intensively to maintain low levels of disease. This situation is especially favorable for a rapid increase in the frequency of fungicide resistance.

Fungicide resistance management provides an interesting new situation for theoretical population biology, because fungal life histories differ in several respects from the model life histories on which most theoretical population biology is based. Most theory has been developed for diploid, randomly-mating organisms, but fungi are generally not diploid and/or are not randomly mating organisms. The vast majority of fungal plant pathogens (and the vast majority of fungal pathogens for which fungicide resistance is important) may be haploid, undergo sexual reproduction rarely if at all, or both. Thus the situation for fungicide resistance differs from the situation for which most theory has been developed, and also differs from insecticide resistance management. Growth of many fungal plant pathogens is essentially clonal, so that inheritance of resistance is not especially important (and maintenance of heterozygosity in the population becomes a much less important strategy than is true for insects and insecticide resistance management).

The evolutionary forces we will discuss include mutation, selection, and migration. Mutation is particularly important in providing sources of resistance, but, in the presence of selection, plays a negligible role in determining the frequency of resistance. Selection is extremely important and is the basis of most fungicide resistance theory and most management strategies. Migration is an important source of resistance for some populations, but in the presence of selection is likely to assume a minor role in the overall evolution.

The Role of Mutation

Mutation has very little effect relative to selection (see below) in changing either the number or frequency of resistant individuals in a population (1). The major role of mutation is as a source of resistance prior to selection by fungicide use. Recurrent mutation, balanced by selection against resistant mutants, results in an initial frequency that can be estimated theoretically.

A basic (conservative) assumption of most fungicide resistance strategy has been that resistant individuals are present in the population prior to use of the "at risk" fungicide. Unfortunately, it is extremely difficult (technically) to determine whether fungicide resistance is present in a population when the frequency is extremely low. Sampling schemes are not adequate to detect extremely low frequencies. For most fungal plant pathogens, there are no estimates of mutation rates for resistance to any fungicide. Such estimates together with selection coefficients might enable the estimation of initial frequencies of fungicide resistance. If the expected frequency is very low, then there is a considerable probability that some finite populations would have no resistant individuals. The importance of this result is that some management strategies (based on the dynamics of selection) may be inappropriate if resistance is not present in a population. Further development of this theory may lead to new possibilities of fungicide use.

For example, consider a finite population ($N=10^{10}$) where the expected frequency of resistance is 10^{-10}. With this frequency, the probability that an individual in a population of 10^{10} is resistant is 0.63 (according to the relationship identified in Table II). The relationshps among population size, frequency of resistance, and the probability of resistance occurring in a finite population are illustrated in Figure 1. Clearly the actual occurrence of resistance in a population increases as the population increases. Therefore, the occurrence of resistance (and its subsequent selection) is less likely in small populations than in large populations.

Table II. Rationale for calculating probabilities for the occurrence of resistance as a function of population size and expected frequency

---

If:
  N = population size
  p_o = expected initial frequency
Then:
  The probability that at least one individual
  in a population (N) is resistant (assuming
  N is very large and p_o is very small)

  = 1 - probability that no individuals are resistant
  = $1-(1-p_o)^N$
  = $1-e^{-Np_o}$

Two important conclusions arise from this analysis. First, resistance may not occur in every population, a fact which might well influence strategy. For example, in isolated locations (no migration) with small areas treated with the "at-risk" fungicide, resistance may be unlikely. Under these conditions, intense use might be a good strategy. Second, it reaffirms our current realization that any technique (cultural, sanitation, host plant resistance, or fungicide) that restricts pathogen population size will be useful in fungicide resistance management.

Unfortunately, we have insufficient information for developing these concepts in a quantitative fashion. Mutation rates for

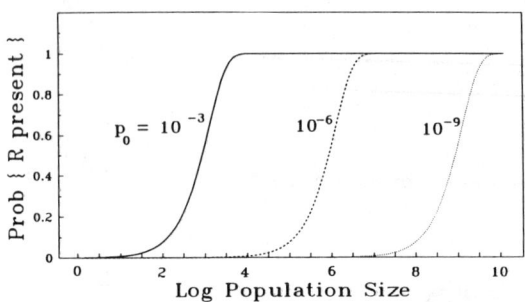

Figure 1. Relationships between population size and the probability that at least one individual in that population is resistant.

specific fungicides are unknown. Accurate estimates of population sizes are typically unavailable. Estimates of expected frequencies of fungicide resistances are generally unknown. Empirical investigations concerning mutation rates, expected initial frequencies of resistance, and methods to estimate initial population sizes are needed before we can take advantage via management procedures, of the fact that resistant individuals don't occur in every population.

The Role of Selection

The dominant role of selection in the evolution of fungicide resistance within a fungal pathogen population is readily apparent (2). Thus, any use of the "at risk" fungicide will influence evolution of the population. And, appropriately, consideration of selection has dominated our thinking. Predictions concerning the influence of selection on plant pathogen populations have been made with the aid of mathematical models (3, 4, 5, 6, 9).

Simple exponential growth models have been useful in several respects. For example, they have predicted that the most rapid absolute change occurs when there are equal numbers of resistant and sensitive types (1). These predictions are consistent with the very few available empirical data (1). Also simple exponential growth models were used to identify two basic strategies for fungicide resistance management (Table III) (1, 6). There are a variety of tactics available for achieving each of these strategies. Some such as use of a second non-selective fungicide, have been evaluated via modeling and experimentation (2, 6, 7, 8, 9).

Table III. Strategies for Managing Fungicide Resistance

---

1) Reduce the growth rates of both sensitive and resistant types.
2) Reduce the growth rate of the resistant type relative to the sensitive type.

Simulation models have been used to evaluate specific approaches in specific agroecosystems (see refs 1, 2, 3, 4, 5, 6, 7). For example, we have used such a system of models in the potato: *Phytophthora infestans* pathosystem. The models describe the dynamics and the effects of chlorothalonil, metalaxyl, and resistant varieties on the dynamics of *P. infestans* in potato foliage. We have then used this system of models in a preliminary analysis of various strategies to achieve the dual goals of maintaining a low frequency of resistance in the population, and of maintaining a low absolute population size (Tables I and IV). From simulation alaysis we learned that a protectant fungicide (chlorothalonil) used in mixture with a selective ("at-risk") systemic fungicide (metalaxyl) or in alternation with the "at-risk" fungicide reduced the selection for fungicide resistance (lower frequency) and also suppressed disease (smaller population) (Table IV). Use of a cultivar with a relatively high level of resistance had an effect similar to that of the protectant fungicide. Both of these tactics employed the same strategy: they reduced the growth rates of both sensitive and resistant types.

Table IV. Predicted relative effects of various management strategies on fungicide resistance and disease suppression using *Phytophthora infestans*: *Solanum tuberosum* and protectant and systemic fungicides via simulation analysis* Data are from Milgroom and Fry (1)

Chemical (kg/ha)	Days of Appl'n	Cv	Relative Increase in Res/Sens*	AUDPC
a) metalaxyl (0.22)	5,19,33,47	S	62,900	2.62
b) metalaxyl (0.22) + chlorothalonil (1.26)	5,19,33,47	S	6,500	0.90
c) metalaxyl (0.22)	5,19,33,47	R	9,900	0.28
d) metalaxyl (0.22) + chlorothalonil (1.26)	5,33 5,19,26,33,47	S	800	0.94
e) metalaxyl (0.22) + chlorothalonil (1.26)	5,33 19, 26, 47	S	9,800	1.34

*Values are averages resulting from simulation analysis of 9 years of diverse weather data in simulations.
Initial ratio of Res/Sens was 10^{-5}.
AUDPC = Area under the disease progress curve.
Cv=cultivar; S=suceptible; R=resistant

Alternations versus mixtures. Plant pathologists have long been intrigued by the question of whether the protectant fungicide should be alternated with the systemic fungicide or should be mixed with the "at-risk" fungicide (4, 7). Both use of mxitures and alternations limit the final selection for resistant biotypes as well as suppress disease (Table V). It is abundantly clear that both mixture and alternations are much preferable to using the "at-risk" fungicide alone (treatments b and e, Table IV). Use of the second fungicide slows pathogen growth rate and therefore satisfies one of the two basic approaches of fungicide resistance management (Table III). Although some authors have felt that mixtures might be somewhat preferable to alternations (7), Doster and Fry (Cornell University, Ithaca, NY, unpublished results) found little difference between alternations and mixtures, when equal amounts of fungicide were used in both strategies (Table V). Alternation of a mixture with a protectant (treatment d, Table IV) was apparently better than use of only a mixture or only an alternation.

Table V. Predicted relative influence of mixtures and alternations on fungicide resistance and disease suppression using constant amounts of fungicide (via simulation analysis)

	Relative Increase in Res/Sens[a]	AUDPC[a]
mixture [b]	181	7.6
alternation [c]	175	10.0

[a] Data are means of results from simulations done with each of 10 years of real weather data.
[b] 5 applications chlorothalonil (1.2 Kg/ha) + metalaxyl (0.15 Kg/ha)
[c] 4 applications chlorothalonil (1.5 Kg/ha) and 3 applications metalaxyl (0.25 Kg/ha)

Growth Rates. Because pathogen growth rates are so important in the evolution of fungicide resistance, plant pathologists should continue to recommend complete, effective systems to curtail growth rates of both resistant and sensitive forms (Table III). These complete systems are likely to include cultural controls, sanitation, resistant varieties, as well as protectant and systemic fungicides.

Opportunities for decreasing the growth rate of the resistant types relative to the sensitive types (Table III) are limited. In some pathosystems negatively correlated cross resistance may enable selective repression of resistant type growth rate. In these systems the second fungicide may be selectively effective against the resistant genotype. A more common possibility for reducing selection for resistance is to use fewer applications or lower doses of the "at-risk" fungicide. Both empirical and simulation (Table IV) experiments support the use of this strategy (2, 8). Restrictions on use of the "at-risk" fungicide illustrate the potential conflict between the goals of disease suppression and resistance suppression (Table I).

If the systemic, "at-risk" fungicide is to be used infrequently, it is important to identify when during the season the application should be made. M. A. Doster, and W. E. Fry

(Cornell University, unpublished data) have begun to investigate timing of the "at-risk" fungicide (metalaxyl), in the potato: *P. infestans* pathosystem. The approach to date has involved both simulation and field experiments. In preliminary simulations with sensitive isolates, applications of metalaxyl early-mid season contributed more to disease suppression than did later applications (Figure 2). The simulation experiments were corroborated by field experiments (Figure 3).

Fitness. Fitness of the resistant isolates may be a potentially important factor in identifying strategies to manage fungicide resistance. If the resistant isolates are considerably less fit than sensitive ones, then reversion to sensitivity in a population becomes a realistic consideration. Unfortunately, if fitness of resistant types is very similar to that of sensitive types, there are few options. However, relative fitnesses of resistant and sensitive biotypes are rarely known, and knowledgeable application of model results awaits empirical evidence.

Empirical investigations of fitness are difficult because of the potentially confounding effects of diverse background genotypes. Differences in fitness between any single pair of field isolates (one resistant and one sensitive) can almost never be attributed with confidence to resistance because the background genotypes of the two isolates can be widely divergent. Fitness differences may be due to background genotype, rather than to resistance.

Role of Migration

Migration may have a variety of roles in the evolution of resistance. An obvious simplistic role is to provide an initial source of resistant types for a neighboring population. The number of migrants is a function of size of the migrating population, the method of migration (dispersal), and the distance between the source and target populations. When there is a mosaic of treated and untreated fields, when fitness costs are very low, and when migration is high, migration will hasten the dominance by the resistant types (1). Unfortunately, the theoretical consideration of migration is still at an initial stage, and development of resistance management strategies based on rigorous application of theory to real situations is preliminary.

Predictions and Explanations

What is the Effect of Curative Sprays? When systemic fungicides were first discovered, plant pathologists were very excited about the potential for using these compounds in a curative manner. However, with recognition of the hazard of resistance, some scientists recommended that these fungicides not be used in a curative manner (10). However, in simulation experiments using the potato late blight pathosystem, we found the counter-intuitive result that curative applications did not necessarily select more

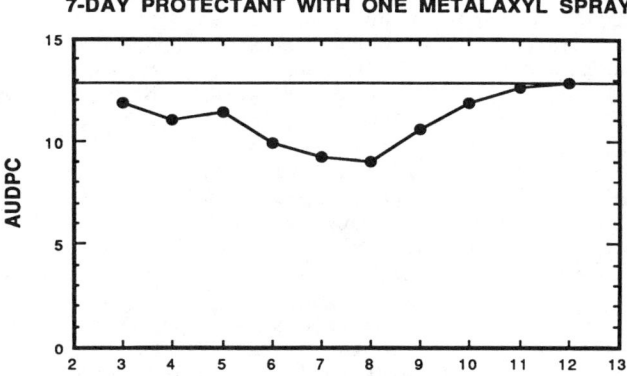

Figure 2. Simulation analysis of value of single metalaxyl sprays in suppressing potato late blight. Data points indicate values for area under the disease progress curve (AUDPC) when metalaxyl was applied on that date. The solid line indicates AUDPC without metalaxyl. Each data point is the mean of values from each of 10 different simulations each done with a distinct season of weather data from central New York. Simulations done by M. A. Doster (Cornell University).

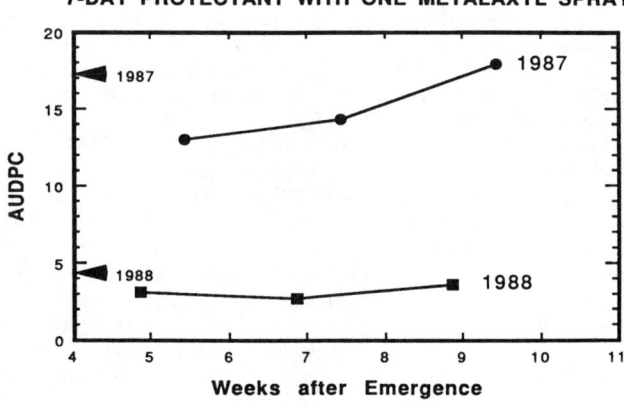

Figure 3. Field experiment analysis of the value of single metalaxyl sprays in suppressing potato late blight. Data points are means of four replications. Arrows indicate the area under the disease progress curve (AUDPC) in the absence of metalaxyl. LSD (P=0.05) for 1987 data was 1.1. There were no significant differences in 1988. Data from M. A. Doster and W. E. Fry (Cornell University).

strongly for resistance than did protective applications (6). One explanation was that the length of time for selection was lessened with curative relative to protective sprays.

Perhaps the major reason that curative sprays were not recommended is related to population size. Once disease is detected, the pathogen population may already be large. When the frequency of resistance is low, large populations are more likely than small populations to contain resistant types. Thus, application of the "at-risk" fungicide in a curative manner may be more likely to expose resistant types to selection than use of fungicide in a protective manner. Of course, this expectation is based on the assumption that protective use will expose the fungicide to smaller populations than will curative use.

Case Histories. The evolution of metalaxyl resistance in the *Bremia lactucae* population in the United Kingdom and in the *Phytophthora infestans* population in Northern Europe illustrate many of the principles discussed in this chapter. Mutation to resistance in the *B. lactucae* population is apparently a very rare event. Metalaxyl was used very effectively for five years (and maintained very low populations) before resistance became a problem (11). However, once established, selection was extremely strong for the resistant type and it quickly dominated *B. lactucae* populations in the lettuce growing areas of the United Kingdom. The available evidence strongly indicates clonal selection: during the first several years of resistance occurrence, resistant individuals collected all over England were indistinguishable from each other (11). Migration was extremely effective in dispersing the resistant phenotype. In addition to aerial dispersal of "conidia", the resistant phenotype was apparently dispersed on infected lettuce seedlings. At the present time, an integrated strategy involving protectant fungicides, use of metalaxyl on certain race specific cultivars, race non-specific resistance, and sanitation are involved in an effective program for managing lettuce downy mildew in England (11).

The more rapid appearance of metalaxyl-resistant individuals in the *Phytophthora infestans* population of Northern Europe suggests multiple mutational events. Presumably the populations of *P. infestans* in the late 70's and early 80's were very large, and several resistant mutants were selected. [Resistant isolates of *P. infestans* from Northern Europe compose a heterogeneous population (Spielman, L. J., Fry, W. E., Cornell University, unpublished data)]. Presumably there were several mutational events in diverse populations, and migration contributed significantly to the dispersal of these individuals. Again, selection was of overriding importance in allowing resistant individuals to dominate *P. infestans* populations in various locations in Europe.

Conclusions

In the United States where metalaxyl resistance is not yet known in the *P. infestans* population, it is important to maintain an integrated approach to disease management. Metalaxyl should be used infrequently because of the extremely strong selection pressure. A variety of effective tactics should be used to maintain low population growth rates and to avoid extreme selection pressure on populations. Analytical models are useful to develop general principles and validated simulation models are useful to predict the relative magnitudes of the effects of various specific tactics.

These conclusions have been derived from and illustrated with, the potato: *P. infestans* pathosystem, but they should have utility with other pathosystems that are epidemiologically similar. Thus wherever a fungicide has an especially strong selection for resistant biotypes and there are potentially many secondary cycles, reduction of the growth rates of both sensitive and resistant populations using an integrated progam of tactics should delay the appearance of problems caused by resistant pathogens.

Acknowledgments

We acknowledge the support of Hatch Project 153430, BARD and NEPIAP funding.

Literature Cited

1. Milgroom, M. G; levin, S. A.; Fry, W. E. In "Plant Disease Epidemiology", Leonard, K. J.; Fry, W. E., Eds.; McGraw Hill Book Publishing Company, New York 1988; Vol. II, pp 340-67.
2. Staub, T.; Sozzi, D. Plant Disease 1984, 68, 1026-31.
3. Barrett, J. A. Phytopathology 1983, 73, 510-12.
4. Kable, P. F.; Jeffery, H. Phytopathology 1980, 70, 8-12.
5. Levy, Y.; Levy, R. S. Phytoparasitica 1986, 14, 303-12.
6. Milgroom, M. G.; Fry, W. E. Phytopathology, 1988, 78, 565-70.
7. Skylakakis, G. Phytopathology 1981, 71, 1119-21.
8. Lalancette, N.; Hickey, K. D.; Cole, H., Jr. Phytopathology 1987, 77, 86-91.
9. Sanders, P. L.; Houser, W. J.; Parish, P. J.; Cole, H., Jr. Plant Disease 1985, 69, 939-943.
10. Urech, P. A.; Staub, T., OEPP/EPPO Bulletin 1985, 15, 539-43.
11. Crute, I. R. In Plant Disease Epidemiology; Leonard, K. J. and Fry, W. E. Eds.; McGraw Hill Book Publishing Company, New York, 1988; Vol. II, P 30-53.

RECEIVED October 16, 1989

Chapter 20

Impact of Fungicide Resistance on Citrus Fruit Decay Control

Joseph W. Eckert

Department of Plant Pathology, University of California, Riverside, CA 92521

> The continuous use of o-phenylphenol, biphenyl, thiabendazole, benomyl, and sec-butylamine for 20 years to control postharvest decay of citrus fruits has resulted in a serious problem of fungicide-resistance in Penicillium digitatum and P. italicum. Fungicide-resistant Penicillium isolates are cross-resistant to structurally-related compounds and, in addition, may be resistant to two or more unrelated compounds. Biotypes of P. digitatum resistant to imazalil have been found recently in 19 packinghouses throughout California. Imazalil treatments (1-2 g/L in water/wax sprays) applied in commercial packinghouses failed to control decay or sporulation of the resistant biotypes. Six imazalil-resistant isolates were cross-resistant in vitro to 8 other sterol demethylation inhibitors, but were sensitive to tridemorph and fenpropimorph. Several imazalil-resistant isolates were resistant also to o-phenylphenol and thiabendazole. The resistant biotypes were as pathogenic as the imazalil-sensitive wild types but were less stable when mixtures of isolates were inoculated into imazalil-free fruit. Resistance control strategies emphasize sequential treatment of fruit with unrelated fungicides and suppression of spore production and dispersal.

Citrus fruits may be stored for weeks or months after harvest in an environment that maintains or improves the market quality of the fruit (1). The usual storage conditions are 3-6°C for oranges and 12-15°C for lemons and a relative humidity of 85-90% for both fruits. Green and blue mold decays of citrus fruits, which are caused by the ubiquitous fungi, Penicillium digitatum Sacc. and P. italicum Wehm., respectively, are the major factor determining the storage life and quality of citrus fruits. Green mold is the most important cause of postharvest decay of citrus fruits grown in areas with low rainfall during the period of fruit development. Blue mold is of lesser overall importance, but may become the major problem under certain environmental conditions or a fungicide regimen that selectively suppresses the development of green mold (1,2). In nature, P. digitatum completes its life cycle only on diseased citrus fruits, whereas P. italicum can infect several fruits and vegetables.

Spores of P. digitatum and P. italicum are produced on the surface of diseased fruit on the ground in citrus groves and in packinghouses. These spores can survive for months under dry conditions and are transported by air currents to healthy fruit. The surface of virtually every citrus fruit is contaminated with these spores at harvest time, but the pathogens are unable to germinate and infect fruit except at injured sites. Potentially, a single spore of P. digitatum that alights in a fresh injury on a citrus fruit may give rise to a progeny of about 10^8 spores in seven days under optimal environmental conditions--20-25°C and a high humidity. The time required to complete the life cycle approximately doubles with every 10°C decrease in storage temperature.

Decay Control Strategies

From the biology of Penicillium molds, it is evident that these diseases can be controlled to at least some extent by: reducing the incidence of fruit injury, decreasing the number of spores in the fruit environment, and lowering the storage temperature. Practices based on these principles were used during the last century and today are major non-selective methods for management of fungicide-resistance problems.

With the advent of long-distance marketing of citrus fruits in the early 1900s, the need for postharvest fungicides was immediately recognized and stimulated relevant research in the USDA and in the agricultural experiment stations of Florida and California (1). It was soon discovered that the germination of Penicillium spores was adversely effected by alkaline conditions, and that dipping citrus fruits in a warm (45°C) solution of borax or sodium carbonate prevented the infection of fresh wounds in the peel. These treatments have been used successfully for more than 60 years to control citrus decay, although their applicability in recent years has been curtailed by water pollution regulations. Although borax and sodium carbonate treatments have encountered practical problems and have virtually no protective or curative action against Penicillium decay, they have never failed to reduce infection because of the tolerance of Penicillium to them.

The development of more effective organic fungicides to control Penicillium decay began with biphenyl in the late 1930s, followed by o-phenylphenol within the next decade (1). Biphenyl was developed as a vapor treatment of packed fruit to suppress sporulation of Penicillium on diseased fruit during shipment and thereby reduce contamination of adjacent fruit in the package. o-Phenylphenol was developed as a washing solution, 0.5%-1.0% sodium o-phenylphenate (SOPP) pH 12, and as a fungicide additive to fruit coating formulations ("waxes"). Since the 1960s, a steady flow of organic fungicides have been introduced to provide improved control of citrus fruit decay (Figure 1). The chemical and biological properties of these compounds has been summarized by Worthing (3).

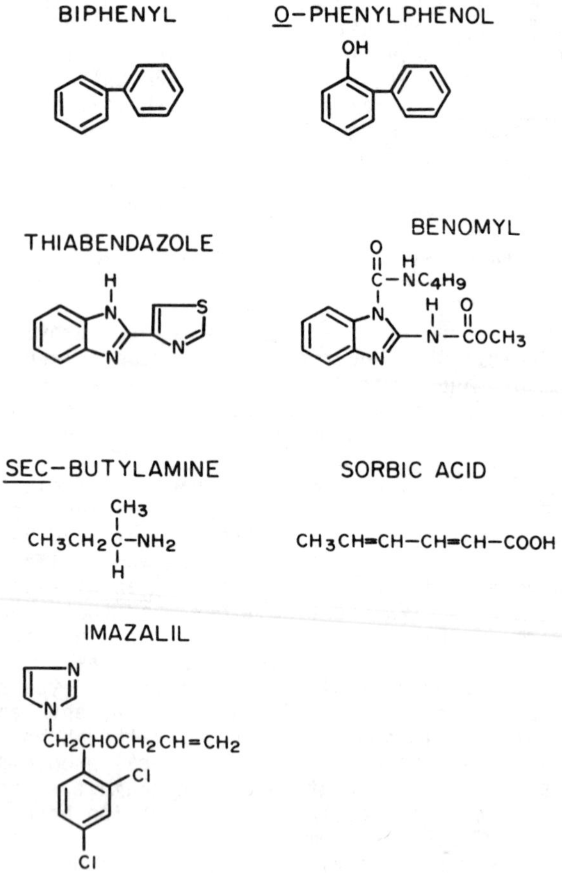

Figure 1. Fungicides applied in commercial packinghouses to control citrus fruit decay.

Fungicide Resistance

Early Fungicides. Several years after biphenyl was recognized as an effective postharvest fungicide, investigators in Israel described mutants of P. digitatum and P. italicum that appeared spontaneously in pure cultures of these fungi that were exposed to biphenyl vapor (4). The existence of biphenyl-resistant biotypes was not regarded as a practical problem until the late 1960s when certain shipments of California lemons arrived in Europe with high levels of Penicillium decay and sporulation. Harding (5,6) observed that many of these problem shipments originated in packinghouses where the lemons had been treated with SOPP before they were stored for several months prior to shipment. The fungal isolates that were resistant to biphenyl were also resistant to SOPP, a correlation that might be anticipated from the structural relationship of the two compounds.

sec-Butylamine (SBA) was developed in the mid 1960s as an aqueous drench and an additive to wax formulations to reduce decay of oranges and lemons during storage (1). By the early 1970s, biotypes of P. digitatum resistant to sec-butylamine were common in California packinghouses, with the result that the effectiveness of the treatment was seriously impaired. sec-Butylamine is no longer registered as a postharvest fungicide in the U.S., but is used in other citrus producing countries.

Benzimidazoles. Thiabendazole, and a few years later, benomyl, were introduced as postharvest fungicides in the late 1960s. These outstanding compounds (benzimidazole fungicides) provided excellent control of both fruit decay and Penicillium sporulation (1,2). Their practical benefits were so apparent that these compounds were adopted throughout the world as the principal postharvest fungicides for citrus fruits. The supremacy of the benzimidazole fungicides was short-lived, however. Benzimidazole-resistant biotypes of P. digitatum and P. italicum were isolated from the atmosphere and fruit in lemon packinghouses about 15 months after thiabendazole was adopted as the standard fungicide treatment for lemons before storage (7). The problem of benzimidazole-resistance in citrus decay control escalated in importance worldwide over the next several years, and by 1979 it was clear that substantial decay losses in major citrus exporting countries could be attributed directly to widespread distribution of benzimidazole-resistant biotypes of P. digitatum and P. italicum (8). In California alone, a loss of 7 million cartons of citrus fruit in 1979 was attributed to this cause (9).

Most observations of benzimidazole-resistant biotypes of Penicillium on citrus fruits have followed the intensive use of thiabendazole and/or benomyl over a period of many months. A survey in Israel revealed that the frequency of benzimidazole-resistance was very low in citrus groves, higher in packinghouses, and highest in fruit storage rooms (10). The frequency of benzimidazole-resistant biotypes in California packinghouses followed a seasonal fluctuation that correlated with the intensity of benzimidazole fungicide use (11). However, a serious resistance problem on satsuma mandarins in Japan resulted from a program of a single spray of thiophanate methyl (a benzimidazole progenitor) on the fruit several days before harvest (12). Furthermore, several investigators have isolated benzimidazole-resistant Penicillium from citrus groves and packinghouses where benzimidazole fungicides have never been used (7,12,13,14), suggesting that mutants resistant to these fungicides arise spontaneously in the Penicillium population.

In genetic studies with other fungi, benzimidazole-resistance has been attributed to the mutation of a single major gene (15,16). Therefore, it is not surprising that resistance in P. digitatum and P. italicum is expressed in a qualitative fashion, i.e., virtually all resistant isolates examined are capable of causing citrus decay, irrespective of the dosage of benzimidazole fungicide applied to the fruit (7,13,14). Furthermore, benzimidazole-resistant isolates, tested as pure cultures, do not differ significantly in virulence from the range which is normal in fungicide-sensitive isolates of P. digitatum and P. italicum.

Virtually all benzimidazole-resistant isolates of Penicillium are cross-resistant to all other benzimidazole fungicides (7,8,13,14). Furthermore, many of these isolates show resistance also to the unrelated postharvest fungicides, SOPP, biphenyl, sec-butylamine, and guazatine (13,14,17,18). Multi-resistant biotypes apparently emerged from a benzimidazole-resistant population that was subjected to the selection pressure of the unrelated fungicides.

Imazalil. This fungicide was introduced in Europe in the mid 1970s coincident with the decline in effectiveness of the benzimidazoles due to Penicillium resistance. Imazalil has been intensively used for over 10 years in Mediterranean citrus production areas with no reports of practical failure due to fungicide-resistant Penicillium spp. (11,18,19). Isolates of P. italicum that are resistant to imazalil in the laboratory have been reported (20,21,22), but their existence was not related to loss of disease control under practical conditions. Laville (23) transferred weekly spores of fungicide-sensitive P. digitatum and P. italicum to culture medium containing benomyl or imazalil. The spores were treated with mutagenizing agents in some instances. Many mutants appeared that were resistant to benomyl but, after 3 years, no imazalil-resistant mutants had been found. De Waard et al. (20) isolated imazalil-resistant biotypes of P. italicum after UV treatment of fungicide-sensitive spores. The resistant mutants were pathogenic to oranges and, when inoculated into fruit, were more difficult to control than the sensitive parent biotype.

Genetic studies with other fungi have shown that imazalil-resistance is controlled by a polygenic systemic rather than a single major gene as in the case of benzimidazole-resistance (15). Mutation of the individual imazalil-resistance genes and their accumulation in the genome of a single biotype results in a progressive increase in the level of imazalil resistance. Thus, a stepwise increase in the level of resistance has been observed after repeated mutagenesis of a P. italicum isolate and continuous cultivation on imazalil containing medium (20,24). Van Tuyl (15) showed that the frequency of mutation for imazalil resistance was relatively high, but resistant mutants often grew poorly in culture and were weakly virulent pathogens. This observation spawned the opinion (hope?) that resistance to imazalil and other sterol demethylation inhibitors was not likely to become a practical problem in disease control (18,19,25), especially in the case of P. digitatum where no imazalil-resistant biotypes had been uncovered despite diligent searches for them for 10 years by several investigators (11,19,23).

Imazalil has been intensively applied to harvested citrus fruits in California packinghouses since 1981. The fungicide was enthusiastically welcomed as the ultimate solution to the serious benzimidazole-resistance problem that had plagued the California citrus fruit industry during the late 1970s (9) and was universally adopted by packinghouses throughout the state. Today, imazalil is applied routinely to harvested oranges and lemons as a water spray (1 g/L) or as a component (2 g/L) of the wax formulation. Correctly applied, these treatments have prevented fruit infection by Penicillium and suppressed sporulation of the fungus on diseased fruit (21).

In mid 1986, several biotypes of P. digitatum with obvious tolerance to imazalil were isolated from imazalil-treated lemons and from the atmosphere of packinghouses where this fungicide had been used for a number of months (26). These isolates formed colonies characteristic of P. digitatum on potato dextrose agar containing 1 µg/ml imazalil (approx. EC_{50}) whereas all isolates collected earlier had EC_{50} values less than 0.1 µg/ml (18). The resistant isolates were strongly pathogenic to citrus fruits and were confirmed to be authenic specimens of P. digitatum (Stoner, M. J., California State Polytechnic University, personal communication, 1988). A survey of 19 California packinghouses yielded 180 isolates of P. digitatum with the same level of resistance as the prototype resistant isolates, but only several imazalil-resistant isolates of P. italicum were found. Very few resistant isolates of P. digitatum or P. italicum were found among several hundred samples of Penicillium collected from fallen fruit in citrus groves throughout the state. All of the isolates from citrus groves were as sensitive to imazalil as stock cultures collected before 1981 when imazalil was first used intensively in California.

Imazalil resistance appeared to be a stable character of the isolates--no change in resistance level was observed after six weekly transfers on imazalil-free culture medium or six disease cycles in untreated lemon fruits. Six imazalil-resistant isolates tested were cross-resistant to propiconazole, fenarimol, flusilazole, penconazole, myclobutanil, prochloraz, diclobutrazol, and bitertanol, all of which inhibit the C-14 demethylation reaction in fungal sterol biosynthesis (27,28). However, the imazalil-resistant isolates were sensitive to tridemorph and fenpropimorph, also inhibitors of sterol biosynthesis but of the Δ^8-Δ^7 isomerization reaction (29). Imazalil-resistant mutants of P. italicum have shown the same pattern of cross-resistance (20,30).

Since the typical imazalil-resistant isolate of P. digitatum showed a relatively low resistance level compared to benzimidazole-resistant isolates, experiments were conducted under practical operating conditions to determine the impact, if any, of these resistant isolates on the effectiveness of typical imazalil fruit treatments. Lemons were inoculated with isolates of P. digitatum, sensitive and resistant to imazalil (300 fruit each), and treated with this fungicide under usual practical conditions in a commercial packinghouse or in a pilot plant. The fruit were analyzed for imazalil to confirm that the treatment had been applied in an acceptable fashion. Table 1 gives the results of a typical imazalil treatment in a water/wax formulation applied to lemons before storage at 14°C for 20 days. Test procedures are described in detail elsewhere (31).

Table 1. Effect of a typical packinghouse application of imazalil upon decay of lemons inoculated with imazalil-sensitive or -resistant biotypes of Penicillium digitatum

P. digitatum biotype	g Imazalil/L wax formulation (fruit residue-mg/Kg whole fruit)			
	0 (0.1)		2 (1.03)	
	% Decay	Penicillium Sporulation	% Decay	Penicillium Sporulation
Imazalil-sensitive	64	4+	28	1
Imazalil-resistant	99	4+	93	4+

In lemons inoculated with an imazalil-sensitive isolate of P. digitatum, the fungicide/wax treatment reduced decay 56% compared to the "wax only" control, and greatly suppressed Penicillium sporulation (0-5 scale) on the treated fruit. In contrast, the imazalil/wax treatment did not significantly reduce decay or Penicillium sporulation on lemons inoculated with an imazalil-resistant isolate of P. digitatum. While the lower level of decay in untreated fruit inoculated with the imazalil-sensitive isolate could reflect a lower virulence of the sensitive strain, prior experiments comparing the virulence of the two strains have shown that this conclusion is not correct. More likely, the difference in decay of the two controls reflects the high sensitivity of the wild-type isolate to imazalil, which was present as a contaminant on the wax applicator brushes in the packinghouse. The untreated fruit contained 0.1 ppm imazalil (whole fruit basis).

Since the residue tolerance for imazalil on citrus fruits is 10 ppm in the U.S. and 5 ppm in most European countries, fruit were treated in a manner that resulted in high residue levels to determine the maximum effect of this fungicide on resistant biotypes of P. digitatum. Lemons were dipped for 30 seconds in aqueous imazalil (1 g/L) and stored at 14°C for 3 weeks. This method of treatment, although impractical for commercial use, resulted in higher residues and improved decay control compared to the imazalil/wax formulation treatment (32). In two experiments, the imazalil residues on dip-treated lemons were 3.0 and 5.8 ppm, respectively. These treatments provided excellent control of Penicillium sporulation on fruit inoculated with two imazalil-sensitive isolates of P. digitatum, but had essentially no effect on the sporulation of six imazalil-resistant isolates on imazalil-treated fruit. In lemons that were wound inoculated 24 hours before treatment, an imazalil dip treatment (1.0 g/L), which resulted in a fruit residue of 3.0 ppm imazalil, reduced infection by two sensitive and two resistant biotypes 99% and 55%, respectively. A 2 g/L dip treatment, which resulted in a fruit residue of 3.4 ppm imazalil, reduced infection by an imazalil-resistant biotype by 85%. These experiments showed that the standard imazalil spray treatment (1-2 g/L imazalil) applied in California packinghouses, which gave residues of 1-2 ppm imazalil on treated fruit, was unlikely to control either infection or sporulation by resistant P. digitatum biotypes that are prevalent in citrus packinghouses throughout the state. Treatment of fruit by dipping in an imazalil solution (1-2 g/L), which resulted in imazalil residues of 3.5 ppm and higher, reduced fruit infection by imazalil-resistant biotypes about 80-90%. However, no imazalil treatment, even those that gave residues in excess of 5 ppm, the legal tolerance in many countries, provided acceptable control of Penicillium sporulation on diseased fruit, a major aspect of citrus decay control.

The fact that high dosages of imazalil controlled fruit infection by resistant biotypes of Penicillium reflects the intermediate nature of resistance in isolates tested to date, compared to the qualitative nature of benzimidazole-resitance in P. digitatum. Moreover, the modest level of imazalil-resistance observed in many isolates of P. digitatum in California may be transient, since De Waard (20,24) showed that higher levels of resistance can be achieved in P. italicum by a repeated process of mutagenizing and selection in a single culture line.

The intensive and indiscriminate commercial use of imazalil and thiabendazole, alternated and in mixture, was followed by the isolation in 1987 of P. digitatum biotypes with multiple resistance to imazalil, benzimidazoles, and SOPP/biphenyl. Treatment of fruit inoculated with these biotypes by dipping in solutions of thiabendazole and imazalil, alone and in mixture, confirmed that the multiply-resistant biotypes could not be controlled by mixtures of these two fungicides, which had been extensively used in an effort to control imazalil-resistant biotypes.

Factors Affecting the Development of Resistance

P. digitatum has demonstrated the genetic capacity for resistance to all organic fungicides that have been used extensively to control citrus fruit decay. Sorbic acid may be an exception, but this fungicide is not highly effective and has not been used as intensively as other organic fungicides. Thus, it is reasonable to assume that P. digitatum will be capable of resistance to selective organic fungicides of the future. In contrast, resistance to sodium carbonate and borax are unknown, despite their continuous use for over 60 years, suggesting that P. digitatum lacks the genetic capability to avert the toxic action of these non-specific fungicides.

In addition to genetic potential, the development of a serious resistance problem involves the interplay of several biological and operational factors as well. The most important of these for Penicillium decay of citrus fruits appear to be: 1) intensity and duration of selection pressure for emergence of fungicide-resistant mutants; 2) rate of inoculum production and efficiency of spore dispersal; 3) relative fitness of sensitive and resistant biotypes and their interaction during pathogenesis.

Selection and Dispersal of Resistant Biotypes. A postharvest fungicide treatment exerts significantly more selection pressure for the emergence of a resistant biotype than a spray application of the same fungicide in the field. The postharvest treatment results in essentially complete coverage of the fruit, and the residues of most postharvest fungicides decrease only slightly during the storage period. The fungicide-treated fruit are often "de-greened" or stored under environmental conditions that permit growth and sporulation of the resistant biotypes on the fruit. The number of spores can increase about 10^8-fold in 7 days at 20°C or in 15 days at 15°C. At still lower temperatures, the rate of reproduction is substantially reduced.

The selection pressure for proliferation of a resistant biotype usually results from a deposit of the same fungicide on the fruit. However, an unrelated fungicide may also select for the same phenotype if a fungus isolate possesses resistance genes for each of the fungicides. For example, biotypes of P. digitatum and P. italicum have been isolated which possess resistance to the benzimidazoles and to one or two unrelated fungicides such as SOPP, SBA, guazatine and imazalil. Treatment of fruit inoculated with a multiply-resistant biotype with SOPP or sec-butylamine resulted in a rapid rise in the level of benzimidazole-resistance in the Penicillium spore population (13,14).

Interaction of Resistant and Sensitive Biotypes During Disease Development. Clearly, resistant biotypes are more fit than sensitive biotypes in the presence of the appropriate postharvest fungicide. Biotypes of Penicillium that possess resistance to a fungicide flourish on fruit treated with that fungicide, whereas fungicide-sensitive biotypes are inhibited and do not reproduce in this environment. Nonetheless, the inhibited biotypes can increase the infectivity of the resistant biotypes by their release of pectolytic enzymes, which increases the probability of fruit infection by spores of the resistant biotypes. Wild and Eckert (33) demonstrated that benzimidazole-sensitive isolates of P. digitatum could increase the infectivity of benzimidazole-resistant biotypes when a mixture of both types was inoculated into a benomyl-treated orange. Benomyl severely stunted the growth of the sensitive biotype, but only slightly reduced germination and pectolytic enzyme production. The dialyzed culture filtrate from the inhibited sensitive isolate also increased the infectivity of spores of the resistant biotype, suggesting the involvement of extracellular pectolytic enzymes from the inhibited sensitive biotype. This observation provides an explanation for the poor performance of benzimidazole fungicides in controlling Penicillium infection even when inoculum (spores) contained only a very small percentage of resistant biotypes (13,14). It also provides a compelling argument for strict packinghouse sanitation as the cornerstone of a program to control benzimidazole resistant isolates of Penicillium.

The relative fitness of resistant and sensitive biotypes in the absence of fungicide selective pressure (untreated fruit) is a matter of considerable practical importance since it could provide a means for reducing the level of resistance in the Penicillium population. Two observations suggest that benzimidazole-resistant biotypes are less fit than sensitive biotypes in the absence of selection pressure. Firstly, fungicide-resistant biotypes are rarely found in citrus groves when benzimidazole fungicides are not applied, although spores of resistant biotypes are carried into these groves regularly on picking boxes and by air currents from packinghouses with high levels of resistance in the Penicillium population. Secondly, the frequency of benzimidazole-resistant spores in a packinghouse characteristically declines sharply after the use of these fungicides is discontinued (11).

Experiments in which untreated citrus fruit were inoculated with standardized mixtures of resistant- and sensitive-biotypes have shown that the frequency of resistant spores in the population decreases in each successive disease cycle and typically is less than 1% of the spore population of the fourth cycle (13,14). By this criterion, virtually all benzimidazole-resistant biotypes of P. digitatum are less fit than sensitive (wild) biotypes. Biotypes of this fungus that are resistant to sterol-demethylation inhibitors also appear less fit than wild types of the species (20,34; Torres, G. P. L. and Eckert, J. W., University of California at Riverside, unpublished). Not all fungal species/fungicides follow the pattern of reduced fitness shown by benzimidazole-resistant biotypes of P. digitatum. Gutter et al. (10) reported that the frequency of resistance to thiabendazole in a population of P. italicum remained constant when a mixture of resistant and sensitive spores was passed through several disease cycles in fungicide-free oranges. Smilanick and Eckert (35) combined spores of nine SBA-resistant isolates of P. digitatum with six SBA-sensitive isolates in 27 combination and injected each pair of isolates into fungicide-free oranges. After four disease cycles, the frequency of the resistant biotypes increased in 70% of the combinations, decreased in 19% and remained constant in 11%. The high degree of fitness of SBA-resistant biotypes is surprising since in most fungus/fungicides combinations investigated, the resistant biotypes were typically less persistant in the population than the wild type in the absence of fungicide selection pressure.

Strategies for Management of Fungicide Resistance

Strategies for control of resistance in P. digitatum and P. italicum should focus on: 1) suppression of resistance buildup in an unselected population of the pathogen which is being controlled by the current fungicide program, or 2) restoration of disease control after the resistant biotypes have exceeded a critical frequency in the population and the current fungicide program does not provide acceptable disease control. The two situations must be handled differently.

An unselected (wild) population of Penicillium containing benzimidazole-resistant biotypes at a frequency of approximately 10^{-8} can be subjected to a number of benzimidazole fungicide treatments over a period of months before the resistant biotypes reach a detectable frequency of about 10^{-2} (36). After resistance detection, several more fungicide applications will produce a level of fungicide resistance (>10%) which can cause a significant loss in disease control. The relentless use of benzimidazole fungicides will result in a Penicillium population that is almost totally resistant to this class of fungicides. Lemon packinghouses with 20% or more benzimidazole-resistance are frequently encountered late in the season and require a different strategy for management than that designed to suppress the buildup of resistance in an unselected population.

Several factors that influence the rate of buildup of resistant biotypes can be manipulated in strategies (Figure 2) to manage the problem: (1) The initial level of the resistant biotypes in the environment; (2) Intensity of fungicide selection pressure; (3) Rate of multiplication of the resistant biotypes; (4) Dispersal of spores of resistant biotypes; (5) Availability of injuries on the fruit surfaces that are susceptible to infection.

Sanitation. Since the probability of infection of a wound on a fungicide-treated citrus fruit depends upon the number of fungicide-resistant spores in the environment (1,33), sanitation of the packinghouse premises with formaldehyde or quaternary ammonium compounds is an essential component of the strategy to prevent the buildup of fungicide-resistant biotypes. Spores of fungicide-sensitive biotypes should also be a target for sanitation treatments, because they can increase the efficiency of infection of fungicide-treated wounds on citrus fruit by resistant spores (33). Ideally, the packinghouse environment should be essentially sterile with respect to P. digitatum and P. italicum spores at the beginning of the fruit harvesting season. If this condition is met, the spore population responsible for fruit infection will be brought in on harvested fruit from the groves where the frequency of resistance is very low, apparently due to the poor fitness of the fungicide-resistant mutants (13,14). Nonselective sanitation treatments decimate the size of the pathogen population but usually do not reduce the frequency of the resistant biotypes in the surviving population. Therefore, when the Penicillium population contains a substantial level of resistance (>20%), the best strategy is thorough sanitation of the packinghouse, followed by discontinuation of the selective fungicide until the level of resistance falls to an undetectable level.

Minimize Fungicide Selection Pressure. Treatment of harvested fruit with a selective fungicide should be limited to situations which require unique action by the treatment (e.g., curative action, sporulation control, etc.). Non-selective treatments such as heat, active chlorine, sodium carbonate, and low temperature should be emphasized whenever possible to discourage the buildup of fungicide-resistant biotypes of Penicillium.

Application of a minimum effective dosage of a selective fungicide is often recommended for field use of fungicides whose residues dissipate rapidly on the plant surface. This strategy is not likely to have a major effect on the selection pressure of postharvest fungicides, which are applied to provide uniform deposition of the fungicide that does not lose its effectiveness during the storage period. A long period of residual activity is generally expected,

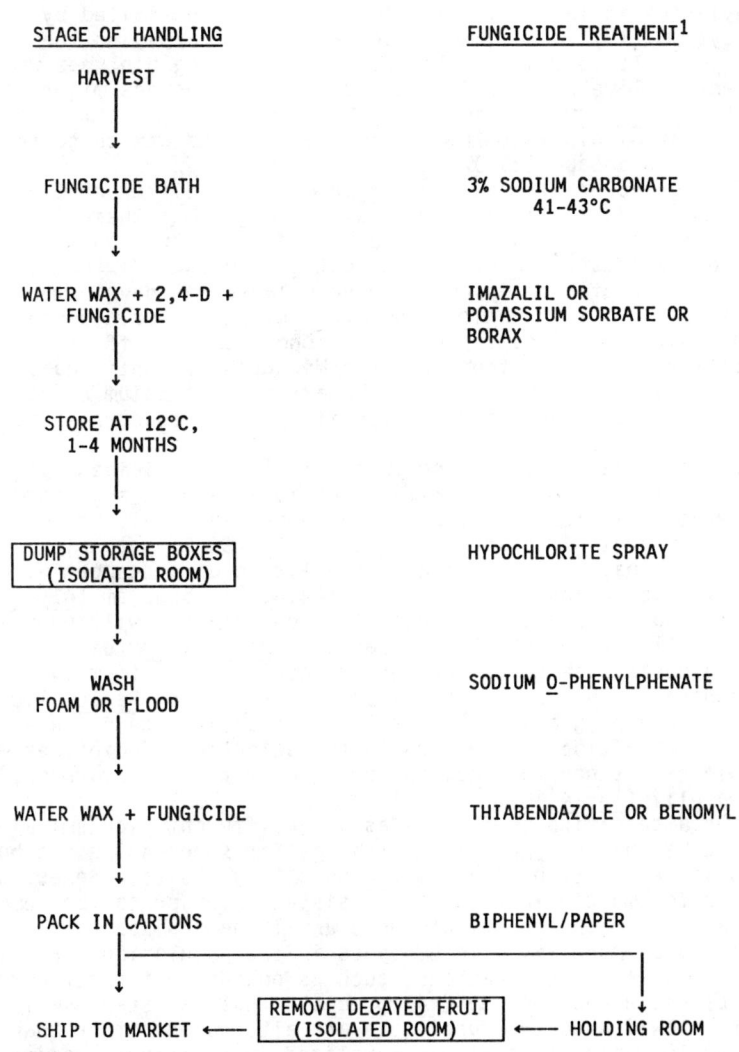

[1] Fungicides listed are registered in the U.S.: Sec-Butylamine, guazatine, and carbendazim can be applied in some other citrus-producing countries.

Figure 2. A strategy for management of fungicide-resistant Penicillium in harvested lemons.

especially for treatments applied to obtain sporulation control or protective action.

In the case of resistance to imazalil and other sterol demethylation inhibitors, which appears to be controlled by a polygenic system (15), an argument can be made for applying the highest dosage possible to minimize the risk of selecting biotypes with intermediate levels of resistance (24). High dosages might also inhibit Penicillium by a secondary biochemical mechanism under genetic control distinct from that governing resistance to the minimum effective dosage (29,37,38).

Mixing and/or alternating fungicides with different biochemical mechanisms is advocated as a general strategy, but there is little direct evidence that this practice is useful in the control of fungicide-resistant Penicillium on citrus fruits. Mixtures of benomyl with captan or iprodione have delayed the development of benomyl-resistance in Penicillium expansum on apples and pears (39,40), but neither captan nor iprodione alone is effective against Penicillium decay of citrus fruits. Mixtures of benzimidazole fungicides with sodium carbonate, borax, and potassium sorbate also failed to prevent the buildup of benzimidazole-resistant Penicillium on oranges (13,14).

The combination of two or more selective fungicides with different biochemical mechanisms of action has been suggested to reduce the selection pressure generated by use of either compound alone. Mixtures of thiabendazole with SBA, SOPP, potassium sorbate and imazalil have been mentioned with little data supporting their effectiveness against the fungicide resistance problem (41,42,43). This strategy should be used only for control of a wild population of Penicillium in which the frequency of Penicillium biotypes carrying resistance genes for both fungicides is extremely rare. Unfortunately, this strategy has been attempted on Penicillium populations containing a significant level of resistance for one of the selective fungicides, resulting in the selection of biotypes with multiple resistance to benzimidazoles, aromatic hydrocarbons, SBA and imazalil (13,14,18).

The alternation of fungicides with different biochemical action seems to be the most practical strategy for suppressing the buildup of fungicide-resistant Penicillium on citrus fruits. Selection pressure for multiplication of a resistant biotype can be completely eliminated from the treatment cycle until the resistant biotype is suppressed by more fit wild biotypes in the population. For complete coverage applications, such as postharvest treatments, the model of Kable and Jeffrey (44) predicts that resistance will build up more slowly when two fungicides are alternated rather than mixed.

Considerable interest has developed in resistance management through the use of: 1) compounds that are more active against fungicide-resistant biotypes than their sensitive counterparts (i.e., the resistant biotypes exhibit negatively-correlated cross resistance) and 2) compounds that interfere with the resistance mechanism (i.e., synergists). Both approaches to resistance control have been reviewed by De Waard (24,45).

Leroux and Gredt (46) described the enhanced sensitivity of benzimidazole-resistant biotypes of Penicillium expansum and Botrytis cinerea to several N-phenylcarbamate herbicides. Following this lead, methyl 3,5-dichlorophenylcarbamate and isopropyl

3,4-diethyoxyphenylcarbamate were reported to control several plant diseases caused by benzimidazole-resistant biotypes of plant pathogens, especially gray mold of grapes caused by Botrytis cinerea (47). Rosenberger and Meyer (48) observed that benzimidazole-resistant biotypes of P. expansum were often more sensitive to diphenylamine than wild-type isolates. Combinations of diphenylamine with benomyl, thiabendazole, and thiophanate-methyl provided better control of benomyl-resistant P. expansum isolates in inoculated apples stored at 2-4°C than either diphenylamine or each fungicide alone.

DeWaard et al. (20) evaluated cross-resistance of fenarinol-resistant mutants of P. italicum to other SBI (sterol biosynthesis inhibitor) fungicides. All fenarimol-resistant mutants were cross-resistant to other SBI fungicides tested, except fenpropimorph, a morpholine derivative (20,49). Biotypes with a high degree of resistance to fenarimol, and cross resistant to other SBI fungicides, exhibited negatively correlated cross resistance to fenpropimorph, tridemorph, and dodine (30,49,50). Fenpropimorph was more effective against decay of oranges inoculated with fenarimol-resistant isolates of P. italicum than oranges inoculated with wild-type isolates (49). Fenarimol-resistant isolates of P. italicum accumulated less of the SBI fungicides than sensitive isolates, except for the morpholine fungicides and imazalil, which were accumulated to the same extent in both resistant and sensitive isolates (30). Cross-resistance between fenarimol and the triazole fungicides was correlated with reduced accumulation in resistant biotypes; negatively correlated cross resistance between the triazoles and the morpholines could not be explained by this mechanism.

The utilization of negatively-correlated cross-resistance as a strategy against practical problems of fungicide resistance is attractive in theory since it offers a rare opportunity to selectively reduce the frequency of fungicide-resistant biotypes in the pathogen population. However, many reported examples of negatively-correlated cross-resistance have indicated that the phenomenon is not universal, but rather it is limited to certain species and biotypes of pathogens (46,48,49). For example, many benzimidazole-resistant isolates of Penicillium spp. are not sensitive to phenylcarbamates and diphenylamine. On the other hand, most benzimidazole-resistant isolates of Botrytis cinerea do exhibit negatively-correlated cross-resistance to phenylcarbamate compounds, and SBI-resistant isolates of Pencillium spp. are more sensitive to morpholine fungicides and dodine than wild-type isolates (26,30,49, 50). Clearly, this approach to fungicide-resistance management deserves further consideration.

Delay the buildup of Resistant Biotypes. The multiplication of recently-selected biotypes is essential to the buildup of a resistant population, because the resistant form is often less fit than the wild biotype and will be lost in the absence of selection pressure. A generation of P. digitatum requires about seven days at 25°C and results in an increase in spores of about 10^7 fold. The rate of multiplication can be reduced substantially by lowering the storage temperature and by treatment of the fruit with imazalil, benomyl, or thiabendazole, all of which suppress sporulation of Penicillium biotypes that are sensitive to these fungicides.

Prevent Dispersal of Resistant Biotypes. The development of a
fungicide-resistance problem can be delayed by isolation of the
diseased fruit to prevent the dissemination of spores of fungicide-
resistant biotypes. Fruit treated with a selective fungicide are
often stored under environmental conditions that permit the development of Penicillium decay, even if slowly, due to low temperature.
The selection pressure of the fungicide residue transforms the
Penicillium population from sensitive to resistant in several weeks
unless an antisporulant treatment has been applied to the fruit.
Eventually, the decayed fruit must be removed from the containers so
that the sound fruit can be sold. It is essential that this operation be carried out in a room that is isolated from the packinghouse, to prevent spores of resistant biotypes from contaminating
recently harvested fruit (51). Lemon fruit are sprayed with a solution of active chlorine as they are emptied from the storage box to
kill fungicide-resistant spores on the fruit surface, so the
resistant biotypes do not spread to other fruit in the packinghouse.

Reduce Incidence of Fruit Infection. The infection of citrus fruit
by Penicillium spores depends upon the inoculation of spores into a
fresh injury 2-3 mm deep that does not contain a deposit of an
effective fungicide. Therefore, a basic method for suppressing the
buildup of fungicide resistance consists of reducing fruit injuries
or their susceptibility to infection until the wound has healed in
three days. Heat (30°C) and volatile amines are examples of non-
selective treatments that temporarily prevent infection while the
wounds are healing (38,52).

Conclusions

Fungicide-resistant biotypes of Penicillium digitatum and P.
italicum are a continuing threat to the fresh citrus fruit business
worldwide. In practical use, all major organic fungicides have
failed to provide satisfactory control of fruit decay when
fungicide-resistant biotypes of Penicillium have reached a moderate
frequency in the packinghouse environment. Management of the
problem can be achieved by minimizing fungicide selection pressure
on the Penicillium population, suppressing the multiplication and
dispersal of spores, and reducing fruit injury to decrease the
probability of infection by Penicillium. Practical implementation
and validation of specific strategies are subjects of on-going
investigations.

Literature Cited

1. Eckert, J. W.; Eaks, I. L. In The Citrus Industry; Reuther,
   W.; Calavan, E. C.; Carman, G. E., Eds.; Univ. Calif. Press:
   Berkeley, CA, 1989, 5, 179-260.
2. Gutter, Y. Phytopathology. 1975, 65, 498-499.
3. Worthing, C. R. The Pesticide Manual 8 Ed.; British Crop
   Protection Council: London, U.K., 1987; pp 1077.
4. Farkas, A.; Aman, J. Palest. J. Bot. (Jerusalem Ser.)., 1940,
   2, 38-45.

5. Harding, P. R., Jr. Plant Dis. Reptr. 1962, 46, 100-104.
6. Harding, P. R., Jr. Plant Dis. Reptr. 1964, 48, 43-46.
7. Harding, P. R., Jr. Plant Dis. Reptr. 1972, 56, 256-260.
8. McDonald, R. E.; Risse, L. A.; Hillebrand, B. M. J. Amer. Soc. Hort. Sci. 1979, 104, 333-335.
9. Anonymous. Citrograph. 1980, 65, 95-96.
10. Gutter, Y.; Shachnai, A.; Schiffmann-Nadel, M.; Dinoor, A. Phytopathology. 1981, 71, 482-487.
11. Kaplan, H. J.; Dave, B. A.; Petri, J. F. Proc. Int. Soc. Citriculture. 1981, 2, 788-791.
12. Kuramoto, T. Plant Dis. Reptr. 1976, 60, 168-172.
13. Eckert, J. W.; Wild, B. L. In Pest Resistance to Pesticides; Georghiou, G. P.; Saito, T., Eds.; Plenum Publ. Corp.: New York, NY, 1983; pp 525-556.
14. Wild, B. L. Ph.D. Thesis, University of California, Riverside, 1980. 89 pp.
15. Van Tuyl, J. M. Meded. Landb. Hogesch. Wageningen. Netherlands. 1977, 2, 1-137.
16. Georgopoulos, S. G.; Skylakakis, G. Crop Protection. 1986, 5, 299-305.
17. Wild, B. L. Ann. Appl. Biol. 1983, 102, 237-241.
18. Dave, B. A.; Kaplan, H. J.; Petri, J. F. Proc. Fla. State Hort. Soc. 1980, 93, 344-347.
19. Eckert, J. W. In Pesticide Science and Biotechnology; Greenhalgh, R.; Roberts, T. R., Eds.; Blackwell Scientific Publ.: London, U.K., 1987; pp. 217-220.
20. De Waard, M. A.; Groeneweg, H.; Van Nistelrooy, J. G. M. Neth. J. Pl. Path. 1982, 88, 99-112.
21. Kaplan, H. J.; Dave, B. A. Proc. Fla. State Hort. Soc. 1979, 92, 37-43.
22. El-Goorani, M. A.; El-Kasheir, H. M.; Kabeel, M. T.; Shoeib, A. A. Plant Dis. 1984, 68, 100-102.
23. Laville, E. Y. Proc. Int. Soc. Citriculture. 1981, 2, 783-784.
24. De Waard, M. A. In Fungicide Resistance in North America; Delp, C. J., Ed.; APS Press: St. Paul, MN, 1988; pp. 98-100.
25. Fuchs, A.; Drandarevski, C. A. Neth. J. Pl. Path. 1976, 82, 85-87.
26. Eckert, J. W. Phytopathology. 1987, 77, 1728.
27. Köller, W.; Scheinpflug, H. Plant Dis. 1987, 71, 1066-1074.
28. Buchenauer, H. In Modern Selective Fungicides; Lyr, H. Ed.; 1987: Longman, U.K., 1987; pp 205-231.
29. Kerkenaar, A. In Modern Selective Fungicides; Lyr, H. Eds.; 1987: Longman, U.K., 1987; pp 159-171.
30. De Waard, M. A.; Van Nistelrooy, J. G. M. Pestic. Sci. 1988, 22, 371-382.
31. Eckert, J. W.; Brown, G. E. In Methods for Evaluating Pesticides for Control of Plant Pathogens; Hickey, K. D. Ed.; APS Press: St. Paul, MN, pp 92-97.
32. Brown, G. E.; Nagy, S.; Maraulja, M. Plant Dis. 1983, 67, 954-957.
33. Wild, B. L.; Eckert, J. W. Phytopathology. 1982, 72, 1329-1332.
34. Van Gestel, J. Proc. Int. Soc. Citriculture. 1988, 1, 1511-1514.

35. Smilanick, J. L.; Eckert, J. W. Phytopathology. 1986, 76, 805-808.
36. Skylakakis, G. In Populations of Plant Pathogens: Their Dynamics and Genetics; Wolfe, M. S.; Caten, C. E., Eds.; Blackwell Scientific Publ.: Oxford, U.K., 1987; pp 227-237.
37. Bartz, J. A., Eckert, J. W. Phytopathology. 1972, 62, 239-246.
38. Eckert, J. W.; Kolbezen, M. J. Phytopathology. 1963, 53, 1053-1059.
39. Rosenberger, D. A.; Meyer, F. W.; Cecilia, C. V. Plant Dis. Reptr. 1979, 63, 1033-1037.
40. Prusky, D.; Bazak, M.; Ben-Arie, R. Phytopathology. 1985, 75, 877-882.
41. Gutter, Y. Crop Protection. 1985, 4, 346-350.
42. Hall, D. J.; Bice, J. R. Proc. Fla. State Hort. Soc. 1977, 90, 138-141.
43. Nelson, P. M.; Wheeler, R. W.: McDonald, P. D. Proc. Int. Soc. Citriculture. 1981, 1, 820-823.
44. Kable, P. F.; Jeffery, H. Phytopathology. 1980, 70, 8-12.
45. De Waard, M. A. British Crop Protection Conf. 1984, 573-584.
46. Leroux, P.; Gredt, M. 6th Reinhardsbrunn Symp. Systemic Fungicides and Antifungal Compounds. 1982, pp. 297-301.
47. Kato, T.; Suzuki, K.; Takahashi, J.; Kamoshita, K. J. Pestic. Sci. 1984, 9, 489-495.
48. Rosenberger, D. A.; Meyer, F. W. Phytopathology. 1985, 75, 74-79.
49. De Waard, M. A.; Van Nistelrooy, J. G. M. Neth. J. Pl. Path. 1982, 88, 231-236.
50. De Waard, M. A.; Van Nistelrooy, J. G. M. Neth. J. Pl. Path. 1983, 89, 67-73.
51. Bancroft M. N.; Gardner, P. D.; Eckert, J. W.; Baritelle, J. L. Plant Dis. 1984, 68, 24-28.
52. Brown, G. E. Phytopathology. 1973, 63, 1104-1107.

RECEIVED October 25, 1989

Chapter 21

Predicting the Evolution of Fungicide Resistance

K. J. Brent, D. W. Hollomon, and M. W. Shaw[1]

Department of Agricultural Sciences, University of Bristol,
AFRC Institute of Arable Crops Research, Long Ashton Research Station,
Long Ashton, Bristol BS18 9AF, United Kingdom

New fungicides are increasingly difficult and costly to discover. It is crucial to ensure that their values to agriculture and to the agrochemical industry do not become lost or diminish through the development of resistance by the target fungi. The ability to predict the risk of resistance would help the selection of candidate chemicals for development, and the establishment of strategies for their durable use.

A knowledge of the potential for genetic variation in fungicide sensitivity in pathogen populations is central to prediction of resistance. Potential variation may be revealed by several methods in the laboratory and field, including mutagenesis, recombination and selection experiments, and monitoring. Benefits and drawbacks of these methods are considered.

Field monitoring is unlikely to detect single-gene resistance early enough for useful resistance forecasting in the region concerned. However, polygenic resistance is likely to develop gradually, allowing time to implement or alter avoidance strategies. In such cases the fitness of the more resistant variants will be of critical importance, and the measurement of both fitness and stabilizing selection in field experiments is discussed.

[1]Current address: Department of Agricultural Botany, University of Reading, Whiteknights, Reading RG6 2AH, United Kingdom

0097-6156/90/0421-0303$06.00/0
© 1990 American Chemical Society

Analysis of past cases of resistance has helped to
identify the many biological, chemical and husbandry
factors that together determine the build-up of
resistance in the field. For a future chemical/use
combination each of these factors can be scored for
risk and the scores summed to give a total risk.
However, uncertainties over the relative importance of
each factor at present limit multifactorial assessment
as a predictive approach.

Predictions based on mathematical modelling, have been
fairly accurate where resistance is controlled by a
single gene. However, where several genes are
involved, modelling is in its infancy, and predictions
of the build-up of quantitative resistance have
greatly underestimated the time needed for it to
develop. Incorporation of additional epidemiological
data has permitted new, testable models for polygenic
fungicide resistance to be developed.

At present a combination of the genetic, monitoring,
multifactorial and modelling approaches should be used
to develop a best judgement of risk before and during
the early use of a fungicide. In this way at least
low, medium and high risk can be distinguished, plus
a shorter or longer-term time-scale. More precise
evaluation of the timing and severity of first
resistance outbreaks will be much harder to achieve.

The discovery and development of a fungicide takes many years.
Costs are high and still increasing (now ca. $50M per compound),
and several years' use is needed before any profit on investment
can be obtained. Moreover many existing compounds have unique
properties of agricultural value which will be extremely
difficult or impossible to replace. The selection of variant
forms of the target fungi with inherited decreases in fungicide
sensitivity can, in some circumstances, rapidly decrease the
performance of a new fungicide to the extent that it must be
withdrawn from use. Against this background, can the chemist
and biologist involved in fungicide discovery rationally predict
whether a newly synthesised fungicide will be long lasting, like
the dithiocarbamates or phthalimides, or whether it will rapidly
succumb to resistance, like benomyl and metalaxyl? Such
prediction will not only guide the selection of candidate
fungicides for further development, but will indicate priorities
in terms of the effort needed to monitor "baseline"
sensitivities, and the need to develop special marketing and use
strategies to ensure sustained performance.
    Even if a new chemical is a member of an existing fungicide
class one cannot be sure of its 'resistance behaviour'. In

field experiments carried out in the UK in 1988 (Hollomon, unpublished results), a new triazole fungicide, tebuconazole ($\underline{1}$) gave excellent control of barley mildew (Erysiphe graminis f.sp. hordei), similar to that given by tridemorph. In the same experiments an older triazole, triadimenol, performed poorly because resistance to it had already developed ($\underline{2}$). Ten years earlier, at the time of its development, tridimenol had given even better disease control than tridemorph. The questions we must attempt to answer are whether the initial lack of resistance to a new fungicide such as tebuconazole could be predicted, and how fast and how widely might it develop in the future? Also could the durability of tridemorph be predicted, bearing in mind its selective biological and biochemical action? We might next ask how effective the strategies of use ($\underline{3}$, $\underline{4}$) intended to combat resistance are likely to be. However, in this paper we will concentrate on how we might predict the evolution of resistance, what information is required to do this, and how the necessary data might be collected.

Fungicide resistance is no longer a new problem facing plant pathologists. Some twenty years' experience of resistance problems, and a great deal of data, have now accumulated. A number of workers have addressed, to varying extents, the problems of predicting the likely development and spread of resistance. Four main approaches can be distinguished: genetic experimentation, monitoring, multi-factorial analysis, and mathematical modelling. These approaches are not separate or exclusive, and the two latter need to incorporate results from the first two.

Genetic Experimentation

Mutagenesis. Mutagenesis by ultra-violet irradiation, or treatment with chemical mutagenic agents, provides a useful and often simple test, which can expose rapidly the biochemical and genetic potential for resistance to any new fungicide, and indicate its possible magnitude. However, without parallel genetic analysis of mutants, it is possible that additional mutations in unrelated genes will have occurred which disturb the "genomic framework", and impair fitness. Consequently, it is rather dangerous to draw practical conclusions, indicating low risk, from an observed association of a laboratory mutation to fungicide resistance with a loss in fitness, as was done for the DMI fungicides ($\underline{5}$). In any case, selection for fitness in the field may well modify any adverse effects mutant genes may have on fitness. Furthermore, similar resistant mutants occurring in field populations at natural mutation rates, are much less likely than laboratory-induced mutants to be linked chromosomally with other deleterious mutations.

The production of 'fit' laboratory-induced resistant mutant can give some positive indication of the likely development of resistance, as for example in the control of Phytophthora megasperma f.sp. medicaginis by metalaxyl ($\underline{6}$). Such mutants cannot be produced in relation to durable fungicides such as copper, dithiocarbamates and phthalimides (7). However, "natural" experiments are incomparably larger than any laboratory mutagenesis program, so that failure to detect a fungicide-resistant mutant cannot be taken, unfortunately, as firm evidence that it will not occur. Also, apparently 'fit' mutants resistant to morpholine fungicides are readily produced in the laboratory (8), and yet resistance has not emerged in practice.

Recombination. Where variation is controlled by many genetic factors, or where monogenic resistance occurs in diploid or dikaryotic pathogens, analysis of progeny of sexual recombination experiments has revealed the extent of inherited variation more effectively than other methods. Recombination tests are less useful than monitoring in detecting very rare single-gene variants, since such mutants are most unlikely to be present in a small sample of parental strains. However, in studies on barley powdery mildew, progeny derived from crosses yielded more variation in sensitivity to the fungicides ethirimol and triadimenol than could be generated through mutagenesis, or exposed by "baseline" monitoring (9). At the same time useful information was obtained on the heritability of this variation. Cleistothecia, often abundant in mildew lesions on ripening barley, can be stored in a refrigerator for years, and yet still produce viable ascospores. Analysis of progeny generated from such cleistothecia would provide a measure of the variation in sensitivity to any new fungicide that was present in the population at the time cleistothecia were collected from the field. Where sexual reproduction has not been found (as with Rhynchosporium secalis), or where it cannot be manipulated in the laboratory (as with Pseudocercosporella herpotrichoides), regeneration of progeny from diploids after protoplast fusion may offer an alternative way of exposing variation (10).

Selection Experiments. The initial low frequency of fungicide resistance alleles in natural populations may be increased to detectable levels by selection. In growth room experiments with populations of P. infestans, initially inoculated to give a frequency of metalaxyl resistance at 1×10^{-4}, Staub and Sozzi (11) increased this frequency to detectable levels after exposure for 3 - 4 generations to metalaxyl sprays. However, in natural populations resistance alleles are likely to initially be at frequencies below 10^{-6}, and possibly much below this if the variation is associated with poor fitness in the absence of the fungicide. For example, starting with a population of one million clones, fungicidal exposure and selection pressure for four generations would be needed to increase the frequency of a resistance allele from 10^{-6} to a detectable level of 10^{-2}, assuming 10% survival in each generation. Where the initial frequency is 10^{-8}, as was measured for MBC resistance in P. herpotrichoides (12), exposure for seven generations might be required. Thus the length of any selection experiment might not need to be greatly increased, but the starting population would need to include 3×10^8 clones if the experiment is to provide a 95% chance of exposing a resistant allele. Consequently, selection experiments need to utilize large starting populations, and where resistance alleles are initially rare, results are likely to be erratic. If selection experiments are carried out in the field, absence of resistance developing in one locality will not mean that it will not develop elsewhere.

Nevertheless, a positive result from a laboratory or field selection experiment (see for example (13) and (14)) signifies a danger that resistance might develop. It suggests that resistant variants are frequent, and fungicide use should be carefully managed to avoid rapid selection. If a resistant variant was identified in a selection experiment, information on its fitness would be particularly useful if it could be gained in a realistic ecological setting. This would pose difficulties, since no one would advocate deliberately inoculating field trials with fungicide resistant variants of a pathogen still effectively controlled in the field.

Where genetic control of fungicide sensitivity is more complex and involves many genes, selection experiments are useful in exposing the extent of variation, the nature of its inheritance, and the strength of any stabilizing selection that might counter the effects of a fungicide. Starting populations must ideally encompass all available variation, and this might be achieved using field plots, rather than laboratory or greenhouse populations. Again, many experiments will probably be needed before any conclusions can be drawn about the possibility of the evolution of resistance to any new fungicide, since local pathogen populations might differ.

Resistance Mechanisms and Cross-Resistance

Mode of action is frequently judged an important element in risk evaluation, and certainly cross-resistance either to benomyl or to metalaxyl extends to other fungicides thought to inhibit tubulin assembly or RNA polymerase, respectively. However, experimental evidence indicates that resistance is not always caused by a change in the target enzyme. Thus mode of action is not necessarily a sound guide to the likely development of resistance. For example, tebuconazole is an inhibitor of sterol 14 α demethylase in target fungi (15), and yet it controls sterol-demethylation-inhibitor-resistant (DMI-resistant) strains of cereal powdery mildews that cause practical resistance problems in field crops (Hollomon, unpublished data, Table I). Although a knowledge of the mechanism of resistance in field variants is undoubtedly useful, critical information on this element is unlikely to be available until it is too late to be of value in any initial prediction scheme.

It is now common to test potential new fungicides at an early stage of development, for in vitro or glasshouse activity against biotypes of the target pathogens that have already acquired practical resistance to one or more existing fungicides (which may or may not be related chemically or by mode of action

to the new one). If these biotypes are resistant also to the
new fungicide, then genetically determined cross-resistance is
indicated. Obviously, such observations can give useful
prediction of commercial risks and guide appropriate use
strategies.

Table I. Cross-resistance patterns between triadimenol and tebuconazole in cereal powdery mildews

	Sensitivity (ED_{50} µg/ml)				
	Triadimenol sensitive		Triadimenol resistant		Resistance factor
Barley powdery mildew					
No. of isolates tested:	10		14		
Triadimenol	0.019	0.002*	1.25	0.084	66
Tebuconazole	0.014	0.002	0.115	0.008	8
Wheat powdery mildew					
No. of isolates tested:	18		8		
Triadimenol	0.087	0.006	1.50	0.112	17.2
Tebuconazole	0.100	0.010	0.293	0.042	2.9

* SED

Detection and Monitoring

Evolution of resistance depends not only on the underlying
capacity for biochemical change, but also on epidemiological and
genetic determinants. How much potential genetic variation
exists within natural populations, how best to reveal it, and
how this variation is likely to respond to selection, are
important questions for the prediction of the development of
resistance. Several approaches have been followed to measure
variation in fungicide sensitivity in a number of plant pathogen
populations, and these have met with different degrees of
success.

Assay methods for fungicide response invariably expose the
pathogen either to a range of fungicide doses, or to a single
discriminating dose if this is considered adequate. These
current methods are imprecise, labour intensive, and time
consuming, especially for obligate parasites, and they detect
only relatively frequent mutants, or continuous variation. The
problem of detecting rare variants is well illustrated in Table
II. Sampling every mildew pustule in one hectare of diseased
crop (assuming 10% leaf area infected) would be needed to
detect, with 95% certainty, a rare (1×10^{-8}) resistant mutant

of a pathogen such as E. graminis. Whilst biochemically based diagnostic tests might improve the chances of detecting known resistance mechanisms, these will not help to detect any unknown resistance mechanism, such as might be selected by a new fungicide. Where resistance is controlled by a single gene, monitoring offers little chance of detecting resistance before it is too late to take avoidance action in the affected area, because of the exponential way in which the frequency of resistance will change (Figure 1). Experience has largely borne this out, for detection of both MBC and phenylamide resistance in several pathogens coincided with disease control failures. In these instances monitoring confirmed the role of resistance, and did provide evidence of the need to take appropriate avoidance action at other locations where these fungicides were being introduced.

Table II. Sample size needed to detect (with 95% confidence) rare resistant mutants in populations of Erysiphe graminis

Mutant frequency	Sample size (number of pustules)	Area of crop (Ha)*
1×10^{-4}	3×10^{4}	0.0001
1×10^{-6}	3×10^{6}	0.01
1×10^{-8}	3×10^{8}	1
1×10^{-10}	3×10^{10}	100
1×10^{-12}	3×10^{12}	10000

* Assumes 10% leaf area infection and every pustule tested separately.

Monitoring for "baseline" sensitivity may reveal significant variation in fungicide sensitivity in some pathogens, especially, it seems, with regard to sensitivity to DMI and 2-aminopyrimidine fungicides in powdery mildews. These populations may include variants that survive exposure to recommended fungicide doses, but which are too infrequent to cause immediate disease control problems. In cases where this type of continuous variation has been identified in natural populations, genetic control is generally polygenic, and changes in response to selection are gradual (Figure 1 and ref. 2). Consequently, monitoring did detect shifts in fungicide sensitivity in cereal mildew populations before difficulties in practical disease control emerged in the field. In these situations it seems likely that resistance mechanisms involve altered metabolic processes that, for some reason unconnected with fungicide use, have been subjected to sufficient selection to establish an optimum sensitivity. As well as the fundamental linear selection process in these cases (Figure 1), associated

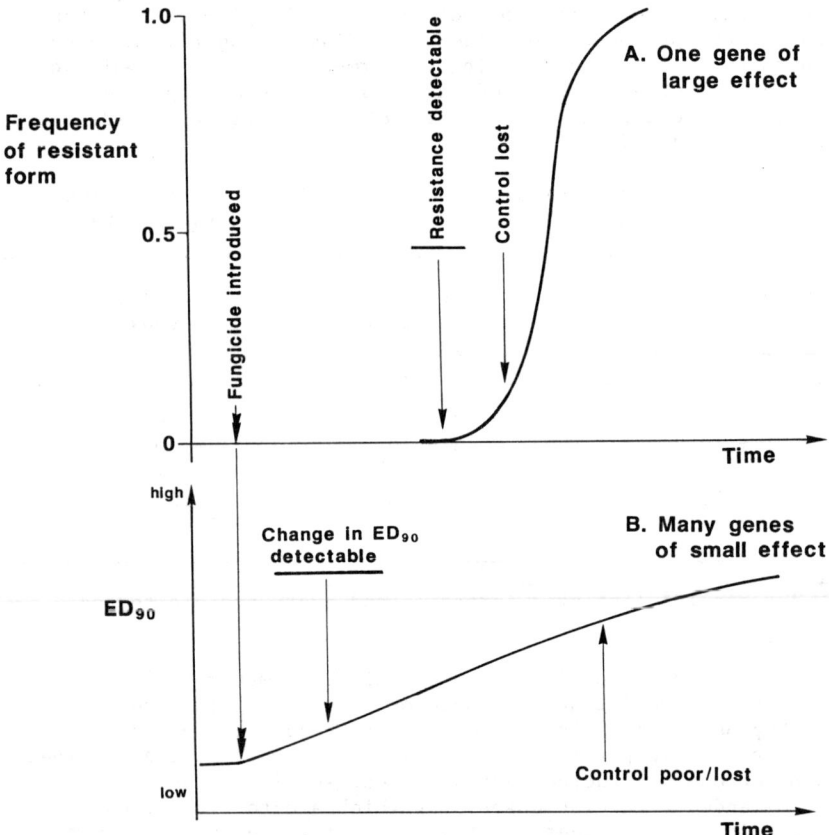

Figure 1. Evolution of fungicide resistance.
A. Exponential increase where only a single gene is involved. B. Gradual increase in resistance where many genes are involved.

pleiotropic effects on fitness may be difficult to disturb by
selection with a new fungicide.

Multi-Factorial Analysis

The analysis of case histories (see ref. 16 for examples) can
provide some indication of the many interacting factors that
determine the rate and severity of resistance build-up. These
factors have been discussed by many authors over the years (see
for example 17 - 21), and at least fifteen, have been included
in a recent scheme prepared by Gisi and Staehle-Cseh (22).
They comprise elements inherent in the biology and biochemistry
of the target pathogen, as well as imposed management factors
reflecting strategies of fungicide use (Table III). A
comprehensive approach to forecasting (23) is to allocate and
add up arbitrary numerical values for the perceived degree of
risk attached to each element.

This scheme provides a useful framework for drawing
together assessments of various critical factors, and could be
further refined. Unfortunately, not all these elements are
easily measured, nor will they be equally important, or differ
consistently in their importance, in determining the outcome
of selection. Consequently, the method of combining the
elements to give a total risk score is extremely arbitrary.
Moreover this approach is not fully predictive. Factors like
'r build- up' and 'field performance' demand years of field
experience on a considerable scale for a proper assessment.
However, we have submitted a few known products to this scheme
as indicated in Table III, and the results do suggest that it
works at a simple 'smaller' or 'larger' risk level, and not
surprisingly gets better as knowledge accumulates during use.
For example, scoring an average '2' when no information exists
or the element is obviously not applicable, we obtained notional
values for mancozeb and metalaxyl used alone of 39 and 75 before
extensive field use, and 39 and 96 after approx. 5 years' use.
This is roughly in line with the actual behaviour of these
materials although the mancozeb score is near the top end of
'low risk', and might have caused some concern whereas in
practice it has proved very low risk.

Modelling

Single-gene models. Many authors have attempted to use
mathematical models to explain the pattern and rate of spread of
resistance, and to indicate effects of different strategies of
use (17 - 21, 24 - 26). Most have concentrated on cases where
two readily identified distinct groups with substantial
differences in resistance, controlled by a single gene, are
present. However, two forms of genetic control of fungicide

TABLE III. Evaluation of elements of inherent and management risks of development of fungicide resistance (adapted, with permission, from ref. 23)

Elements of inherent risk	Management factors
1. Cross-resistance (to existing fungicides with known resistance problems).	Number and timing of applications.
2. Detection of r-strains in field.	Spray intervals. Size of treated area.
3. Selection of r-strains in laboratory.	Use of fungicide alone or in mixtures or rotations.
4. Production of r-strains by laboratory mutagenesis.	Scoring:
5. Selection of r-strains in heavily treated field plots.	The candidate fungicide is given scores of 1 (low-) 2 (medium-) and 3 (high-risk) for each of the 12 inherent characters.
6. Decrease of field performance.	
7. Unimodal or bimodal distribution of r and s forms (mono- or polygenic change).	Management risk is assessed as a whole at 1, 2 or 3 (low, medium or high risk).
8. Segregation of progeny in crossing experiments.	Multiplication of management risk (1 - 3) by inherent risk (12 - 36) gives overall risk (low 12 - 43, medium 44 - 75, high 76 - 108).
9. Fitness of r strains.	
10. Laboratory build-up of r forms on treatment of r - s mixtures.	
11. Length of fungicide persistence of action.	
12. Fungus biology (short generation time, low migration rate).	

resistance are now recognised (27): the first where only a single gene is responsible; and the second where many genes, each with a small but additive effect, control resistance. Where a single gene is involved, the fundamental population dynamics of the evolution of resistance to levels of practical significance are exponential. However, where many genes are involved, change will be more gradual and roughly linear over a period of time (Figure 1).

Published models make many different assumptions, but all attempt to estimate how fungicide application and pathogen epidemiology combine to influence the rate of selection. The general conclusions from single gene models are broadly similar, and in keeping with "common sense" predictions. They predict that resistance will develop most rapidly in rapidly reproducing pathogens exposed to highly effective, persistent fungicides. The rarer that resistant mutants are in the initial population, the longer that resistance will take to develop. Poor fitness may prevent the development of resistance, but fungicide selection pressures are sometimes extremely strong, especially in enclosed environments such as packing-sheds and glass-houses, so that relative unfitness of resistant forms will only slow down the evolution of resistance substantially if the fungicide is not used continuously.

For the limited number of cases where sufficient data are available, and where resistance is due to a single major gene, model predictions seem to fit the outcome in practice fairly well (Table IV). However, agreement may sometimes be fortuitous. For instance evidence suggests that dimethirimol resistance in Sphaerotheca fuliginea is not monogenic (28), but involves additive effects of many genes, so that published models do not apply.

TABLE IV. Predicted and observed duration of selection pressure required for resistant subpopulations of selected pathogens to cause a practical resistance outbreak

Pathogen	Chemical	Standard selection time* (days)	Duration of selection pressure (d = days y = years)	
			Predicted	Observed
Sphaerotheca fuliginea	Dimethirimol	8.5-16.5	98-236d	112-224d
Phytophthora infestans	Metalaxyl	3.7- 3.8	57- 70d	200-400d
Cercospora beticola	Benomyl	9.5-14.3	130-263d	140-200d
Ustilago nuda	Carboxin	158	5-7y	11y

* Time for proportion of resistant sub-population to increase by e (2.7 times).
SOURCE: Reprinted with permission from ref. 32. Copyright 1988 The British Crop Protection Council. Additional data from ref. 8.

Polygenic Models

A polygenically controlled character is one where variation is caused by many genes, each with a small effect. Change in such characters is usually modelled assuming that the difference between the mean value of the character before and after selection within one generation is known. The change between generations is related to the heritability of the character. This approach is suitable for artificial selection studies in higher organisms, but for plant pathogens reproducing continuously in time, it is hard to estimate the change in fungicide sensitivity within generations. A preliminary attempt to do this for S. fuliginea and E. graminis f.sp. hordei overestimated this change between generations, and predicted a faster evolution of resistance than that which actually occurred in practice (29). Two recent papers (30, 31) discuss the treatment of epidemiological and population genetic parameters needed to extend models to cover quantitative evolution of fungicide resistance.

In Europe, control of powdery mildew in small grain cereals has relied greatly on fungicides that either inhibit sterol 14α-demethylase (DMI's) or adenosine deaminase (ethirimol). Extensive use of both these fungicides in some regions has led to a gradual selection for resistance, with the result that performance has declined, especially on mildew-susceptible cultivars of wheat and barley (2, 33). Variation in sensitivity to both these fungicides is controlled by many genetic factors (34, 35) and, although selection with DMI's and ethirimol decreased the mean sensitivity of the population, selection also reduced the total variation in the population, suggesting that stabilizing selection might be operating in both instances (2, 36). As mentioned above, existing models of the evolution of resistance in polygenic systems are inadequate (29, 37), and require additional epidemiological data which are hard to obtain in practice. We have, therefore, attempted to produce more suitable models of the evolution of polygenic resistance, and to develop techniques which permit the measurement of critical epidemiological parameters in field experiments.

By comparison of geographically separate epidemics of barley mildew, caused by populations with a given level of fungicide resistance, we have attempted to measure how fitness changes with respect to fungicide sensitivity. By plotting the difference in population growth rate ("r" of VanderPlank, 38) between treated and untreated plots against mean ED50 of the mildew population at each site, the slope of the line (β, Figure 2) relating change in fitness in the presence of the fungicide to fungicide sensitivity can be estimated. Other factors will influence the slope of β, including the degree of fungicide coverage. The most reliable estimates of β are likely to come from field, rather than glasshouse experiments, and though more difficult, such field experiments do not appear impracticable.

Where selection operates strongly to stabilize the pathogen population about some optimum fungicide sensitivity, effects of fungicide sensitivity are likely to be minimized. The further the sensitivity of any variant clone departs from this optimum sensitivity (Figure 3), which may have been established through prior selection processes unconnected with any fungicide, the faster the growth rate of the clone will decline on untreated plants. The steeper this decline (γ), the stronger the effect of stabilizing selection. One way in which it may be possible to measure stabilizing selection is by analysis of variation in fungicide sensitivity before and after sexual reproduction in the absence of any fungicide selection. Recombination will tend to restore any variation lost in previous asexual generations. Another way of measuring stabilizing selection could involve observing changes in population variance after a fungicide is withdrawn from use.

In addition to epidemiological measurements of fitness and stabilizing selection, genetic information is also required on the heritability (h^2) of any fungicide resistance, and on the genotypic variation (σ^2) in sensitivity to any new fungicide. Although estimates of heritability are obtained from analysis of progeny from sexual crosses, epidemics of many plant pathogens, including powdery mildews, develop primarily through asexual reproduction. In this case heritability should be unity, unless variation in sensitivity is controlled by non-nuclear genes, in which case heritability between successive asexual generations may depart from unity. Shaw (30) has developed a model which describes both the evolution of quantitative resistance in terms of β, γ, h^2 and σ^2, and the theoretical implications of this model on fungicide use. We are now evaluating it against field data already available for ethirimol and the DMI fungicide triadimenol, as well as applying it to new fungicides.

Attempts to predict the evolution of fungicide resistance by measurement of fitness, and other parameters in field experiments, assume that an effectively closed population is being studied. Although this may be true for "rain splash" pathogens (e.g., R. secalis, P. herpotrichoides), and enclosed environments such as greenhouses and packing-sheds, others such as cereal mildews are wind-borne pathogens that can travel considerable distances (39). This long distance transport is known to influence the deployment of effective host-plant resistance genes in barley cultivars in different regions in Europe (40). Ingress of mildew from outside the experimental area offers a possible explanation for changes in triadimenol sensitivity reported for wheat mildew in an experiment in Northern Germany (41), which is one of the few relevant experiments so far published in the literature. Clearly, any conclusions drawn from field trials about the evolution of resistance to a new fungicide must, at least for powdery mildews, allow for the fortuitous ingress of either more sensitive, or more resistant inoculum. At present, it is not clear how such ingress can be effectively measured.

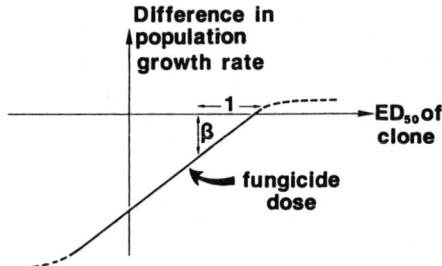

Figure 2. Estimation of the effect of fungicide sensitivity on fitness in the presence of fungicide.

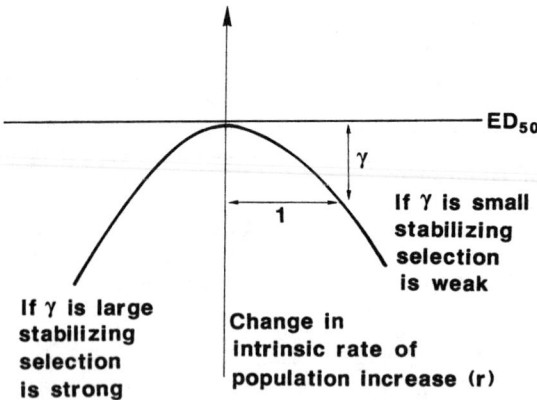

Figure 3. Relationship between fungicide sensitivity of a clone and stabilizing selection. Clones with ED_{50}'s further from the optimum grow more slowly than those closer to it.

Conclusions

It is impossible to avoid all risks of error in predicting the likelihood of resistance developing to a new fungicide if they were newly discovered. We could not identify with full confidence as low-risk the fungicides such as copper, sulfur, dithiocarbamates, phthalimides or morpholines, which have been widely used for many years without resistance problems. Identification of high risk situations can be made with a greater degree of confidence. The combined use of multi-factor and modelling approaches, incorporating genetic and monitoring data whenever they become available, and continuous re-appraisal, do permit useful judgements at low, medium and high risk level plus an indication of shorter or longer term timescale.

Refinements in modelling, backed by more validation, will enable us to recognise instances where significant variation in sensitivity already exists within target populations, but where selection is unlikely to occur rapidly, and where sensible use strategies will permit a new compound to make a worthwhile contribution to disease control. Identification of these situations, and providing the agrochemical industry with the evidence it needs to develop new products with confidence, remain important objectives of our work on fungicide resistance.

Acknowledgments

Our thanks are due to ICI Agrochemicals, Ciba-Geigy Agrochemicals, Bayer AG and the UK Home-Grown Cereals Authority, for financial support for our research on fungicide resistance, which has provided much of the background for this contribution.

Literature Cited

1. Reinecke, P.; Kaspers, H.; Scheinpflug, H.; Holmwood, G. Proc. 1986 Br. Crop Prot. Conf. - Pests and Diseases, 1986, 41-6.
2. Hollomon, D. W.; Brent, K. J. Tagungs. Akad. Landwirtschaft. DDR. 1987, 253 S, 45-53.
3. Staub, T. H.; Diriwaechter, G. Proc. 1986 Br. Crop Prot. Conf. - Pests and Diseases, 1986 pp 771-80.
4. Locke, T.; Fletcher, J. T.; Griffin, M. J. In Combating Resistance to Xenobiotics: Biological and Chemical Approaches, Ford, M. G.; Hollomon, D. W.; Khambay, B. P. S.; Sawicki, R. M., Eds.; Ellis Horwood: Chichester UK; 1987; pp 63-73.
5. Fuchs, A.; de Ruig, S. P.; van Tuyl, J. M.; de Vries, F. W. Neth. J. Plant Pathol. 1977, 83 S, 189-205.
6. Davidse, L. C.; Neth. J. Plant Pathol. 87, 11-24.
7. Dekker, J. Proc. 1981 Br. Crop Prot. Conf. - Pests and Diseases, 1981, 850-860.
8. Leroux, P. Agronomie, 1986, 6, 225-6.

9. Butters, J.; Clark, J.; Hollomon, D. W. Proc. 1986 Br. Crop Prot. Conf. - Pests and Diseases, 1986, 561-5.
10. Hocart, M; Lucas, J. A.; Peberdy, J. F. J. Phytopath. 1987, 119, 193-205.
11. Staub, T. H.; Sozzi, D. Proc. 10th Int. Congr. Plant Prot. 1983, Vol 2, 291-8.
12. Fehrmann, H, Phytopath. Z. 1976, 86, 144-85.
13. Thind, T. S.; Clerjeau, M.; Olivier, J. M. Proc. 1986 Br. Crop Prot. Conf. - Pests and Diseases, 1986, 2, 491-498.
14. Hunter, T.; Jordan, V. W. L.; Kendall, S. J. Proc. 1986 Br. Crop Prot. Conf. - Pests and Diseases, 1986, 523-30.
15. Berg, D.; Born, L.; Buchel, K. -H.; Holmwood. G.; Kaulen, J. Pflanzenschutz Nachrichten Bayer, 1987, 111-32.
16. Dekker, J.; Georgopoulos, S. G. (Eds.) Fungicide Resistance in Crop Protection, Pudoc: Wageningen; 1982.
17. Kable, P. F.; Jeffrey, H. Phytopath. 1980, 70, 8-12.
18. Delp, C. J. Plant Dis. 1980, 64, 652-7.
19. Skylakakis, G. Phytopath. 1981, 71, 1119-21.
20. Skylakakis, G. Proc. 1984 Br. Crop Prot. Conf. - Pests and Diseases, 1984, 565-72.
21. Levy, Y.; Levi, R.; Cohen, Y. Phytopath. 1983, 73, 1475-80.
22. Gisi, U.; Staehle-Cseh, U. In Fungicide Resistance in North America, Delp C. J., Ed.; American Phytopathological Society: St. Paul; 1988; pp 101-106.
23. Gisi, U.; Staehle-Cseh, U. Proc. 1988 Br. Crop Prot. Conf. - Pests and Diseases, 1988, pp 359-366.
24. Josepovits, G.; Dobrovolszky, A. Pestic. Sci. 1985, 16, 17-22.
25. Chin, K. M. Phytopath. 1987, 77, 666-9.
26. Milgroom, M. G.; Fry, W. E. Phytopath. 1988, 78, 565-70.
27. Skylakakis, G. EPPO Bull. 1985, 15, 519-25.
28. Schepers, H. T. A. M. Meded. Landbouwhogesch. Wageningen, 1985, pp 1-56 (PhD thesis).
29. Skylakakis, G.; Hollomon, D. W. In Combating Resistance to Xenobiotics: Biological and Chemical Approaches; Ford, M. G.; Hollomon, D. W.; Khambay, B. P. S.; Sawicki, R. M., Eds.; Ellis Horwood:Chichester UK; 1987; 94-103.
30. Shaw, M. W. Plant. Pathol. 1989, 38, 44-55.
31. Josepovits, G. Crop Protection, 1989, 8, 106-113.
32. Skylakakis, G. Crop Protection, 1982, 1, 249-62.
33. Brent, K. J.; Hollomon, D. W. In Sterol Biosynthesis Inhibitors: Pharmaceutical and Agrochemical Aspects; Berg, D.; Plempel, M., Eds.; Ellis Horwood:Chichester UK; 1988; pp 332-346.
34. Hollomon, D. W. Phytopath. 1981, 81, 536-40.
35. Hollomon, D. W.; Butters, J.; Clark, J. Proc. 1984 Br. Crop Prot. Conf. - Pests and Diseases, 1984, 477-82.
36. Brent, K. J. In Fungicide Resistance in Crop Protection; Dekker, J; Georgopoulos, S. G. Eds.; Pudoc : Wageningen; 1982; pp 219-30.

37. Via, S. In Pesticide Resistance: Strategies and Tactics for Management, National Academy Press:Washington DC; pp 222-35.
38. VanderPlank, J. E. Epidemics and Control, Academic Press; New York; 1963; pp 349.
39. Hermansen, J. E.; Torp, U.; Prahm, L. Yearbook Royal Veterinary and Agricultural University Copenhagen, 1975, 17-30.
40. Limpert, E. J. Phytopath. 1987, 298-311.
41. Schulz, U.; Dutzmann, S.; Scheinpflug, H. Pflanzenschutz Nachrichten Bayer, 1986, 39, 209-45.

RECEIVED December 11, 1989

Chapter 22

The Fungicide Resistance Action Committee

An Update on Goals, Strategies, and North American Initiatives

M. Wade[1] and C. J. Delp[2]

[1]Shell Research, Ltd., Sittingbourne Research Centre, Sittingbourne, Kent, ME9 8AG, England
[2]Consultant, 145 Kentucky Avenue, SE, Washington, DC 20003

> FRAC is an inter-company committee dedicated to prolonging the effectiveness of fungicides liable to encounter resistance problems and to limit crop damage due to resistance. Through educational and research programs, FRAC communicates information on resistance and fungicide-use strategies to agricultural, academic, industrial and regulatory sectors worldwide and promotes cooperative action to solve resistance problems. FRAC coordinates four Working Groups for fungicides at risk (phenylamides, dicarboximides, demethylation inhibitors, benzimidazoles). Through resistance monitoring and research programs, Working Groups develop and recommend technical resistance management strategies. FRAC stays alert to developments in the field of fungicide resistance and is constantly reviewing situations and updating strategies and educational programs. Current FRAC strategies for resistance management are described along with a review of recent FRAC educational and policy initiatives in North America.

FRAC is an inter-company committee dedicated to prolonging the effectiveness of fungicides liable to encounter resistance problems and to limit crop damage during the emergence of resistance. For 8 years, FRAC and its Working Groups have assumed an important international role in effective fungicide resistance management. The Working Groups have been successful in developing and promulgating TECHNICAL strategies designed to delay or prevent the onset of resistance to the four major classes of fungicides at risk--the phenylamides, benzimidazoles, dicarboximides, and demethylation inhibitors (DMIs).

0097–6156/90/0421–0320$06.00/0
© 1990 American Chemical Society

The FRAC Steering Committee coordinates activities of the Working Groups and takes a leading role in educational activities in and policy initiatives with respect to resistance. At all levels, FRAC enjoys the close collaboration and support of university and government scientists throughout the world. Only through cooperation may sound strategies be developed and implemented.

Objectives

FRAC was formed in 1981 in responsee to a growing awareness by the Industry that the complex problems posed by fungicide resistance could only be tackled effectively through cooperation. Of particular concern were issues relating to cross-resistance. One company's efforts to safeguard the effectivenesss of a product, by promoting sensible usage, could be nullified by the abuse of a related product through cross-resistance. Other concerns included the lack of education or awareness of the causes and consequences of resistance with users, distributors, marketing managers and registration officials. This lack of awareness was a major obstacle to the adoption of anti-resistance strategies in practice. There was a need to standardise definitions of resistance and monitoring methods so as to avoid confusion and unnecessary alarm and problems to growers, advisors and manufacturers. FRAC was established to tackle these problems and set the general objective: to prolong the effectivenesss of fungicides liable to encounter resistance problems, and to limit crop damage during the emergence of resistance.

To do this FRAC established Working Group for four fungicide types considered to be at risk, identifies existing and potential resistance problems , collects and generates information, communicates to those involved in fungicide research, distribution and use.

Principal Functions of FRAC are to:

- Initiate, stimulate and monitor the Working Groups.
- Provide guidance and coordination of Working Groups.
- Help Working Groups communicate their conclusions.
- Publicize guidelines on procedures/definitions of practical resistance research.
- Provide technical counsel for resistance courses and research.

FRAC had set itself and the Working Groups a very ambitious set of objectives. However, it is pleasing to report that by and large the objectives are being achieved.

Accomplishments of Working Groups

A common feature of the Working Groups is the close colloboration with advisory officers and institutes. This collaboration is an essential feature of effective resistance management. Each of the Groups has been successful in coming to grips with their particular problems by establishing trust, pooling information, and formulating use recommendations that are reported below.

Phenylamides (formerly acylalanines). The phenylamides are recognised as having a high potential for inducing resistance. They have a specific mode of action, and resistant strains can have a high levels of resistance and fitness. The pre-packaged mixtures of products based on phenylamide and residual fungicides have given encouraging results in recent years allowing the reintroduction of phenylamides where they had been withdrawn. The Working Group recommendations are to use:

- Only pre-packed mixtures for foliar use.
- Three quarters to full rate of the partner applied at intervals not to exceed 14 days.
- Limited number of sprays per season.
- No curative use.
- No soil treatments for control of airborne pathogens (with the exception of Tobacco Blue Mold).

Benzimidazoles. The benzimidazoles have a high resistance potential with many pathogens because:

- They have a specific mode of action that can lead to the selection of strains with high resistance levels.
- Pathogen populations have naturally occurring resistant strains.
- Resistance is not usually associated with a significant loss of fitness of the pathogen.

When benzimidazoles are abused, resistance is likely to occur and often has. However, when used judiciously as part of well considered strategies, benzimidazoles continue to give excellent control. Mixtures and combinations of mixtures and alternations with multi-site contact fungicides have proved very effective, especially if used before problems become apparent.

To cope with problems of cereal eyespot resistance, the Benzimidazole Working Group recommends the following:
- Use a mixture of a benzimidazole and non-benzimidazole in fields that have a high risk from eyespot or have reeceived benzimidazole treatments for several years.
- A benzimidazole should not be used where previous disease control failure has been due to resistance.

Dicarboximides. Acute problems of resistance to dicarboximides can arise when the products are intensively and exclusively used over many seasons. Resistant isolates are moderately resistant and tend to be less fit than

sensitive strains in the absence of the fungicides. The Dicarboximide Working Group recommends the following:
- Grapes. Two sprays of dicarboximides, applied as a mixture with a downy mildew fungicide having additional activity against Botrytis, may be applied once at bunch closing and once at beginning of ripening. In seasons when bad weather extends the flowering period and provides conditions suitable for infection, the first application can be brought forward.
- Glasshouse crops. Growers are advised to restrict the number of dicarboximide treatments to no more than three per crop in situations where resistance is present. Where resistance has not been a problem previously, a fourth application may be used if infection pressure is very high. When infection pressure is high, ensure that dicarboximides are used either in alternations or in mixtures with contact fungicides like chlorothalonil, captan, or thiram.
- Strawberries. A limited number (3-4) of applications per season is recommended in mixtures and alternated with contact fungicides such as thiram.
- Stone/pome fruit. For the control of Sclerotinia and Monilinia, a limited number of treatments is recommended in mixtures and alternations with conventional products and demethylation inhibitors.

Demethylation Inhibitors (DMIs). Generally with the DMIs, the need for early pre-emptive strategies is not so clear for many crop-pathogen combinations as with some other classes of fungicides. Less sensitive strains appear to be at a selective disadvantage in the absence of the fungicides. Gradual "step-by-step" shifts in the pathogen population toward decreased sensitivity occurs. This shift can be reversed by limiting the use of DMI's in favor of non-DMI fungicides.

In the case of cucurbit powdery mildew, a clear correlation has been established between the intensity of DMI use and reduced efficacy of this class of fungicides. This is the only crop-pathogen combination for which this correlation has been clearly shown. For these situations, the FRAC DMI Working Group recommends the use of alternations or mixtures. In grape powdery mildew, isolates with lower levels of sensitivity have been detected in Portugal. Recommended strategies include the use of DMI and sulphur mixtures, limiting DMI input to critical periods in the growing season and avoiding curative applications of DMI's wherever possible.

No problems in the field performance of DMIs have been observed yet for the control of apple scab, although strains of Venturia inaequalis with reduced sensitivity to DMIs have been observed in a few field plots.

Based on these findings the FRAC DMI Working Group recommends that repeated application of DMI fungicides alone for the control of cereal diseases in the same season should be avoided. The use of mixtures (DMI/non-DMI's) for reducing the shift to lower sensitivity should be practised.

Recently, the DMI Group held its first Banana subgroup meeting in Miami. At this meeting, the agrochemical manufacturers along with the major banana growers of Central and Southern America and government scientists agreed on a monitoring method and the following policy for the use of DMI's on bananas for Sigatoka control:

- Limit the total number of DMI sprays to 8 per year.
- Schedule applications in blocks of 2-4 consecutive DMI treatments.
- Intersperse the blocks with non-DMI treatments.

FRAC Accomplishments

Industry takes the lead in establishing strategies, in advertising them, and in organizing educational programs to foster an awareness of the causes and consequences of resistance. Companies are working hard to ensure good communications and clearer labeling of products to reduce the chance of cross-resistance. Industry bears the major responsibility for monitoring of resistant strains in pathogen populations - an expensive overhead. But industry cannot and should not face the common problems of resistance alone. Only through cooperation with a shared will to use fungicides intelligently can effective strategies for resistance management be successful.

The FRAC Steering Committee meets regularly to review the activities of the Working Groups and to deal with the wider aspects of resistance. These include publishing definitions of resistance, guidelines for containment of resistant strains, conclusions and recommendations of Working Groups, minutes of FRAC meetings, articles and posters on industry's responsee to resistance, and a book and brochure based on proceedings of the North American workshop. FRAC also provides advice and funds for research (via companies). and cooperates with the FAO and ISPP in running courses on resistance for developing nations. It developed a code of ethics to avoid inappropriate use of resistance in advertisements.

FRAC is alert to fungicide resistance developments in the field and laboratory and to issues relating to the use, reporting, regulation and advertising etc. FRAC considers education to be most important. Without a firm understanding of the problem and the rationale behind usage strategies it is impossible to gain acceptance within companies, by advisors, distributors and users. For this reason, FRAC has a concerted educational policy to take the "message" to all people involved in fungicide use. This is being done through active support of friends and colleagues in universities, advisory services and government in promulgating sound technical strategies. Together we can continue to make real progress toward preserving the invaluable option of chemical disease control for crops.

NORTH AMERICAN INITIATIVES

FRAC, in cooperation with the Department of Plant Pathology at Pennsylvania State University, organized and sponsored the North American Fungicide Resistance Workshop and Conference, held September 20-25, 1987, at University Park, Pennsylvania. The objectives of the Workshop and Conference were to:

- Develop North American fungicide resistance research goals.
- Prolong the usefulness of fungicides at risk from resistance in North America.
- Promote antiresistance use strategies for fungicides in North America.
- Stimulate North American research potential in the area of fungicide resistance.
- Improve collaboration among industry, government, academia, and private organizations for determining and implementing appropriate fungicide use strategies.
- Encourage and demonstrate the commitment and leadership of Industry in the accomplishment of the foregoing objectives.

Research and Management Goals

A major objective of the 1987 North American Fungicide Resistance Workshop was to develop goals for fungicide resistance research and management and to ensure that the goals are implemented in North America. Specific goals were developed for each of the four fungicide groups, in addition to general goals. The 60 participants were encouraged to contribute their suggestions, to modify proposed goals, and to arrive at a consensus. General goals listed below are followed by goals for each of four fungicide groups at risk from resistance. Neither goals nor recommendations are presented in a specific order implying importance or sequence of execution since many activities are related and interdependent. Non-North American experiences have been incorporated into these goals and should be given consideration in the future. Many of the goals for one group apply to other fungicide classes as well. Their successful implementation relies on the constant awareness and cooperation of all involved with fungicide use, from the manufacturing company to the end user. The ultimate solution is careful and responsible use of fungicides in an effective manner that will preserve their usefulness for future needs.

General Goals

I. Provide Educational and Training Aids. There are many misconceptions concerning resistance problems, even among agricultural professionals. Educational and training aids should be readily available to a wide variety of audiences to overcome these difficulties.
- Use FRAC video tapes, slide sets, and RESISTAN-type resistance development models
- Develop special audiovisuals and handouts for extension agents and consultants.
- Promote and use aids developed by industry for users of individual products.

II. Improve Communication of Resistance Strategies at the User Level. Improved mechanisms are needed for reaching a consensus on use strategies. In addition, strategies are often misunderstood by users and, therefore, inadequately applied. There are also problems of abuse despite well- defined strategies.
- Use FRAC Working Groups to attain agreements.
- Obtain concurrence of industry management (marketing/ sales and research) and those who make recommendations.
- File written documents (recommendations, labels, advertisements) with FRAC Working Group.
- Report concerns of misuse to FRAC Working Groups.
- Promote strategies and educate users.

III. Strengthen Basic Research on Fungicide Resistance Basic research on fungicide resistance has been insufficient in North America, and greater support is needed.
- Conduct or contract basic research through industry as cooperative industry/academic projects.
- Develop and coordinate grants from multiple sources (including U.S. Department of Agriculture, National Science Foundation, and industries) through North American subgroups of FRAC Working Groups.
- Define and study epidemiological factors critical to the development of resistance.
- Determine genetic control and biochemical modes of action/resistance for DMIs and dicarboximides.
- Study biochemical interactions of candidate fungicides with other fungicides, negatively cross-resistant fungicides, and nonfungicide compounds with physiological actions including effects on host resistance.
- Develop new biotechnology tools to study resistance.
- Encourage nonchemical approaches to manage resistance.

IV. Avoid Establishment of Resistant Strains in New Areas. It may be necessary to experiment with introduced or created strains, but only with proper precautions.
- Avoid the introduction of resistant strains in a manner where they may be released into a virgin environment.

- Ensure the containment of experimental field strains when testing resistance management strategies.

V. Anticipate the Impact of New Candidate Fungicides on Resistance Problems. The methods (especially field studies) for evaluating resistance potential of new compounds must be improved and used.
- Determine the resistance and cross-resistance potential of each new candidate with major fungicide groups.
- Determine the influence of new candidates on pathogen populations resistant to previously used fungicides.
- Determine mode of action/resistance of new candidates.
- Develop and validate reliable monitoring methods.
- Establish baseline sensitivity levels in wild strains.
- Build on the "Gisi model" for risk assessment.

VI. Coordinate Industry, Government, and Academic Activities.
- Establish North American subgroups of the FRAC DMI and dicarboximide working groups.
- Develop cooperation among industry, government, and academia, including entomologists and weed scientists.

VII. Improve Understanding and Cooperation of EPA to Implement FRAC Resistance Management Strategies.
- Regulation and labeling can provide more useful functions in resistance management.
- Coordinate labeling for cross-resistant products.
- Persuade EPA of the need for the retention of products for use as companions in resistance management strategies, the use of resistance factors in risk-benefit analysis for compounds under special review, and the consideration of mode of action as well as efficacy in registration of new products.

VIII. Discover and Develop Fungicides with New Modes of Action. The discovery of new fungicide target sites is required to compensate for the loss of products due to de-registration and resistance.
- Provide accurate market value data.
- Promote "orphan drug" legislation to encourage development of compounds with insufficient economic potential for unsubsidized development.

Benzimidazole Resistance Research and Management Goals and Their Implementation in North America

I. Determine Optimal Benzimidazole Use Strategies. Strategies for fungicide use are critical in preventing and managing resistance. Evaluation of use strategies can be facilitated through field research experiments and computer models. Investigation of strategies will focus on two economically important diseases.

Apple Scab. The computer simulation model RESISTAN, with the use of currently available field data, can aid development of management recommendations tailored to regional situations. The following procedure can be used to translate the data into recommendations and evaluate the management strategies:
- Set parameters of the model with available field data.y Validate and refine model by comparing field data with simulation output.
- Run model to select best management options for specific situations.

Stone Fruit Brown Rot (Fruit Rot and Blossom Blight). Field and laboratory investigations can provide useful information in selecting disease management tactics:
- Develop rapid resistance monitoring systems.
- Investigate biology of resistant fungal isolates.
- Evaluate integrated disease management strategies.

II. Implement Optimal Fungicide Use Strategies Through Cooperation and Education. Cooperation and education are needed to distribute information and implement appropriate fungicide use strategies.
- Promote technical cooperation within industry to recommend appropriate use of fungicides through FRAC.
- Promote strategies with growers through cooperation among industry, academia, extension, and regulatory agencies.

III. Use Resistance Monitoring to Study Resistant Populations. Sensitivity monitoring used to determine the level and frequency of resistant strains is needed in guiding fungicide use strategies.
- Improve, exchange, and standardize monitoring techniques and data interpretation among industry, academia, and extension through FRAC.

Dicarboximide Resistance Research and Management Goals and Their Implementation in North America

I. Identify the Most Effective Use Strategies to Delay or Prevent Resistance Development This is the guiding goal for all resistance research.
- Determine the effects of rate, timing, and spray deposition on resistance development on strawberry, stone fruit, grape, and greenhouse crops.
- Determine the efficacy of dicarboximides in the presence of various proportions of resistant strains. Degrees of continued efficacy have been noted despite the occurrence of substantial resistance.
- Explore the effects of companion materials on resistant population dynamics and disease control.
- Develop and implement distinct regional strategies for strawberry, stone fruit, grape, and greenhouse crops.

- Develop models of population dynamics to integrate the above factors and help anticipate the effects of resistance management strategies.

II. <u>Characterize Field Populations of Dicarboximide-Resistant Strains of Botrytis and Monilinia</u>. A thorough understanding of dicarboximide resistance behavior would permit more accurate prediction and interpretation of the effects of potential use strategies. It could also direct efforts toward the most fruitful areas of research. Note that despite the advantages of generating resistance in the laboratory, these studies become more meaningful to subsequent interpretation for strategy development if field-resistant isolates are employed. It is prudent, however, for the responsible researcher to avoid developing field resistance in experimental plots or introducing laboratory or greenhouse-derived resistant isolates into the field.
- Identify the genetic control and biochemical mode of resistance.
- Determine the levels of resistant populations and their relative virulence.
- Further define the overall fitness and fitness factors (sporulation, latent period, pathogen survival, pathogenicity, etc.) of resistant populations selected from the field relative to wild type (sensitive) populations.
- Determine the effects of dicarboximides and/or companion fungicides on the pathogenicity of selected resistant isolates.

III. <u>Establish Comprehensive Monitoring Programs</u>. Without timely, efficient, and reliable monitoring, no resistance program can be properly developed or evaluated.

- Simplify monitoring techniques to be more rapid, less expensive, and less labor-consuming.
- Establish resistance monitoring programs for <u>Botrytis</u> in strawberry, grapes, and greenhouse crops and <u>Monilinia</u> in stone fruit. These can be implemented by industry, regulatory agencies, U. S. Department of Agriculture, grower groups, academia, or private advisors to validate resistance cases, assess risk, and estimate efficacy before a treatment is applied. This includes expanding resistance "indexing" programs that monitor for resistant pathogens on propagation material to be broadly distributed.
- Validate methods. It is very important that any monitoring method be reproducible and that the results relate directly to the actual field situation.
- Link some product uses to monitoring programs. Label restrictions could prevent the use of dicarboximides in high-resistance areas.

IV. Incorporate Nonfungicide Disease Control Measures with Dicarboximide Use. Methods other than chemical control can be used to reduce disease pressure and subsequent likelihood that resistance will develop.
- Follow recommended cultural practices and use resistant varieties to reduce disease pressure. An example is the practice of pruning and leaf removal around grape clusters to increase air movement and reduce gray mold.
- Integrate resistance models with other systems that predict disease, pests, plant growth, etc. (i.e., an IPM system incorporating resistance management).
- Integrate a resistance monitoring system into other management/monitoring practices.

V. Promote Interindustry Cooperation and Communication Among Industry, Institutions, Regulatory Agencies, and Academia. Without cooperation and communication, existing strategies will rarely be effectively implemented, and the development of new ideas will be greatly hindered.
- Establish guidelines for cooperation. Traditionally competitive companies must determine how and when to act in concert and must agree to adhere to consensus decisions.
- Establish common use patterns for dicarboximides on all crops for which more than one dicarboximide is or may be registered. This includes: number of applications permitted, rates used, timing of applications, and possible use of companions and alternatives.
- Develop common label statements to warn of resistance and restrict the use of other dicarboximides with cross-resistance.
- Establish a North American dicarboximide FRAC Working Group.
- Communicate to regulatory agencies the importance of cooperation in resistance management.
- Provide educational material about resistance management through technical information bulletins, brochures for grower meetings, and promotional material.
- Promote presentations and discussion sessions on resistance at professional meetings.

Phenylamide Resistance Research and Management Goals and Their Implementation in North America

I. Prioritize Attention to Two Diseases.
Lettuce Downy Mildew (caused by *Bremia lactucae*).
- Make growers aware of the threat of resistance through California farm advisors and Ciba-Geigy.
- Fund research to study the genetic resistance in California lettuce cultivars to pathovar III, the metalaxyl-resistant strain of *Bremia lactucae*, at the University of California - Davis.

- Explore the value of alternating fosetyl Al and metalaxyl plus maneb sprays in limiting resistance.
- Investigate the factors (especially use pattern of metalaxyl) that lead to the development of resistance.

Pythium Blight on Turfgrass
- Introduce a prepack of metalaxyl and mancozeb to limit the development of resistance.
- Breed perennial ryegrass with greater resistance to *Pythium*, as most fungicide resistance has been in ryegrass fairways.
- Continue to promote the alternation of metalaxyl with a non-cross-resistant *Pythium* fungicide. Ciba-Geigy held a turf symposium on December 4, 1987.

II. <u>Standardize Monitoring Programs.</u> It takes different concentrations of metalaxyl to control different stages in the life cycles of *Phytophthora* spp. and *Pythium* spp. There is also a marked difference between the ED50 and ED95. Most current monitoring techniques do not detect the proportion of resistant individuals within a population. The possibility to make such detections with DNA probes is an exciting challenge.
- Organize and standardize monitoring programs.
- Develop and validate rapid (on-site if possible) monitoring techniques.
- Determine the impact of carryover inoculum.

III. <u>Implement Use of Management Strategies Developed by FRAC.</u> Select use strategies complementary to the following FRAC recommendations:
- Introduce prepacked mixtures with residual or systemic fungicides with a different mode of action. Use three-fourths of the full rates of the mixing partners.
- Limit the number of sprays to two to four per crop and season.
- Keep applications intervals under 14 days.
- Prohibit curative/eradicative use.
- Prohibit soil application for control of airborne pathogens. (The exception is where there are no available companion fungicides, e.g., tobacco blue mold.)
- Provide educational materials about the philosophy and strategy of resistance management for improved communication with government agencies, professional societies, trade associations, growers, and academia.
- Strengthen basic studies of pathogen population dynamics as influenced by cultural systems and resistant cultivars. Determine criteria for epidemic development.
- Develop and use <u>RESISTAN</u>-type models to plan anti-resistance strategies.
- Study the molecular genetics of target fungi in an attempt to develop diagnostic tools.

Demethylation-Inhibitor Resistance Research and Management Goals and Their Implementation in North America

I. Organize Input to Determine Use Strategies for DMIs on Selected Crops in North America. Registrations of DMI fungicides in North America are increasing, and the number of EPA-approved active ingredients is expected to continue to increase. The biochemical mode of action of DMIs leads to the expectation of cross-resistance. A consensus use strategy for DMIs on a crop could maintain product usefulness and longevity.
- Establish a U. S. subgroup of the FRAC DMI Working Group.
- Focus efforts on grapes, cereals, apples, turfgrass, peanuts, and stone fruit.
- Build on and upgrade previously formulated strategies.
- Communicate recommendations through FRAC.

II. Determine Baseline Sensitivity to DMIs of Selected Pathogens on Selected CROPS. Management strategies and use patterns of DMIs in certain markets can be influenced by sensitivity changes in the target fungus. Adequate warning of significant sensitivity changes allows for prudent strategy decisions.
- Define "baseline sensitivity."
- Emphasize powdery mildew on grapes, apple scab, *Septoria* on wheat, late leaf spot of peanuts, dollar spot of turfgrass, and brown rot of stone fruit.
- Standardize monitoring methods.
- Share data and communicate results through FRAC.
- Cooperate with industry, academic institutions, and government agencies to communicate existing and future monitoring techniques and interpretation of results.

III. Determine Mechanisms of Resistance to DMIs and Possible Differences Among Compounds Within the Class Is cross-resistance automatic across all members of the DMI class, and if not, why not?
- Document and detail methodology to allow substantiation of claims of "new" or "different" modes of action.
- Encourage and support university research projects.

Action Items for the Implementation of Fungicide Resistance Research and Management Goals in North America

As a result of the recommended goals, the following actions should be taken:
- Establish North American subgroups of the FRAC DMI and Dicarboximide Working Groups.
- Organize California lettuce downy mildew action group.

- Designate members of the following groups to represent and promote North American Workshop (FRAC) goals:
    APS Chemical Control Committee
    APS Industry Committee
    CPS Chemical Control Committee
    ISPP Chemical Control Committee
    ACS American Chemical Society
    ESA Insecticide Resistance Committee
    USDA Agricultural Research Service
    WSSA Weed Science Society of America
- Organize field demonstrations and seminars for the U.S. Department of Agriculture and EPA.
- Publish standardized monitoring methods.

Workshop Proceedings

The FRAC sponsored book, FUNGICIDE RESISTANCE IN NORTH AMERICA, published by APS Press, is based on papers presented at the North American Workshop. Although the focus of the book is on North America, the inclusion of international experiences expands the state of the art and science of fungicide resistance. The book updates the activities of academia, government, and extension and also documents the active involvement of industry scientists in fungicide resistance research and management and their exemplary cooperative efforts to preserve the effectiveness of fungicides. It remains to be seen if the necessary cooperation will be developed to carry out the tough tasks of resistance management in the future.

Sponsorship

FRAC is sponsored by GIFAP the international group of national associations of the agrochemical manufacturers located in Brussels, Belgium.

Acknowledgments

This article is adapted from references (1) and (2).

References

1. Wade, M. Plant Dis. 1987, 71, 652-653.

2. Wade. M. Tagungsber. Akad. Landwirtschaftswiss. D.D.R. 1987, 253, 411-416.

RECEIVED September 1, 1989

' # HERBICIDES

Chapter 23

Herbicide Resistance in Weeds and Crops

An Overview and Prognosis

Homer M. LeBaron[1] and Janis McFarland[2]

[1]New Technology and Basic Research Department, Ciba-Geigy Corporation, Greensboro, NC 27419
[2]Metabolism Department, Ciba-Geigy Corporation, Greensboro, NC 27419

Weeds resistant to herbicides are rapidly becoming important factors in crop production and agricultural technology. Resistance to triazine herbicides has been confirmed in 55 species with one or more resistant biotypes in 31 states in the U.S., four provinces of Canada and 18 other countries. In addition, there has been a serious spread of weeds having multiple or cross resistances to various classes of herbicides and a recent development of weed biotypes resistant to the herbicides that inhibit acetohydroxyacetate synthase. The need for research on the prevention and management of herbicide resistance is obviously urgent. Herbicide resistant weeds may become a more serious economic problem within five to 10 years than pest resistances to insecticides and fungicides due to the greater use of herbicides in agriculture. This is almost certain to be the case if we depend too much on only a few of the newer herbicides and discard the older ones. We will need all the tools we currently have, as well as those that modern technology can provide, to manage our weed pests while further reducing or eliminating soil tillage, and to conserve essential soil and water for future crop production and public use. Research on herbicide resistant weeds should complement biotechnology research aimed at developing herbicide resistant crops, but the strategy and objectives of the biotechnology research must be altered to some extent. In particular, efforts should be aimed at developing major crops resistant to many herbicides, rather than one or two. This would provide greater flexibility in rotating or alternating herbicides to prevent resistant weeds from evolving, and controlling those resistant populations that appear.

0097-6156/90/0421-0336$06.00/0
© 1990 American Chemical Society

The impact of herbicides on our modern agricultural technology has been phenomenal. It is almost impossible to measure or to overemphasize the importance of herbicides in their potential for overcoming famine, pestilence, poverty and crop losses due to weeds that were prevalent throughout most of the history of mankind prior to the 1950s. Herbicides have been essential, not only for their direct effects on weed control and crop yield, but also in providing more efficient use of water and fertilizers for the optimum development of improved crop varieties and hybrids. They have also been essential tools in the recent development of conservation tillage methods, which have reduced topsoil erosion and moisture losses in many areas of the world.

In terms of improved agricultural efficiency and crop productivity, we maintain that never in our history have so few researchers done so much for agriculture in such a short time as weed scientists in the past 40 years. Most weed science societies in the U.S. have existed for only 40 years. In other countries, they have been organized for a shorter time. The Entomological Society of America is celebrating its centennial in 1989, and the American Phytopathological Society is now 80 years old. In spite of their great contributions, even today, most weed scientists must obtain their training in other related departments because few universities have departments of weed science. Conversely, almost all agricultural schools have departments of entomology and plant pathology, with these specialists outnumbering weed scientists about 10 to 1 on many university facultues.

Although the discovery and development of herbicides came after the discovery and development of organic insecticides and fungicides, herbicide use has far surpassed all other pesticides combined in the marketplace. As can be seen in Table I, herbicides account for about 65% of all agricultural pesticide sales in the U.S. Herbicides also represent 85% of the total pesticide use on major crops in the U.S. (see Table II). For a number of reasons, including increasing registration costs, toxicological and environmental concerns, and the evolution of resistance in the target pests, the total effort by industry to discover and develop new insecticides and fungicides is tending to decrease. This is not yet the case with herbicides, but we predict that it will be in the near future for some of the same reasons. Any decreased development effort will make the threat of herbicide resistant weeds an even greater concern.

Mechanisms of Herbicide Action and Plant Selectivity

One of the most important characteristics contributing to the value of herbicides has been plant selectivity. The fact that some plants (weeds) are killed by herbicides while others (crops) are naturally resistant has been very important to selective weed control in crop production. A few herbicides that have been useful without natural or biological resistance, such as glyphosate and paraquat, have found great success in crop production due to the use of other methods of selective weed control (e.g., directed applications, applying prior to crop emergence or applying to dormant perennials).

Table I. U.S. Pesticide Sales in 1986*

	Agriculture Uses		Total Uses	
	($ Millions)	%	($ Millions)	%
Herbicides	2,775	63	3,625	56
Insecticides	1,050	24	1,980	30
Fungicides	310	7	515	8
Other	290	6	370	6
Total	4,425	100	6,490	100

Worldwide total sales of all pesticides in 1986 was $16,150,000,000 (estimated).

*From Economic Analysis Branch, Office of Pesticide Programs, EPA.

Table II. Amounts of Pesticides Used on Major Crops in U.S.*

	1987		1988	
	Pounds (Millions)	%	Pounds (Millions)	%
Herbicides	365.0	85.2	372.0	84.7
Insecticides	56.6	13.2	59.7	13.6
Fungicides	7.0	1.6	7.6	1.7
Total	428.6	100.0	439.3	100.0

*From USDA-AR II, July 1988, Agriculture Resources Inputs: Situation and Outlook Report.

Our knowledge about herbicide sites of action has been essential in our research and understanding of herbicide resistance mechanisms. On the other hand, herbicide resistant weeds have proven to be valuable scientific tools that have contributed greatly to our understanding of herbicide action, plant biochemical and physiological processes, molecular genetics, plant morphology, and plant anatomy. We believe that herbicide resistant weeds have contributed more benefits to society in the past through scientific advances and to our understanding of basic botany and plant physiology than they have cost. However, herbicide resistant weeds are rapidly becoming serious economic problems.

Triazine Resistance

Intraspecific triazine resistance was first discovered in common groundsel (Senecio vulgaris L.) in western Washington in the late 1960s. The subsequent widespread and frequent occurrence of other triazine resistant weeds over the past 20 years, has made triazine resistance the best known and most studied example of herbicide resistance (1). Triazine resistance has also been of great interest because of the importance and extensive use of this group of herbicides.

Biotypes of at least 40 broadleaf and 15 grass weed species are known to have evolved resistance to triazine herbicides somewhere in the world. Only 21 of these resistant biotypes have been found in the U.S., and one or more of these resistant biotypes have invaded 31 states, four provinces of Canada, and 18 other countries. The confirmed triazine resistant weed species and their distribution by years are summarized in Tables III and IV from data collected in our recent worldwide survey. Details of this extensive survey, mostly conducted in 1988, have not yet been published. Data in the summary Tables III, IV and V are published here for the first time.

Past experience has shown that weeds resistant to triazines can be managed or confined within a reasonable limit. In the U.S., the total area of land or crops infested with triazine resistant weeds is still relatively small and does not seem to be expanding rapidly. In most areas of the U.S. where triazine resistant weed populations have evolved, it has not been necessary or desirable to discontinue the use of the triazine herbicide of choice, due to the many triazine susceptible weeds that are usually prevalent. In a few cases, the resistant biotypes have even disappeared (e.g., some triazine resistant biotypes of Setaria sp. in Nebraska). However, in other countries, especially when resistance strategies were not followed, resistant biotypes quickly became serious problems.

Weed Resistance to Other Herbicides

Over the past years since the extensive use of modern herbicides began, cases of intraspecific resistance to a few other herbicides have developed, and there have been many examples of evolution toward interspecific herbicide tolerance (i.e., the more tolerant species gradually take over, such as 2,4-D tolerant broadleaf weeds and dalapon tolerant grasses). Within the last two years, there

Table III. Distribution of Triazine Resistant Dicot
(Broadleaf) Weeds
(as of December 1988)

Scientific Name	Common Name	Location	Number of States, Provinces or Countries
Abutilon theophrasti Medik.	velvetleaf	U.S.	1
		Europe	1
Amaranthus spp. (8)[a]	pigweed or amaranth	U.S. (4)[b]	21
		Canada (2)	3
		Europe (5)	10
		Israel (1)	
Ambrosia artemisiifolia L.	common ragweed	U.S.	1
		Canada	1
Arenaria serpyllifolia L.	thymeleaf sandwort	Europe	1
Atriplex patula L.	spreading orach	Europe	2
Bidens tripartita L.	bur beggarticks	Europe	2
Brassica campestris L.	birdsrape mustard	Canada	1
		Europe	1
Capsella bursa-pastoris (L.) Medik.	shepherdspurse	Europe	1
Chenopodium spp. (6)	lambsquarter or goosefoot	U.S. (2)	19
		Canada (2)	3
		Europe (4)	9
		New Zealand (1)	
Epilobium spp. (2)	willowweed	Europe (2)	6
Erigeron (or Conyza spp.) canadensis L.	horseweed	Europe	6
Galinsoga ciliata (Raf.) Blake	galinsoga		
Kochia scoparia (L.) Schrad.	kochia	U.S.	12
Matricaria matricarioides (Less.) C. L. Porter	pineappleweed	Europe	1
Myosoton aquaticum (L.) Moench.	starwort	Europe	2
Physalis longifolia Nutt.	longleaf groundcherry	U.S.	1
Polygonum spp. (5)	smartweed or knotweed	Europe (5)	4
Senecio vulgaris L.	common groundsel	U.S.	3
		Canada	2
		Europe	9
Sicyos angulatus L.	burcucumber	U.S.	1
Sinapis arvensis L.	wild mustard	Canada	1
Solanum nigrum L.	black nightshade	U.S.	1
		Europe	7
Sonchus asper (L.) Hill	spiny sowthistle	Europe	1
Stellaria media (L.) Vill.	common	Europe	2

Total known to date = 40 species in 23 genera.

[a]Numbers after genera indicate the number of species within that genus.

[b]Numbers after location indicate the number of species within the genus found within U.S., Canada, Europe, or other countries.

[c]These numbers indicate the number of states, provinces or countries where the species or genus has been found.

Compiled by authors.

Table IV. Distribution of Triazine-Resistant Monocot
(Grass) Weeds
(as of December 1988)

Scientific Name	Common Name	Location	Number of States, Provinces or Countries
Alopecurus myosuroides Huds.	slender foxtail	Israel	
Brachypodium distachyon (L.)P. Beauv./R.et S.	bromegrass	Israel	
Bromus tectorum L.	downy brome	U.S.	5
		Europe	1
Digitaria sanguinalis (L.) Scop.	large crabgrass	Europe	3
Echinochloa crus-galli (L.) Beauv.	barnyardgrass	U.S.	2
		Canada	1
		Europe	4
Koeleria phleoides (Vill.)Pers. [or Lophochloa cristata (L.)Hyl.]	annual cat's tail	Israel	
Lolium rigidum Gaud.	Swiss or annual ryegrass	Israel Australia	
Panicum capillare L.	witchgrass	U.S.	1
		Canada	1
Phalaris paradoxa L.	hood canarygrass	Israel	
Poa annua L.	annual bluegrass	U.S. Europe	1 7
Polypogon monspeliensis (L.) Desf.	rabbitfootgrass	Israel	
Setaria spp. (4)[a]	foxtail	U.S. (2)[b]	2
		Canada (1)	1
		Europe (3)	1

Total known to date = 15 species in 12 genera.

Grand total of all known triazine resistant weeds = 55 species and 35 genera.

[a]Numbers after genera indicate the number of species within that genus.

[b]Numbers after location indicate the number of species within the genus found within U.S., Canada, Europe, or other countries.

[c]These numbers indicate the number of states, provinces or countries where the species or genus has been found.

Compiled by authors.

TABLE V. Occurrence and Distribution of Weed Biotypes Resistant to Various Nontriazine Herbicides (as of December 1988)

Herbicide	Scientific Name	Common Name	Year Found	Location
Aminotriazole	Lolium rigidum Gaud.	annual ryegrass	1988	Australia
	Poa annua L.	annual bluegrass	1986	Belgium
Bromoxynil	Chenopodium album L.[a]	common lambsquarter	1988	West Germany
Carbamates	Amaranthus hybridus L.[ab]	smooth pigweed	1988	Hungary
	Amaranthus retroflexus L.[ab]	redroot pigweed	1988	Hungary
Chlorbromuron, chlorotoluron, isoproturon, & some other substituted ureas	Alopecurus myosuroides Huds.[b]	slender foxtail	1983 1984	West Germany United Kingdom
	Amaranthus hybridus L.[ab]	smooth pigweed	1988	Hungary
	Amaranthus retroflexus L.[ab]	redroot pigweed	1988	Hungary
	Chenopodium album L.[a]	common lambsquarter	1988	Hungary
	Erigeron canadensis L.[a]	horseweed	1988	Hungary
2,4-D and phenoxys	Cirsium arvense (L.) Scop.	Canada thistle	1985	Hungary
	Daucus carota L.	wild carrot	1962	Ontario Canada
	Sphenoclea zeylanica Gaertn.	gooseweed	1982	Philippines
Diclofop methyl	Alopecurus myosuroides Huds.	blackgrass	1988	United Kingdom
	Avena fatua L.	wild oat	1985 1987	South Africa Australia
	Lolium multiflorum Lam.[c]	Italian ryegrass	1987	Oregon
	Lolium rigidum Gaud.[d]	annual ryegrass	1982	Australia
Diuron	Amaranthus bouchonii Thell.[a]		1988	Hungary
Mecoprop	Stellaria media (L.) Vill.	common chickweed	1985	United Kingdom
MSMA & DSMA	Xanthium strumarium L.	common cocklebur	1984 1987	South Carolina North Carolina
Paraquat and Diquat	Arctotheca calendula (L.) Levyns	capeweed	1986	Australia
	Epilobium adenocaulon Hausskn.	American willow-herb	1984 1983	Belgium United Kingdom
	Conyza bonariensis (L.) Cronq.	hairy fleabane	1979 1984	Egypt Hungary
	Erigeron canadensis L.	horseweed	1980 1988	Japan Hungarya
	Erigeron philadelphicus L.	Philadelphia fleabane	1980	Japan
	Erigeron sumatrensis Retz.	oharechinoqiku	1986	Japan
	Hordeum glaucum Steud.	wall barley	1983	Austral
	Hordeum leporinum Link	hare barley	1988	Austral
	Lolium perenne L.	perennial ryegrass	1976	United
	Poa annua L.	annual bluegrass	1978	United Kingdom
	Youngia japonica(L.) DC	onitabirako		
Propanil	Echinochloa colonum (L.) Link	junglerice	1986	Japan
	Echinochloa crus-galli (L.)Beauv.	barnyardgrass	1988	Columbia

Continued

been an increasing number of weeds which have evolved
resistance to other types or classes of herbicides, especially to
sulfonylureas and other acetohydroxyacetate synthase (AHAS)
inhibitors (also referenced as acetolactate synthase or ALS
inhibitors). These weed species and their distributions are presented in Table V.

Mechanisms of Resistance

Before Radosevich and De Villiers found in 1975 that isolated chloroplasts of resistant common groundsel were insensitive to atrazine and simazine (2), it had been erroneously assumed that all living plants would die if the herbicides could reach their target site intact. We now know that mechanisms of selectivity in crops can be due to differences in metabolism rates, uptake, translocation, site of action or avoidance mechanisms. However, the mechanisms of herbicide resistance that have evolved in weeds are usually different from the mechanisms of herbicide selectivity in most crops. This is certainly true with the most prevalent and thoroughly studied cases of herbicide resistance, including the triazines, dinitroanilines, and AHAS inhibitors.

For example, in the goosegrass [Eleusine indica (L.) Gaertn.] weed biotype resistant to trifluralin, the tubulin is apparently altered so that dinitroaniline herbicides are not effective in preventing tubulin polymerization into microtubules, which is the mechanism of action of these herbicides (3). However, selectivity in most crops to dinitroaniline herbicides is believed to be due to the high lipid content of the seed, which compartmentalizes the herbicide separately from dividing cells, and to the ability of their tap roots to rapidly grow through the treated soil layer, thereby avoiding significant herbicide exposure.

The mechanism of resistance to the sulfonylurea herbicides and other AHAS enzyme inhibitors in many resistant biotypes also differs from crop selectivity mechanisms. Based on research to date, resistance mechanisms in weed biotypes usually appear to be due to an alteration in the gene coding for AHAS, resulting in variable forms of insensitive AHAS enzymes, which is the main target site of these herbicides (4). It is likely that all weed species possess some ability to evolve resistance to these herbicides by mutation of the target site. Crop selectivity, on the other hand, seems to be most frequently dependent on differential metabolism. The annual ryegrass (Lolium rigidum Gaud), which first evolved resistance to diclofop-methyl and poses a very serious problem in Australia, is an important exception. It is cross-resistant to various herbicides, including AHAS inhibitors, possibly due to an increased rate of herbicide metabolism (5). Research to date indicates that in most triazine resistant biotypes, the normal triazine binding properties in their chloroplasts are altered, whereas crop selectivity is due mainly to metabolism differences. Differential translocation is also responsible for crop selectivity with prometryn in cotton. However, some weed species have evolved biotypes with resistance mechanisms similar to those in crops and a few have evolved more than one resistance mechanism in the same species. For example, while most triazine resistant biotypes of barnyardgrass [Echinochloa crus-galli (L.) Beauv.] have insen-

Table V. Continued

Herbicide	Scientific Name	Common Name	Year Found
Pyrazon	Chenopodium album L.[a]	common lambs-quarter	1986 1978 1984
Sulfonylureas, imidazolinones & triazolopy-, rimidine	Ixophorus unisetus (Presl) Schult./Schlecht. Kochia scoparia (L.) Schrad.	hatico kochia	1984 1988 1987 1987 1988 1988 1988 1988 1988 1988
	Lactuca serriola L. Lolium rigidum Gaud.[d] Salsolo iberica Sennen & Pau Stellaria media (L.) Vill.	prickly lettuce annual ryegrass Russian thistle common chickweed	1987 1986 1988 1988 1988
Trifluralin and other dinitroanilines	Eleusine indica (L.) Gaertn.	goosegrass	1973 1982 1987 1987
Uracils (e.g. bromacil)	Setaria viridis (L.) Beauv. Amaranthus hybridus L.[b]	green foxtail smooth pigweed	1988 1988
	Amaranthus retroflexus L[ab]	redroot pigweed	1988

[a] This biotype evolved a secondary cross-resistance after evolving atrazine resistance.

[b] This biotype usually shows varying degrees of cross-resistance to several classes of herbicides, including diclofop-methyl, pendime triazines and sulfonylureas.

[c] This biotype is cross-resistant to other polycyclic alkanoic acid herbicides.

[d] This biotype evolved resistance to diclofop methyl after several years of use, and was then found to often be cross-resistant to m sulfonylurea herbicides.

Compiled by authors.

sitive chloroplasts, one biotype was found in France to have another type of resistance, apparently due to metabolism (6). Triazine-resistant velvetleaf (Abutilon theophrasti Medik.) in Maryland has been found to be resistant due to enhanced glutathione transferase activity, which is similar to the major protection mechanism in corn (7).

While most plants are susceptible to paraquat, some paraquat-resistant horseweed (Erigeron sp. and Conyza sp.) biotypes are apparently insensitive to the herbicide due either to elevated levels of superoxide dismutase and other enzymes in a pathway detoxifying oxygen radicals or to differential sequestration of paraquat in the weed (8, 9). Data on the mechanism of most other types of herbicide resistance in weeds are still not complete.

Prediction of Future Herbicide Resistance

Based on our knowledge of herbicide modes of action and other characteristics, we are able to predict, with some confidence, which herbicides will have a high risk for resistance. There are wide margins for error due to uncertainty about genetic variability of weeds (i.e., the relative frequency of occurrence of different resistance mutations), the lack of knowledge on the primary sites or modes of action for certain herbicides, and the lack of understanding of cross-resistance between classes of herbicides.

In general, we can say that the following characteristics of herbicides, listed in their approximate order of importance, will contribute to a high risk for resistance:

1. A single target site of action.

2. Extremely active and effective in killing a wide range of weed species.

3. Provide long-term soil residual and season-long control of germinating weeds, or applied many times throughout the year.

4. Applied frequently and over several growing seasons without rotating, alternating, or combining with other types of herbicides.

Other factors may be important in some cases, but based on the above criteria, our experience to date, and the general use of the herbicides, we have listed some of the more common herbicides that are considered low risk (Table VI) and high risk (Table VII) for weed resistance.

The Need for a Variety of Herbicides

Resistance to the AHAS inhibitors and other newer herbicides (e.g., diclofop methyl) has the potential to be a serious issue in the future. This new generation of herbicides can usually be used at much lower rates, with what is considered to be lower human, animal and environmental exposure and risks. They have also greatly expanded the crops and use areas where problem weeds can be effectively controlled. They are being introduced at a time when many

Table VI. Herbicides With Low Risk for Weed Resistance

Herbicide Classification	Common Name	Trade Name
Acetanilides or Amides	alachlor	Lasso
	diphenamid	Enide
	metolachlor	Dual
	napropamide	Devrinol
	propachlor	Ramrod
	propanil	Stam
Aliphatics	dalapon	Dowpon
	glyphosate	Roundup
	TCA	Sodium TCA
Benzoics	chloramben	Amiben
	dicamba	Banvel
Carbamates	asulam	Asulox
	barban	Carbyne
	chlorpropham	Chloro IPC
	phenmedipham	Betanal
Nitriles	bromoxynil	Brominal
Organic Arsenicals	cacodylic acid	Rad-E-Cate
	DSMA	several
	MSMA	Ansar, Daconate
Phenoxys	2,4-D	several
	MCPA	several
Thiocarbamates	butylate	Sutan
	diallate	Avadex
	EPTC	Eptam
	molinate	Ordram
	vernolate	Vernam

Table VII. Herbicides With High Risk for Weed Resistance

Herbicide Classification	Examples	
	Common Name	Trade Name
Bipyridiliums	diquat	Diquat
	paraquat	Paraquat
Dinitroanilines	benefin	Balan
	oryzalinte	Surflan
	pendimethalin	Prowl
	trifluralin	Treflan
Diphenyl Ethers	acifluorfen	Blazer
	diclofop methyl	Hoelon
	lactofen	Cobra
	oxyfluorfen	Goal
Imidazolinones	imazapyr	Arsenal
	imazaquin	Scepter
Sulfonylureas	chlorsulfuron	Glean
	chlorimuron	Classic
	sulfometuron-methyl	Oust
	triasulfron	Amber
Triazines	atrazine	AAtrex
	cyanazine	Bladex
	metribuzin	Sencor, Lexone
	prometon	Pramitol
	prometryn	Caparol
	simazine	Princep
Uracils	bromacil	Hyvar
	terbacil	Sinbar
Ureas	diuron	Karmex
	fluometuron	Cotoran
	linuron	Lorox
	tebuthiuron	Spike

of the earlier herbicides are being discontinued or are in trouble because of economics, reregistration requirements, toxicology questions, and environmental concerns. Benbrook and Moses (10) recently stated the following relative to industrial policies on developing herbicide resistant crops:

"Another significant public policy issue surrounding the development of the new herbicides and resistant cultivars (crops) is the marked divergence in prospective public health outcomes. On the one hand, little progress toward safer herbicides can be expected if resistance research is focused on and successful in sustaining or expanding the market share of older compounds, particularly those that are relatively toxic and applied at higher rates per acre. Conversely, resistance research, and the emergence of resistant cultivars, could speed adoption of new proprietary compounds. When this outcome involves low application rates of more fully tested and relatively safer compounds, environmentalists should rejoice. The environmental attributes of the new products generally are markedly improved over the older products."

These and other conclusions made by the authors seem to be premature, but a very strong perception exists among government agencies and policymakers that the new herbicides will replace the major herbicides in current use, and that we should not try to save or justify keeping the older herbicides. A major justification for not dropping older herbicides is the need for a maximum number of herbicides in rotation as a resistance prevention and management tool (11). It is essential to understand that the main reason that small amounts of atrazine and some of the other older herbicides have been detected in some groundwater samples is not because they readily leach or move in soils, but because they are relatively persistent and used in large quantities. It is also important to understand that most of the herbicides found in groundwater, especially at concentrations above 1 to 2 ppb, are due to point source contaminations or correctible errors. Atrazine has virtually made conservation tillage possible in corn and other major crops, resulting in tremendous environmental, as well as economic benefits. Even though newer herbicides are effective at much lower rates, they are currently unable to fill all agricultural needs. We maintain that there will be a great need for many of the older herbicides and other tools of agricultural technology to be preserved for future use.

Just as pharmaceuticals have contributed much to increased life expectancy, hunger and famine have been greatly alleviated by increased food production due to pesticides. Pesticides must certainly be a major and increasing part of the agricultural technology in the decades ahead, which will be needed to provide the constantly greater demand for food, fiber and shelter, with adequate cost effectiveness. But we must not become complacent, discard other means of pest control, or depend too much on pesticides alone. Herbicide resistance is acting as a self-limiting system of nature. Nature sets the rules. We have to adapt or we lose.

In spite of the evolution of resistance to pesticides and other limitations or problems associated with their use, no one who understands the adverse consequences to humanity in its present numbers can seriously recommend that chemicals be immediately or in the foreseeable future replaced by alternative means of pest control. Weed scientists are very supportive of research on alternative methods of weed management, such as biological control, integrated pest management (IPM), biotechnology, sustainable agriculture, and any other method of pest or crop management that will work, and, thereby, keep our agriculture competitive, strong and safe. However, idealism must be tempered with realism. Furthermore, it is incorrect to assume that all biological controls are safer than pesticides. If we have learned anything in the past 40 years, it is that we will need all the help we can get to keep ahead of pests. To depend on only one tool or method against major pests is a sure road to failure and scientific heresy. Certainly, a few chemicals should and will go, but chemicals will continue to be essential and a major line of defense against most pests, especially weeds, and will help to produce the crops and pay for the research on biological controls, biotechnology, and sustainable agriculture while these tools are being developed.

Urgent Need for Implementation of Resistance Prevention and Management Strategies

With the first invasion of resistant weed populations, prompt action has been essential to avoid serious and permanent problems. Preventive action to avoid herbicide resistant weeds from evolving in the first place is definitely the best strategy. It is virtually essential in all cases of herbicide resistance to have other classes or types of herbicides, with alternate sites, modes of action and degradation available. In some countries, because of failure to respond promptly, the lack of suitable alternatives, or for other reasons, triazine resistant weeds have not been controlled, resulting in rapid spread, leading to the almost total loss of effective weed control with these herbicides (1).

In 1989, there are great concerns and doubts whether we can be as successful in avoiding or managing some of the more recent resistant biotypes as we have been with triazines in the past. Table VIII shows some interesting trends in the first occurrences of herbicide resistant biotypes. Not only are herbicide resistant weeds appearing after fewer repeat annual applications of some of the newer herbicides (e.g., 4 to 5 years of treatment with sulfonylures herbicides), but there seem to be more species that have potential for resistance, as shown by the 26 new cases in 1988. In addition, the resistant biotypes are more fit and competitive than most biotypes resistant to triazine herbicides.

The lower relative fitness in most triazine resistant weeds is a very important reason why they have been fairly easily controlled, and why more problems of cross-resistance or multiple resistance have not occurred where both a triazine and another type of herbicide have been used together repeatedly. However, some cases of such cross-resistance are now beginning to appear, consistent with the predictions by Gressel and Segel (12).

We must apply all the wisdom and understanding we have gained on pest resistance to pesticides, as well as exercise a greater degree of marketing control and self-restraint than has ever been demonstrated in U.S. agriculture to protect, preserve or prolong the use of the newer sulfonylurea and other herbicides with a single site of action and, therefore, at high risk for resistance.

Table VIII. Number of States, Provinces and Countries Reporting the First Appearance of Herbicide Resistant Biotypes by Year

Year	Triazines	Non-triazine Herbicides
1962	0	1
1968	3	0
1969	0	0
1970	2	0
1971	0	0
1972	1	0
1973	5	1
1974	1	0
1975	4	0
1976	7	1
1977	15	0
1978	24	2
1979	10	1
1980	12	2
1981	22	0
1982	19	3
1983	20	3
1984	23	7
1985	17	2
1986	11	6
1987	9	8
1988	5	26
Total	210	63

The following are some of the changes or strategy rules we believe will be required:

1. Crop and herbicide rotations should be used wherever possible. Try to avoid crop rotations with the same weed spectra.

2. Use of long residual herbicides with a single site of action should be avoided or minimized.

3. Use the lowest rates possible.

4. Minimize the number of applications per season, and use the same herbicide only every two or three years.

5. These herbicides should be used in crops only where other mixing partners, cultivation, or other weed control options are available.

6. Especially in major crops, these herbicides should be marketed only in combinations if other types of herbicides are available as suitable partners.

7. Cultivators or other mechanical weed control options should remain available. Conservation tillage systems may not always be an option in situations where resistance is a problem.

8. Don't yet throw away the hoe, but rogue out the weeds that escape if resistance occurs or is suspected, or use management systems that preferentially control resistant weeds.

9. Industry should not develop and market AHAS resistant crops or crops resistant to only one herbicide with a high risk for resistance for the purpose of greatly expanding their use. This approach should be used to enhance herbicide selectivity in crop varieties, to avoid carry-over injury, for specific and limited special problems, and for minor acreage and high value crops. A major objective of developing herbicide resistant crops should be to provide more flexibility in control of resistant weeds.

10. If possible, industry should continue to try to develop chemicals that will inhibit all types of AHAS enzymes.

11. Develop other herbicides that do not have a single site of action and are not as likely to induce resistance.

12. Education and cooperation of industry management, marketing, sales, extension, farmers, and others is essential.

13. Government agencies and policymakers must realize that all possible herbicides must be retained as potential mixing partners.

Literature Cited

1. LeBaron, H. M. and J. Gressel, Ed.; Herbicide Resistance in Plants; John Wiley & Sons: New York, 1982.

2. Radosevich, S. R.; DeVilliers, O. T. Weed Sci. 1976, 24, 229.

3. Vaughn, K. C.; Vaughn, M. A., This Volume.

4. Thill, D. C.; Mallory, C. A.; Saari, L. L.; Cotterman, J. C.; Primiani, M. M. WSSA Abstract #297, 1989, in press.

5. Powles, S. B.; Holtum, J. A. M.; Matthews, J. M.; Liljegren, D. R. This volume.

6. Gasquez, J., INRA Lab. de Malherbologie, Dijon, France, 1988, Personal communication.

7. Anderson, R. N.; Gronwald, J. W. Weed Sci. 1987, 35, 496-8.

8. Fuerst, E. P.; Nakatani, H. Y.; Dodge, A. D.; Penner, D.; Arntzen, C. J. Plant Physiol. 1985, 77, 984-9.

9. Shaaltiel, Y.; Chva, N.-H.; Gepsteins, S.; Gressel, J. Appl. Genet. 75, 850-856, 1988.

10. Benbrook, C. M.; Moses, P. B. In Bio Expo 86 Proceedings, Butterworth Publ., Boston, 1986; p. 27-54.

11. Gressel, d; Segel, L. A., This Volume.

12. Gressel, J.; Segel, L. A. In Herbicide Resistance in Plants; LeBaron, H. M.; Gressel, J., Ed.; John Wiley & Sons: New York, 1982; p. 325.

RECEIVED October 31, 1989

Chapter 24

Fate of Herbicide Resistance Genes in Weeds

H. Darmency and J. Gasquez

Laboratoire de Malherbologie, Institut National de la Recherche Agronomique, BV 1540, 21034 Dijon, Cedex, France

The appearance of herbicide resistance genes in weed populations may originate from mutations or genetic exchanges with other organisms. Using triazine resistance as an example, we attempted to answer several questions about the origin and the spread of genes for resistances in weed populations.

We found that triazine resistant mutants do not appear at random, but are the result of particular genomes capable of producing a high number of mutants, which we refer to as a founder effect in each resistant population. However, the presence of the mutated chloroplast gene is not believed responsible for high levels of resistance in these weeds. We also believe that because of the high selection pressure conferred upon the resistant genes in herbicide treated areas, mutation events even at very low frequencies have high probabilities of evolving resistant plants, and that spread of resistance from engineered crop plants will likely occur in spite of cytoplasmic inheritance.

During the last two decades, weed control techniques in Europe have been marked by four major features: the release of antigramineae herbicides in cereals; an increased selective control of perennials; a trend to use lower quantities of active ingredients; and the appearance of herbicide resistant weeds which is becoming increasingly important in crop production (1). The appearance of herbicide resistant weeds indicates a clearly adapted response of weeds to agricultural practices and has aided research in the development of herbicide resistant crops. This paper discusses the origins of herbicide resistance genes.

HERBICIDE RESISTANCES

Occurrence: The appearance of herbicide resistant weeds was predicted as early as 1950. Blackman (2) then pointed out the likenesses between mass selection for a given character in crops and selection of weeds after continuous herbicide treatments in fields. Herbicide resistance has been reported worldwide in a wide variety of crops that involve herbicide families with different modes of action and several weeds species (1).

However, contrary to the prediction that assumes the involvement of many genes, the genetic control of the following resistances is determined by only one or a few genes: triazines (loss of herbicide binding due to a chloroplast mutation (3)), bipyridilium (single gene control of the amount of three detoxication enzymes (4)), and dinitroaniline resistances (unaffected microtubule formation due to tubulin mutation (5)). The genetics for diclofop methyl (6), chlorotoluron (7), mecoprop (8), and triallate resistances (9) are still not clear. Other cases of differential tolerance in response to herbicides are known among different biotypes in several species (10) but are not clearly related to the selection of resistant plants in the field.

Problems and Model: If the development of herbicide resistance continues, the lifespan of several herbicides will probably be shortened, leaving farmers to face several impossible weed control problems. Moreover, weeds resistant to herbicides are costly because alternative herbicides are not always available. In some cases, supplementary treatments must be applied in addition to the standard treatment that remains efficient for controlling numerous "ordinary" weeds. Of the 3 million hectares of corn grown in France, 1.2 million hectares are now subjected to a postemergence treatment (e.g., pyridate) specifically directed against triazine resistant weeds. Thus, this so-called "remedial treatment" is becoming a standard treatment that is carried out after the preemergence atrazine treatment. This creates a four-fold increase in the cost of weed control in these corn fields.

The reduced efficacy of certain herbicides and the increased cost of weed management are two reasons to study the appearance of herbicide resistant plants. In addition, fundamental knowledge of plant population biology is also a concern. In contrast with most ecological factors, herbicide treatment is one of the rare selection pressures that is easy to study within a genuine environment. Biologists can understand and test by experimental methods the long-term effects of field applied herbicides on weed populations. Hence, the weed population/herbicide model is an excellent example for studying adaptation and evolution.

Therefore, all of the operative factors contributing to the appearance of resistant plants in a population or a species should be precisely defined. The genetics of weed species are certainly the first areas that must be investigated. Questions to be raised to study weed genetics include:

1. Does mutation occur at random?
2. Does a special genetic structure of weed populations exist in those populations evolving toward resistance?
3. How does the regulation of resistance genes work?
4. What are the risks of gene transfer associated with the release of genetically improved resistant crops?

Triazine Resistance: We attempted to answer the previous four questions using data and examples derived from the study of the best documented case of herbicide resistance, triazine resistance. Two kinds of mechanisms may be responsible for this triazine resistance; first is the presence of detoxification metabolic pathways, as seen in corn (11). This also may occur in weed populations, especially Panicoideae, but a low heritability makes its study complex. The second mechanism of triazine resistance is the loss of herbicide binding at the level of the chloroplast.

This second mechanism is due to a point mutation at the chloroplast gene encoding for a Photosystem II 32 KD membrane protein (3). This point mutation is the result of a change of one amino-acid residue and the subsequent loss of the protein-herbicide affinity. Resistant chloroplasts are 1,000-fold more resistant than susceptible chloroplasts, when assayed in vitro. Whole plants also show at least a 200-fold increased resistance, which confers a clear, selective advantage in triazine treated fields, orchards and vineyards (13). Evidence that a specific gene is responsible for resistance was proven by the production of tolerant transgenic tobacco (12). Now, nearly 50 species are known to have at least one triazine resistant population and several million hectares in more than 15 countries are infected with triazine resistant weed populations (1).

SPONTANEOUS MUTANTS

It is intriguing that all triazine resistant biotypes in which chloroplast DNA has been analyzed are mutated at the same position on the psbA gene on the chloroplast DNA (Table I). This mutation corresponds to the change of one amino-acid at Position 264 of the psbA gene product, the 32 KD protein. The fact that the mutation occurs on the psbA gene is not surprising because atrazine is a competitor of quinones binding to the 32 KD protein. However, the reason for the exclusive involvement of Position 264 in this resistance in higher plants is not known.

Other mutations at the psbA gene are known to confer atrazine resistance in some organisms e.g., in the algae Chlamydomonas (14, 15). This indicates that a diversity of mutants could have been expected. From the data now available, it appears that 14 mutants are altered at Position 264 out of 21 DNA analyses of triazine resistant organisms (Table I). Therefore, mutants at other positions may have theoretical frequencies between 15% and 59% (probability of 95%). To explain the higher frequency of mutants at Position 264, in spite of the lower fitness associated with it (16), we investigated whether some genetic mechanism increased the chance of this mutation. Such a mechanism, was illustrated in Chenopodium album and may provide some light on the appearance of chloroplast mutants.

Precursor Plants: Rare events in populations generally cannot be detected unless 1,000 or more individuals are analyzed from each location. However, we proceeded in a slightly different manner. Seedlings of C. album were collected in areas that have never been treated with chemicals. When transplanted in a greenhouse, one leaf per plant was cut off and left to absorb atrazine in the dark and then the amounts of chlorophyll fluorescence were recorded. The results of this test indicated that all plants were susceptible to atrazine (Figure 1). The plants set seeds and the seeds from one plant formed a family. Then at least 100 seeds of each family were sown and analyzed for fluorescence.

Of the 869 families derived from plants collected in five populations (Table II), 33 showed at least one seedling with an intermediary fluorescence curve (Figure 1 (21)). These seedlings proved later to have the mutated psbA gene at Position 264 (19) and were moderately resistant to atrazine. They were called Type I (for intermediate level of resistance). The 33 mother plants and their corresponding seed families were called Sp. because they were special susceptible plants that produced mutant plants.

TABLE I: MUTATION FOR ATRAZINE RESISTANCE

Species	Resistance Level	Mutation	Reference
Higher Plants			
Amaranthus bouchonii	>500	psbA 264	17
A. cruentus	>500	psbA 264	17
A. hybridus	>500	psbA 264	17
A. retroflexus	>500	psbA 264	3
Brassica napus	>500	psbA 264	18
Bromus tectorum	>500	psbA 264	17
Chenopodium album	>500	psbA 264	19
Phalaris paradoxa	>500	psbA 264	20
Poa annua	>500	psbA 264	21
Senecio vulgaris	>500	psbA 264	17
Solanum nigrum	>500	psbA 264	22, 23
Others			
Chlamydomonas reinhardtii	84	psbA 264	24
C. reinhardtii	15	psbA 256	25
C. reinhardtii	15	psbA 255	25
C. reinhardtii	25	psbA 251	15
C. reinhardtii	2	psbA 219	25
Synechococcus sp.	10	psbA 264	26
S. sp.	2	psbA 219	27
S. sp.	7	psbA 211	27
Synechocystis	70	psbA 264	*
Synechocystis	100	psbA 211-251	*

*G. Ajlani; C. Astier, personal communication.

TABLE II: OCCURRENCE OF MUTANT PRECURSOR PLANTS IN POPULATIONS

Population	Origin	No. of Families	No. of Sp. Families	Mean % of 1 Mutants within Sp. Families
Burgundy B	Garden	180	8	5.3
Burgundy C	Garden	191	17	3.2
Burgundy D	Garden	166	1	1.1
Alpes A1	Garden	187	5	1.6
Alpes A2	Garden	145	2	1.7

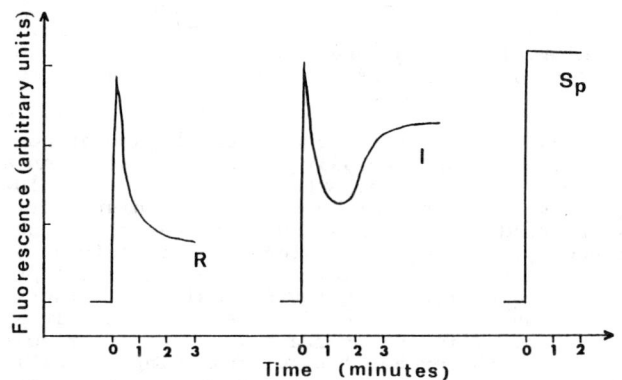

Figure 1. Fluorescence curve of whole leaves of the three phenotypes of *Chenopodium album* after a one night incubation with 30 ppm atrazine

A Mutator System: The Sp. families probably display a peculiar genome because they show an average Type I mutation frequency as high as 3.3%. This is much greater than the mutation rate expected due to chance alone. Arntzen and Duesing (29) previously suggested that a mutation for atrazine resistance could have resulted from a specific mutator system. In their system, a high frequency of variegated plants were obtained. The suggested mutator system could probably be responsible for other types of mutations at the chloroplast level, but appropriate screenings (e.g., the wide use of atrazine in fields) have not been used. If true, this mutator system could be the starting point for numerous resistances to different herbicides. The evolution of cross-resistances could therefore be expected to occur within a short time.

In addition, we observed that Type I mutants have all of their psbA gene copies mutated (99.9% probability) while no heteroplasmicity was found in Sp. plants (19). This indicates that a mechanism other than a high mutation rate must work to allow the release of individual plants that have their entire chloroplast population mutated. This could be due to the control of chloroplast DNA replication directed by mitochondrial or nuclear genomes. Indeed, a sequence homology has been found between mitochondrial DNA and part of the chloroplast psbA gene. Moreover, this homologous sequence is expressed as an RNA transcript in atrazine resistant C. album only and not in the susceptible plants (30).

Randomness: Predictive models focusing on the appearance and spread of resistant weeds generally assume that mutation events fit a uniform statistical law, i.e., that is they occur in low frequencies at every location. At best, a normal distribution or normal laws are used. In contrast, weed research on populations teaches that weed distributions instead fit contagious (aggregative) laws (31). In addition, the presence of special precursor plants that produce a high frequency of resistant mutants leads us to imagine a patchy distribution of resistance genes in spite of a random repartition. This is important in the case of polygenic resistances because the probability of combining two or more genes in one plant by outcrossing may widely vary from location to location because the frequency of single mutants may be very different.

POPULATION STRUCTURE

Monomorphism: An additional piece of evidence in favor of a peculiar genome for Sp. plants is provided by isozyme analyses of two populations from the Burgundy region of France. The Sp. plants showed the same electrophoretic pattern while 36 other patterns were detected in the plants that do not contain the Sp. genome in these two populations (32). Therefore, the potential to produce mutants seems to be restricted to only one type of plant within a given population. Of course, this could be simply coincidental and due to a founder effect. In the populations from the other regions, the Sp. plants showed other electrophoretic patterns, indicating that there is no relationship between mutation frequency and isozymes production.

The fact that Type I mutants originate from only one type of plant within a population is corroborated by the monomorphic structure of all resistant populations of C. album studied in France and Canada (33, 34). This is certainly the result of a founder effect by one plant only. Thereafter, a quick buildup of a population occurred due to the high fitness value of the resistant plant in a herbicide treated area. Monomorphism has been maintained thanks to low allogamy (5% in corn fields (35) and the isolation of pollen outside the field. In contrast, if the location is suitable for a high seed migration rate and a high allogamy, a high polymorphism level can be expected, as found in a resistant population of Alopecurus myosuroides near a roadside in Israel (Chauvel, INRA Dijon, Personal communication).

Polymorphism: We previously studied polymorphic resistant populations of Poa annua. In this case, allogamy was low (10% maximum in the greenhouse), but a peculiar feature greatly enhanced the hybrid output between the resistant plants and the adjacent susceptible ones. The selection pressure due to habitat ecology was hidden for a time due to the drastic selection pressure due to the herbicide. The resistant population developed in the central strip of a city avenue, an area open to seed migration and previously colonized by the annual erect ecotype. By chance, the former resistant plant in this area was a perennial prostrate ecotype. Because leaching of atrazine through the soil was rapid, immigrant susceptible plants (mainly annual erect types) grew and flowered among resistant plants 5 or 6 months after a herbicide treatment. Isozyme analysis indicated that many hybrids developed (36). The population of resistant Poa annua colonizing this area gradually evolved as an annual erect. The drastic herbicide selection pressure was hidden for a time due to the habitat ecology. This disequilibrium was the source of the polymorphism.

Origin or Consequence: In light of these studies, it appears that the genetic structure of a resistant population is not the image of the ancestral structure of the population from which resistant plants originated. Therefore, it is difficult to draw a conclusion about a peculiar genetic structure promoting the appearance of atrazine resistant plants.

REGULATION OF RESISTANCE

The fact that populations or plants display resistance genes is not necessarily synonymous with the expression of resistance at the whole plant level. This is clear for polygenic resistances for which the accumulation of two or more genes is necessary for a sufficient level of resistance. This situation also occurs for atrazine resistance.

The Search for Resistant Plants: The search for Sp. and Type I mutant plants in populations from cultivated fields failed. These plants, as well as many other types of mutants may have been eliminated because cultivation and other cultural practices have reduced the polymorphism of field populations (32). Moreover, Type I mutants probably have a low fitness value in the absence of triazines (16). The fact that mutants originate from only one or very few types of individuals in a population represents a bottleneck to the spread of resistance. Hence, resistance certainly evolved in fewer locations than could have been expected. For instance, no resistant C. album plants were found in Burgundy while Sp. mother plants having Type I descendants were found in gardens in Burgundy.

A second bottleneck to the spread of the resistance was found in C. album populations. The Type I mutants were resistant to 0.5 kg a.i./ha only and 100% mortality was reached with 1 kg a.i./ha (19), while the actual dose used in corn fields was at least 1.5 kg a.i./ha (the lethal dose for susceptible plants is 0.1 kg a.i./ha).

Apart from the chance of having a heterogeneous herbicide treatment leaving micro-niches with less atrazine applied in some area, and unless germination occurred late in the season after partial leaching or degradation of the atrazine, Type I mutants are likely to be killed. This incomplete resistance of a psbA-mutated plant brings up the question of how the true resistant plants (those tolerating more than 40 kg a.i./ha) evolved.

Lamarkism for Resistance: When growing Type I mutants of C. album free of any chemical, their characteristics (lethal dose and fluorescence curve) are maternally inherited (28). However, when pesticides, including herbicides at sublethal doses, are sprayed on Type I seedlings, the seeds obtained from these plants give individuals showing lethal doses of 40 kg a.i./ha and fluorescence curves characteristic of the very resistant plants found in infested corn fields (19). Thus, a change of phenotype occurs within one generation after a chemical stimulus. This seems to be one of the very few cases of environmentally induced genetic changes cited in the literature (37, 38).

Up to now, we have not found relevant differences of growth, reproduction, psbA sequence, photochemical characteristics, in vitro chloroplast response to atrazine and membrane lipids (19, 39) between Type I mutants and induced resistant mutants (Ri). Crosses using S, I, Ri, and treated Type I mutants (Ii) were studied to determine the nature and the inheritance of the induction. Preliminary results showed the presence of I and R phenotypes at the F_2 generation of I x R crosses. Some exceptional crosses with paternal inheritance were studied in detail and showed peculiar features in subsequent F_1 to F_5 generations (40).

Therefore, even with an apparently simple maternal inheritance, the gene for atrazine resistance may be expressed in different ways according to the genetic background.

SPREAD OF RESISTANCES FROM CROPS

With the release of crops, in which foreign genes for antibiotic or herbicide resistances (e.g., bacterial genes) have been introduced using recombinant DNA techniques, the risk of dispersal of foreign genes become actual. The possible spread of such genes through insertion in viruses, bacterial, and nematodes that live in close contact with transgenic plants should be studied. However, outcrossing with related wild species should be of greater concern. Outcrossing was known for a long time under the terminology of introgression.

Introgressions: Introgression or outcrossing between cultivated and wild plants often results in the release of more aggressive weeds. Examples of this feature can be found in Raphanus (41), Secale (42), Sorghum (43), and Setaria (44).

A good example of introgression related to weed control practices was the use of red pigmented rice in India because it made hand weeding of wild plants easier. Ten years later, wild plants with a purple coloration in their leaves were found in a weed population (45), thus indicating hybridizations between cultivated and wild rice. The purple characteristic would most likely spread in weed populations because it escaped hand weeding.

Similarly, one can ask whether the diclofop-methyl resistant wild oat populations in Australia are the result of crosses between wild plants and some tolerant oat cultivar. Indeed, an oat cultivar that has been shown to be tolerant to diclofop has been grown for 85 years. Because this trait appeared to be simply inherited (2 nuclear loci (46), the transfer of tolerant genes into the wild oat could have occurred long before any herbicide spray.

The risk of passing resistance genes from a crop to a wild related species is not inconceivable. A number of crops, (e.g., rice, millets, sorghum, oats, rapeseed, sugar beets, sunflower, alfalfa, peas, and potatoes) could be involved in introgression. Therefore, as herbicide resistant crops are engineered, a genetic barrier between crops and weeds must be devised.

Limitation of the Spread: The spread of genes from transgenic plants is generally interpreted as resulting from pollen dispersal of the crop toward wild plants. Therefore, it could be that maternal inheritance of a resistance trait, as for atrazine resistance, may greatly delay the time needed to release resistant weeds. The gene flux between an atrazine resistant foxtail millet (Setaria italica) and its wild relative, the green foxtail (S. viridis) is now being investigated.

Triazine resistance found in one plant of the wild green foxtail was transferred to the cultivated foxtail millet to improve weed control in this crop (47). Growing foxtail millet with atrazine resistant cytoplasm in areas infested with green foxtail may result in the fertilization of some florets of cultivated spikes by pollen from wild plants. This is likely to occur close to the border of a field, where wild plants grow, as well as within a field where some plants survived herbicide applications. Spontaneous hybridizations lead to less than 0.01% hybrids of the crop (unpublished data), and nearly 0.2% of the wild plants (45). Because harvest machines leave a small proportion of grain on the soil and foxtail millet yield is more than 10^9 grains/ha, it is highly probable that resistant hybrid seeds will be released in the soil.

The behavior of hybrid plants and their descendents will depend on seed viability, predation and germintion, plant survival from herbicide or crop competition, fertility, and gene exchanges within the weed population. Preliminary estimates lead us to think that the risk of having resistant weeds is not acceptable. Therefore, it is necessary to start a breeding program to produce a tetraploid resistant foxtail millet that reduces the chance of having viable hybrids between the crop and the weed.

CONCLUSION

Several aspects of the appearance of resistance genes and plants in weed populations were illustrated using data from triazine resistance studies. The following applies to weeds resistant to triazines; some peculiar genomes appeared to be able to produce a high number of mutants and, therefore, mutants do not appear at random in fields; this results in a founder effect in each resistant population; the presence of the mutated chloroplast gene is not sufficient to bring about highly resistant plants; spread of the resistance genes from engineered crop plants to wild plants is likely to occur in spite of a cytoplasmic inheritance.

These characteristics are peculiar to triazine resistance, but this example may be a good study model. This study revealed bottlenecks that have reduced the spread of triazine resistant plants. It will be important to discover and analyze similar phenomena in future cases of herbicide resistance (48). This knowledge will be helpful in elaborating mathematical models and determining weed control strategies to prevent herbicide resistant plants from appearing and spreading.

However, each new resistance that develops will produce a different genetic situation. The solutions found by weeds to escape herbicide selection pressures may be varied. Due to the high selective value conferred by the resistance genes in herbicide treated areas, mutation events at very low frequencies have high probability to lead to the appearance of resistant plants. In addition, weeds will certainly display after a short delay the bacterial genes transferred to crops for herbicide resistance, and some wild plants could be expected to become new weeds because of resistant genes.

LITERATURE CITED

1. LeBaron, H. M.; McFarland, J., This Volume.
2. Blackman, G. E., J. Roy. Soc. Arts, 1950, 499-517.
3. Hirshberg, J.; McIntosh, L., Science, 1983, 22, 1346-1349.
4. Shaaltiel, Y.; Chua, N. H.; Gepstein, S.; Gressel, J., Theor. Appl. Genet., 1988, 75, 850-856.
5. Vaughn, K. C.; Vaughan, M. A., This Volume.
6. Powles, S. B., Proc. 8th Aust. Weed Conf., 1987.
7. Kemp, M. S.; Caseley, J. C., Proc. Brit. Crop Prot. Conf. Weeds, 1987, 895-899.
8. Lutman, P. J. W.; Snow, H. S., Proc. Brit. Crop Prot. Conf. Weeds, 1987, 901-908.
9. Jana, S.; Naylor, J. M., Can. J. Bot., 1982, 60, 1611-1617.
10. LeBaron, H. M.; Gressel, J., Herbicide Resistance in Plants, Wiley: New York, 1982, Appendix.
11. Schimabukuro, R. H., Plant Physiol., 1968, 43, 1925-1930.
12. Cheung, A. Y.; Bogorad, L.; Van Montagu, M.; Schell, J., Proc. Natl. Acad. Sci., 1988, 85, 391-395.
13. Ryan, G. F., Weed Sci., 1970, 18, 614-616.
14. Erickson, J. M.; Rahire, M.; Rochaix, J. D.; Mets, L., Science, 1985, 228, 204-207.
15. Johanningmeier, U.; Bodner, U.; Wildner, G. F., Febs Letters, 1987, 211, 221-224.
16. Holt, J. S. Proc., This Volume.
17. McNally, S.; Bettini, P.; Sevignac, M.; Darmency, H.; Gasquez, J.; Dron, M., Plant Physiol., 1987, 83, 248-250.
18. Reith, M.; Straus, N. A., Theor. Appl. Genet., 1987, 73, 357-363.
19. Bettini, P.; McNally, S.; Sevignac; Darmency, H.; Gasquez, J.; Dron, M., Plant Physiol., 1987, 84, 1442-1446.
20. Schönfeld, M.; Yaacoby, T.; Ben-Yehuda, A.; Rubin, B.; Hirschberg, J., Z. Naturforsch, 1987, 42c, 779-782.
21. Barros, M. D. C.; Dyer, T. A., Theor. Appl. Genet., 1988, 75, 610-616.
22. Hirschberg, J.; Bleecker, A.; Kyle, D. J.; McIntosh, L.; Arntzen, C. J., Z. Naturforsch, 1984, 39c, 412-420.
23. Goloubinoff, P.; Edelman, M.; Hallick, R. B., Nucleic Acid Res., 1984, 12, 9489-9496.
24. Erickson, J. M.; Rahire, M.; Bennoun, P.; Delepelaire, P.; Diner, B.; Rochaix, J. D., Proc. Natl. Acad. Sci., 1984, 81, 3617-3621.

25. Rochais, J. D.; Erickson, J. M., Trends Biochem. Sci., 1988, 13, 56-59.
26. Golden, S. S.; Haselkorn, R., Science, 1985, 229, 1104-1107.
27. Gingrich, J. C.; Buzby, J. S.; Stirewalt, V. L.; Bryant, D. A., Photosyn. Res., 1988, 16, 83-99.
28. Gasquez, J.; Al Mouemar, A.; Darmency, H., Pestic. Sci., 1985, 16, 392-396.
29. Arntzen, C. J.; Duesing, J. H., In Advances in Gene Technology Molecular Genetics of Plants and Animals; Ahmed, F.; Downey, K.; Schulz, J.; Voellmy, R. W., Eds.; Academic Press: New York, 1983, 273-294.
30. Bettini, P.; McNally, S.; Sevignac, M.; Dron, M., Theor. Appl. Genet., 1988, 75, 291-297.
31. Chauvel, B.; Gasquex, J.; Darmency, H., Weed Res., 1989, 29, 213-219.
32. Al Mouemar, A.; Gasquez, J., Weed Res., 1983, 23, 141-149.
33. Gasquez, J.; Compoint, J. P., Agroecosystem, 1981, 7, 1-10.
34. Warwick, S. I.; Marriage, P. P., Can. J. Bot., 1982, 60, 483-493.
35. Gasquez, J., In Genetic Differentiation and Dispersal in Plants; Jacquard, P.; Heim, G.; Antonovics, J., Eds.; NATO ASI Series, Springer-Verlag: Berlin, 1984, 57-66.
36. Darmency, H.; Gasquez, J., New Phytol., 1983, 95, 299-304.
37. Durrant, A., Heredity, 1962, 17, 27-61.
38. Al Saheal, Y. A.; Larik, A. S., Genome, 1987, 29, 643-646.
39. Tremolieres, A.; Darmency, H.; Gasquez, J.; Dron, M.; Connan, A., Plant Physiol., 1988, 86, 967-970.
40. Gasquez, J.; Al Mouemar, A.; Darmency, H., VII° Col. Int. Ecol. Biol. Syst. Mauvaises Herbes, 1984, 281-286.
41. Panestos, C. A.; Baker, H. G., Genetica, 1967, 38, 243-274.
42. Suneson, C. A.; Rachie, K. O.; Khush, G. S., Crop Sci., 1969, 9, 121-124.
43. Baker, H. G., In The Genetics of Colonizing Species; Baker, H. G.; Stebbins, G. L., Eds; Academic Press: New York, 1965, 147-172.
44. Darmency, H.; Zangre, G. R.; Pernes, J., Genetica, 1987, 75, 103-107.
45. Oka, H. I.; Chang, W. T., Phyton., 1959, 13, 105-117.
46. Warkentin, T. D.; Marshall, G.; Mckenzie, R. I. H.; Morrison, I. N., Weed Res., 1988, 28, 27-35.
47. Darmency, H.; Pernes, J., Weed Res., 1985, 25, 174-179.
48. Gressel, J.; Segel, L. A., This Volume.

RECEIVED December 20, 1989

Chapter 25

Structural and Biochemical Characterization of Dinitroaniline-Resistant *Eleusine*

Kevin C. Vaughn and Martin A. Vaughan

Southern Weed Science Laboratory, Agricultural Research Service, U.S. Department of Agriculture, Stoneville, MS 38776

Dinitroaniline-resistant goosegrass (Eleusine indica) biotypes were first discovered in South Carolina and have now been found in several southern states where trifluralin is used for weed control in cotton. Two kinds of resistant biotypes have been noted: one is highly-resistant (R) and is unaffected even by saturated solutions of dinitroaniline herbicide, whereas an intermediate-resistant (I) biotype is only 50X resistant to trifluralin and less than 10X resistant to oryzalin compared with the susceptible (S) biotype. Both R and I biotypes are cross-resistant to phosphoric amide herbicides. Tubulin from the R is able to polymerize into microtubules even in the presence of oryzalin, whereas that of the S biotypes cannot. Western blots of tubulin from the R biotype reveal two β-tubulin isotypes whereas only one form is noted in the S biotype. Because the R biotype is hypersensitive to the microtubule-stabilizing agent taxol, it is likely that the R biotype is resistant by having hyperstabilized microtubules. The I biotype has no gross alteration in tubulin nor extreme sensitivity to taxol, indicating that this biotype has a different resistance mechanism than the R.

The dinitroaniline herbicides, trifluralin and pendimethalin, have been utilized in greater than 80% of the cotton acreage in the Southern United States because of their very effective weed control in this crop (1). Many of these fields are essentially in cotton monoculture and hence the continued use of these herbicides has constantly selected out those weeds most tolerant of these herbicides. Under such a selection pressure, the appearance of weed biotypes resistant to dinitroaniline herbicides is expected (2). The first report of a resistant biotype did not appear until 1984, Mudge et al. (3) described the occurrence of dinitroaniline-resistance in Eleusine indica in counties in South Carolina where cotton is extensively cultivated. Since that initial report, dinitroaniline-resistant Eleusine has been detected throughout the midsouth (H. LeBaron, personal communication).

The occurrence of dinitroaniline-resistance in Eleusine is quite alarming, because it is one of the world's ten worst weeds (4) and dinitroaniline herbicides are a cost-effective way of controlling this weed in cotton. Even more alarming is

This chapter not subject to U.S. copyright
Published 1990 American Chemical Society

the apparent fitness of this biotype: Murphy et al. (5) found virtually no significant difference between the two biotypes when grown under non-competitive conditions. This makes the potential of this resistance problem much greater than triazine resistance, in which the resistant mutant is much less competitively fit in the absence of herbicide.

Mechanisms of Tolerance/Resistance to Dinitroaniline Herbicides

Crop plants in which dinitroaniline herbicides are used are tolerant of the herbicide by a number of mechanisms. Some researchers have turned their attention to these in the hope of finding a mechanism for dinitroaniline resistance in Eleusine.

Lipids. Dinitroaniline herbicides are effective on small-seeded, lipid-poor species. Hilton and Christiansen (7) examined the level of seed lipid and the susceptibility of a plant to trifluralin and found a good correlation between the two. These authors concluded that the herbicides would be compartmentalized into the lipid bodies of the seed and away from the growing tip of the plant. Upadhyaya and Nooden (8) even found that there is a differential between susceptible and tolerant species in the uptake of oryzalin into the membrane system, indicating that more than the seed lipids may be involved in determining dinitroaniline sensitivity. Chernicky (9) investigated the possibilities that alterations in the amount of lipid is involved in the resistance of Eleusine to dinitroaniline herbicides. Both susceptible (S) and resistant (R) biotypes had less total lipid than tolerant crop species and even most sensitive weed species. The S biotype had actually 36% more total lipid in the roots than the R biotype (a result opposite to what one would expect if higher lipid content correlates dinitroaniline resistance).

Translocation and Metabolism. All of the dinitroaniline herbicides are poorly translocated from the roots to the shoots in both susceptible and tolerant species and little or no metabolism of the herbicide occurs in the plant (10). For example, carrot, a species highly tolerant of dinitroaniline herbicides, 89% of the C^{14}-trifluralin applied to the soil was still in the form of the parent compound after one month. All of the metabolites were found in the soil as well as the plant, indicating that all of the metabolites were either chemically or microbially produced. Chernicky (9) found that there were some slight differences between the R and S biotypes in translocation of C^{14}-trifluralin to shoots of root-fed Eleusine seedlings, whereas C^{14}-oryzalin was equally translocated in both R and S biotypes. Because the R biotype is resistant to both herbicides, it is likely that the differences in trifluralin translocation have relatively little to do with dinitroaniline resistance. However, these data might suggest that a small tolerance to trifluralin may have allowed selection of progenitors of the R biotype as tolerant survivors from initial herbicide treatment. A similar case scenario was reported by Gressel et al. (11) in a triazine-resistant Brachystylon: the biotype had developed resistance both by metabolism of the herbicide and by an alteration of the site of action (the 32 kD protein). Because field selection of herbicide-resistance generally is noted after many years of use, the probability of a biotype acquiring several tolerance/resistance mechanisms is high. To our knowledge, no one has investigated herbicide metabolism in the dinitroaniline-resistant Eleusine and due to the low metabolism of these herbicides in general, even in tolerant plants, it is unlikely that much metabolism is occurring. Powles et al. (12) have found that a diclofop-methyl resistant biotype of Lolium rigidum shows an amazing range of cross-tolerance to sulfonylurea, other propionic acid herbicides, and trifluralin. From studies with cytochrome P450 inhibitors, these workers concluded that the mechanism of this cross resistance is

due to metabolism of these herbicides by the plant mono-oxygenase system. Because the dinitroaniline resistant Eleusine does not have a similar pattern of cross-resistances to the Lolium [see the study of Mudge et al. (3) for example], it is doubtful that, if metabolism is involved, it is the same type as the Lolium.
Calcium. A number of mitotic disrupter herbicides cause an efflux of calcium from the mitochondria into the cytoplasm (13). Calcium is known to destabilize microtubules and is thought to be involved in the catostrophic loss of microtubules during mitosis so that existing arrays are destroyed and new conformations of microtubules may be made. Although these herbicides may disrupt the calcium efflux into mitochrondria, the calcium effects occur at 100-1000 X more herbicide than is required to cause mitotic disruption (14), indicating that this may be a secondary consequence of the herbicide. Calcium localization may be easily made using the electron microscope, using pyroantimonate to precipitate calcium into an electron opaque precipitate (15). When the pyroantimonate procedure (15) was utilized to detect calcium distributions in control and trifluralin-treated Eleusine, loss of calcium from the mitochondria was noted in both R and S biotypes at 10 M (Fig. 1) but not at lower trifluralin concentrations (not shown). Because both biotypes were affected and then only at this high concentration, we conclude that the efflux of calcium from the mitochondria is not related to herbicide resistance. Moreover, this efflux does not even seriously disrupt mitosis in the R biotype, as no mitotic disruption is noted at this level of herbicide (16).
Microtubules. Tubulin, the major protein constituent of the microtubule, is the primary site of action of dinitroaniline herbicides. Strachan and Hess (17) and Morejohn et al. (18) have isolated tubulin from Chlamydomonas and higher plants respectively and have established a binding of herbicide to the tubulin dimer and an inhibition of tubulin polymerization into microtubules in vitro. These data are the strongest evidence that tubulin is the principle site of action of the dinitroaniline herbicides.

Microtubules have a number of functions in the higher plant cell, the most important of these being the movement of chromosomes during cell division, the formation of the cell plate, and the determination of cell shape (19). As a consequence of dinitroaniline herbicide treatment, mitosis is arrested at prometaphase, cells become isodiametric, and cell plates are either misshapened or non-existant (13). Microtubule loss after treatment also has a dramatic effect on the gross morphology of the seedling root (13, 15). The loss of cortical microtubules in the zone of elongation makes these cells isodiametric rather than elongate, whereas the inhibition of cytokinesis makes these cells larger. As a consequence, a characteristic club-shaped root is formed, similar to the classic disrupter colchicine. Each of the tubulin genes is found in multiple copies in most organisms (e.g. 19), which makes the probability of all (or most) of the tubulin genes mutating to a new resistant form rather low. Because naturally-selected triazine resistant weeds and lab-selected animal cell line resistant to microtubule disrupters are unfit, one would expect mutants resistant to herbicides that act similarly also to be selectively unfit. In the case of the Eleusine, however, the resistant biotype is comparatively as fit in a number of growth parameters as the susceptible biotype (5). Despite this, our research shows that there apparently is an alteration in β-tubulin present in the R biotype that may account for dinitroaniline resistance in Eleusine.

Structural and Cross-Resistance Studies
In an effort to determine the magnitude of the resistance to dinitroaniline herbicides, we (15) utilized root tip squashes to determine changes in the mitotic indices of the R and S biotypes. Dinitroaniline herbicides disrupt mitosis at prometaphase, due to a loss of microtubules. Chromosomes fail to arrange at the

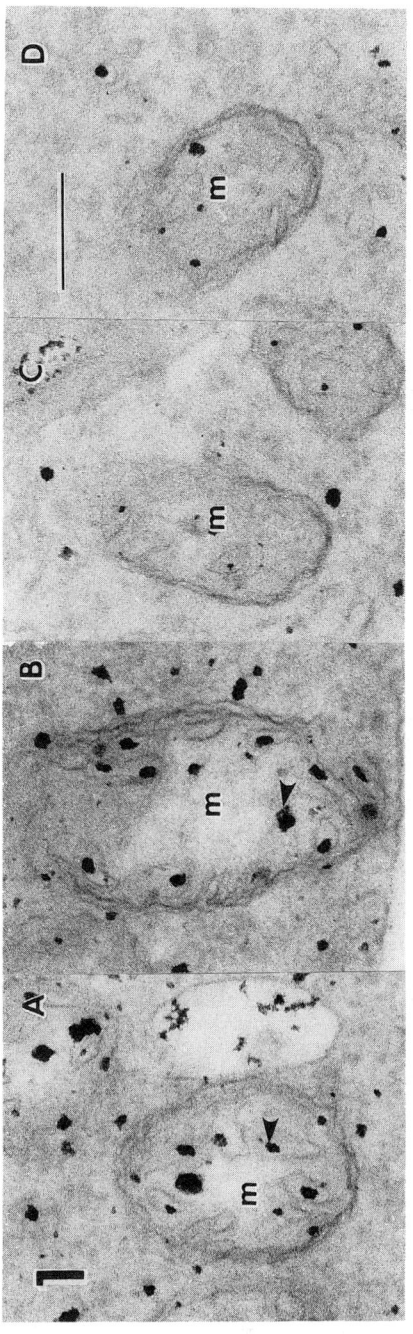

Figure 1. Electron micrographs of pyroantimonate precipitation of calcium (arrow) in Eleusine root tip mitochondria in the presence and absence of trifluralin. A. S biotype. B. R biotype. C. S biotype grown 24 h in 10^{-5} M trifluralin. D. R biotype grown 24 h in 10^{-5} M trifluralin. Bar = 0.5 μm. m = mitochrondrion.

metaphase plate and do not undergo anaphase movement. As a consequence, cells spend a longer period in mitosis so that the chances of observing a given cell in mitosis increases. When the R biotype was treated with up to 10^{-5} M trifluralin for 24 h, no effect on the mitotic indices of the treated seedlings was noted, whereas there was a significant increase in the mitotic indices of the S biotype even at 10^{-8} M trifluralin (15). No disruption of mitosis was noted in electron microscopic observations of the R biotype, although the S biotype appeared to be greatly affected even at 10^{-8} M trifluralin. These differences in sensitivity as measured by mitotic indices indicate a 1000-10,000-fold increase in resistance to these herbicides.

Root tips squashes were also used to examine the cross-resistance of the R biotype to other mitotic disrupters. Mudge et al. (3) had previously shown that field applications of all of the dinitroaniline herbicides were ineffective in controlling the R biotype, and this was confirmed using root tip squashes as well (20). The phosphoric amide herbicide amiprophosmethyl also inhibits polymerization of tubulin into microtubules (21) and causes the same kinds of mitotic disruption as the dinitroaniline herbicides. The R biotype is also cross-resistant to this herbicide (20). Because the structure of amiprophosmethyl and trifluralin are quite different, it is likely that mechanisms of resistance based upon translocation and/or metabolism of herbicides is unlikely (but see 16 for an exception).

The R biotype is sensitive to many other mitotic disrupters including the classic disrupters, colchicine, vinbastine, and podophyllotoxin (20). Actually, the R biotype appears more sensitive (~10-fold) to compounds that cause multipolar mitosis (propham, chloropropham) or disturb cell plate formation (caffeine). This hyper-sensitivity may be due to the already misarranged cell plate formation in the R biotype that is more easily exacerbated than those of the S biotype. As noted below, there may be another reason for this difference that is related to the stability of the microtubules in the R biotype.

Because dinitroaniline herbicides are one of the most effective methods of weed control in cotton, the development of resistance to dinitroaniline herbicides in Eleusine was especially serious. Moreover, many of the herbicides that are effective on the R Eleusine [see Mudge et al. (3)] are much more costly or more detrimental to the growth of the cotton than the dinitroaniline herbicides. In an attempt to find a solution to this problem, Figliola et al. (22) isolated two pathogenic fungi,
Bipolaris setariae Saw. and Piricularia grisea (Cke.) Sacc. from infected Eleusine seedlings and found that those pathogens could be used as biocontrol agents of both the S and R biotypes of Eleusine. Neither of these pathogens infected any of the major crops of the Carolinas (soybean, tobacco, peanut, and cotton). So, when released, biocontrol agents should be an important part of the weed control program.

Comparisons of Tubulin and Microtubules in R and S Biotypes

With the elimination of many of the other possible mechanisms for resistance, we began an investigation of the site-of-action, tubulin, for differences that might be responsible for dinitroaniline resistance.

Extracts of the R and S biotypes of Eleusine were fractionated by stepwise increases in polyethylene glycol concentration and the various fractions were monitored for the presence of tubulin and other proteins by electrophoresis and Western blotting. By making small increases in PEG concentration, one fraction contained virtually all the recognizable tubulin and was > 85% pure as determined by Coomassie blue staining of denaturing gels (23, 24). This is comparable in purity to the protocols of Morejohn et al. (25), but is a much faster method and allows concentration of tubulin even from a relatively poor source like the 1-month-old Eleusine plants.

Tubulin may be polymerized into microtubules in vitro by incubation of a relatively concentrated protein extract in the presence of GTP and high concentrations of dimethylsulfoxide (25). When extracts of the R and S biotypes were subjected to this protocol in the presence of the dinitroaniline herbicide oryzalin, only the R biotype formed recognizable microtubules whereas the S biotype formed only small fragments or ring-like structures that may be small aggregates of tubulin (Fig. 2). Both biotypes were able to form microtubules in the absence of oryzalin (23, 24). These data indicate that the tubulin of the R biotype is different than the tubulin of the S biotype in its sensitivity to dinitroaniline herbicides.

To further examine differences between the R and S biotypes, we ground root tips from 5-day-old seedlings under liquid nitrogen and then homogenized in Laemmli (26) solubilization buffer in the presence of the three protease inhibitors recommended by Morejohn et al. (27). These precautions were taken to assure that the tubulin was not degraded by proteases during the isolation. The extracts were centrifuged in microfuge tubes and the supernatant was applied to either Laemmli (26) or Blose (28) formulation gels. The extracts were blotted to nitrocellulose and probed with commercially available monoclonal antisera to α-tubulin and β-tubulin, as well as to a polyclonal antiserum that recognizes both tubulin forms (29, 30). On one-dimensional gels, only one α-tubulin form is found in both biotypes and only one β-tubulin form in the S biotype. In the R biotype, however, two β-tubulin bands are noted (Fig. 3). These two β-tubulin bands vary in mobility, depending on the buffer and gel system utilized, but in virtually every system two are resolved. This does not appear to be a simple geographical variant, as R biotypes obtained from North Carolina, Arkansas, and South Carolina have the β-tubulin doublet even though they are geographically widely separated. The only exception to this point is the so-called intermediate-resistant (I) biotype found in a few counties of South Carolina (see below), which has similar tubulin profiles to the S biotype. Although equal numbers of seedlings were ground and these were of nearly equal fresh weight and protein content, the extracts of the R biotype always give a more intense staining in Western blots, indicating that there may be more tubulin, as well as the new form. Southern blot analysis indicates a novel set of β-tubulin genes in the R biotype, not found in the S biotype (M. D. Marks and D. P. Weeks, unpublished). These data indicate that there is a the novel form of β-tubulin present in the R biotype and it may be responsible for the insensitivity of the tubulin of the R biotype to dinitroaniline herbicides.

Mechanism for the Resistance

Although our original thought was that the alteration in β-tubulin would prevent binding of the dinitroaniline herbicide, it appears that a more complex mechanism is responsible for the resistance. In animal cell lines resistant to microtubule disrupters such as colchicine, the cell lines have hyperstabilized microtubules (31) so that the turnover of subunits at both plus and minus ends of the microtubule does not occur as frequently and the destabilizing effects of the disrupter are also minimized. Along with this increase in hyperstability, the cell lines are more sensitive to agents that hyperstabilize microtubules, such as taxol, than the disrupter-sensitive lines. This has been explained using a model of stability and instability (31).

When the R and S biotypes were treated with taxol, the R biotype was 100 X

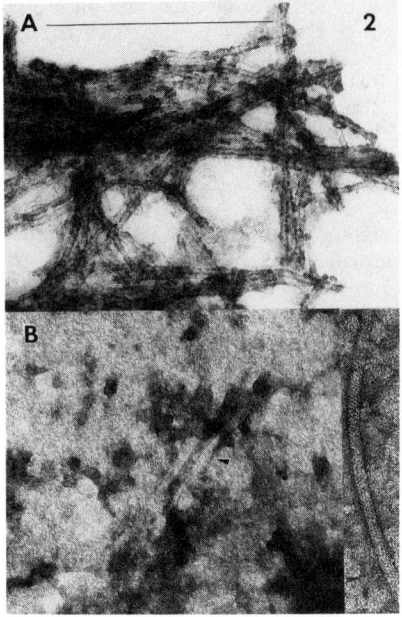

Figure 2. Electron micrographs of in vitro polymerized microtubules from R (A) and S (B) biotypes of Eleusine in the presence of 10^{-5} M oryzalin. Although good microtubules are formed in the extracts of the resistant biotype, only fragments or structures that represent limited polymers of tubulin were found in the S. Inset shows a higher magnification of an individual microtubule (polymerized without oryzalin) revealing the "beaded" structure typical of microtubules. Bar = 0.5 μm

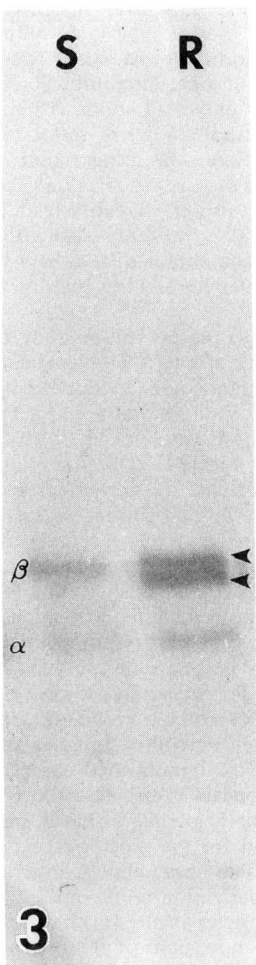

Figure 3. Example of Western blots of R and S Eleusine extracts probed with anti-sea urchin tubulin that recognizes both α- and β-tubulins. The R biotype has two β-tubulin bands, but only one α-tubulin in one dimensional gel extracts.

more sensitive to this compound than the S biotype (32). This was determined by examining the response of microtubules (by immunofluorescence) and other cellular structures (by electron microscopy) to taxol over 10^{-8} to 10^{-5} M taxol treatments. In the taxol-treated R biotype, the microtubule arrays included star-like structures, where the microtubules radiated from a central region, to multipolar spindles, and extensive cortical microtubule arrays (Fig. 4). The S biotype was affected but generally only wall abnormalities were noted, similar to that noted previously in ultrastructural studies of the R biotype grown on water (15). Because of the similarity between the ultrastructure of the taxol-treated S biotype and the untreated R biotype, it is conceivable that the taxol-treated S-biotype was also dinitroaniline resistant. Treatment of the S biotype with taxol and subsequently with trifluralin or oryzalin revealed little or no mitotic disruption. Similar results have been obtained by others who used taxol as a protectant for other mitotic disrupter herbicides, such as the phosphoric amide herbicide cremart (33).

These studies established that (a) the microtubules of the R biotype are hyperstabilized and (b) phenocopies of the R biotype in structure and in insensitivity to microtubule disrupters could be induced by treating the S biotype with taxol. From this, one could conclude that a major factor in dinitroaniline-resistance in Eleusine is the hyperstability, presumably caused by the novel β-tubulin form in the R biotype. This does not eliminate the possibility that the binding of dinitroaniline herbicides is altered in the R biotype and we hope to investigate this possibility as well.

The dinitroaniline herbicides are relatively insoluble in water (and consequently probably in the cytoplasm) so that a mechanism of resistance based upon hyperstability would probably allow for a rather complete resistance to these compounds. Other compounds, like colchicine, are much more water-soluble and, at the high concentrations of these compounds required to elicit effects in all plants (34), little difference in selectivity between the biotypes is noted (20). When small differences were noted, the R biotype was always slightly more resistant (unpublished). In the case of the carbamate herbicides, the initial effect is to depolymerize microtubules and then new mini-spindles are formed directing the chromosomes to many poles (S. Wick, unpublished observations). The hyperstability in the tubulin of the R biotype appears to promote the assembly at many sites, as the frequency of multipolar mitosis is greater in the R biotype than the S (20). A similar explanation may be offered for the greater effect of caffeine on the R biotype than the S biotype (20). The phragmoplast microtubules arrays require a rapid polymerization and depolymerization of the microtubules to orient and allow for fusion of the golgi-derived vesicles in the forming cell plate. Hyperstabilized microtubules may prevent proper movement of the microtubules and hence orientation of the cell plate. This may be the reason for the abnormal extensions of the cell wall, similar to partial cell plates, sometimes found in the R biotype grown on water (15). Caffeine affects the fusion of vesicles in the cell plate so that the combined effects of hyperstabilized microtubules and lack of vesicle fusion result in binucleate cells in the R biotype after caffeine treatment (20). Thus, the hyperstability hypothesis can explain the patterns of cross-resistance and sensitivity as well as the resistance to dinitroaniline herbicides.

Tissue cultures of both biotypes of Eleusine can be initiated by a simple combination of plant hormones. Surprisingly, when the cultures were tested for resistance, there was no difference in the susceptibility (as measured by growth) even though the regenerated plants retained the characteristics of the parental line (35). Because hormone concentrations greatly affect the stability of microtubules, it is possible that the hormones used in these cultures have altered microbubule

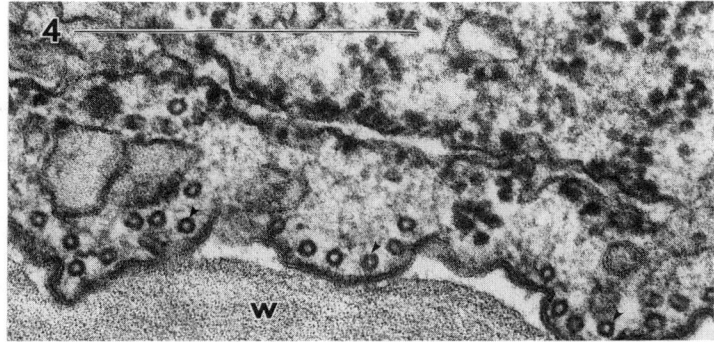

Figure 4. Electron micrograph of the effects of taxol on the R biotype of Eleusine. The walls in this figure have extensive groups of microtubules (arrows) along the wavy wall (w). Bar = 0.5 μm.

stability so the herbicides have no additional effect, i.e. 2,4-D reduces the stability of the microtubules. We are currently investigating the tissue cultures using electron microscopy to determine what the effect of tissue culture has upon the microtubule display as well as what effects dinitroaniline herbicides have on the tissue culture.

Intermediate-Resistant Biotype

In the initial description of dinitroaniline-resistance in Eleusine, Mudge et al. (3) reported that biotypes from some counties could be controlled by high rates of dinitroaniline herbicides although normal field rates were relatively ineffective at control.

Recently, we have investigated this intermediate-resistant (I) biotype in more detail (36). The level of resistance to trifluralin based both upon seedling growth and changes in mitotic indices is much less than the R biotype, around 50 X. The pattern of cross-resistance is also much different from the R biotype. While the I biotype is highly resistant (1000 X) to isopropalin, it is only slightly (10X) resistant to oryzalin. There is no consistent structural or solubility characteristic of the dinitroaniline herbicide that correlates with the resistance of the I biotype (36). The I biotype does not make abnormal cell walls nor is it more sensitive to taxol than the S bioype, indicating that the microtubules of the I biotype are not hyperstabilized. Similarly, there is no obvious β-tubulin doublet in this biotype. At present, we are not sure why the I biotype is resistant to these compounds. This biotype is very similar in level of resistance and cross-resistance to the Chlamydomonas mutants of James et al. (37) that appear not to be tubulin mutants but rather map to loci involved with flagellar function, possibly microtubule-associated proteins. Thus, resistance to dinitroaniline herbicides does not necessarily involve alterations in tubulin, although this is one potential mechanism. When the I biotype was self-pollinated and the progeny were only of the I phenotype (unpublished), so these individuals are not a hybrid of S and R biotypes.

So far, the resistant biotypes obtained from other areas have all been of the highly-resistant (R) rather than I type (unpublished). Although all of the biotypes arose under the same situations (cotton monoculture and dinitroaniline herbicide as a pre-plant herbicide treatment) only the populations from two counties in South Carolina apparently have this lower level of resistance. As our case scenarios on other occurrences of dinitroaniline resistance increase, we may be able to relate this to certain sets of unique environmental conditions that may have allowed selection of one type of resistance phenotype over another.

Acknowledgments

Thanks are extended to B. J. Gossett, N. D. Camper, D. Marks and D. Weeks who supplied published and unpublished reports for preparation of this manuscript. Ms. Ruth Jones provided excellent technical assistance during these experiments. This work was supported by a USDA Competitive Grant No. 86-CRCR-1933 to K. C. Vaughn.

Literature Cited

1. Parka, S.J.; Soper, Q.F. Weed Sci. 1977, 25, 79-87.
2. Gressel, J.; Segel, L.A. J. Theor. Biol. 1978, 75, 349-371.
3. Mudge, L.C.; Gossett, B.J., Murphy, T.R. Weed Sci. 1984, 32, 591-594.

4. Holm, L.G.; Pluckett, D.L.; Pancho, J.V.; Herberger, J.P. The World's Worst Weeds Distribution and Biology. Univ. Hawaii Press, Honolulu. 1977.
5. Murphy, T.R.; Gossett, B.J., Toler, J.E. Weed Sci. 1986, 34, 704-710.
6. Gressel, J.; Ben-Sinai, G. Plant Sci. 1985, 38, 29-32.
7. Hilton, J.L.; Christiansen, M.N. Weed Sci. 1972, 20, 290-294.
8. Upadhyaya, M.K.; Nooden, L.D. Ann. Bot. 1987, 59, 483-485.
9. Chernicky, J.P. Ph.D. Thesis, University of Illinois, Urbana, 1985.
10. Probst, G.W.; Golab, T.; Herberg, R.J.; Holzer, F.T.; Parka, S.T.; Van der Schans, C.; Tepe, J.B. J. Agric. Food Chem. 1967, 15, 592-599.
11. Gressel, J.; Shimabukuro, R.H.; Drupen, M.E. Pestic. Biochem. Physiol. 1983, 19, 361-370.
12. Hertel, C.; Quader, H.; Robinson, D.G.; Marme, D. Eur. J. Cell. Biol. 1979, 20, 121-130.
13. Hess, F.D. Rev. Weed Sci. 1987, 3, 183-203.
14. Duke, S.O., Vaughn, K.C.; Wauchope, R.D. Pestic. Biochem. Physiol. 1985, 24, 384-394.
15. Vaughn, K.C. Pestic. Biochem. Physiol. 1986, 26, 66-74.
16. Powles, S.B.; Liljegren, D. Weed Sci. Soc. Am. Abst. 1988, 20, 67.
17. Strachan, S.D.; Hess, F.D. Pestic. Biochem. Physiology 1983, 20, 141-150.
18. Morejohn, L.C.; Bureau, T.E.; Mole-Bajer, J.; Bajer, A.S.; Fosket, D.E. Planta 1987, 172, 252-264.
19. Oppenheimer, D.G.; Haas, N.; Silflow, C.D.; Snustad, D.P. Gene 1987, 63, 87-102.
20. Vaughn, K.C.; Marks, M.D.; Weeks, D.P. Plant Physiol. 1987, 83, 956-964.
21. Morejohn, L.C.; Fosket, D.E. Science 1984, 224, 874-876.
22. Figliola, S.S.; Camper, N.D.; Ridings, W.H. Weed Sci. 1988, in press.
23. Vaughn, K.C.; Vaughan, M.A. Plant Physiol. 1986, 80s, 67.
24. Vaughn, K.C. Weed Sci. Soc. Am. Abst. 1986, 26, 77.
25. Morejohn, L.C.; Fosket, D.E. Nature 1982, 297, 426-428.
26. Laemmli, U.K. Nature 1970, 227, 680-685.
27. Morejohn, L.C. Bureau, T.E.; Fosket, D.E. Cell Biol. Int. Rep. 1985, 9, 849-857.
28. Blose, S.H. Cell Motil. 1981, 1, 417-431.
29. Kilmartin, T.V.; Wright, B.; Milstein, C. J. Cell Biol. 1981, 93, 576-582.
30. Vaughan, M.A.; Vaughn, K.C. Pestic. Biochem. Physiol. 1987, 28, 182-193.
31. Cabral, F.R.; Brady, R.C.; Schibler, M.J. Ann. N.Y. Acad. Sci. 1986, 466, 745-756.
32. Vaughan, M.A.; Vaughn, K.C. Plant Physiol. 1987, 83s, 107.
33. Doonan, J.H.; Cove, D.J.; Lloyd, C.W. J. Cell Sci. 1988, 89, 533-540.
34. Morejohn, L.C.; Bureau, T.E.; Tocchi, L.P.; Fosket, D.E. Planta 1987, 170, 230-241.
35. Figliola, S.S.; Camper, N.D. Bull. S. Carolina Acad. Sci. 1986, 48, 121.
36. Vaughan, M.A.; Vaughn, K.C.; Gossett, B.J. Plant Physiol. 1987, 86s, 98.
37. James, S.W.; Ranum, L.P.W.; Silflow, C.D; Lefebre, P.A. Genetics 118, 141-147.

RECEIVED September 14, 1989

Chapter 26

Herbicide Resistance in *Alopecurus myosuroides*

Malcolm S. Kemp, Stephen R. Moss, and Tudor H. Thomas[1]

Weed Research Department, AFRC Institute of Arable Crops Research, University of Bristol, Long Ashton Research Station, Long Ashton, Bristol, BS18 9AF, United Kingdom

> Resistance to chlorotoluron has been identified in black-grass (*Alopecurus myosuroides*) populations at several sites in the UK. The degradation and detoxification of chlorotoluron is more rapid in resistant than in susceptible populations. These processes are inhibited by the P450 mixed function oxidase (MFO) inhibitor 1-aminobenzotriazole (ABT) which enhances the phytotoxicity of chlorotoluron in resistant populations. The synergistic effects of other MFO inhibitors have been investigated as potential herbicide adjuvants to overcome resistance. In general, chlorotoluron-resistant stocks demonstrate cross-resistance to a wide range of herbicides of differing modes of action which suggests a common mechanism of herbicide degradation and detoxification. However, in contrast to previous findings, a chlorotoluron-resistant stock has now been identified which does not show cross-resistance to diclofop-methyl or pendimethalin, indicating that a number of oxidases may operate selectively in herbicide degradation, or additional resistance mechanisms might occur in black-grass.

In the UK *Alopecurus myosuroides* Huds. ('black-grass', 'slender foxtail') is a major annual grass weed of winter cereal crops. The phenylurea herbicides chlorotoluron and isoproturon are mainly used for its control (1) and many fields in England and Wales have received successive, annual applications of these herbicides since their introduction approximately 15 years ago. Under these

[1]Current address: Broom's Barn Experimental Station, AFRC Institute of Arable Crops Research, Highham, Bury St Edmunds, Suffolk, IP28 6NP, United Kingdom

circumstances the discovery of resistance to phenylurea herbicides in black-grass was not surprising (2).

Partial resistance to chlorotoluron was first detected in the UK at Faringdon, Oxfordshire, in 1982 and more pronounced resistance at Peldon, Essex, in 1984 (3). Chlorotoluron-resistant black-grass is now known to occur on at least five farms within 4 km of the original site at Peldon (4), but has also been detected at sites 11 km, 30 km and 140 km from this region. We believe that resistance may be present, or is developing elsewhere, but has yet to be detected. Black-grass resistant to substituted-urea herbicides has also been detected in Germany (5).

Significantly, the chlorotoluron-resistant black-grass from Peldon shows cross-resistance to other herbicides of different chemical groups and differing modes of action (Table I, and refs. 4, 6, 7). The degree of resistance varies between herbicides but is not related in any direct way to particular groups of herbicides. For example, there is cross-resistance to pendimethalin but not to trifluralin, despite both herbicides being dinitroanilines. The degree of resistance is less than that which commonly occurs with triazine resistance, but nevertheless it is sufficient to cause substantial reductions in the efficacy of herbicides at normal field rates (4,6). Similar patterns of cross-resistance have been observed in all chlorotoloron-resistant lines of black-grass examined to date, with one exception which does not show cross-resistance to diclofop-methyl or pendimethalin (Moss, unpublished results).

Comparisons of the degree of resistance of plants grown from seeds collected over several years from one site has indicated that the development of resistance has been gradual. The most likely explanation for this is through selection pressure due to continued annual applications of chlorotoluron or isoproturon. However, it is not yet known why, following such treatments, resistance should occur in some fields but not in others.

Mechanisms of herbicide resistance in black-grass

The broad-spectrum, moderate, cross-resistance of black-grass is quite unlike the almost complete immunity to a single herbicide or specific group of herbicides which can arise as a result of structural modifications to sites of action, as is observed in cases of triazine and simazine resistance (8). Rather, it indicates a deficiency of herbicide arriving at the site of action. This could be due to restricted uptake and/or inhibited translocation to the site of action, or rapid degradation and detoxification of the herbicide within the resistant plant.

Black-grass plants show considerable morphological variation in root and leaf structure. Those individuals with narrower foliage, more prostrate habit and a shallow, well-developed, fine, secondary root system might be expected to be more vulnerable to residual and translocated herbicides since these features provide a larger target area and, consequently, more efficient herbicide uptake. However, no significant morphological differences are apparent between resistant "Peldon" black-grass and a standard susceptible biotype collected from Rothamsted, Hertfordshire. Also, the resistance to pre-emergence applications of residual herbicides, e.g. pendimethalin, is difficult to explain in terms of morphological variation. Furthermore, the difference between the two populations in their tolerance to chlorotoluron and isoproturon is evident when seedlings are grown in a hydroponic system in which the roots are exposed to herbicide incorporated in the liquid medium (6,9). Under these conditions, the Peldon population withstands concentrations of chlorotoluron and isoproturon approximately ten-fold and three-fold greater, respectively, than the maximum concentration sustained by the susceptible Rothamsted population (Figure 1). Thus resistance to chlorotoluron and isoproturon is intrinsic within the tissues of the "Peldon" plants rather than being attributable to reduced contact with the herbicide caused by a different root or foliage morphology.

The uptake, translocation and degradation of chlorotoluron and isoproturon have been compared in Peldon and Rothamsted populations of black-grass using C^{14}-radiolabelled herbicides (Kemp *et al.*, in preparation). The concentration of radioactivity in the foliage of hydroponically grown plants after incorporation of labelled herbicide (~0.25 mg/l) in the liquid nutrient for 24h was similar for the two biotypes, and remained so for four days after treatment. No further measurements were made after this period. Resistance in the Peldon population, therefore, cannot be explained in terms of restricted herbicide movement into the leaf. The rates of uptake of chlorotoluron and isoproturon by the roots and subsequent acropetal movement into the foliage were similar in resistant and susceptible populations. These observations do not eliminate the possibility of resistance in the Peldon black-grass being attributable to compartmentalisation of the herbicide within the leaf tissues, thus restricting its access to the site of action, although the cross-resistance to a variety of herbicides which possess very different modes of action (e.g. inhibitors of electron transport in PSII and disruption of meristematic activity) suggests that this mechanism is improbable.

There is no evidence of any substantial loss of radioactivity from the leaf tissues via volatile or gaseous, low molecular weight metabolites, but detoxification through a less complete degradation may take place. Analysis of the composition of the radioactivity within the foliage revealed that there was relatively little herbicide degradation in the susceptible Rothamsted population. Immediately after the 24 h treatment, 75% of the chlorotoluron and 86% of the isoproturon remained unchanged, and substantial amounts were still present four days later, i.e. 40% of the chlorotoluron and 58% of the isoproturon. In the resistant Peldon population, however, chlorotoluron was degraded rapidly (Figure 2). Only 31% of the radioactivity in the foliage remained as chlorotoluron immediately after treatment, and this was reduced to 11% over the following two days. Most activity was in the form of conjugated metabolites, with free metabolites and tissue-bound residues as minor components. Isoproturon was also degraded by the resistant Peldon population, but at a slower rate than chlorotoluron; 68% of the isoproturon remained immediately after treatment, and 41% and 33%, respectively, remained one and four days later. Therefore, resistance to chlorotoluron and isoproturon in the Peldon population of black-grass may be attributable to degradation and

Table I Herbicides assessed for activity on chlorotoluron-resistant ('Peldon') black-grass

Herbicide group	Resistance demonstrated	No evidence of resistance
phenylureas	chlorotoluron, isoproturon, metoxuron, methabenzthiazuron	
sulfonylureas	chlorsulfuron	
triazines	terbutryne, simazine, cyanazine	
triaziones	SMY 1500	
aryloxyphenoxy-propanoates	diclofop-methyl	fluazifop-P-butyl quizalofop-ethyl
cyclohexeneones		sethoxydim
thiocarbamates	tralkoxydim tri-allate	
carbamates	barban	carbetamide
imidazolinones	imazamethabenz	
dinitroanilines	pendimethalin	trifluralin
amides		propyzamide
anilides	metazachlor	
organophosphates		glyphosate
bipyridyliums		paraquat
unclassified		ethofumesate

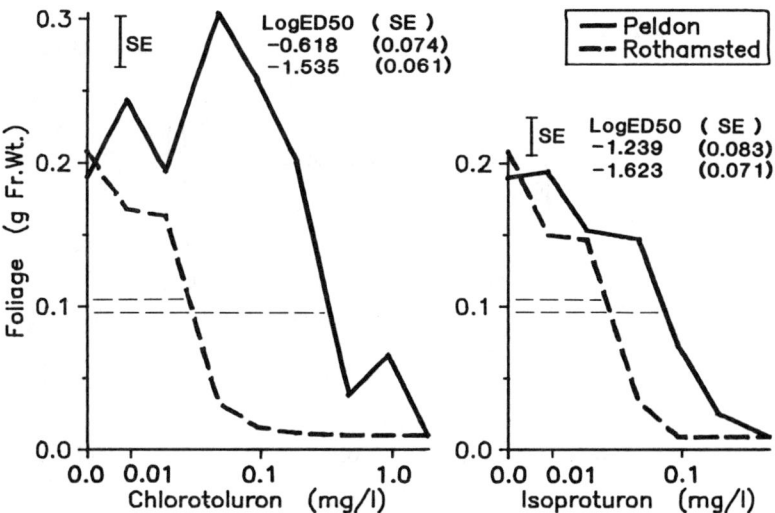

Figure 1. Phytotoxicity of chlorotoluron and isoproturon incorporated in liquid medium of hydroponically grown resistant "Peldon" and susceptible "Rothamsted" black-grass. Standard error bars represent pooled errors of means, each determined from 10 measurements. Log ED50 values (plus standard errors) have been determined from a logistic curve fitted to the log-transformed data using a maximum likelihood program (67).

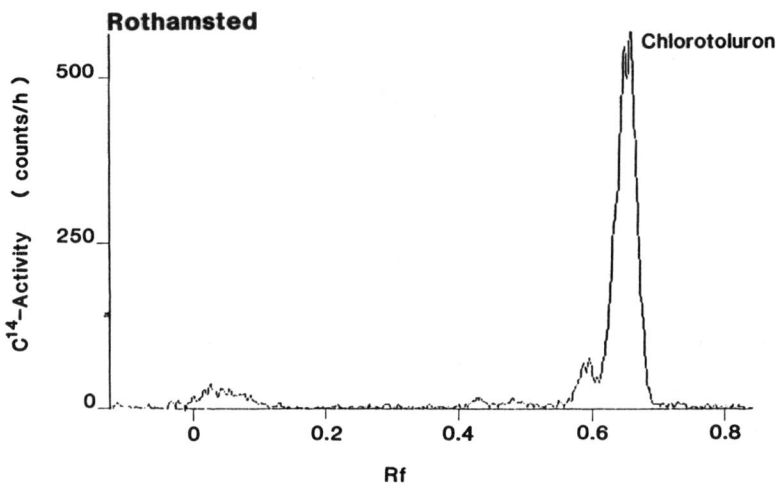
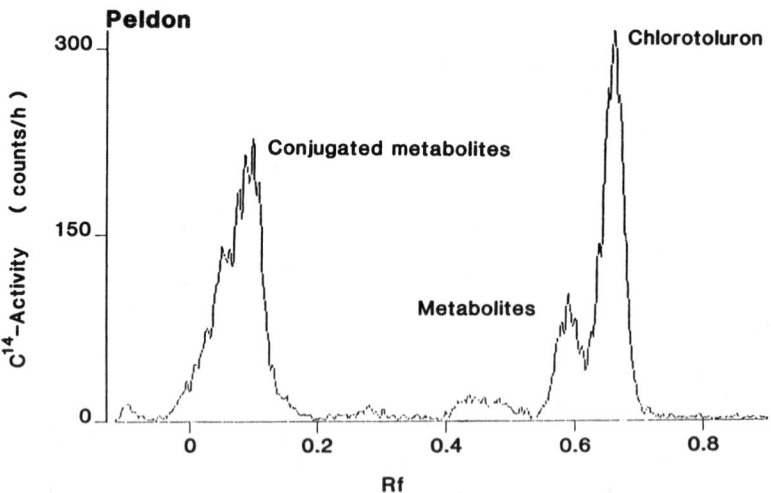

Figure 2. Radio-TLC- analysis of components extracted by 20% aqueous MeOH from foliage of susceptible "Rothamsted" and resistant "Peldon" black-grass immediately after 24h treatment with C^{14}-chlorotoluron. Eluent 20% EtOH/CHCl$_3$.

detoxification of the herbicide by the plant tissues. The lower resistance of Peldon black-grass to isoproturon compared with chlorotoluron appears to be the result of differential rates of degradation of the two herbicides.

Evidence that herbicide degradation takes place in leaf tissue has been obtained using a leaf section assay (Kemp, Moss, Hartnoll and Newton, in preparation). When 4mm leaf sections were sunk by vacuum infiltration in potassium phosphate buffered (pH 6.5) aqueous solutions of chlorotoluron (0-100 mg/l), then illuminated in the same solutions containing freshly dissolved sodium carbonate (0.2% w/v), photosynthesis, which was manifest by leaf sections becoming buoyant (cf. 10 and Underwood & Clay, Long Ashton Research Station, unpublished), occurred only in those treatments containing less than 5ppm chlorotoluron. The responses of the resistant Peldon and susceptible Rothamsted plants were comparable. In similar experiments where the leaf sections were transferred to herbicide-free medium immediately after vacuum infiltration, the response of the susceptible Rothamsted tissue was little changed, photosynthesis being again inhibited by treatments of > 10 mg/l chlorotoluron. In the case of the resistant Peldon biotype, however, treatments of 100 mg/l or more chlorotoluron were now required to prevent photosynthesis. Lower concentrations only delayed the process, indicating that the small amount of chlorotoluron incorporated in the leaf sections following impregnation with these dilute solutions is rapidly degraded by the "resistant" tissue.

Phenylurea herbicides are degraded and detoxified in plants via primary or phase I oxidative processes of N-dealkylation, N-dealkoxylation, oxidation of ring substituents and rarely by ring hydroxylation (11-14). Phytotoxicity is only partially removed by N-monodemethylation; both N-methyl groups must be removed for complete detoxification of the free catabolite. The resulting phase I catabolites then undergo secondary or phase II processes in which they are conjugated with endogenous plant constituents, including glucoside formation, association with amino-acids, and binding to cell-wall membranes (12,13,15).

In the case of chlorotoluron, N-demethylation, ring-methyl oxidation and subsequent conjugation with glucose are the principal processes of degradation, but the extent of N-mono and N-didemethylation and whether these processes are predominant over ring-methyl oxidation are dependent upon the plant species and cultivar (16-25).

Attempts to identify which of the phase I catabolic processes predominate in resistant "Peldon" black-grass, as opposed to the susceptible "Rothamsted" biotype, by comparing the relative quantities of free catabolites in each one are unreliable because the majority of the catabolised material exists as conjugates. Results obtained from conjugate hydrolysis with β-glucosidase and subsequent measurement of the released free metabolites were also inconclusive. Treatment with β-glucosidase revealed that after four days 32%, 10% and 20% of the chlorotoluron in Peldon plants had undergone N-monodemethylation, N-didemethylation and ring-methyl oxidation, respectively, compared with 21%, 4% and 8% in the susceptible Rothamsted population. Isoproturon was degraded less than chlorotoluron in both biotypes, but N-demethylation and ring-alkyl oxidation were again more extensive in Peldon plants compared with Rothamsted plants. However, more than 20% of the chlorotoluron and isoproturon translocated into the leaves of Rothamsted plants and 30-40% in the leaves of Peldon plants remained in the form of conjugated materials after treatment with β-glucosidase. Until this β-glucosidase-resistant metabolic fraction has been characterised, the relative importance of N-demethylation and ring-alkyl oxidation in the detoxification of chlorotoluron and isoproturon in resistant black-grass cannot be determined precisely. However, the results discussed here suggest that both processes are substantial contributors to the resistance mechanism.

Support for these findings is provided by comparing the phytotoxicity of phenylureas in "Peldon" and "Rothamsted" populations of black-grass with those observed in chlorotoluron-susceptible and tolerant cultivars or isogenic lines of wheat. In wheat, degradation is mainly by ring-methyl hydroxylation and is particularly well developed in some chlorotoluron-tolerant varieties (16,18-21) where it imparts a tolerance typically 5-fold above that of susceptible cultivars (26, Blair, A.M.; Martin, T.D., Broom's Barn Experimental Station, Institute of Arable Crops Research, unpublished results; Kemp *et al.*, unpublished results). However, varietal differences in the tolerance of wheat to isoproturon is not so well defined, being less than two-fold (26, Blair, A.M.; Martin, T.D., unpublished results); and there is no evidence of any differential tolerance in wheat cultivars to diuron (19) or linuron (Kemp *et al.*, unpublished results). These observations might be predictable in those plants which, like wheat, detoxify phenylureas predominantly via ring-alkyl oxidation. In black-grass there is a broader spectrum of resistance to the phenylureas. The "Peldon" population of black-grass, in comparison with the "Rothamsted" population, is resistant to chlorotoluron (~10-fold), isoproturon (~3-fold) (9) (Figure 1), linuron and diuron (~5-6-fold) (Kemp *et al.*, unpublished results).

Resistance to chlorotoluron and isoproturon in the "Peldon" black-grass, therefore, appears to be attributable to rapid herbicide degradation via oxidative processes of N-dealkylation and ring-alkyl oxidation. The observed cross-resistance to a wide variety of other herbicides including pendimethalin and diclofop-methyl may be the result of detoxification via similar oxidative degradations. Two points are of particular interest in this regard. Firstly, cross-resistance is not extended to trifluralin, which is structurally related to pendimethalin but lacks available ring-methyl substituents. Secondly, the tolerance of wheat to diclofop-methyl has been attributed to its ability to hydroxylate this compound in the chloro-substituted aryl-ring (27,14 and refs. therein).

Synergistic effects of cytochrome P450 oxidase inhibitors

1-aminobenzotriazole (ABT) There is accumulating evidence that some phase I oxidative processes of N-dealkylation, N-dealkoxylation, ring-alkyl oxidation and ring-hydroxylation, and analogous processes in secondary plant metabolism are enzymically controlled by cytochrome P450 mono-oxygenases (25,28-35). These enzymes have a haem group at their active centre, and can be "poisoned" by suicide substrates which bind to the haem group and so make it unavailable for oxidation (36-40). Such a compound is ABT (1-aminobenzotriazole) (41-43) which inhibits the degradation of chlorotoluron in wheat, cotton and other plants (25,44-47) and so interacts synergistically with herbicide activity (48). If the resistance to chlorotoluron and isoproturon in "Peldon" black-grass is attributable to rapid herbicide degradation via P450 mono-oxygenases, then application of ABT could reduce herbicide resistance.

This synergism between ABT and phenylurea herbicides has been demonstrated in hydroponically grown black-grass plants (Figure 3) (9). Incorporation of 10 and 5 mg ABT/l in the nutrient medium eliminated resistance to chlorotoluron and isoproturon, respectively, in the "Peldon" population but had little effect on phytotoxicity of these herbicides in the susceptible "Rothamsted" population (Figure 4). Similar results have been observed for linuron and diuron (Kemp et al., unpublished results). In comparable experiments with chlorotoluron-tolerant and susceptible isogenic lines of wheat, a strong synergistic interaction between chlorotoluron and ABT was observed in the tolerant line. The two lines were equally susceptible to linuron and both showed weak synergistic interaction between this herbicide and ABT.

The proposed mechanism of these synergistic interactions has been confirmed in black-grass by measuring the effects of ABT on the uptake, translocation and degradation of chlorotoluron and isoproturon (Kemp et al., in preparation). Incorporation of ABT (7.5 mg/l) in the liquid medium of hydroponically-grown plants during and after a 24 h treatment with C^{14}-radiolabelled herbicide (~0.25 mg/l) severely suppressed herbicide degradation in "Peldon" and "Rothamsted" plants (Figure 5). The amount of chlorotoluron remaining undegraded in the foliage four days after treatment was raised from 9% to 35% in "Peldon" plants and from 40% to 72% in "Rothamsted" plants. The corresponding figures for isoproturon are 34% to 70% in "Peldon" and 58% to 89% in "Rothamsted" biotypes. Both N-demethylation steps, and particularly ring-alkyl oxidation in the Peldon plants, appeared to be inhibited. However, the effects of ABT upon these degradative pathways could not be determined precisely since the β-glucosidase-resistant fraction again represented a large proportion of the conjugates and its chemical composition remains to be determined. ABT also suppressed uptake and translocation of chlorotoluron and isoproturon by approximately 20-25%. However, this reduced uptake is insufficient to overcome the effects of reduced herbicide degradation, and there is an overall induced increase in herbicide accumulation within the plant foliage which becomes phytotoxically critical in the resistant "Peldon" tissue.

The accumulation of abscisic acid (ABA) is regulated by its metabolism to phaseic and dehydrophaseic acids via 8'-hydroxymethyl-ABA (49). The ABA-hydroxylating enzyme isolated by Gillard and Walton from *Echinocystis lobata* possesses properties characteristic of cytochrome P450 mono-oxygenases, i.e. it is microsomal, O_2 and NADPH-dependent, and is inhibited by CO. There is circumstantial evidence that ABA-hydroxylation is inhibited by cytochrome P450 mono-oxygenase inhibitors (50) leading to increased ABA levels and, consequently, closed stomata and reduced transpiration. The reduced uptake of chlorotoluron

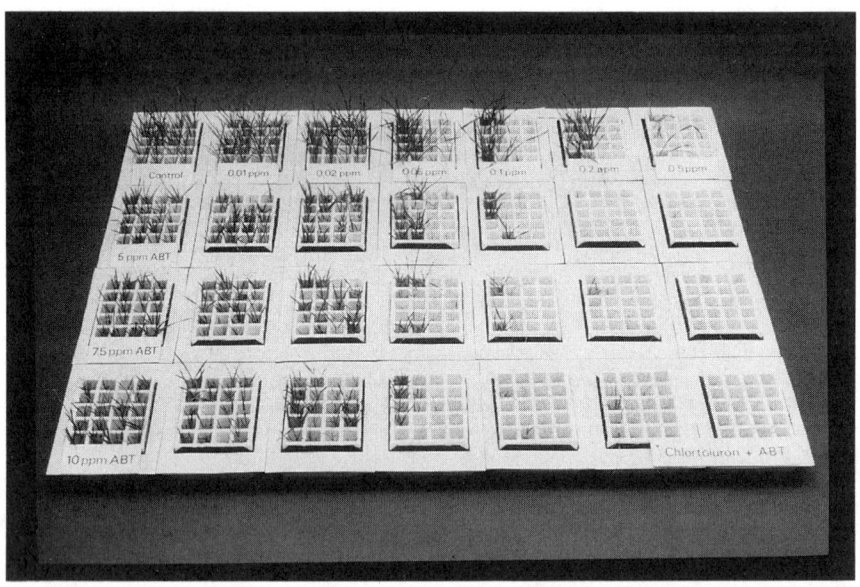

Figure 3. Effects of incorporating 1-aminobenzotriazole (ABT) (0, 5, 7,5 and 10 mg/l) with chlorotoluron (0, 0.01, 0.02, 0.05, 0.1, 0.2 and 0.5 mg/l) in the liquid medium on hydroponically grown black-grass. In each tray resistant "Peldon" plants are on the left and susceptible "Rothamsted" plants are on the right.

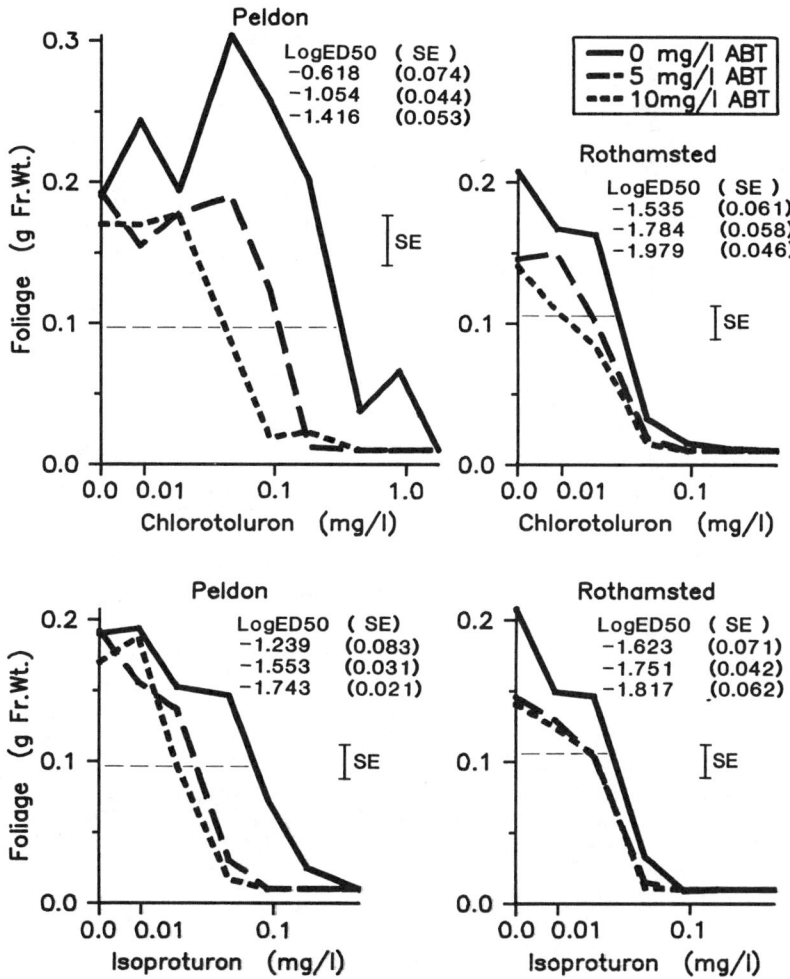

Figure 4. Synergistic effects of 1-aminobenzotriazole (ABT) on phytotoxicity of chlorotoluron and isoproturon in Peldon and Rothamsted populations of black-grass (*Alopecurus myosuroides*). Determination of standard error bars and log ED50 values are as described for Figure 1.

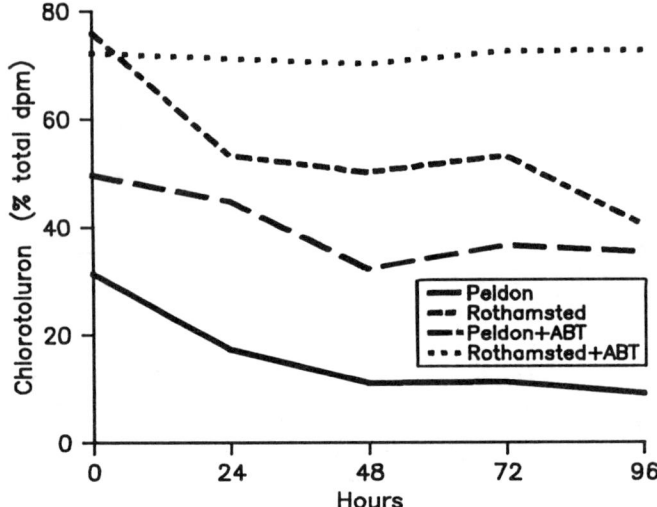

Figure 5. Degradation of chlorotoluron in foliage of hydroponically grown "Rothamsted" and "Peldon" black-grass and its inhibition by 1-aminobenzatriazole (ABT). Concentration of chlorotoluron is expressed as a percentage of total radioactivity found in the foliage following 24 h application of C^{14} chlorotoluron via the roots.

and isoproturon induced by ABT in black-grass may be the result of a similar inhibition of transpiration.

Other physiological effects of ABT, including stunting (shorter, broader leaves and stems with shorter internodal distances and darker pigmentation) and mild phytotoxicity (yellowing of leaves), may also be attributable to the inhibition of microsomal cytochrome P450-dependent processes. The stunting effects are characteristic of gibberellic acid deficiency, the biosynthesis of which is dependent upon a cytochrome P450 mediated oxidation of *ent*-kaurene to ent-kaurenoic acid (31), although it is unclear whether all three steps in this oxidation (alcohol, aldehyde, acid) are catalysed by a single or multi-enzyme system. Demethylation of phytosterols at C-14α by oxidation of the C-32 methyl group is also cytochrome P450-dependent (35). Inhibition of this process results in the accumulation of 14α-methyl steroids and an associated inhibition of growth and development of phytotoxic symptoms (40,51,52) possibly resulting from the disruption of cell membrane functionality. These processes, like the N-demethlyation and ring-methyl oxidation of chlorotoluron, involve the oxidation of a methyl group. Their inhibition by ABT may account for the growth-regulating and phytotoxic properties of this triazole, but it is yet to be established whether these or other cytochrome P450 enzymes are responsible for the exceptionally rapid degradation of phenylurea herbicides in the resistant "Peldon" black-grass.

Anti-kaurene-oxidase plant growth regulators and ergosterol-biosynthesis-inhibiting fungicides   The plant growth regulating triazoles and related growth retardants operate through the inhibition of cytochrome P450 kaurene-oxidase, thereby inhibiting gibberellic acid biosynthesis (53,54). The closely related triazole fungicides and many other ergosterol biosynthesis-inhibiting fungicides inhibit the cytochrome P450 dependent C-14α demethylation of 24-methylenedihydrolanosterol to ergosterol in fungi (55). They do not, however, necessarily inhibit the analogous process of C-14α demethylation of obtusifoliol in higher plants (38,56). Several of these plant growth regulators and fungicides are available commercially and more are being tested for use in agriculture (57). The majority are nitrogen heterocyclic compounds of diverse chemical structures including pyridines, pyrimidines, imidazoles and triazoles. Their mode of action and interaction with other P450 systems is the subject of much recent research (38,53,54). These compounds are potential herbicide synergists. Indeed, the kaurene-oxidase inhibitor, tetcyclasis, is at least 100 times more active in preventing chlorotoluron breakdown than is ABT in cell suspension cultures of cotton (*Gossypium hirsutum*) and maize (*Zea mays*) (47).

The effects of some of these nitrogen heterocyclic PGRs and fungicides on chlorotoluron phytotoxicity in hydroponically grown resistant "Peldon" and susceptible "Rothamsted" black-grass have been examined by incorporating them in the liquid medium with the herbicide (58). As with ABT, synergistic responses were observed only in the resistant "Peldon" plants. The magnitude of the responses varied considerably between compounds and appeared to be independent of phytotoxicity and growth inhibition associated with the heterocyclic compounds which affected both resistant and susceptible plants, (Figure 6). Of the compounds tested, some derivatives of 3,3-dimethyl-1-(1H-1,2,4-triazol-1-yl)butan-2-ol (e.g. triadimenol) and the keto-analogue triadimefon were the most active, the latter compound irradicated chlorotoluron resistance in the Peldon population at 1 mg/l.

The resolved enantiomers of triadimefon were equally active (Kemp *et al.*, in preparation), as was to be expected because they are quickly racemised in aqueous solution (59). However, the two pairs of enantiomers of triadimenol, an active metabolite of triadimefon, are stable (60) and possess independent properties (Kemp *et al.*, in preparation). The *SR*, *RR* and *SS* configurations all synergised chlorotoluron phytotoxicity in "Peldon" black-grass, the *SR* form being marginally the most active, and the SS the least active of the three. However, the RS form, had no synergistic activity. If the synergistic response is attributable to the binding of a "suicide substrate" to the haem group or groups in the cytochrome P450 which catalyses the oxidative degradation of chlorotoluron, then these observations suggest that this binding is stereochemically controlled. Furthermore, the cytochrome P450 responsible for the degradation of chlorotoluron may not be the same as that responsible for *ent-*kaurene oxidation or 14α-demethylation of phytosterols since the relative synergistic activities of the triadimenol enantiomers do not correspond with those observed for anti-kaurene oxidase or anti-14α-demethylase activity (56). However, these interpretations should be treated with caution since the observed activities may be influenced by differences in the uptake, translocation and metabolism of enantiomers, and there may be interspecific variations in the haem groups of P450 *ent-*kaurene -oxidase and 14α-demethylase.

Activity of triadimefon under semi-field conditions Triadimefon was further tested under semi-field conditions (Kemp, Moss, Hartnoll and Newton, in preparation). Resistant Peldon black-grass and wheat were grown together from seed in loam-filled containers in plunge-beds. When the black-grass had attained growth-stage 1.1-1.3 (61), the seedlings were sprayed with chlorotoluron (3.5 kg/ha) or isoproturon (2.5 kg/ha) each mixed with triadimefon (0, 125 and 625 g/ha). Phytotoxicity was assessed by measuring the fresh weight of foliage four weeks after treatment. Treatments of chlorotoluron, chlorotoluron plus triadimefon and chlorotoluron plus the higher application of triadimefon gave 38%, 45% and 60% control of black-grass, respectively. With isoproturon, the synergism was greater, the corresponding effects of these treatments being 67%, 73% and 88% control, respectively. Treatment with isoproturon alone inhibited growth of wheat plants, reducing the foliage of individual plants by 33%. Treatments incorporating triadimefon were generally less damaging, probably due to the removal of competition from black-grass.

Conclusions

Herbicide resistance resulting from enhanced degradation is generally of a lower magnitude than that observed in other forms of resistance which tend to be specific to a single herbicide or

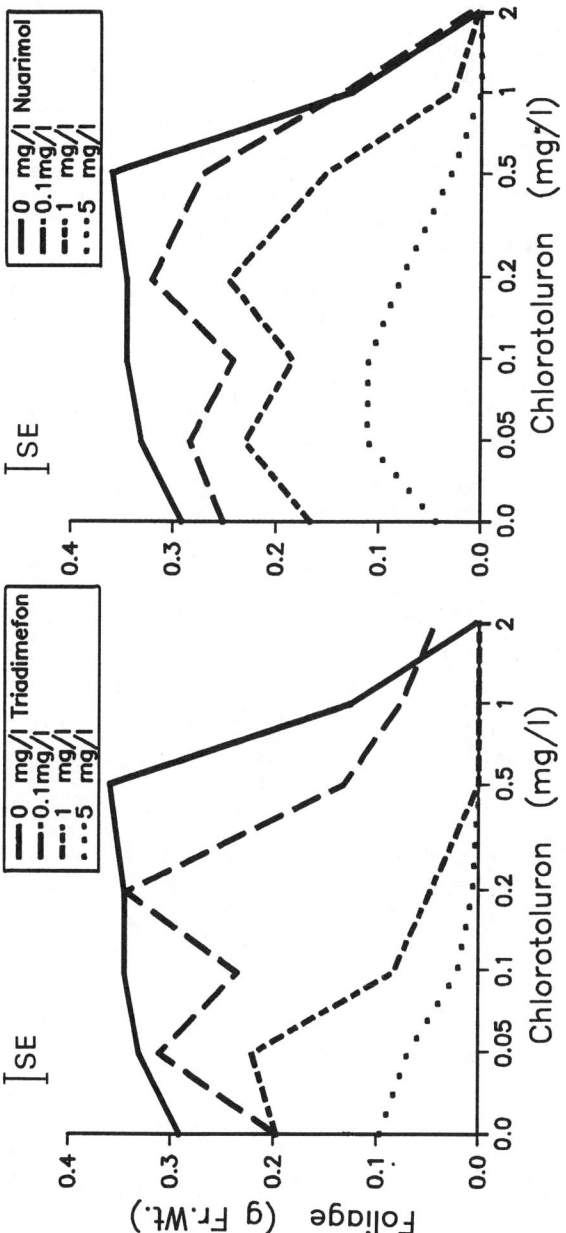

Figure 6. Comparison of predominantly synergistic effects of triadimefon and additive phytotoxic effects of nuarimol on chlorotoluron phytotoxicity in resistant "Peldon" black-grass. Standard errors bars represent pooled errors of means each determined from 10 measurements.

group of herbicides. Nevertheless, the former type of resistance can be a serious agricultural problem when it arises which cannot necessarily be overcome by simply applying an alternative herbicide. The black-grass from Peldon shows cross-resistance to all herbicides currently recommended for the efficient control of *Alopecurus myosuroides* in winter cereals (Table 1, cf. 62,63). Solutions to problems of resistance based on reducing the tolerance of resistant weeds, and so maintaining their control with lower levels of herbicides, have environmental and economic advantages. The results presented here encourage further development of this approach. The synergistic activity of inhibitors of cytochrome P450 monooxygenases may be exploited in the practical control of resistance arising from enhanced oxidative degradation of herbicides. Some of the most useful compounds for this purpose might be PGRs or fungicides already approved for agricultural use.

At a more fundamental level, there is much to be discovered about these degradative processes and the enzyme systems involved. Rapid degradation of chlorotoluron and isoproturon, and synergistic responses resulting from the inhibition of these processes have been demonstrated in the resistant Peldon black-grass, but similar mechanisms and synergistic responses, though implied, have yet to be demonstrated for cross-resistance to other herbicides.

The accumulating evidence indicates that cytochrome P450 monooxygenases control these degradations, but this remains to be established unequivocally. Confirmation of binding to the haem groups by the synergist and some quantification of cytochrome P450 content in resistant and susceptible plants might be achieved by the measurement of difference spectra induced by the herbicide synergist in isolates of the corresponding plant tissue, though this may not be specific for the haem groups of interest. Additional enzymic studies, using cell-suspension cultures and purified cell-free systems, are required.

A recently identified chlorotoluron-resistant population of black-grass does not show exactly the same pattern of cross-resistance to herbicides as the Peldon population. There are also differences between the cross-resistance in black-grass and that found in resistant strains of annual ryegrass (*Lolium rigidum*) (cf. 64-66). Should the resistance mechanisms in these plants also prove to be associated with herbicide degradation, then a comparison of the enzyme systems involved and their interaction with potential synergists could provide valuable insight into inter- and intra-specific variations in mechanisms of xenobiotic degradation which give rise to resistance and their selective inhibition.

More information is needed on the factors determining the development of resistance caused by enhanced herbicide degradation. At present we do not know whether the potential to develop resistance occurs in all black-grass populations. Investigation of the genetics of resistance in black-grass will provide information of value in the development of effective control and containment strategies. There is some evidence to indicate that resistance in individual fields is developing at a relatively slow rate. This may be difficult to detect in the field as resistance can be masked by the many soil and climatic factors which influence herbicide performance. Slowly evolving resistance may however cause increasing problems in the future, especially if resistance develops despite the use of herbicides with differing modes of action. If resistance develops slowly, despite high selection pressure, then it may be possible to reduce selection pressure enough to prevent resistance development. This might be achieved by applying effective herbicides in non-cereal crops and by adopting integrated control strategies.

Acknowledgments

We thank the Winston Churchill Memorial Trust, Ciba-Geigy (Basle) and the American Chemical Society for supporting this work, and Lesley Newton and Jill Hartnoll for experimental assistance. Long Ashton Research Station is financed through the Agricultural and Food Research Council.

Literature cited

1. Sly, J.M.A. Arable farm crops and grass 1982. Preliminary Report of Pesticide Usage England and Wales No. 35. Ministry of Agriculture, Fisheries and Food: London, 30pp, 1984.
2. Putwain, P.D. Proc. British Crop Protection Conference - Weeds 1982. 2, 719-728.
3. Moss, S.R.; Cussans, G.W. Aspects of Appl. Biol., 1985, 9, 91-98.
4. Moss, S.R. Proc. British Crop Protection Conference - Weeds, 1987. 3, 879-886.
5. Niemann, P.; Pestemer, W. Nachricht. Deutschen Pflanzenschutz. 1984, 36, 113-118.
6. Moss, S.R.; Cussans, G.W. In Combating Resistance to Xenobiotics: Biological and Chemical Approaches; Ford, M.; Hollomom, D.; Khambay, B.; Sawicki, R., Eds.; Ellis Horwood: Chichester, 1987; 200-213.
7. Orson, J.H.; Livingston, D.B.F. Proc. British Crop Protection Conference - Weeds, 1987. 3, 887-894.
8. LeBaron, H.M.; Gressel, J., Eds. Herbicide Resistance in Plants; Wiley, New York, 1982.
9. Kemp, M.S.; Caseley, J.C. Proc. British Crop Protection Conference - Weeds, 1987. 3, 895-899.
10. Hensley, J.R. Weed Sci. 1981, 29, 70-73.
11. Frear, D.S.; Hodgson, R.H.; Shimabukuro, R.H.; Still, G.G. Advances in Agronomy. 1972, 24, 327-378.
12. Geissbühler, H.; Martin, H.; Voss, G. In Herbicides, Chemistry, Degradation and Mode of Action, Vol. I. Kearney, P.C.; Kaufman, D.D., Eds.; Marcel Dekker: New York and Basel, 1975; Chapter 3, 209-291.
13. Shimabukuro, R.H.; Lamoureux, G.L.; Frear, D.S. In Biodegradation of Pesticides. Matsumura, F.; Murti, C.R.K., Eds.; Plenum Press: New York, 1982; Chapter 2, 21-66.
14. Cole, D. In Progress in Pesticide Biochemistry and Toxicology, Vol. 3. Hutson, D.H.; Roberts, T.R., Eds.; Wiley, Chichester, 1983; Chapter 4, 199-254.
15. Lamoureux, G.L; Rusness, D.G. In Xenobiotic Conjugation Chemistry. Paulson, G.D.; Caldwell, J.; Hutson, D.H., Menn, J.J., Eds.; American Chemical Society: Washington, DC, 1986, Chapter 4, 62-105.

16. Müller, F.; Frahm, J.; Sonad, A. Mitt. Biol. Bund. Anst. Ld. - Forstw. Berlin-Dahlem. 1977, 178, 241-251.
17. Müller, F.; Frahm, J.; Med. Fac. Landbouww. Rijksuniv. Gent. 1980, 45, 1017-1036.
18. Gross, D.; Laanio, T.; Dupuis, G.; Esser, H.O. Pestic. Biochem. Physiol. 1979, 10, 49-59.
19. Ryan, P.J.; Gross, D.; Owen, W.J.; Laanio, T.L. Pestic. Biochem. Physiol. 1981, 16, 213-221.
20. Ryan, P.J.; Owen, W.J. Proc. British Crop Protection Conference - Weeds, 1982. 1, 317-324.
21. Ryan, P.J.; Owen, W.J. Aspects of Appl. Biol. 1983, 3, 63-72.
22. Cabanne, F.; Gaillardon, P.; Scalla, R. Pestic. Biochem. Physiol. 1985, 23, 212-220.
23. Cole, D.J.; Edwards, R.; Owen, W.J. In Progress in Pesticide Biochemistry, Vol. 6, Herbicides. Hutson, D.H.; Roberts, T.R., Eds.; Wiley, Chichester, 1987; Chapter 2, 57-104.
24. Owen, W.J. Proc. British Crop Protection Conference - Weeds, 1987. 1, 309-318.
25. Gonneau, M.; Pasquette, B.; Cabanne, F.; Scalla, R. Weed Res. 1988, 28, 19-25.
26. Tottman, D.R.; Holroyd, J.; Lupton, F.G.H.; Oliver, R.H.; Barnes, T.R.; Tysoe, R.H. Proc. European Weed Research Society Symposium, Status and Control of Grass Weeds in Europe, 1975, 360-368.
27. Gorbach, S.G.; Kuenzler, K.; Asshauer, J. J. Agr. Food Chem. 1977, 25, 507-511.
28. Frear, D.S.; Swanson, H.R.; Tanaka, F.S. Phytochemistry. 1969, 8, 2157-2169.
29. Russell, D.W. J. Biol. Chem. 1971, 246, 3870-3878.
30. Sandermann, H.; Diesperger, H.; Scheel, D. In Plant Tissue Culture and Its Bio-technological Application; Barz, W.; Reinhard, E.; Zenk, M.H., Eds.; Springer-Verlag: Berlin, 1977, 178-196.
31. Hasson, P.J.; West, C.A. Plant Physiol. 1976, 58, 473-484.
32. West, C.A. In The Biochemistry of Plants, Volume 2, Metabolism and Respiration; Davies, D.D., Ed.; Academic Press: New York and London, 1980, Chapter 8, 317-364.
33. Fujita, M.; Asahi, T. Plant Cell Physiol. 1985, 26, 389-395.
34. Fonne-Pfister, R.; Simon, A.; Salaun, J-P.; Durst, F. Plant Sci. 1988, 55, 9-20.
35. Rahier, A.; Taton, M. Biochem. Biophys. Res. Commun. 1986, 140, 1064-1072.
36. Coolbaugh, R.C.; Hirano, S.S.; West, C.A. Plant Physiol. 1978, 62, 571-576.
37. Wiggins, T.E.; Baldwin, B.C. Pestic. Sci. 1984, 15, 206-209.
38. Vanden Bossche, H.; Marichal, P.; Gorrens, J.; Bellens, D.; Verhoeven, H.; Coene, M-C.; Lauwers, W.; Janssen, P.A.J. Pestic. Sci. 1987, 21, 289-306.
39. Katagi, T.; Mikami, N.; Matsuda, T.; Miyamoto, J. J. Pestic. Sci. 1987, 12, 627-633.
40. Taton, M.; Ullmann, P.; Benveniste, P.; Rahier, A. Pestic. Biochem. Physiol. 1988, 30, 178-189.

41. Ortiz, de Montellano, P.R.; Mathews, J.M. Biochem. J. 1981, 195, 761-764.
42. Reichart, D.; Simon, A.; Durst, F.; Mathews, J.M., Ortiz de Montellano, P.R. Arch. Biochem. Biophys. 1982, 216, 522-529.
43. Ortiz de Montellano, P.R.; Mathews, J.M.; Langry, K.C. Tetrahedron. 1984, 40, 511-519.
44. Gaillardon, P.; Cabanne, F.; Scalla, R.; Durst, F. Weed Res. 1985, 25, 397-402.
45. Cabanne, F.; Gaillardon, P.; Scalla, R.; Durst, F. Proc. British Crop Protection Conference - Weeds, 1985. 3, 1163-1170.
46. Cabanne, F.; Huby, D.; Gaillardon, P.; Scalla, R.; Durst, F. Pestic. Biochem. Physiol. 1987, 28, 371-380.
47. Cole, D.J.; Owen, W.J. Plant Sci, 1987, 50, 13-20.
48. Morse, P.M. Weed Sci. 1978, 26, 58-71.
49. Gillard, D.F.; Walton, D.C. Plant Physiol. 1976, 58, 790-975.
50. Asare-Boamah, N.K.; Hofstra, G.; Fletcher, R.A.; Dumbroff, E.B. Plant Cell Physiol. 1986, 27, 383-390.
51. Burden, R.S.; Clark, T.; Holloway, P.J. Pestic. Biochem. Physiol. 1987, 27, 289-300.
52. Burden, R.S.; James, C.S.; Cooke, D.T.; Anderson, N.H. Proc. British Crop Protection Conference - Weeds, 1987. 1, 171-178.
53. Rademacher, W.; Fritsch, H.; Graebe, J.E.; Sauter, H.; Jung, J. Pestic. Sci. 1987, 21, 241-252.
54. Hedden, P. In Plant Growth Substances. 1988.
55. Henry, M.; Sisler, H. Pestic. Biochem. Physiol. 1984, 22, 262-275.
56. Burden, R.S.; Carter, G.A.; Clark, T.; Cooke, D.T.; Croker, S.J.; Deas, A.H.B.; Hedden, P.; James, C.S.; Lenton, J.R. Pestic. Sci. 1987, 21, 253-267.
57. Worthing, C.R.; Walker, S.B., Eds. In The Pesticide Manual, A World Compendium. 8th Edition; British Crop Protection Council: Thornton Heath, 1987.
58. Kemp, M.S.; Newton, L.V.; Caseley, J.C. In Proc. European Weed Research Society Symposium, Factors Affecting Herbicidal Activity and Selectivity. 1988.
59. Deas, A.H.B.; Carter, G.A. Proc. British Crop Protection Conference - Pests and Diseases, 1986, Vol. 2, 835-841.
60. Clark, T.; Vogeler, K.; Ishikawa, I. Proc. British Crop Protection Conference - Pests and Diseases, 1986. 2, 475-482.
61. Tottman, D.R.; Broad, H. Ann. Appl. Biol. 1987, 110, 441-454.
62. Ivens, G.W. (1988) The U.K. Pesticide Guide 1988. CAB International, British Crop Protection Council, Lavenham Press, Lavenham, Suffolk.
63. Flint, C. In Crop Guide to Herbicides. Reed Business Publishing, Wallington, Surrey, 1987.
64. Heap, I.; Knight, R. Aust. J. Agric. Res. 1986, 37, 149-156.
65. Heap, I.M. Proc. 8th Aust. Weed Conf. 1987, 114-118.
66. Powles, S.B. Proc. 8th Australian Weeds Conference. 1987, 109-113.
67. Ross, G.J.S. MLP (3.08) Maximum Likelihood Program, Rothamsted Experimental Station, 1987.

RECEIVED October 27, 1989

Chapter 27

Herbicide Cross-Resistance in Annual Ryegrass (*Lolium rigidum* Gaud)

The Search for a Mechanism

S. B. Powles, J. A. M. Holtum, J. M. Matthews, and D. R. Liljegren

Waite Agricultural Research Institute, University of Adelaide, Glen Osmond, 5064, South Australia

Currently, biotypes of some 100 weed species have been documented as herbicide resistant. In all but two cases the resistance is confined to an individual chemical type and mode of action. Cross-resistance to a range of dissimilar herbicides has now been documented in biotypes of the prominent grass weeds *Alopecurus myosuroides* and *Lolium rigidum*. Biotypes of *Lolium rigidum* are cross-resistant to a range of aryloxyphenoxyproprionate, cyclohexanediones and sulfonylurea herbicides and the dinitroaniline herbicide trifluralin. Studies to identify the mechanism of cross-resistance in *Lolium rigidum* indicate that resistance is not the result of any barrier to herbicide uptake or translocation. There is no major difference in the sensitivity of acetyl coenzyme-A carboxylase, the enzymic target site for the aryloxyphenoxypropionate and cyclohexanedione herbicides. Similarly there are no differences in acetolactate synthase, the target site of the sulfonylurea herbicides. Activity of the conjugating enzyme, glutathione-S-transferase, is similar in the resistant and susceptible biotypes. Cross-resistance in *Lolium rigidum* may be related to enhanced metabolic de-toxification of a range of dissimilar herbicide types.

Herbicides account for more than 60% of the tonnage of pesticides used worldwide. Their use has increased as more selective herbicides have become available and as they replace mechanical tillage as a method of weed control. An undesirable outcome of the widespread use of herbicides has been the appearance and proliferation of resistant weed biotypes (1). There are currently some 100 weed biotypes with resistance, the great majority of which have acquired resistance following the continued use of a particular class of herbicide (LeBaron H. M. & McFarland, J., this volume). Resistance has emerged because the persistent use of herbicides acts as a selection pressure which results in increases in the frequency of resistant genes in a weed population. Examples include biotypes resistant to the soil-residual triazine herbicides which have been used over long periods for weed control in maize monoculture and for total vegetation control (Holt, J. S. & LeBaron H. M. Weed Technol., in press). Resistant weed biotypes are often cross-resistant to other herbicides of similar chemical type and modes of action. However, the resistant biotypes are usually susceptible to other herbicides with very different structures or with different modes of action. Thus the agricultural impact of the resistant weed biotypes has been limited as alternative herbicides can be used, albeit sometimes at greater cost.

A major practical problem in plant protection is the appearance of pests with resistance to a range of pesticides. This phenomenon, termed cross-resistance, has developed in a number of prominent insect pests (2). Until recently there had been no reports of weeds with cross-resistance to herbicides. Cross-resistance to herbicides has now been detected in at least three cases. In Australia, biotypes of the important cereal and grain legume grass weed, annual ryegrass (*Lolium rigidum*), originally identified as resistant to the graminicide, diclofop-methyl (3), exhibit resistance to other aryloxyphenoxy-propionate herbicides, to several cyclohexanediones, to at least two sulfonylureas and to the dinitroaniline herbicide, trifluralin (4). In Western Australia a biotype of *Lolium rigidum* is resistant to triazine, triazole and dimethylurea herbicides (Hildebrand, O., Powles, S.B. & Holtum, J.A.M., unpublished). In the United Kingdom a population of the grass weed *Alopecurus myosuroides* Huds. (blackgrass) is resistant to the dimethylurea herbicides, chlortoluron and isoproturon, and to the chemically unrelated herbicides chlorsulfuron, diclofop-methyl and pendimethalin (5). A population of *Conyza canadensis* L. Cronq. in Hungary is resistant to the dissimilar herbicides paraquat and atrazine (6).

The clear practical and scientific importance of cross-resistance is reflected in the considerable research attention that the resistant biotypes are being accorded. Work on the mechanism of cross-resistance in *Alopecurus* is reported elsewhere in this volume by Kemp et al. (see also 7). In our laboratory, research is underway to establish the mechanism(s) which endow cross-resistance in *Lolium rigidum*. Here we report on the current state of our investigations.

POSSIBLE MECHANISMS FOR CROSS-RESISTANCE IN *LOLIUM RIGIDUM*

In our experimental program to date, studies have been conducted with a single biotype of *Lolium rigidum* resistant to the herbicides diclofop-methyl, fluazifop-butyl, haloxyfop-methyl, quizalofop-ethyl, sethoxydim, tralkoxydim, chlorsulfuron and trifluralin (unpublished data). This biotype is susceptible to herbicides such as paraquat, glyphosate, propham and carbetamide. The herbicides against which there is resistance belong to four structurally distinct groups of chemicals, examples of which are shown in Figure 1.

An understanding of the mechanistic basis of cross-resistance in *Lolium rigidum* requires recognition of the fact that the plants are resistant to herbicides which act differently within the plant. Any mechanism, or mechanisms, of resistance must therefore be general enough to account for the resistance to a number of dissimilar herbicides, yet specific enough to account for the herbicide susceptibility that is still observed. There are at least five general, not necessarily mutually exclusive, mechanisms which could account for cross-resistance :

1. Reduced uptake of herbicides
2. Reduced translocation of herbicides to their active sites
3. Reduced sensitivity of the herbicide target sites
4. Enhanced detoxification of herbicides
5. Rapid repair of damage

We are examining whether cross-resistance in *Lolium* is due to one or more of these possible mechanisms. Our results and interpretations to date are as follows.

UPTAKE OF HERBICIDES. It is possible that a reduction in the permeability of the epidermis of plants to herbicides could endow resistance. Preliminary

ARYLOXYPHENOXYPROPIONATES

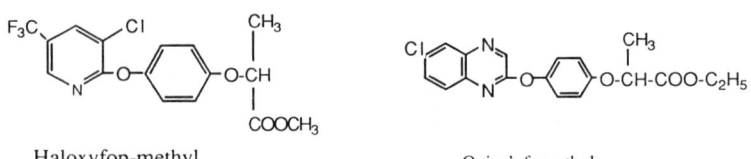

Diclofop-methyl

Fluazifop-butyl

Haloxyfop-methyl

Quizalofop-ethyl.

CYCLOHEXANEDIONES.

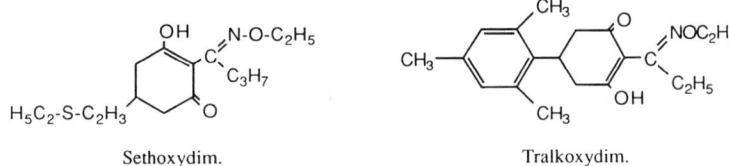

Sethoxydim.

Tralkoxydim.

DINITROANILINES

SULFONYLUREAS

Trifluralin

Chlorsulfuron.

Figure 1. Examples of structurally distinct herbicides against which biotypes of *Lolium rigidum* are resistant.

unpublished work by us and others indicates that the rates of uptake of ^{14}C-diclofop-methyl by biotypes of *Lolium* susceptible and resistant to diclofop-methyl, applied post-emergent, are not appreciably different.

REDUCED TRANSLOCATION OF HERBICIDES TO THEIR ACTIVE SITES. To be effective, an herbicide, or a toxic metabolite of the herbicide, must reach a target site at a concentration and for a time period sufficient to exert the desired herbicidal effect. Any reduction in the rate of herbicide transport to the active site, in its biotransformation to an active form, or an increased diversion to another non-target compartment, can reduce the active concentration at the herbicide active site and thus reduce efficacy.

Diclofop-methyl is a herbicide which, upon entry into the plant, undergoes minimal acropetal and basipetal transport (8). In our experiments transport to the roots or to the leaves of [^{14}C] diclofop-methyl applied to the axils of two-leaved susceptible and resistant *Lolium* plants does not appear to differ. No data are presently available as to whether proplastids or chloroplasts from the two biotypes exhibit differential permeability to diclofop-methyl or to the active acid derivative diclofop. Similarly, there is no evidence for a differential capacity to convert diclofop-methyl to diclofop nor differential sequestration of the ester or the acid in some secondary compartment such as the vacuole or within membranes.

REDUCTION IN THE SENSITIVITY OF HERBICIDE TARGET SITES. Cross-resistance may involve changes in the sensitivity of herbicide target sites. However, if this is the only mechanism for resistance in *Lolium* there must be changes in more than one target site since there is resistance to more than one herbicide class.

a) Aryloxyphenoxypropionate and cyclohexanedione herbicides.
Cross-resistant biotypes of *Lolium* are invariably resistant to a variety of, but not all, aryloxyphenoxypropionate and cyclohexanedione herbicides. These selective grass herbicides probably share at least one common target site (9-12). They inhibit, at nanomolar concentrations *in vitro*, the plastidic enzyme, acetyl coenzyme-A carboxylase (ACC), an enzyme involved in an early step in fatty acid biosynthesis. ACC levels are greatest in differentiating meristematic tissue. ACC from dicotyledons, and possibly some tolerant grasses, is less sensitive to these compounds. Pyruvate decarboxylase is also inhibited by at least some of these compounds but the I_{50} values are about 100-fold higher than for ACC (13). A number of aryloxyphenoxypropionates and cyclohexanediones, including diclofop, also function as protonophores and antagonize some auxin-mediated processes (14). Although the importance of these responses for the herbicidal activity of diclofop is not yet clear (14) it is likely that these effects add to the stress imposed by the herbicide on the target plant.

ACC was extracted from hydroponically-grown susceptible and resistant biotypes of *Lolium*. The extracts were subjected to $(NH)_2SO_4$ fractionation followed by gel filtration. The ATP-, acetyl Co-A- and protein- dependent incorporation of radioactivity from $H^{14}CO_3$ into acid-stable products was monitored. The sensitivity of ACC from both biotypes to diclofop-methyl, diclofop-acid, fluazifop-acid, sethoxydim and tralkoxydim was similar (Table I). The small differences observed are unlikely to account for resistance at the whole plant level. ACC was not affected by either chlorsulfuron or trifluralin, two herbicides against which there is resistance but which have different modes of action.

The extractable activities of ACC from shoot bases of *Lolium* changes during ontogeny (Figure 2). The activity was greatest when the plants were at the two-leaf stage, about 15 to 18 days after planting. Similar changes in the extractable

Table I. Inhibition of ACC from susceptible and resistant *Lolium rigidum*.
Values are the concentrations required for 50 % inhibition

Herbicides	Susceptible	Resistant
	I_{50} [μM]	
Diclofop methyl	0.6	1.4
Diclofop acid	0.2	0.3
Haloxyfop acid	0.3	0.7
Fluazifop acid	0.6	1.9
Sethoxydim	2.7	2.5
Tralkoxydim	0.3	0.4
Trifluralin *	-	-
Chlorsulfuron *	-	-

* At a concentration of 10 μM these herbicides have no effect on ACC activity *in vitro*.

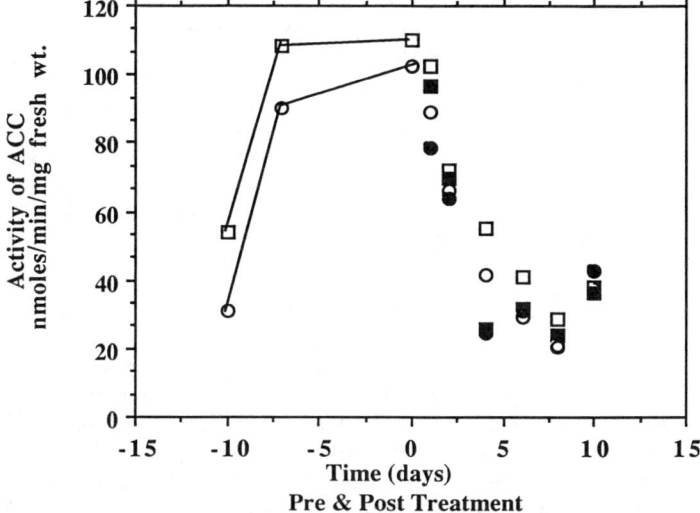

Figure 2. The activity of ACC extracted from susceptible [○, ●] and resistant [□, ■] biotypes of *Lolium rigidum* exposed to 0 [○, □] or 500 nM [●, ■] diclofop-methyl in the hydroponic solution. Plants were treated at the full 2 leaf stage 18 days after planting. Tillering began about 9 days post-treatment.

activity of ACC have been observed in wheat (15). At any comparable stage of development, the activities of ACC in the shoot base of resistant and susceptible *Lolium*, exposed at the two-leaf stage to either 0 or 500 nM diclofop-methyl in the nutrient medium, were similar (Figure 2). The sensitivity of the enzyme from both biotypes to diclofop-methyl and diclofop-acid and the affinities of the enzymes for the substrate acetyl Co-A did not change following exposure to 500 nM diclofop-methyl in the nutrient medium.

We conclude that, in *Lolium*, neither intrinsic nor induced differences in the amounts or the characteristics of ACC are responsible for cross-resistance to the aryloxyphenoxy-propionate and cyclohexanedione herbicides.

b) Sulfonylurea herbicides
Lolium biotypes exist which have resistance to the sulfonylurea herbicides chlorsulfuron and metsulfuron methyl (4). The biotype used in the studies presented here is resistant to both these sulfonylurea herbicides. Sulfonylurea herbicides inhibit the chloroplastic enzyme acetolactate synthase (ALS), also known as acetohydroxyacid synthase (AHAS) (16). Inhibition of this enzyme results in disruption of the synthesis of the branched-chain amino acids valine and isoleucine (16). The imidazolinone herbicides also inhibit ALS (17). In some species auxins can protect against chlorsulfuron inhibition (S. Frear, USDA North Dakota, personal communication); the mechanistic basis for this protection is not known. We have measured the ALS activity in the resistant and susceptible *Lolium* and have also checked for any induction of ALS activity following treatment with the sulfonylurea herbicide chlorsulfuron.

In these experiments, and in the majority of experiments reported here, *Lolium rigidum* biotypes with and without cross-resistance, were grown in a hydroponic system. Pre-germinated seedlings were placed in individual cells of polystyrene planter trays (72 plants/ 0.08 m^2 tray). The trays, containing seedlings, were floated on nutrient solution (4 liters) in a growth room (temperature 20°C/16°C, photon flux density (400-700 nm, 300 µmol quanta $m^{-2} s^{-1}$). At the two-leaf stage of growth, herbicide was added to the nutrient solution. Enzyme was extracted from leaf tissue at various times throughout an 8-day period after treatment. These experiments showed that the inhibition kinetics of ALS extracted and assayed (18) from resistant and susceptible biotypes grown in nutrient solution containing 0 or 100 nM chlorsulfuron were virtually identical, with 50% inhibition occuring at a herbicide concentration of 12 nM (Figure 3). Clearly the resistance to chlorsulfuron, evident in the cross-resistant biotype, does not result from an ALS mutation displaying altered kinetics. However, our initial experiments showed that the levels of enzyme activity extractable from the susceptible biotype decreased 70% within two days of chlorsulfuron treatment, whereas the amount from the resistant plants remains constant. This result may be interpreted in two ways. Either the resistant biotype is capable of more rapid *de novo* synthesis to replace the ALS inhibited by chlorsulfuron or it more rapidly metabolizes the herbicide to a metabolically inactive form.

It appears that resistance to the aryloxyphenoxypropionate herbicides, the cyclohexanedione herbicides and to the sulfonylurea herbicides is unlikely to be due to reductions in the sensitivity or increases in the amounts of their respective target enzymes (Figures 2 & 3). Studies have not yet been performed to examine if the resistance to the dinitroaniline herbicide trifluralin is associated with any change at the tubulin polymerization site.

DETOXIFICATION OF HERBICIDES. Almost all herbicides are metabolized to some extent by higher plants. The selective use of herbicides is often possible because the rates and the mechanisms by which plants detoxify herbicides differ

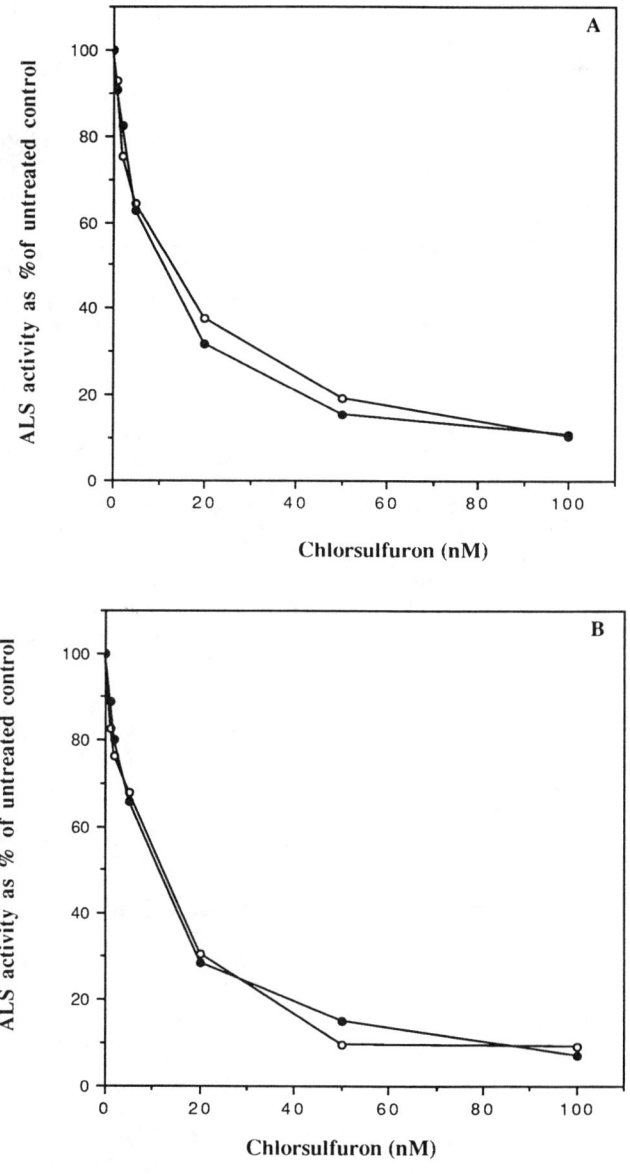

Figure 3. Effect of chlorsulfuron on acetolactate synthase (ALS) activity in extracts from leaves of resistant (A) and susceptible (B) *Lolium rigidum*. ALS was extracted from plants following 7-8 days growth in nutrient solution in the presence (●) or absence (○) of 100 nM chlorsulfuron.

(cf.19). The phytotoxicity of a herbicide is often closely associated with the rate at which it is metabolized. It is possible that cross-resistance in *Lolium rigidum* is related to an enhanced capacity for the selective metabolism of herbicides.

One method of detoxification involves the glutathione S-transferase-catalyzed conjugation of a herbicide with glutathione. Conjugation is often followed by sequestration of the conjugated metabolite in a physiologically inactive compartment such as a vacuole. Glutathione S-transferase-catalysed conjugation, which is a relatively common mechanism by which insects detoxify insecticides (20), is well documented in plants and is responsible for the tolerance of maize and sorghum to triazine-type herbicides (cf. 19). Although an unlikely candidate for the mechanism of cross-resistance in *Lolium rigidum*, we considered it necessary to test for the levels of this enzyme in a resistant and a normal population. The concentrations of reduced glutathione (data not shown) and the extractable activities of glutathione S-transferase were similar in both susceptible and resistant *Lolium* (Figure 4). In both biotypes the levels of glutathione S-transferase, determined using chlorodinitrobenzene as substrate, increased in a similar manner for 3 days following spraying with 572 g a.i. ha^{-1} diclofop-methyl (Figure 4). Cross-resistance cannot be explained by differential glutathione S-transferase activity.

In insects with cross-resistance to insecticides the most common mechanism for the primary detoxification of insecticides involves oxidation by mixed-function oxidases (MFOs) (cf. 21-23). These enzymes oxidise a wide range of substrates with over 1,000 having been catalogued to date (22). MFOs and MFO-type reactions have been detected in plants (cf. 24). In insects, cross-resistance can sometimes be overcome by combining an insecticide with an MFO inhibitor. Compounds such as methylenedioxyphenyl (1,3-benzodioxole) derivatives act as insecticide synergists by retarding the MFO-catalysed detoxification of the insecticide thereby improving the efficacy of the insecticide (cf. 23). Piperonyl butoxide (PBO) is a common commercial methylenedioxyphenyl synergist used to overcome some forms of resistance (23).

Evidence for the involvement of MFOs in the detoxification of herbicides is restricted either to the demonstration that the breakdown products of radiolabelled herbicides are those expected if MFO-catalysed reactions had occurred, or to the demonstration that microsomal preparations, containing cytochrome P-450, catabolize herbicides only under conditions propitious for the operation of MFOs. To our knowledge no MFO protein with a capacity to detoxify a herbicide has been purified from a resistant or a tolerant plant source. The only successful purifications of plant MFOs have been confined to the extraction of MFOs from etiolated or storage tissues of a few plant species (25). Despite these technical difficulties, it has been established that MFOs are involved in the metabolism of dimethyl-urea herbicides (26-29). There are data showing that, in some species, MFOs are involved in the metabolism of at least some of the herbicides against which biotypes of *Lolium* are resistant. MFOs have been implicated in the metabolism of the aryloxyphenoxypropionate herbicide diclofop-methyl in wheat (30) and the sulfonylurea herbicide chlorsulfuron (31-33). We know of no publications addressing involvement of MFOs in the degradation of cyclohexanediones or trifluralin.

In considering whether MFO-catalyzed metabolism of herbicides is associated with cross-resistance in *Lolium rigidum* we have taken, initially, an indirect approach. Growth experiments with wheat showed that the MFO inhibitors aminobenzotriazole (ABT) and PBO synergized chlortoluron when the herbicide and inhibitors were added as a soil drench (34). This synergism was presumed to be due to the inhibition of MFO-catalysed metabolism of chlortoluron (27-29). Significantly, Kemp and Caseley (7) have shown that cross-resistant *Alopecurus* are relatively more susceptible to chlortoluron in the presence of ABT and other

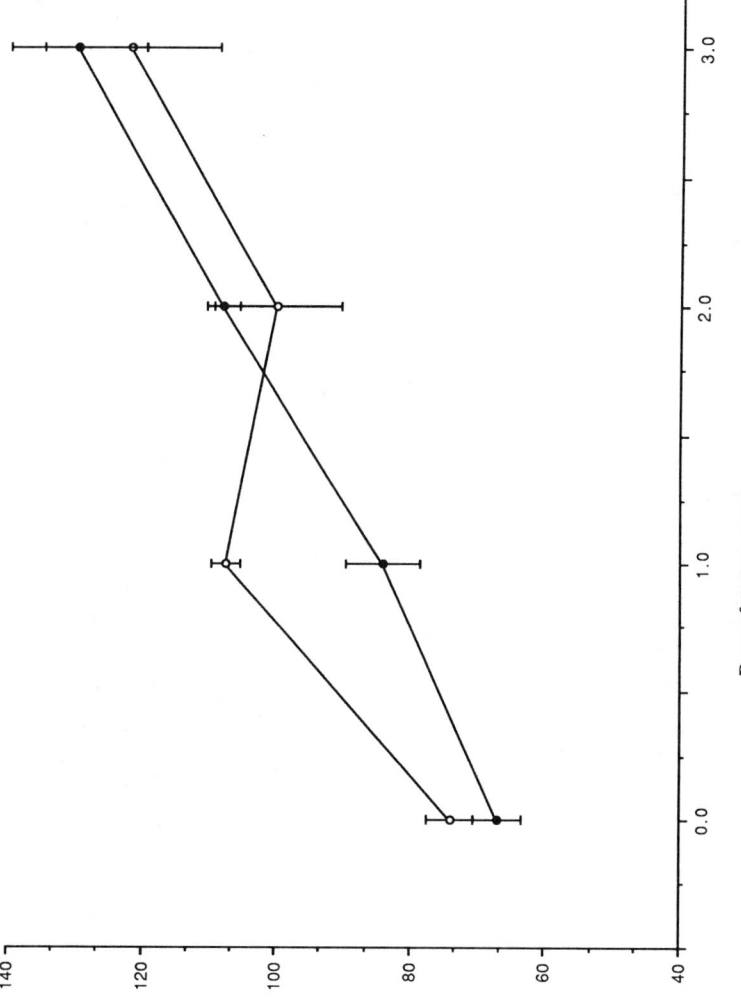

Figure 4. Glutathione S-transferase activity measured on extracts from leaves of resistant (○) and susceptible (●) *Lolium rigidum* after treatment (2-leaf stage) with diclofop-methyl at 572 g a.i. ha^{-1}.

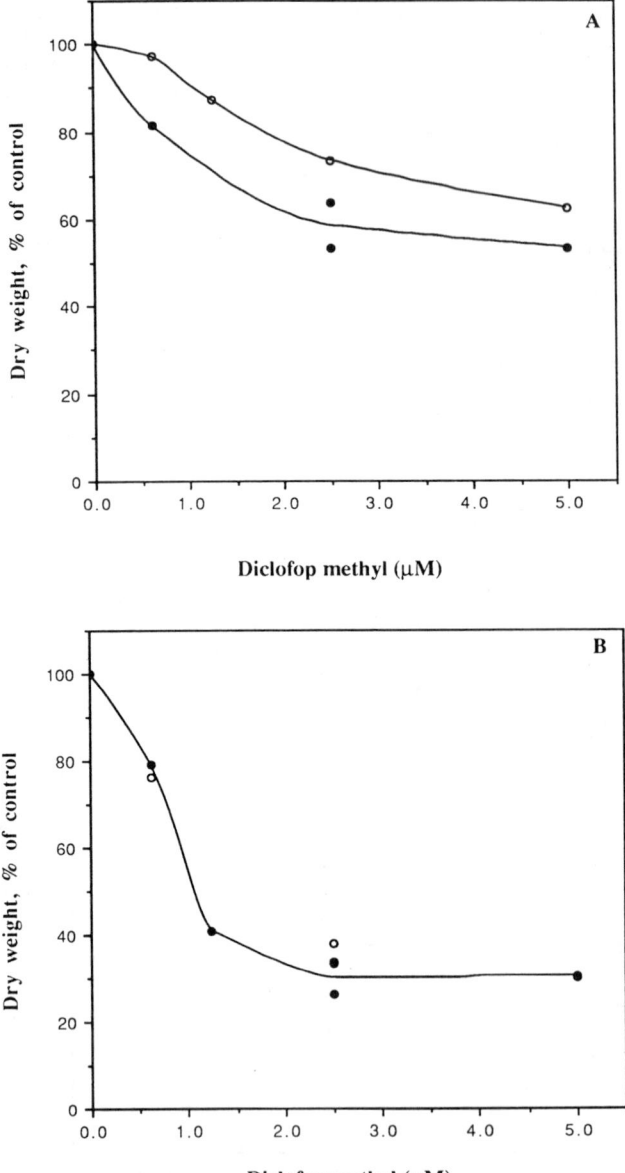

Figure 5. Effect of diclofop-methyl on growth of resistant (A) and susceptible (B) *Lolium rigidum* in the presence (●) or absence (○) of 10 µg/ml aminobenzotriazole. Each point is the mean dry weight of 36 plants determined 7 days after commencement of the treatment.

MFO inhibitors. Further experiments are reported in these proceedings. We have been conducting similar experiments with *Lolium rigidum*.

In a resistant biotype of *Lolium rigidum* we have found synergistic interactions between diclofop-methyl and the MFO inhibitor ABT (Figure 5). Similar synergism has been observed between diclofop-methyl and PBO and between chlorsulfuron and PBO (data not shown). With diclofop-methyl or chlorsulfuron used alone there is minimal effect upon the dry weight of resistant plants, but substantial effect on the susceptible plants. The addition of the MFO inhibitors PBO or ABT consistently reduced the dry weights of the resistant plants. As PBO and ABT are inhibitors of some forms of MFO activity the synergism evident in Figure 5 is, in our opinion, the only evidence available which is consistent with the involvement of MFOs in the metabolism of diclofop-methyl and chlorsulfuron in the cross-resistant *Lolium rigidum* (see also 7; 34). It is stressed, however, that any rigorous proof of the involvement of MFOs in the resistance response requires the detection of MFO-catalysed herbicide breakdown products and the demonstration that the biotypes have different capacities to produce these compounds. In the case of PBO, synergism is only evident when PBO is added to the nutrient solution in hydroponic experiments. When applied with herbicide to the leaves of pot- or field-grown plants the synergistic effect of PBO is not observed. We assume that there is limited uptake of PBO through the foliage or soil.

CONCLUSIONS

To date our studies of possible mechanisms for cross-resistance in *Lolium rigidum* have provided no evidence that resistance is the result of:
1. Any barrier to herbicide uptake.
2. Any differential translocation of herbicide within the plant.
3. Any major differences in the sensitivity of the enzymic target sites of the aryloxyphenoxypropionate, cyclohexanedione or sulfonylurea herbicides.
4. Any differential activity of glutathione S-transferase.

We have no evidence for, or against, any differential sequestration of herbicide or any different pathway for herbicide detoxification in the resistant biotype.

The mixed-function oxidase inhibitors aminobenzotriazole and piperonyl butoxide can synergize herbicide activity in resistant *Lolium* growing in a hydroponic system. This indicates that at least one aspect of cross-resistance in *Lolium rigidum* may be related to enhanced metabolic activity of mixed-function oxidazes acting to detoxify herbicides. We are now concentrating on direct studies of herbicide metabolism in resistant biotypes.

ACKNOWLEDGMENTS

Partial support for this research has been provided by a grant from the Wheat Research Council of Australia. We thank Dr. I. Heap and Dr. R. Knight for provision of the seed population on which this study has been based. The expert technical assistance of Ms. B. Furness and Mr. N. Charman is gratefully acknowledged.

LITERATURE CITED

1. LeBaron, H. M.; Gressel, J. Herbicide Resistance In Plants. 1982, John Wiley & Sons, New York.

2. Georghiou, G. In. Pesticide Resistance: Strategies And Tactics For Management; National Acad. Press: Washington, 1986; pp 14-45.
3. Heap, J.; Knight, R. J. Aust. Inst. Ag. Sci. 1982, 48, 156-7.
4. Heap, I.; Knight, R. J. Aust. J. Ag. Res. 1986, 37, 149-56.
5. Moss, S. R. British Crop Prot'n Conf. 1987, 8C-3, 879-86.
6. Polos, E.; Mikulas, Z.; Szigeti, Z.; Matkovics, B.; Do Quy Hai.; Parducz, A.; Lehoczki, E. Pest. Biochem & Physiol. 1987, 30, 142-54.
7. Kemp, M. S.; Casely, J. C. British Crop Prot'n Conf. 1987, 8C-5, 895-98.
8. Walter, H.; Koch, W.; Müller, F. Weed Research 1980, 20: 325-331.
9. Secor, J.; Cseke, C. Plant Physiol. 1988 86, 10-12.
10. Burton, J. D.; Gronwald, J. W.; Somers, D. A.; Connelly, J. A.; Gengenbach, B. G. Biochem. Biophys. Res. Commun. 1987, 148, 1039-1044.
11. Focke, M.; Lichtenthaler, H. K. Z. Naturforsch. 1987, 42c, 1361-1363.
12. Rendina, A. R.; Felts, J. M. Plant Physiol. 1988, 86, 983-986.
13. Cho, H-Y.; Widholm, J. M.; Slife, F. W. Plant Physiol. 1988, 87, 334-340.
14. Wright, J. P.; Shimabukuro, R. H. Plant Physiol. 1987, 85, 188-193.
15. Hawke, J.C.; Leech, R.M. 1987, 171, 489-495.
16. Ray, T. B. Plant Physiol. 1984, 75, 827-31.
17. Shaner, D. L.; Anderson, P. C.; Stidham, M. A. Plant Physiol. 1984, 76, 545-6.
18. Huppatz, J. L; Casida, J. E. Z. Naturforschung. 1985, 40c, 652-56.
19. Shimabukuro, R. H. In. Weed Physiology Vol 2. Duke, S. O., Ed.;, 1985, pp 215-40, C.R.C. Press.
20. Motoyama, N.; Dauterman, W. C. In. Reviews In Biochemical Toxicology; Hodgson, E.; Bend, J. R.; Philpot, R. M., Eds.; Elsevier: Amsterdam, 1980; Vol. 2, pp 49-69.
21. Oppenoorth, F. J. In. Comprehensive Insect Physiology, Biochemistry And Pharmacology; Kerkut, G. A.; Gilbert, L. I., Eds.; Permagon Press: Oxford, 1984; Vol. 12, pp 731-73.
22. Hodgson, E. In. Comprehensive Insect Physiology, Biochemistry And Pharmacology; Kerkut, G. A.; Gilbert, L. I., Eds.; Permagon Press: Oxford, 1985; Vol. 2, pp 226-321.
23. Wilkinson, C. F. In. Pest Resistance To Pesticides; Georghiou, G. P.; Saito, T., Eds.; Plenum Press: N.Y. & London, 1983; pp 175-205.
24. West, C. A. In. The Biochemistry Of Plants; Davies, D. D., Ed.; Academic Press: New York, 1980; Vol. 2, pp 317-364.
25. Benveniste, I.; Gabriac, J.; Durst, F. Biochem. J. 1986, 235, 365-73.
26. Frear, D. S.; Swanson, H. R.; Tanaka, F. S. Phytochem. 1969, 8, 2157-69.
27. Cabanne, F.; Huby, B.; Gaillardon, P.; Scalla, R.; Durst, F. Pest. Biochem. & Physiol. 1987, 28, 371-80.
28. Cole, D. J.; Owen, W. J. Plant Sci. 1987, 50, 13-20.
29. Gonneau, M.; Pasquette, B.; Cabanne, F.; Scalla, R. Weed Res. 1988, 28, 19-25.
30. McFadden, J.J.; Frear, D.S.; Mansager. E.R. 196th ACS National Meeting. 1988, (Abstract).
31. Erbes, D. L. Pest Biochem & Physiol.,
32. Sweetser, P. B. British Crop Prot'n Conf. 1985, 9B-1, 1147-54.
33. O'Keefe, D. P.; Romesser, J. A.; Leto, K. J. In. Recent Advances In Phytochemistry; Saunders, J. A.; Kosak-Channing, L.;.Conn, E. E., Eds.; Plenum Press: N.Y., 1986; Vol. 21, pp 152-173.
34. Gaillardon, P.; Cabanne, F.; Scalla, R.; Durst, F. Weed Res. 1985, 25, 397-402.

RECEIVED August 23, 1989

Chapter 28

Peroxidizing Herbicides

Some Aspects on Tolerance

Gerhard Sandmann and Peter Böger

Lehrstuhl für Physiologie und Biochemie der Pflanzen, Universität Konstanz, D–7750 Konstanz, Federal Republic of Germany

Destruction of plant membranes, pigments and other cell constituents occurs in the presence of "peroxidizing herbicides" like p-nitrodiphenyl ethers (1), lutidine derivatives (2) or cyclic imides (3); some structures are given in Figure 1. Such compounds are grouped as a particular class of xenobiotics because – in contrast to bipyridylium ions like paraquat – they do not operate as (artificial) terminal electron acceptors in photosynthetic electron transport. Secondly, they do not cause a light-induced oxygen uptake stoichiometrically linked to electron transport and thirdly, their mode of action is connected with interference of chlorophyll biosynthesis.*)

Measurement of ethane formation has been used to quantitate peroxidation especially with higher-plant cell cultures and microalgae. Molecular oxygen reacts with polyunsaturated fatty-acid radicals to form radical intermediates and peroxides. Figure 2 explains assumed chains of events leading to short-chain hydrocarbons from polyunsaturated fatty acids. The chain length of peroxidatively formed alkane species (together with small amounts of alkenes) depends on the nature of the fatty acids occurring in the membranes. As was established in our laboratory the chain length of short-chain (gaseous) hydrocarbons formed is $\omega - 1$ (7).

Radicals and/or activated oxygen initiate the herbicide-induced degradations in the cells (8). Electron spin-resonance signals can be found by illuminating isolated thylakoids from higher plants treated with peroxidizing compounds provided spin-trap techniques are applied. Photosystem-II inhibitors may alleviate peroxidation in photosynthetic cells (e.g. [1,9]). Some light-induced oxygen uptake was observed with higher concentrations of a peroxidizing diphenyl ether (10), but neither substantial formation of peroxide (11) nor superoxide was

---

*) Only selected references are cited. For more detailed overviews the reader may consult (4,5,6).

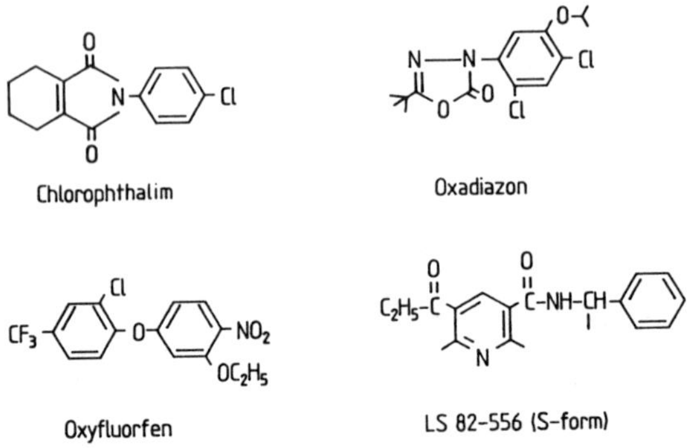

Figure 1. Chemical structures of four typical peroxidizing compounds: chlorophthalim, N-(4-chlorophenyl)-3,4,5,6-tetrahydrophthalimide; oxadiazon, 3-(2,4-dichloro-5-isopropoxyphenyl)-5-tert-butyl-1,3,4-oxadiazol-2(3H)-one; oxyfluorfen, 2-chloro-4-(trifluoromethyl)phenyl-3-ethoxy-4-nitrophenyl ether; LS 82-556, (S)3-N-(methylbenzyl)-carbamoyl-5-propionyl-2,6-lutidine. Also 2,4,5-phenylsubstituted pyrimidinediones exhibit peroxidizing activity, e.g. 3-(4-chloro-5-ethoxy-2-fluorophenyl)-1-methyl-6-(trifluoromethyl)-2,4(1H,3H)pyrimidinedione (34).

detected (1). Singlet oxygen was shown to be generated in the light by isolated thylakoids and apparently no photosynthetic electron transport was required (12). Generally, light is needed for "activation" of peroxidizing herbicides. There is, however, also phytotoxic activity taking place in the dark (13), and the basis for this is not yet understood. As shown by a recent study, cells can evolve substantial ethane in the dark after preillumination for some hours with peroxidizing compounds present (unpubl. results).

Using various tetrahydrophthalimides a comparative study demonstrated that growth and chlorophyll content of both heterotrophic (dark-grown) and autotrophic (light-grown) Scenedesmus acutus were inhibited showing the same structure-activity relationship. Furthermore, this relationship was found identical with a root-growth inhibition assay (Echinochloa utilis) and with the greenhouse pot test as well using different weeds (13). Obviously, an identical basic phytotoxic mechanism is instrumental in the light and the dark which has not been clarified at the moment.

The light-dependent mechanism(s) as to how the starting radicals for peroxidation are formed is not yet elucidated. In aerobic alkaline solutions photoreduction by visible light of nitrodiphenyl ethers to ESR-detectable radicals using β-carotene as sensitizer was reported recently (14). With protoporphyrin IX and an appropriate reductant it could be demonstrated that only nitrodiphenyl ethers, not the p-chloro analogs, led to light-induced radicals although the chloro analogs exhibited strong peroxidation activity (unpubl. results of our laboratory). Accordingly, direct formation of diphenyl ether radicals is not an essential step. The action spectra for phytotoxicity were reported with maxima at 450 and 670 nm (using Chlamydomonas and diphenyl ethers) indicative of carotenoids and chlorophylls as possible photoreceptors (15). Peaks at 550 and 650 nm and a minor one at 450 nm were found when Cucumis sativus was used, which were also seen in the presence of diuron. In non-chlorophyllous tissue (grown in far-red light) the peak at 450 nm was lost (16). The authors suggest a multiple photoreaction, one photoreceptor being chlorophyll or a related pigment. This latter action spectrum reported was observed by treatment with the tetrahydrophthalimide S-23142, N-(4-chloro-2-fluoro-5-propargyloxyphenyl)-3,4,5,6-tetrahydrophthalimide, as well as with the diphenyl ether acifluorfen-methyl, methyl 5-[2-chloro-4-(trifluoromethyl)-phenoxy]-2-nitrobenzoate. However, the intensities of the maxima of the action spectra reported do not match with the absorbance peaks of protoporphyrin IX (17). Using non-chlorophyllous soybean cells the maximum peroxidation activity was found between 350 and 450 nm, with less activity between 450 and 700 nm (18). It should also be mentioned that autotrophic herbicide-treated Scenedesmus cells with no detectable tetrapyrroles present need substantially less light intensity for peroxidation than heterotrophic cells containing tetrapyrroles (19).

The peroxidative response against herbicides can be different in certain species. For example, we find tolerance against oxyfluorfen (20). As shown in Figure 3, the microalgae Scenedesmus

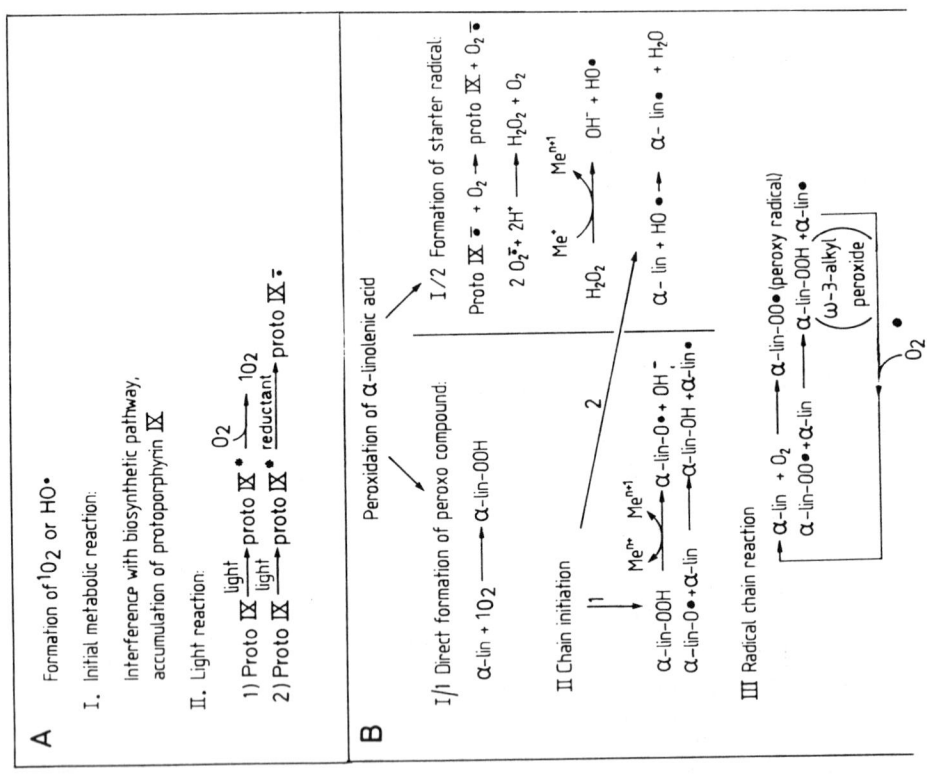

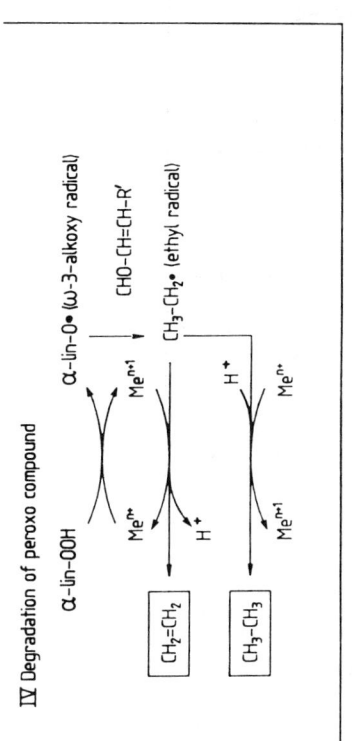

Figure 2: Peroxidation of polyunsaturated fatty acids exemplified with α-linolenic acid (=α-lin). In this scheme radicals are produced by the presence of protoporphyrin IX. In the light, either singlet oxygen may be generated or a superoxide anion provided a suitable redox reaction occurs in case an appropriate reductant is available (part A II, reactions 1,2). Singlet oxygen may lead to formation of a peroxo compounds (of α-linolenic acid, part B I/1); the protoporphyrin IX-radical may form a hydroxy radical HO• via superoxide, see part I/2 (41,42). Both, peroxo linolenic acid and the hydroxy radical will initiate a radical chain reaction (part B, II, III) with linolenic acid leading to the α-lin radical possibly in two ways as indicated by the arrows numbered (1) and (2). This radical supports the chain reaction. The α-linolenic peroxo compound (ω-3-alkyl peroxide) can be degraded to aldehydes and alkanes with some alkenes (R IV) depending on the valence state of the metal ions (Fe) in the cell (43,44).

The formation of the starter radical(s) and the radical chemistry at the photosensitized porphyrin still has to be experimentally proven and is hypothesized by analogous reactions.

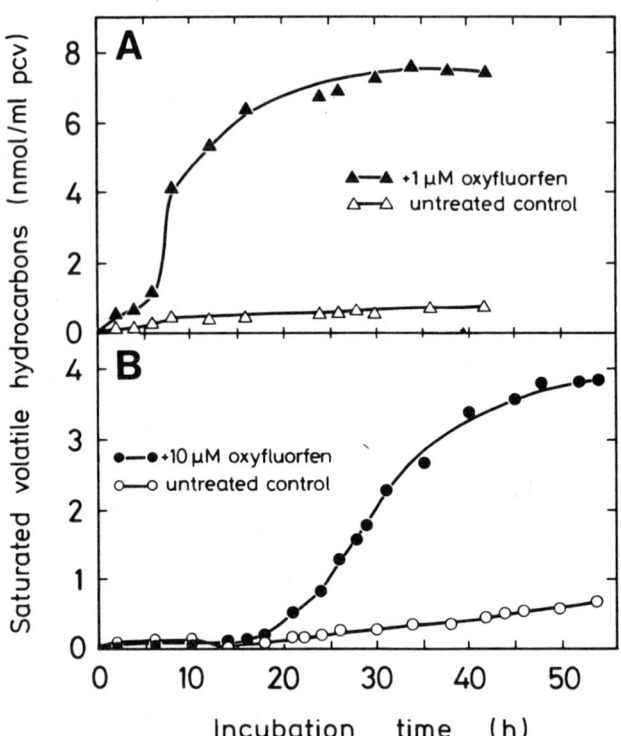

Figure 3. Peroxidative formation of short-chain hydrocarbons induced by oxyfluorfen in cultures of Scenedesmus (A) and Bumilleriopsis (B); modified after (20). pcv, the packed cell volume, refers to the amount of biological material.

and Bumilleriopsis exhibit different kinetics in peroxidative formation of hydrocarbon gases initiated by oxyfluorfen. After oxyfluorfen application both response curves showed a lag phase of about 5 h for Scenedesmus or 20 h for Bumilleriopsis, although the oxyfluorfen concentration was 10-fold higher in the latter culture. Saturation of hydrocarbon formation was reached approx. 30 h after the onset of gas production in both cases. Uptake and incorporation of oxyfluorfen into the cells was the same after 2 and 5 h in Scenedesmus and 3 and 24 h in Bumilleriopsis. Therefore, it can be concluded that the lag phase is not caused by slow oxyfluorfen uptake but rather accounts for that period an antioxidative system can cope with peroxidation before it is overtaxed. Determination of endogenous ascorbate has shown that the tolerant Bumilleriopsis cells contain 0.25 to 0.3 µg ascorbate/µl packed cell volume, that is a 15-fold higher level than in Scenedesmus (unpubl. data). Therefore, the oxyfluorfen tolerance, expressed as an extended lag phase of peroxidation with Bumilleriopsis, can be explained by its more efficient protection system against peroxidative damage. Vice versa, peroxidation can be speeded up by weakening the antioxidative system (see ref. 21 for further details). It should be noted, however, that a different peroxidative response may also be due to different radical formation mechanisms. These mechanisms are still unclear (see Fig. 2). No fluorescent tetrapyrroles could be found in Neurospora or Phycomyces when grown in the dark with chlorophthalim present (up to 10 µM, 4 days).

It has been demonstrated that peroxidizing herbicides interfere with the chlorophyll biosynthetic pathway (22) by accumulation of tetrapyrroles (17,18,23,24). It has also been shown that soluble plastidic cytochrome decreases when oxadiazon is present in a concentration when no peroxidation is yet apparent (25). For the identification of the tetrapyrrole the tolerant Bumilleriopsis was helpful. This species excretes large amounts of a tetrapyrrole into the liquid culture medium when treated with peroxidizing herbicides like oxadiazon, chlorophthalim, oxyfluorfen, or LS 82-556. The product found in the medium could be identified by mass and NMR spectrometry as protoporphyrin IX (17).

About 85% of the protoporphyrin IX accumulated by autotrophic cultures of Bumilleriopsis was excreted (Table I). The amount in the cell increased 5-fold vs. control cells. The effect of peroxidizing herbicides on protoporphyrin-IX accumulation by the sensitive alga Scenedesmus was also determined. This species offers the advantage to grow either autotrophically in the light or heterotrophically in the dark forming chlorophyll. In cells treated with 0.08 µM chlorophthalim protoporphyrin-IX levels were 15-times higher than in the untreated control. No protoporphyrin IX in the culture medium was found. In (autotrophic) cells grown in the light protoporphyrin IX could neither be detected in the control nor in treated cultures (Table I), although these cells are very sensitive against peroxidizing compounds. Tetrapyrroles may be degraded by photodestruction but this explanation remains tentative as yet (19).

Seemingly, under the influence of peroxidizing herbicides in higher plants protoporphyrin IX acts as photosensitizer initiating peroxidation as has been proposed ([18]; see e.g. ref. 6 for overview). Treatment of mung-bean seedlings with chlorophthalim + gabaculine (the latter inhibits δ-aminolevulinic acid formation) prevents the toxic effect of the herbicide. Malondialdehyde formation is similar to the control while the sample treated with chlorophthalim alone exhibited a 100-fold higher level (unpubl. results).

Tappel (26) suggested that ascorbate will reduce tocopheroxyl radicals formed by free-radical reaction(s). Consequently, one α-tocopherol molecule may scavenge many radicals and subsequently stop peroxidation (4). The ratio of ascorbate to α-tocopherol is important for a plant to perform tolerance against peroxidizers. A ratio of ascorbate/α-tocopherol between 10:1 and 15:1 (wt/wt) is effective (see ref. 27 and Table II, first three species). It appears that any ratio above or below this level is correlated with reduced tolerance against peroxidative attack. The capacity of the antioxidative system is influenced by reduction of dehydroascorbate through glutathione, which in turn is reduced by NADPH generated photosynthetically ([21], see the data of Halliwell assembled in [28]). On the other hand, biosynthesis of ascorbate is important, which is closely linked to the assimilatory part of photosynthesis. Treating bean leaves (Phaseolus vulgaris) with acifluorfen, sodium 5-[2-chloro-4-(trifluoromethylphenyl)phenoxy]-2-nitrobenzoate, increased production of glutathione and ascorbate as well as the activities of glutathione reductase and galactonolactone oxidase (the latter catalyzing the last step in the ascorbate-biosynthesis pathway) (29). Apparently, peroxidation will become toxic, namely leading to irreversible degradations of cell constituents, if the antioxidative system cannot cope with the amount of herbicide-induced radicals. Such a situation may explain controversial findings, namely that reduced glutathione and the reductase activity (of cucumber disks) decreased after treatment with acifluorfen in the light (30). Increase of ascorbate, glutathione and α-tocopherol was observed when spruce needles were treated with sulfur dioxide (31) which has been shown to induce peroxidation (32, 33). Elevated levels of both glutathione reductase and superoxide dismutase were also reported in a mutant of Conyza bonariensis, tolerant against paraquat and acifluorfen as well (34).

It needs to be pointed out that in case of paraquat resistance increases in glutathione reductase and superoxide dismutase levels do not alone adequately explain tolerance. Glutathione levels and superoxide dismutase activity was found even lower in a paraquat-resistant Conyza biotype (35). Compartmentalization appears to be an important factor since paraquat apparently was neither transported into the leaves of resistant biotypes (36) nor did it reach its site of action in the chloroplast (37,38).

As demonstrated by Figure 4 (comp. [33]) the glutathione content of bean leaves can be manipulated either by applying different concentrations of oxothiazolidine carboxylate - leading to increase - or buthionine sulfoximine, leading to decrease of

Table I. Formation of protoporphyrin IX by Bumilleriopsis and Scenedesmus in the presence of chlorophthalim

Species/culture type	Protoporphyrin IX (nmol/ml pcv)	
	Control	+ Chlorophthalim
(1) Bumilleriopsis, autotrophic cells	15.0	76.8
", excretion into medium	0	504.2
(2) Scenedesmus, autotrophic cells	0	1
", heterotrophic cells (dark)	3.7	53.7
", excretion into the medium	0	0

Herbicide concentration in (1) was 20 µM and in (2) 0.08 µM; cultivation was for 2 days; pcv indicates packed cell volume (see ref. 17 for details).

Table II. Content of α-tocopherol and ascorbate (mg/100 g dry weight) and the ratio ascorbate/α-tocopherol in different plant species demonstrating relation to peroxidative cell damage

Species	α-Tocopherol	Ascorbate	Ratio	Cell damage *) (%)
Mustard	50	469	9.4	2
Sicklepod	60	861	14.4	3
Alfalfa	10	143	14.3	12
Lambsquarter	12	58	4.8	25
Velvetleaf	50	92	1.8	31
Jimson weed	83	114	1.4	59
Morning glory	10	2	0.2	68
Buckwheat	28	537	19.2	41
Pigweed	10	504	50.4	49

*) Increase of the ratio dry wt/fresh weight in % of untreated control. Plants were sprayed with oxyfluorfen (1 kg/ha) and cultivated in vermiculite; after Finckh and Kunert (27).

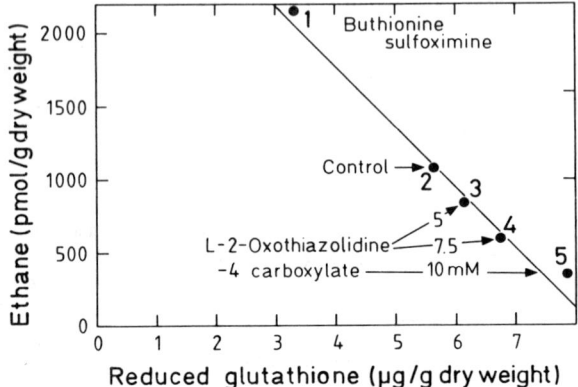

Figure 4: Correlation between glutathione level and peroxidative ethane formation in bean leaves. The glutathione level was changed either by an enhancer (no. 1) or by inhibitors (nos. 3,4,5) of glutathione formation (see text).

glutathione (39). Peroxidative activity, which could be induced by appropriate compounds, correlated exactly with the amount of glutathione in the leaves (Fig. 4). It appears that manipulation of the antioxidative enzymes either by appropriate inhibitors (see some details in ref. 27) or genetic manipulations (for example gene amplification or modifying the ascorbate pool [40]) should be investigated as a means to adjust sensitivity of plants towards this class of herbicides.

Literature cited

1. Kunert, K. J.; Böger, P. Weed Sci. 1981, 29, 169.
2. Matringe, M.; Dufour, J. L.; Lherminier, J.; Scalla, R. Pestic. Biochem. Physiol. 1986, 26, 150.
3. Sato, R.; Nagano, E.; Oshio, H.; Kamoshita, K. Pestic. Biochem. Physiol. 1987, 28, 194.
4. Kunert, K. J.; Sandmann, G.; Böger, P. Rev. Weed Sci. 1987, 3, 35-55.
5. Gillham, D. J.; Dodge, A. D. In Herbicides; Hutson, D. H; Roberts; T. R., Eds.; Wiley Publ., Chichester, 1987; Vol. 6, p. 147.
6. Böger, P.; Sandmann, G. In Chemistry of Plant Protection; W.S. Bowers; W. Ebing et al., Eds.; Springer Publ., Berlin-Heidelberg, 1989, in press.
7. Sandmann, G.; Böger, P. Lipids 1983, 18, 37.
8. Sandmann, G.; Böger, P. In Photosynthesis III, Encyclopedia Plant Physiology, New Series; Staehelin, L.A.; Arntzen, C..; Springer Publ., Berlin-Heidelberg-New York-Tokyo, 1986; Vol. 19, pp. 595-602.
9. Nurit, F.; Ravanel, P.; Tissut, M. Pestic. Biochem. Physiol. 1988, 31, 67
10. Ridley, S. M. Plant Physiol. 1983, 73, 461.
11. Wettlaufer, S. H.; Alscher, R.; Strick, C. Plant Physiol. 1985, 78, 215.
12. Haworth, P.; Hess, F. D. Plant Physiol. 1988, 86, 672.
13. Wakabayashi, K.; Sandmann, G.; Ohta, H.; Böger, P. J. Pestic. Sci. 1988, 13, 461.
14. Rao, D.N.; Mason, R.P. Photochem. Photobiol. 1988, 47, 791.
15. Hess, F. D. Plant Physiol. 1985, 77, 503.
16. Sato, R.; Nagano, E.; Oshio, H.; Kamoshita, K.; Furuya, M. Plant Physiol. 1987, 85, 1146.
17. Sandmann, G.; Böger, P. Z. Naturforsch. 1988, 44c, 699.
18. Matringe, M.; Scalla, R. Plant Physiol. 1988, 86, 619.
19. Nicolaus, B.; Sandmann, G.; Wakabayashi, K.; Böger, P. Pestic. Biochem. Physiol. 1989, submitted.
20. Lambert, R.; Sandmann, G.; Böger, P. Z. Naturforsch. 1987, 42c, 819.
21. Böger, P. In Target Sites of Herbicide Action; Böger, P.; Sandmann, G., Eds.; CRC Press, Boca Raton, FLA, USA, 1989, in press.
22. Wakabayashi, K.; Matsuya, K.; Teraoka, T.; Sandmann, G.; Böger, P. J. Pestic. Sci. 1986, 11, 635.

23. Witkowski, D.A.; Halling, B.P. Plant Physiol. 1988, 87, 632.
24. Lydon, J.; Duke, S. O. Pestic. Biochem. Physiol. 1988, 31, 74.
25. Sandmann, G.; Reck, H.; Böger, P. J. Agric. Food Chem. 1984, 32, 868.
26. Tappel, A. L. Geriatrics 1968, 23, 97.
27. Finckh, B. F.; Kunert, K. J. J. Agric. Food Chem. 1985, 33, 574.
28. Halliwell, B. Chloroplast Metabolism, Clarendon Press, Oxford, 1984, chapter 8.
29. Schmidt, A.; Kunert, K. J. Plant Physiol. 1986, 82, 700.
30. Kenyon, W. H.; Duke, S. O. Plant Physiol. 1985, 79, 862.
31. Mehlhorn, H.; Seufert, G.; Schmidt, A.; Kunert, K. J. Plant Physiol. 1986, 82, 336.
32. Mottley, C.; Trice, T. B.; Mason, R. P. Molec. Pharmacol. 1982, 22, 732.
33. Sandmann, G.; Gamez, H. Environ. Poll. 1989, in press.
34. Shaaltiel, Y.; Glazer, A.; Bocion, P. F.; Gressel, J. Pestic. Biochem. Physiol. 1988, 31, 13.
35. Pölös, E.; Mikulás, J.; Szigeti, Z.; Matkovics, B.; Hai, Do Quy; Párducz, Á.; Lehoczki, E. Pestic. Biochem. Physiol. 1988, 30, 142.
36. Tanaka, Y.; Chisaka, H.; Saka, H. Physiol. Plant. 1986, 66, 605.
37. Fuerst, E.P.; Nakatani, H.Y.; Dodge, A.D.; Penner, D.; Arntzen, C.J. Plant Physiol. 1985, 77, 984.
38. Vaughn, K.C.; Vaughan, M.A.; Camilleri, P. Weed Sci., 1989, in press.
39. Meister, A. Science 1983, 220, 472.
40. Schmidt, A.; Kunert, K. J. In Molecular Strategies for Crop Protection, Arntzen, C. J.; Ryan, C., Eds.; A.R. Liss Inc., New York (UCLA Symp. Molec. Cell. Biol., New Series), 1987; Vol. 48, pp. 401-413.
41. Hopf, F.R.; Whitten, D.G. In Porphyrins and Metalloporphyrins; Smith, K.M., Ed.; Elsevier Publ., Amsterdam, 1975; pp. 667-780.
42. Bachowski, G.J.; Girotti, A.W. Free Rad. Biol. Med. 1988, 5, 3.
43. Kajiwara, T.; Nagata, N.; Hatanaka, A.; Naoshima, Y. Agric. Biol. Chem. 1980, 44, 437.
44. Tappel, A. L. In Free Radicals in Biology, Pryor, W.A., Ed.; Academic Press, New York, 1980; Vol. 4, pp. 1-47.

RECEIVED June 27, 1989

Chapter 29

Fitness and Ecological Adaptability of Herbicide-Resistant Biotypes

Jodie S. Holt

Department of Botany and Plant Sciences, University of California, Riverside, CA 92521

> Natural selection for a particular trait incurs an initial cost to the organism in terms of fitness, or its ability to survive and reproduce. In weeds selected for herbicide resistance, this generalization holds true for biotypes possessing the maternally inherited trait of triazine resistance. This mutation has a detrimental effect on photosynthesis that results in decreased biomass production and reproductive output. However, compensatory interactions of the chloroplast and nuclear genomes may partially overcome reduced productivity. Expression of reduced productivity also appears to be regulated by environmental conditions. Whether similar trends in relative fitness will be found in weeds resistant to other herbicides remains to be examined.

Resistance to pesticides raises many questions about the population dynamics and evolution of resistant types. These questions are important to the pest manager interested in preventing and managing the spread of resistance, as well as to the crop breeder or genetic engineer interested in developing commercially usable resistant crops. The many factors that influence the rate of evolution of resistance in the field have been divided for convenience into three categories, genetic, biological/ecological, and operational factors. These have been reviewed extensively by Georghiou and Taylor ([1](#)). In plants, biological and ecological factors have been particularly important in delaying the evolution of resistance to herbicides, in contrast to the rapid rate of evolution of pesticide resistance in other organisms ([2](#)). The objective of this paper is to explore some of the biological factors that regulate the evolution of herbicide resistance in weeds.

Fitness

Fitness is generally defined as reproductive success, or the proportion of genes an individual leaves in the gene pool of a population ([3](#), [4](#)). The two fundamental components of fitness are survival and reproduction, two alternate ways in which a plant may use limited resources ([4](#)). The relative allocation of resources to survival (maintenance and growth) and reproduction throughout the life cycle determines the fitness of an organism. Since

0097–6156/90/0421–0419$06.00/0
© 1990 American Chemical Society

natural selection favors those genotypes that leave the most descendants, the organism that leaves more offspring (the most fit) is the one whose genes come to dominate the gene pool (3, 4).

While fitness, or reproductive success, is the "currency" of natural selection, many separate characteristics of a plant combine to determine its fitness. These characteristics include seed germination and dormancy, establishment, the physiological processes that result in growth rate, Parkinsonian plasticity, seed size, and seed yield per plant (5). In addition, the fitness of a particular phenotype is determined in the context of prevailing environmental conditions and relative to the survival and reproductive success of other neighboring phenotypes (3, 4,). Therefore, both adaptation, or conformity between the plant and its environment, and competition with neighboring plants will determine fitness. Competition reduces the amount of resources available for maintenance, growth, and reproduction and thus may reduce the fitness of competing organisms (3). Complicating factors that may also determine fitness include other herbicides or weed control measures that may affect resistant and susceptible genotypes differently. All of these factors combine to determine the relative fitness, and thus, frequency of resistant and susceptible plants in a mixed population.

When one gene replaces another in a population, the new gene is usually less adaptive because it has some physiological disadvantages relative to the original gene. These disadvantages keep the new gene at a very low frequency in the population (6). Eventually, if changes in the environment or in the residual genotype occur, the new genotype may become advantageous after a few generations. An increase in the advantage of the new type may occur when changes in the environment take place that favor the new type. In addition, the physiological disadvantages of the new gene may be gradually compensated by the selection of modifiers (6). Until established in the population, the new gene usually causes increased mortality or reduced fertility of the organism. Thus, the cost that accompanies a mutation can be decreased fitness.

In the case of resistance to pesticides, the presence of the chemical selecting agent will favor a new resistant genotype. In the absence of the chemical, the resistant genotype theoretically should suffer some cost in fitness relative to the original dominant susceptible genotype. Since selection of modifying traits is an evolutionary process that occurs over a long period of time, a recently-selected pesticide resistant genotype theoretically should possess some physiological disadvantages over the short term that are expressed as reduced fitness, before compensatory traits are acquired. Over the long term, in the absence of the pesticide selector, the resistant genotype would be replaced by the susceptible one, due to decreased fitness. Efforts by Gressel and Segel (2) to model the dynamics of resistance to herbicides demonstrate the importance of fitness among the factors that regulate the rate of appearance of resistant genotypes in a population.

Biomass production is often assumed to be a good indicator of fitness. While this assumption may not hold true under all conditions, high crop yield is one of the most desirable plant attributes in agricultural systems where short-term productivity is the goal. Weeds with a high biomass productivity are often the most competitive, and therefore, are successful in agricultural systems. Understanding the extent to which herbicide resistance is correlated with reduced fitness in the form of reduced biomass production is essential for the development of herbicide resistant crops, as well as for the prediction of the appearance of resistant weeds. If the trait of resistance directly limits plant growth, its introduction into agromomically important species will be of questionable value.

Fitness of Herbicide Resistant Plants

It should be possible to assess the potential and the reason for decreased fitness in resistant biotypes in cases where the mechanism of resistance is known (7). Table I lists the herbicides or classes of herbicides to which resistance has been found in field situations and for which the mechanism of resistance is known or suspected. (A complete list of weeds with herbicide resistant biotypes as of December, 1988 is presented in the chapter by LeBaron and McFarland in this book.)

Table I. Herbicides or Classes of Herbicides With Resistant Weed Biotypes and the Known or Suspected Mechanism of Resistance

Herbicide or Class	Mechanism of Resistance
Diclofop-methyl	Detoxification
Dinitroanilines (Trifluralin)	Altered tubulin binding site
Paraquat and Diquat	Detoxification and sequestration
Substituted ureas (chlortoluron)	Detoxification
Sulfonylureas (chlorsulfuron)	Modified enzyme binding site
Triazines	Altered PSII binding site

(Source: LeBaron, H. M., McFarland, J., this volume; references 8-13)

In instances where resistance is due to increased amounts of detoxifying enzymes (e.g., paraquat, diclofop-methyl, and chlortoluron) (10-12), the cost of resistance might be a concomitant decrease in other essential proteins (7). Where resistance is due to a modification in substrate affinity of the target enzyme (e.g., sulfonylureas) (13), the organism may produce increased amounts of the enzyme to compensate for its altered activity (7). Both of these mechanisms of resistance could theoretically result in slower growth of the organism. Unfortunately, at the present, no data is available on the relative fitness of plants resistant to these four groups of herbicides.

Trifluralin and other dinitroaniline herbicides inhibit the formation of microtubules and thereby block mitosis in susceptible plants. Resistance to trifluralin in goosegrass (*Eleusine indica* (L.) Gaertn.) is conferred by an altered form of tubulin that results in microtubule insensitivity to the dinitroanilines (9). The alteration that causes resistance also results in aberrations in some of the functions of microtubules in resistant plants in the absence of the herbicide (9). The effect of trifluralin resistance on fitness has been examined to a limited extent. Growth and development were evaluated using plants from several resistant and susceptible populations of goosegrass, grown under noncompetitive conditions (14). Most measured characters varied within and among populations but did not vary consistently between biotypes. However, inflorescence dry weight was significantly greater in susceptible than in resistant plants (14). In a similar experiment, measured and calculated growth parameters were similar between biotypes of goosegrass, while reproductive weight was lower in the resistant biotype (15). In addition, resistant plants were less competitive than susceptible plants and responded to competition by reduced reproductive output (15). These data suggest that dinitroaniline resistance is correlated with reduced fitness. Whether this reduction is due to impaired growth caused by altered tubulin has not been determined.

Triazine Resistance

The most extensively studied type of herbicide resistance is that of triazine resistance, first documented in 1970 (16). In resistant plants, a mutation in the chloroplast gene that encodes the 32 kDa herbicide binding protein, Q_B, of photosystem II (PSII) markedly decreases the affinity of the binding protein for triazine molecules, causing resistance (8). As a consequence of this mutation, the rate of $Q_A \rightarrow Q_B$ electron transfer is at least 10-fold slower in chloroplasts of resistant plants (17). Many reports of reduced plant vigor in triazine resistant biotypes compared to their susceptible counterparts are found in the literature (18). In the absence of triazine herbicides, susceptible biotypes of *Senecio vulgaris* L. and *Amaranthus retroflexus* L. demonstrated more vigorous growth, greater seed production, and greater competitiveness than resistant ones (19-21). Several populations of *S. vulgaris* from various parts of the world also demonstrated greater biomass production by susceptible biotypes, while seedling dry weight, flowering time, and reproductive effort were more variable (22). Greater vigor and productivity of the susceptible biotype was found for other species possessing triazine resistant biotypes, including *Amaranthus hybridus* L. (23), *A. powellii* S. Wats. (24), *Brassica campestris* L. (25), and *Chenopodium album* L. (26, 27). An example of the type and magnitude of whole plant differences between triazine resistant and susceptible biotypes of several species is shown in Table II.

If the triazine resistance mutation imposes a direct limitation on plant growth potential, as indicated by data in the literature, yields of engineered triazine resistant crops would be expected to be reduced. Such crops have been developed in Canada, using conventional breeding techniques to cross resistant weed species and susceptible crop cultivars of *Brassica* (28). Yields of these resistant crop cultivars are reduced between 10 and 40 % compared to susceptible cultivars (29, 30). Research using *Chlamydomonas reinhardtii* mutants with triazine resistance of a magnitude comparable to that of higher plants also showed that decreased growth accompanied this trait (31, 32). These results all suggest that triazine resistance caused by a mutation in the chloroplast genome is correlated with decreased productivity.

Mechanism of Reduced Productivity

The physiological basis for the reduced productivity of triazine resistant biotypes has been investigated extensively (33-35). Net CO_2 fixation differed between biotypes of *Senecio vulgaris* at all light levels (35). Lower photosynthetic rates were also measured in resistant biotypes of *Amaranthus retroflexus* (36) and *Brassica* spp. (37, 38), relative to susceptible biotypes. The quantum yield of CO_2 fixation was significantly lower in resistant than in susceptible biotypes of *Amaranthus hybridus* (39), *Brassica campestris* (40), and *Senecio vulgaris* (35, 41), suggesting that the thylakoid mutation conferring triazine resistance is accompanied by an intrinsic inefficiency in photosynthetic light reactions. This conclusion was supported by lower rates of PSII electron transport measured at all light levels in resistant relative to susceptible thylakoid membranes of *S. vulgaris* (35), *Amaranthus hybridus* (17, 39), *Brassica campestris* (42), and its backcross progeny *B. napus* L. ssp. *Rapifera* (Metzg.) Minsk. (42). An analysis of leaf anatomy and physiological characteristics that may limit photosynthesis revealed some differences between biotypes of *S. vulgaris*, but none that explained reduced photosynthetic capacity in resistant plants (33). Examples of physiological differences

Table II. Whole-plant differences between triazine resistant and susceptible biotypes[a]

Species	Parameter	Response Res.	Susc.	Ref.
Senecio vulgaris	Total dry wt (g)	2.9	3.9	19
Amaranthus retroflexus	Total dry wt (g)	9.8	16.1	
Senecio vulgaris	Total dry wt (g)	2.0	6.1	20
	Veg. dry wt (g)	1.4	4.8	
	Repro. dry wt (g)	0.6	1.3	
Senecio vulgaris	Total dry wt (g)	0.7	1.1	21
	Veg. dry wt (g)	0.6	0.9	
	Repro. dry wt (g)	0.1	0.2	
Senecio vulgaris	Total dry wt (g)	0.9	1.4	22
	Veg. dry wt (g)	0.8	1.1	
	Repro. dry wt (g)	0.2	0.3	
Amaranthus retroflexus	Top dry wt (g)	63.1	74.3	24
Brassica campestris	Total dry wt (g)	1.6	2.5	25
Chenopodium album	Total dry wt (g)	28.3	40.8	27
	Veg. dry wt (g)	16.6	25.2	
	Repro. dry wt (g)	11.6	15.6	
Chenopodium strictum	Total dry wt (g)	21.7	24.6	27
	Veg. dry wt (g)	14.3	17.2	
	Repro. dry wt (g)	7.4	7.4	

[a]Plant age at the time of harvest varied from 4 to 11 weeks among these references; in each case, the maximum dry weight reported is presented in the Table.

between triazine resistant and susceptible biotypes of several species are shown in Table III.

The mechanism by which productivity of triazine resistant biotypes is reduced may be a combination of the direct effects of altered PSII functioning and the indirect or compensatory effects that result in the inefficient expenditure of fixed carbon. Since electron flow through PSII limits overall electron transport in thylakoid preparations of *Senecio vulgaris* and many other species (35, 43), it is likely that the compensatory mechanism operative in triazine resistant biotypes is a higher ratio of PSII/PSI reaction centers. Evidence supporting this hypothesis is the lower ratio of chl a/b in resistant biotypes compared to susceptible in *Senecio vulgaris* and a number of other species (33, 44, 45). In addition, resistant chloroplasts are characterized by increased grana lamellae and stacking relative to susceptible chloroplasts (40, 45-47), suggesting a greater proportion of PSII reaction centers. Finally, some of the differences in thylakoid membrane lipid composition between resistant and susceptible chloroplasts indicate that resistant chloroplasts contain higher PSII/PSI ratios (44, 47, 48). The synthesis of increased PSII reaction centers represents a diversion of carbon resources from other areas of growth and may partially explain reduced productivity of triazine resistant biotypes.

Genetic Basis of Fitness Differences

While the triazine resistant biotypes of virtually all weed species studied differ substantially from their susceptible counterparts in photosynthetic efficiency at the level of the light reactions, data describing whole plant performance is actually quite variable. For example, in one report, relative differences in CO_2 assimilation rate between susceptible and resistant biotypes varied among six species examined (36). In other reports, resistant biotypes were actually more productive and competitive than susceptible under some conditions (49-51). It is difficult to interpret such data since most research to date has been conducted with weed populations of uncertain genetic backgrounds.

The many documented differences in structure and function between triazine resistant and susceptible biotypes may occur as a consequence of reduced PSII in resistant plants and thus, be due to pleiotropic effects of the chloroplast mutation. It is also possible that differential productivity between biotypes of the same species is the result of variation in other traits that are under nuclear, not chloroplast, control (39, 52). The role of the chloroplast genome in regulating growth and productivity of resistant plants is best determined by examining nuclear-isogenic susceptible and resistant biotypes, rather than field-collected ones, since differences between biotypes in nuclear genome-controlled traits may mask or even compensate for detrimental effects of the chloroplast mutation.

A limited number of comparisons have been made using triazine resistant and susceptible biotypes with hybrid or nearly identical nuclear genomes. In comparisons of nearly nuclear-isogenic resistant and susceptible cultivars of canola (*Brassica napus*), dry weight, number and weight of seed pods, and seed yield of the resistant biotype were all less than those of the susceptible biotype (30). Comparisons of the agronomic performance of F_1 reciprocal hybrids of triazine resistant and susceptible *B. napus* were also made (53). These plants possess either resistant or susceptible chloroplasts in a common, intermediate nuclear background. Resistant F_1 hybrids exhibited 21 to 25% lower yields, flowered later, and were shorter in height than susceptible hybrids (53). These results all indicate that triazine resistance and reduced productivity are indeed linked in some way.

To clarify the role of the chloroplast and nuclear genomes in regulating productivity in triazine resistant plants, F_1 hybrids were developed from purebreeding susceptible (S) and resistant (R) biotypes of *Senecio vulgaris* (43). In F_1 hybrids, atrazine response, carbon fixation, and electron transport rates were nearly identical to those of their respective maternal parents (43). The two susceptible (S, SxR) biotypes had significantly higher rates of CO_2 fixation, PSII, and whole-chain electron transport than did the two resistant (R, RxS) biotypes. PSI rates were identical in all four biotypes. In contrast, hybrid biotypes were not identical to their maternal parents in biomass production. S, SxR, and RxS plants all achieved greater biomass than R plants (43). These results, summarized in Figure 1, suggest that the resistance mutation has a detrimental effect on thylakoid performance that is translated into decreased plant yields. Recent investigations of photosynthetic characteristics of nearly isonuclear resistant and susceptible *B. napus* have confirmed that decreased quantum yield and decreased maximum rate of photosynthesis are consequences of the triazine resistance mutation (54). However, this deleterious effect may be partially overcome at the whole plant level by the interaction of the plastid and nuclear genomes (43).

Environmental Effects on Fitness

There is increasing evidence that the relative productivity of susceptible and resistant biotypes of a single species may depend upon environmental conditions, in particular, temperature. For example, resistant biotypes of *Setaria* spp. produced more biomass under cooler than warmer temperatures and yield was inversely related to increasing temperature, while susceptible plants grew best at higher temperatures (50). Growth was also related to temperature in *Brassica napus* cultivars grown in Canada. On most sites, yields of the resistant cultivar were 80% those of susceptible cultivars, while in hot, dry, summer conditions, the resistant cultivar produced 73% of typical susceptible rapeseed yields (29). A resistant biotype of *Phalaris paradoxa* L. also displayed more vigorous growth than the susceptible biotype when grown under conditions of early winter (17 C days, 12 C nights, 8 hour photoperiod) (51). These data indicate that growth temperature will influence the relative productivity of triazine resistant weeds and crop cultivars.

Temperature sensitivity may be directly related to the triazine resistance mutation. In five different species, high temperature (>38 C) inhibited to a greater extent electron transfer in resistant leaves than in susceptible leaves, perhaps through a conformational change in the herbicide binding protein (55). Furthermore, between 25 C and 35 C, electron transport activity in resistant chloroplasts deteriorated to a greater extent than it did in susceptible ones. These results suggest that the mutation in the binding protein in triazine resistant plants may increase the thermal sensitivity of the oxygen-evolving system (55). In *Conyza canadensis* (L.) Cronq., electron transport activity was also temperature-dependent. Resistant plants had lower rates than susceptible ones at 25 C but higher rates at 20 C (47). Strikingly similar results were found for *Polygonum lapathifolium* L. (56). Changes in membrane lipid composition observed in chloroplasts of resistant plants may be responsible for these differences in temperature sensitivity between susceptible and resistant biotypes (47, 48, 55). Recent investigations using nuclear-isogenic biotypes of *Solanum nigrum* L. resistant or susceptible to the triazines confirmed that resistant types were less tolerant of high temperature (57). However, other recent results have been less conclusive (58).

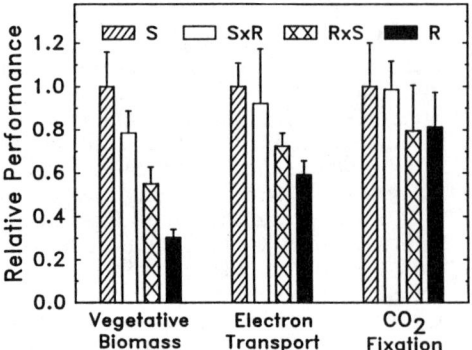

Figure 1. Relative performance of parental and F_1 reciprocal hybrid triazine resistant and susceptible *Senecio vulgaris* biotypes. Data modified from 43.

Table III. Physiological differences between triazine resistant and susceptible biotypes[a]

Species	Parameter	Response Res.	Susc.	Ref.
Amaranthus hybridus	Ps_{net} ($mgCO_2/dm^2h$)	66.0	82.5	23
Senecio vulgaris	Ps_{net} ($nmolCO_2/cm^2s$)	1.5	1.8	35
	O_2 evolution (μmol O_2/mg chl·h)	17.8	41.7	
Amaranthus retroflexus	Ps_{net} ($mgCO_2/dm^2h$)	30.0	35.0	36
Polygonum lapathifolium	Ps_{net} ($mgCO_2/dm^2h$)	41.0	48.0	
Brassica campestris	Ps_{net} ($nmolCO_2/cm^2s$)	1.7	2.0	38
Brassica napus	Ps_{net} ($nmolCO_2/cm^2s$)	1.7	1.9	
Brassica juncea	Ps_{net} ($nmolCO_2/cm^2s$)	2.3	2.6	
Amaranthus hybridus	Ps_{net} ($\mu molCO_2/m^2s$)	49.0	56.0	39
Brassica campestris	PSII (μmol DPIP/mg chl·h)	106.7	153.3	40

[a]Experimental conditions varied among these references. In some cases, data were extrapolated from graphs. Ps = photosynthesis.

Environmental conditions may thus significantly influence differential vigor, and possibly fitness, between triazine resistant and susceptible plants.

Concluding Remarks

In the field, natural selection acts to optimize the adaptations of a plant so as to maximize individual fitness. In triazine resistant weeds, coevolution of the less productive chloroplast with the nuclear genome could have resulted in compensatory mechanisms not present in susceptible plants. These mechanisms could optimize productivity and maximize the fitness of those individuals, within the constraints imposed by impaired PSII. Even slight differences in the chloroplast genome between susceptible and resistant biotypes could result in different responses to the same selection pressures in the environment, perhaps in the direction of overcoming limitations caused by the resistance mutation. Thus, after several generations of selection, many nuclear-genome controlled traits are likely to appear that could mask intrinsic differences between biotypes due to the chloroplast mutation. It is also possible that the alteration in PSII in resistant plants triggers developmental events that compensate in some way. Effects of the resistance mutation may be compensated for by other aspects of plant performance such as carbon allocation or rate of development. Resistant plants may not necessarily be less productive than susceptible ones when traits not directly linked with triazine resistance are considered.

Few inferences are possible about overall fitness, or reductions in fitness, of a resistant biotype relative to a susceptible one from growth and biomass data alone. Fitness can be defined and understood only in the context of an organism's total environment (3). To understand fully the effect of the mutation conferring resistance on fitness, the genetic background of the plants should be evaluated, as well as their phenology, germination, growth, productivity, and reproductive output. Since these life history traits may vary widely in response to environmental conditions and biotic interactions, comparisons among several populations of resistant and susceptible biotypes over a range of conditions may be necessary to evaluate the variability expressed by resistant plants in the field. Furthermore, the significance of differences in growth and reproduction between resistant and susceptible biotypes of any species is the magnitude of those differences under competition in the field. Species shifts may result from competition and alter reproductive success in the form of seed output. This information will be required for predicting the field performance of newly-developed resistant crops and for predicting potential problems to be encountered with newly-selected resistant weeds.

Literature Cited

1. Georghiou, G. P.; Taylor, C. E. In Pesticide Resistance. Strategies and Tactics for Management; National Academy Press: Washington, D. C., 1986; pp 157-169.
2. Gressel, J.; Segel, L. A. J. Theoret. Biol. 1978, 75, 1-23.
3. Pianka, E. R. Evolutionary Ecology; Harper & Row: New York, 1978; Chapter 1, 5.
4. Silvertown, J. W. Introduction to Plant Population Ecology; 2nd Ed.; Longman Scientific & Technical: New York, 1987; Chapter 1,7.
5. Gressel, J.; Segel, L. A. In Herbicide Resistance in Plants; John Wiley & Sons: New York, 1982; Chapter 17.
6. Haldane, J. B. S. J. Genetics 1960, 57, 351-360.

7. Gressel, J. In Molecular Form and Function of the Plant Genome; van Vloten-Doting, L., Groot, G. S. P., Hall, T. C., Eds.; Plenum Press: New York, 1985; pp 489-504.
8. Hirschberg, J.; Bleeker, A.; Kyle, D. J.; McIntosh, L.; Arntzen, C. J. Z. Naturforsch 1984, 39c, 412-419.
9. Vaughn, K. C. Pestic. Biochem. Physio. 1986, 26, 66-74.
10. Shaaltiel, Y.; Gressel, J. Plant Physiol. 1987, 85, 869-871.
11. Powles, S. B.; Liljegren, D. Weed Sci. Soc. Amer. Abstr. 1988, 28, 67.
12. Moss, S. R.; Cussans, G. W. In Combating Resistance to Xenobiotics. Biological and Chemical Approaches; Ford, M. G., Holloman, D. W., Khambay, B. P. S., Sawicki. R. M., Eds.; Ellis Horwood: New York, 1987; Chapter 17.
13. Chaleff, R. S.; Mauvais, C. J. Science 1984, 224, 1443-1445.
14. Murphy, T. R.; Gossett, B. J.; Toler, J. E. Weed Science 1986, 34, 704-710.
15. Valverde, B. E.; Radosevich, S. R.; Appleby, A. P. West. Soc. Weed Sci. Proceedings 1988, 41, 81.
16. Ryan, G. F. Weed Science 1970, 18, 614-616.
17. Bowes, J.; Crofts, A. R.; Arntzen, C. J. Arch. Biochem. Biophys. 1980, 200, 303-308.
18. Radosevich, S. R.; Holt, J. S. In Herbicide Resistance in Plants; LeBaron, H. M., Gressel, J., Eds.; John Wiley & Sons: New York, 1982; Chapter 9.
19. Conard, S. G.; Radosevich, S. R. J. Appl. Ecol. 1979, 16, 171-177.
20. Holt, J. S.; Radosevich, S. R. Weed Science 1983, 31, 112-120.
21. Holt, J. S. J. Appl. Ecol. 1988, 25, 307-318.
22. Warwick, S. I. Weed Res. 1980, 20, 299-303.
23. Ahrens, W. H.; Stoller, E. W. Weed Science 1983, 31, 438-444.
24. Weaver, S. E.; Warwick, S. I. New Phytol. 1982, 92, 131-139.
25. Mapplebeck, L. R.; Souza Machado, V.; Grodzinski, B. Can. J. Plant Sci. 1982, 62, 733-739.
26. Marriage, P. B.; Warwick, S. I. Weed Res. 1980, 20, 9-15.
27. Warwick, S. I.; Black, L. Can. J. Bot. 1981, 59, 689-693.
28. Weiss, J.; Beversdorf, W. D. Can. J. Plant Sci. 1981, 61, 723-726.
29. Beversdorf, W. D.; Hume, D. J. Can. J. Plant Sci. 1984, 64, 1007-1009.
30. Gressel, J.; Ben-Sinai, G. Plant Sci. 1985, 38, 29-32.
31. Erickson, J. M.; Rahire, M.; Bennoun, P.; Delepelaire, P.; Diner, B.; Rochaix, J.-D. Proc. Natl. Acad. Sci. 1984, 81, 3617-3621.
32. Galloway, R. E.; Mets, L. J. Plant Physiol. 1984, 74, 469-474.
33. Holt, J. S.; Goffner, D. P. Plant Physiol. 1985, 79, 699-705.
34. Holt, J. S.; Radosevich, S. R.; Stemler, A. J. Biochim. Biophys. Acta 1983, 722, 245-255.
35. Holt, J. S.; Stemler, A. J.; Radosevich, S. R. Plant Physiol. 1981, 67, 744-748.
36. van Oorschot, J. L. P.; van Leeuwen, P. H. Z. Naturforsch. 1983, 39c, 440-442.
37. Donnelly, M. J.; Hume, D. J. Can. J. Plant Sci. 1983, 64, 432.
38. Hobbs, S. L. A. Can. J. Plant Sci. 1987, 67, 457-466.
39. Ort, D. R.; Ahrens, W. H.; Martin, B.; Stoller, E. W. Plant Physiol. 1983, 72, 925-930.
40. Burke, J. J.; Wilson, R. F.; Swafford, J. R. Plant Physiol. 1982, 70, 24-29.
41. Ireland, C. R.; Telfer, A.; Covello, P. S.; Baker, N. R.; Barber, J. Planta 1988, 173, 459-467.
42. Ali, A.; Fuerst, E. P.; Arntzen, C. J.; Souza Machado, V. Plant Physiol. 1986, 80, 511-514.

43. Stowe, A. E.; Holt, J. S. Plant Physiol. 1988, 87, 183-189.
44. Chapman, D. J.; De-Felice, J; Barber, J. Planta 1985, 166, 280-285.
45. Vaughn, K. C.; Duke, S. O. Physiol. Plant. 1984, 62, 510-520.
46. Lemoine, Y.; Dubacq, J. -P.; Zabulon, G.; Ducruet, J. -M. Can. J. Bot. 1986, 64, 2999-3007.
47. Polos, E.; Laskay, G.; Szigeti, Z.; Pataki, Sz.; Lehoczki, E. Z. Naturforsch. 1987, 42c, 783-793.
48. Pillai, P.; St. John, J. B. Plant Physiol. 1981, 68, 585- 587.
49. Jansen, M. A. K.; Hobe, J. H.; Wesselius, J. C.; van Rensen, J. J. S. Physiol. Veg. 1986, 24, 475-484.
50. Ricroch, A.; Mousseau, M.; Darmency, H.; Pernes, J. Plant Physiol. Biochem. 1987, 25, 29-34.
51. Schonfeld, M.; Yaacoby, T.; Michael, O.; Rubin, B. Plant Physiol. 1987, 83, 329-333.
52. Gasquez, J.; Compoint, J. P. Agro-ecosystems 1981, 7, 1-10.
53. Beversdorf, W. D.; Hume, D. J.; Donnelly-Vanderloo, M. J. Crop Science 1988, 28, 932-934.
54. Jursinic, P. A.; Pearcy, R. W. Plant Physiol. 1988, 88, 1195-1200.
55. Ducruet, J. M.; Lemoine, Y. Plant Cell Physiol. 1985, 26, 419-429.
56. Darmency, H.; Gasquez, J. Plant Sci. Lett. 1982, 24, 39-44.
57. Ducruet, J. M.; Ort, D. R. Plant Sci. 1988, 56, 39-48.
58. Jacobs, B. R.; Duesing, J. H.; Antonovics, J.; Patterson, D. T. Can. J. Bot. 1988, 66, 847-850.

RECEIVED June 27, 1989

Chapter 30

Herbicide Rotations and Mixtures

Effective Strategies to Delay Resistance

J. Gressel[1] and L. A. Segel[2]

[1]Department of Plant Genetics, Weizmann Institute of Science, Rehovot 76 100, Israel
[2]Department of Applied Mathematics, Weizmann Institute of Science, Rehovot 76 100, Israel

Herbicide resistant populations have evolved only in mono-culture and/or mono-herbicide conditions at the rates we have previously predicted. Contrary to our original model, no populations of atrazine-resistant weeds have appeared in corn where rotations of crops and herbicides or herbicide mixtures were used. This is due to the previous lack of information about the greatly reduced fitness of the resistant individuals, which could be expressed only during rotational cycles, and also to the greater sensitivity of resistant individuals to other herbicides, pests and control practices. Our new model, presented here, describes how these factors reduce the resistant individuals to extremely low frequencies during rotation.

It is not yet clear whether such factors will delay the rate of appearance of other types of resistance; e.g. the multiple resistances to grass-killing herbicides in wheat, or to inhibitors of acetolactate synthase, where the fitness of resistant biotypes may be higher.

Weeds have only evolved resistant populations to the herbicides described in the previous chapters where there was mono-culture with a single family of herbicides. The only exceptions have been where different herbicides having the same site of action were used (e.g. a rotation of photosystem II inhibiting herbicides). It is expected that resistant populations will soon evolve in wheat monoculture where high selection pressure herbicides with different sites of action but the same mode of degradation are rotated (1, 2). The appearance of resistance in mono-culture and/or mono-herbicide usage followed a simple population genetics model that integrated: (a) the selection-pressure of the herbicide (based on the rate used and its persistence); (b) the germination dynamics of the weeds (over the season and from the soil seed-bank); (c) the initial frequency of resistants due to mutation of susceptible individuals; (d) the fitness of the resistant mutants in competition with the wild type under field conditions and (e) the number of generations (seasons) the herbicide was used (3, 4). This model was helpful in understanding why resistance evolved to the high selection pressure s-triazines in corn but not to lower selection pressure thiocarbamates, chloroacetamides and phenoxy herbicides in this crop. It is consistent with our model that there was no evolution of resistant populations of weeds to 2,4-D and MCPA in wheat. The model led one to expect the appearance of weed biotypes resistant to the sulfonylureas, chlorotoluron, diclofop-methyl and mecoprop in weeds of wheat. Resistant biotypes evolved first in those weeds where the herbicides exert the highest selection pressure. For example diclofop-methyl exerts a much higher selection pressure on *Lolium* spp than on *Avena* spp; in agronomic terms, this herbicide is more effective on *Lolium*, and *Lolium* has indeed evolved resistance with greater rapidity (Powles et al., this volume).

Because of a lack of field ecology data, the early models inaccurately predicted evolution of resistance where herbicide mixtures and rotations were used. Vast areas of corn have received such rotations with s-triazine herbicides for 30 years, and resistant populations have not appeared. The models did not consider the then unknown extreme lack of fitness in many resistant biotypes in the seasons the triazines were not used. The model (3, 4) correctly predicted that mixtures could considerably delay the evolution of resistance but the lack of field data on selection pressure of mixtures left it to the reader to insert the correct data into the equations and the accompanying figures. The revised model presented below considers what happens during the "off" years when a given herbicide is not used, and predicts that some high selection pressure herbicides can be sparingly used, in rotation, possibly even after resistance has appeared. The data show the urgent need of further research concerning the physiological-ecology of resistant weed populations.

The Original Model

The original model (3, 4) described different possible rates of enrichment of resistant individuals in populations, until the populations were predominantly resistant. It was assumed that (different) constant proportions of susceptibles and resistants germinated, survived to the end of the season, etc., and that susceptible and resistant individuals had (different) constant seed yields. It was further assumed that resistants initially formed an exceedingly small fraction of the population, certainly a well-warranted assumption during the years of resistance enrichment, until there were sufficient numbers of individuals for resistance to be evident. The following simplified equation described the proportion of resistant individuals of a given weed species, N_n in the nth year following continuous application of herbicide (3,4):

$$N_n^{(R)} = N_0^{(R)} \left[1 + \frac{f\alpha}{n}\right]^n \qquad \text{(Eq. 1)}$$

Strictly speaking, $N_n^{(R)}$ gives the proportion of resistant seeds deposited per unit area in year n. However, other quantities such as the number of resistant *plants* that germinate or that become established can be obtained from $N_n^{(R)}$ by multiplication with suitable constants. What is essential is that all these various measures of resistance change by a constant factor each year, giving rise to an exponential increase in resistant individuals (Fig. 1A). In the case of Eq. 1 this factor is $1 + (f\alpha/\bar{n})$.

Various parameters appear in the simple algebraic Eq. 1 for the frequency (proportion) of resistant weeds in a species that is preponderantly sensitive to a given herbicide. $N_0^{(R)}$ is the very low frequency of resistant individuals in the population before it is exposed to the herbicide. In the absence of herbicide, resistance is sustained in the population by a balance between mutation and depletion of the proportion of resistant individuals by their lesser fitness in the absence of selection, resulting in a resistance fraction somewhat lower than the mutation frequency. If mutants were more fit than the wild type, they would be the wild type. Mutant fitness can be near neutrality, and the mutants would then be found in different proportions at various geographical areas due to "drift".

The factor f is a reproductive fitness measure that compounds the relative ability of resistants, compared to susceptibles, in germination, establishment, growth, seed production and survival. Fitness is thus always measured when the resistant and susceptible individuals are in competition with each other in the absence of selection for the mutants. When they are grown separately the resistant individuals are often less 'productive', and in competitive situations the resistant individuals are more affected (Table I). (See Appendix 1 for exact mathematical definitions.)

The factor α describes the selection pressure. It is the ratio of the fraction of resistants that survive herbicide application during a season to the corresponding fraction of susceptibles. Thus, early or late germinating susceptible individuals that produce seeds are considered in this "effective-kill", in contrast with the initial control usually measured by weed scientists. Thus, for example, if the herbicide kills 95% of the susceptibles and none of the resistants, then $\alpha = 20$.

The factor $\bar{n}$ is the average number of years that a seed remains viable in the seed bank.

According to Eq. 1, the frequency of resistance in the population starts at a very low value and then increases by a constant factor each year. In spite of the exponential increase, it will be hard to detect resistant individuals in a field until resistance is at a level of 10–30% of the population.

Eq. 1 allowed the plotting of rates of enrichment of resistance, with different scenarios of selection pressure, seed bank size, fitness, and initial mutation frequency (Fig. 1A). The values that could be "plugged" into the equation to generate the scenarios were based on a very limited data-base, mostly from corollary systems, such as heavy-metal tolerance. We knew too little about weed-herbicide interactions at that time to make precise estimates. With the experience of hindsight, we can see where the model was clearly correct, and where it needed modification.

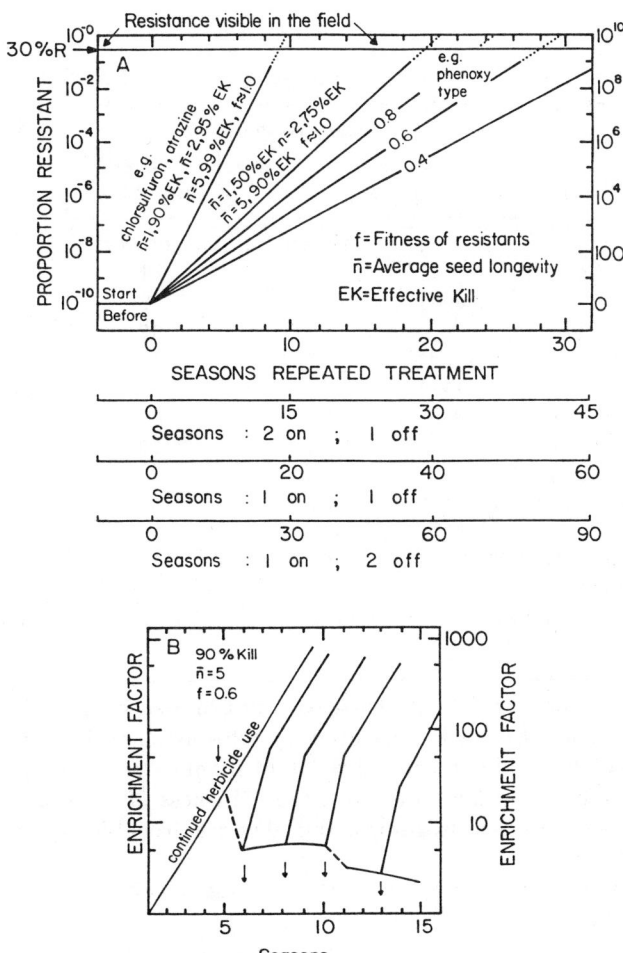

Figure 1. Prediction of the effect of herbicide rotation in the original model. A. Overall average effect of scenarios with different selection pressures ($\alpha = 0.1 = $ 90% effective kill (EK); $\alpha = 0.01 = $ 99% EK, $\alpha = 0.33 = $ 75% EK, $\alpha = 0.5 = $ 50% EK), seed bank dynamics ($\bar{n}$; See Table III) and differential fitness. The different scales give the different rotational scenarios, from mono-herbicide to one treatment in 3 seasons. B. Calculated effect of herbicide stoppage, and restarting at various intervals. Source: Redrawn from equations and figures in refs. (3) and (4).

Table I. Lower Productivity and Competitive Fitness of Resistant Biotypes

Herbicide Species resistant	Productivity	Competitive Fitness (1:1)[b]	Notes	Ref
s-triazines	resistant ÷ susceptible			
Amaranthus hybridus (seed)[a]	–	.11		(5)
(seed)	.90	.18		(6)
Senecio vulgaris	–	.32		(5)
(seed)	.47	.43		(7)
(DW)	.60	–	isogenic[c]	(8)
Chenopodium album (seed)	.75	.08		(9)
C. strictum (seed)	1.00	1.78		(9)
Brassica napus (seed)	.76	.28	isogenic	(10)
Solanum nigrum (seed)	.67	–	isogenic	(11)
Trifluralin				
Eleucine indica (DW)	.68	–		(12)
Diclofop-methyl				
Lolium rigidum (DW)	0.59–0.74	–	density dependent	(13)

Productivity is measured by growing resistant and susceptible biotypes separately; competitive fitness is measured by growing them in a mixture. [a]DW is the dry weight of above ground parts of plants; "seed" may include fruit or whole flower depending on study cited. [b]Fitness of 1:1 mixture; [c]isogenic means that nearly nuclear isogenic material or reciprocal F_1 hybrids were used.

Affirmation of the model

(a) *The importance of selection pressure.* The atrazine and chlorsulfuron levels required to kill different weeds vary over more than an order of magnitude. The selection-pressure of both herbicides is greatest on broadleaf species requiring the lowest rates for weed-control. The selection pressure is least for those weeds requiring the highest herbicide levels, i.e. the grasses. The first weeds to evolve atrazine resistant populations were *Senecio, Amaranthus* spp and *Chenopodium* spp; the last were the grasses, as the model predicted. At least two broad-leaved weeds and one crop have already evolved chlorsulfuron resistance in the field, under selection pressure of this recently introduced and highly persistent herbicide.

The herbicides with the greatest persistence will exert the highest selection pressure. The triazines, dinitroanilines, and sulfonylureas meet this criterion with season-long control, and resistance has evolved to all these groups. By contrast, 2,4-D and other phenoxy herbicides, thiocarbamates, etc. have much shorter biological-persistence in the soil and resistance has not evolved to them. Paraquat resistance has evolved, and may seem to contradict the theory, as paraquat loses biological activity upon touching the soil, due to binding to colloids. The lack of residual persistence was balanced by farmer persistence. All cases of paraquat resistance occurred where this herbicide was employed 5-10 times during each season in mono-herbicide usage.

(b) *Seed bank life.* The longer the life in the seedbank, the greater the buffering effect of susceptible seed from previous years, decreasing the rate of resistance. *Senecio vulgaris* has evolved resistance in orchards, nurseries and roadsides where there was no mechanical cultivation, but not in cornfields. The *Senecio* seed is incorporated into the soil seed bank in cultivated cornfields, where it is viable for many years (14). All *Senecio* seed falling on undisturbed soil on roadsides or orchards germinates or dies during the following season (15). Resistance thus evolved where there was the lowest average seed bank life time $\bar{n}$, as predicted. Such information must be considered in formulating strategies for resistance management. Many other species do not have a seedbank, i.e. in no-till agriculture where $\bar{n} = 1$.

The model predicted that the rotation or mixing of herbicides with the same site of action (and thus similar mode of resistance) will result in the same effect as using a single herbicide. It was therefore suggested that use of corn/soybean; atrazine/metribuzin or sulfonylurea/imidazolinone rotations with various crops would be contraindicated (1, 2, 4). For the same reason, the genetic engineering of atrazine-resistant soybean for use in monoherbicide culture would be a mis-directed project, unless the atrazine usage in corn were replaced by other herbicides.

Atrazine-resistant lambsquarters appeared in corn/sugarbeet; atrazine-/pyrazon rotations (16) following this prediction. A case of decreased sensitivity to all oxidant generating herbicides has appeared in *Conyza canadensis* in orchards treated with paraquat after this weed evolved atrazine resistance (17).

Cases where the model was or may be inadequate

(a) *Use of other photosystem II herbicides against atrazine-resistant weeds.* The first atrazine-resistant *Chenopodium* appeared after 5-10 years of continuous herbicide use. According to the model, it should take 5-10 years for resistance to occur to each other photosystem II herbicide not having cross resistance (e.g. diuron, pyridate, phenolic types), if used on atrazine-resistant weeds, as further mutations would be required. Instead there is evidence that resistances to other photosystem II herbicides evolve more quickly.

When the model was formulated, the precise mutation frequency of chloroplast genome mutations was unknown and could only be estimated as being an exceedingly low number, much lower than nuclear mutation frequencies. There are now some data to indicate that the mutation frequency of chloroplast genes is orders of magnitude higher in certain individual plants than the normal plastid mutation frequency. This is due to a nuclear "plastome mutator" gene which increases the frequency of chloroplast mutations (18).

Triazine resistance seems to have evolved in the individuals bearing the plastome mutator gene, and triazine-resistant populations will carry a much higher frequency of this mutator gene than susceptible populations. This would allow other chloroplast inherited photosystem II herbicide resistances to evolve at much higher rates in triazine-resistant populations than in wild type populations (19). If all cases of triazine resistance occurred where there was an active plastome mutator gene, it is highly likely that in most cases where there is a mutation conferring triazine resistance, there will be other mutations in the chloroplast genome as well. These would also lower fitness.

(b) *Multiple-resistances to herbicides with vastly different modes of action.* Multiple-resistances to insecticides and drugs are rather common, and are documented to the level of molecular biology. They are a rather recent occurrence with herbicides. *Lolium rigidum* that evolved resistance to diclofop-methyl was found to have a surprising cross resistance to chlorsulfuron as well as to all other wheat selective herbicides (20). *Alopecurus myosuroides*, which evolved resistance to chlorotoluron, was cross resistant to chlorsulfuron, pendimethalin and diclofop-methyl (21). Similar surprising multiple cross resistances to insecticides and drugs were often traced to the evolution of higher levels of nonspecific esterases, hydrolases or monooxygenases. The resistance in *Alopecurus* to chlorotoluron can be abolished by adding specific monooxygenase inhibitors along with the herbicide (22). Thus, we may have problems similar to those of the entomologists and pharmacologists. We can learn from their experience that multiple resistance can be both alleviated and delayed or prevented from occurring by the addition of synergists, to the herbicides (22) in our case.

(c) *Lack of triazine resistance in rotation with herbicides that do not directly inhibit photosynthesis.* The original model did not adequately account for events in the "off-years" during rotations. Resistance was held to evolve at a fixed rate that was a function of the number of generations or seasons a weed was treated with a particular herbicide. According to the old model, this meant that if it would take 6–10 years for resistance to occur in monoculture corn with atrazine as the sole herbicide, it would take 9–15 years in a corn/corn/wheat (or soybean) rotation where atrazine was used 2 of 3 years; or 12–20 years in a corn/wheat (or soy) rotation where atrazine was used every other year; or 18–30 years in a corn/wheat/soy rotation where atrazine was used once in 3 years (Fig. 1A). When the model was formulated 10 years ago, it appeared that the time was just about ripe for triazine resistance to break out in vast areas of the cornbelt where such rotations were used, as there had been 6–10 years use of atrazine since it was introduced. Resistant populations did not appear.

Very few farmers in the cornbelt grow monoculture corn, unlike in areas to the east of the cornbelt where atrazine resistance appeared. Some farmers use rotations. Some employ mixtures of atrazine with chloroacetamide herbicides, which allow the use of less atrazine (lowering the selection pressure) and which also kill *Amaranthus* and *Chenopodium*, as well as grass weeds. Such mixtures substantially delay resistance both according to the old model and from practice, although the magnitude is yet unclear. With the chloroacetamide herbicides used in these mixtures under attack and their use restricted in many areas and forbidden in others, herbicide rotation may be the only strategy remaining to delay the evolution of triazine resistance in corn.

Why has rotation been a better strategy than predicted? We now think we know some of the answers and can incorporate them into an updated model, using similar equations, but with better data inserted. The newer data mainly consider the highly reduced fitnesses of the resistant biotypes, which are of greater magnitude and importance than we had expected. We discuss below some of these indications; more data are clearly needed, especially on intraspecific competition.

Revised model: problems to be considered

The original model used an average fitness differential for all the generations treated (Fig. 1A). The fitness differential between resistant and susceptible individuals essentially disappears with herbicides such as triazines, giving season long control, as there is no time for this differential to be expressed. Only the resistant biotypes can survive when the herbicide is present. Thus, the fitness differential seemed to be unimportant with triazines, but was possibly an important factor in delaying resistance to other herbicides with more ephemeral action. The fitness differential was only considered important when herbicide usage was stopped for a long duration and was responsible for the decay in the resistant populations (Fig. 1B) (3, 4). We now think we know more about how fitness should be measured, and we have learned that resistant biotypes are often more susceptible to some of the herbicides and cultivation procedures used in the rotational years (negative cross-resistance). Based on this we now have to modify the model to consider what happens to resistant individuals in the "off" years when the herbicide in question is not used.

Fitness of resistant weeds

The very low frequencies of resistant individuals in the field must compete with the crop, with resistant members of other weed species, and (when the herbicide is not present) with susceptible members of the same and other species. If triazine resistance evolved only in populations with plastome mutator genes (see above), there is the strong possibility of multiple mutations in these plastid genomes, including deleterious mutations giving unfit alleles of other genes. These may explain much of the unfitness of resistant plants, as well as the published variabilities of plastid fitness. Deleterious nuclear mutations can be bred or selected out of populations, because of chromosomal segregation and through somatic and meiotic crossing over. This is not as easy with chloroplasts, where such recombinations as negligible. Thus, the unfitness in atrazine-resistant plants may not be due to the *psbA* gene mutation, as has already been argued on biophysical grounds (e.g. 7, 24–26).

During the evolution of resistant populations only intraspecific competition has been considered, except for one study (9). We await more data from the agro-ecologists on the importance of interspecific competition.

The earliest and classical studies on competitive fitness were performed by pregerminating seedlings of resistant and susceptible individuals, interplanting them at fixed distances, and allowing them to grow to maturity (5). The yields of the resistant and susceptible biotypes were measured. In almost all cases where this was done, the susceptible individuals outyielded the resistant ones (Table I). Some of the triazine-resistant grasses have been reported to be more productive (27), but this must be checked under more rigorous conditions.

Most competition experiments have not been made with material that has nuclear isogenicity, where resistant and susceptible alleles are in otherwise identical nuclear backgrounds. Nuclear isogenicity is easy to achieve by using reciprocal hybrids in cytoplasmically inherited triazine resistance. There is no way at present to guarantee an identical plastome (except for the resistance gene) in such crosses until plastomes are engineered by site directed mutagenesis to have only psbA gene mutations. Repeatedly backcrossed material also provides near-nuclear isogenicity with a large differential in competitive fitness (Table I).

In *Chenopodium strictum* there seems to be no fitness difference between resistant and susceptible individuals (9). *C. strictum* is a very slow-growing species and it is unlikely that photosynthetic electron transport limits its growth. Other cases are more complicated and may be due to various interrelated functions: (a) fitness was not measured from germination on; (b) there may be density dependent functions; (c) there may be different reactions to environmental conditions; (d) there may be different germination characters and seed bank dormancies; (e) the narrow genetic base of resistant individuals (28–30) is probably detrimental compared to the broad base in the susceptible wild type.

(a) *Fitness from germination.* Weeds produce hundreds to thousands of seeds to replace one plant. Clearly most must perish before maturity, many during overwintering, early germination, and establishment. The competition prior to establishment can well be the fiercest. None of the reported competitive fitness studies measured competition at this stage. The simplest way would be to plant mixtures of resistant and susceptible seed and assay (using small leaf discs) which plants are resistant and susceptible. The seeding should be done at a variety of depths and densities (see below), and preferably under field conditions to best mimic the natural environment. As seeds from resistant plants are often smaller than those from sensitive plants, there surely should be a definite competitive disadvantage to resistance when the selecting herbicide is not present.

(b) *Density dependence of fitness* has not been adequately measured. For example, the reasons that *Poa annua* biotypes resistant to triazines evolved only in genotypes that were prostrate and not in those that are erect (31) are not clear. Possibly, the lack of competition allowed the prostrate types to spread. There are variations in the density dependence of diclofop-methyl resistant *Lolium* (13). The resistant plants were more fit under sparse than dense spacing. The fitness also varies with ratio of resistant to susceptible individuals in competition. In Table I, only the data from a 1:1 ratio are given. A perusal of much of the original data show that the resistant biotypes are even less fit than in the 1:1 mixture when they are in a lower proportion in the population. The data suggest that fitness should be measured at various densities and ratios of sensitive to resistant individuals.

(c) *Environment and fitness.* A lowered optimal temperature for growth and photosynthesis for resistant biotypes is one of the common (32) but not universal (32) pleiotropic (33) effects concomitant with triazine resistance. Nuclear isogenic lines of the same species were used in (11, 32) but not in (33). The triazine-resistant biotypes often germinate at lower temperatures (34, 35). Interpretation of these findings can be complicated: the earlier germination and 'head start' can be highly advantageous under many greenhouse conditions but in the field it can be devastating. A late frost will decimate the earlier germinating resistant population and leave the later germinating susceptible population. This shows that fitness must be measured under field conditions to be indicative of what happens in the field.

(d) *Changing seedbank characters.* Repeated and strong selection for resistant weeds under monoculture can easily abolish the "spaced out" germination typical of weeds during the season and over many seasons. The selection pressure being applied is similar to that used by our ancestors who turned weeds into crops by selecting for germination uniformity, thus abolishing weed germination characters. Under many fitness tests, this higher immediate germination of resistant individuals versus the susceptibles (e.g. 36) may give real but misleading results that will not mimic long term fitness properties under field conditions.

(e) *Narrow genetic base.* Electrophoretic studies of resistant populations have always shown that they possess a much narrower base than adjacent susceptible populations (28–31). This is due to the 'founder effect' of mutants in diverse populations measured soon after evolution. In genetic evolutionary terms this suggests that under certain narrow conditions the resistant populations may be more fit than the wild type, but under broad and varying environmental conditions they will be less fit.

In many cases this might mean that the fitness will slowly increase given the chance of repeated crossing with the wild type susceptible population due to 'nuclear compensation'. Because of the fundamental biochemical lesion caused by triazine resistance it is doubted that fitness could really increase. One can increase the yield of triazine-resistant species by intercrossing. Still the reciprocal intercrosses with the sensitive biotype as female parent always outyielded the resistant offspring in such crosses (37). The effect of interbreeding increasing fitness will have to be checked with other types of herbicide resistance.

Negative cross resistance

One of the most important factors that the early model did not consider, because it was then unknown, is that triazine-resistant individuals were often less fit under the agronomic procedures used during the 'off' years when triazines were not used. Unfortunately we have no accurate counts of the demise of resistant populations when triazine usage was stopped due to high resistance levels; these would have been most telling. What has been shown is as follows:

(i) Standard mechanical cultivations of mixed resistant and sensitive *Senecio vulgaris* populations reduced the resistant individuals by a far greater extent than the susceptible ones (14).

(ii) The lack of fitness often can be due to other biotic factors. It was found that triazine-resistant rutabagas, which nominally produced yields as high as the near iso-nuclear susceptible biotype, were totally and selectively decimated by a viral infection (Souza-Machado, 1987, pers. comm.). Triazine-resistant *Amaranthus hybridus* was selectively eaten by beetle larvae, and triazine-resistant *Chenopodium album* was more susceptible to fungal disease than the wild-types (R. Ritter, 1988, pers. comm.).

(iii) Many herbicides are more toxic to resistant individuals than to susceptible ones (Table II). Table II only contains data for negative cross-resistances, but these are not the preponderant case and are clearly not universal. Still, they can clearly be elucidated and incorporated into strategies for managing resistant weeds.

The negative cross-resistances in atrazine-resistant weeds include herbicides that act at or near the same site in photosystem II (DNOC and dinoseb) as well as herbicides acting on other photosystems (paraquat) or at totally different sites. There was negative cross-resistance to other tubulin binding herbicides in dinitroaniline resistant *Eleucine indica* (Table II), but not to six commercial herbicides on this weed (12). The negative cross-resistance to imazaquin (Table II) occurred in only one of 21 chlorsulfuron resistant mutants. The other mutants had varying levels of co-resistance to imazaquin.

Resistant biotypes sometimes grow better in the presence of the herbicide than without the herbicide. For instance, Lipecki (pers. comm.) found that a triazine-resistant biotype of *Amaranthus hybridus* had double the dry weight per plant at 5 kg/ha simazine than without the herbicide. It would be useful to have more such quantitative experiments. This lower resistant biotype productivity when the herbicide is not present results in a stronger lack of competitive fitness in the "off" years.

How the model must be modified

In the seasons when herbicides are used, the enrichment for resistance will be as in Fig. 1A for lines with no fitness differential $(f=1)$. In the "off" years (when the herbicide is not used) the fitness of the resistant individuals may be exceedingly low. The overall fitness is a compounded fitness of each stage or condition where measured: germination, establishment, growth, effects of environment, agronomic procedures and herbicides. Both inter- and intraspecific

Table II. Negative Cross-Resistance of Herbicide Resistant Biotypes

Primary resistance species	Negative cross-resistance	parameter measured	I_{50} R/S	ref
Dinitroaniline			[b]	
Eleucine indica	chlorpropham			(38)
triazines				
Amaranthus retroflexus	dinoseb	FW	.27	(39)
	fluometuron	thylakoids	.22	(40)
	DNOC	thylakoids	.5	(40)
Chenopodium album	dinoseb		.27	(39)
Brassica napus	dinoseb		.66	(39)
Senecio vulgaris	dinoseb		.21	(39)
Conyza canadensis	DNOC	thylakoids	.1	(41)
Kochia scoparia	2,4-D	FW	[b]	(42)
Epilobium ciliatum	oxyfluorfen	FW	[b]	(43)
	paraquat	FW	[b]	(43)
	pyridate	FW	[b]	(43)
	chlorpropham	FW	.46	(44)
mecoprop				
Stellaria media	benazolin	FW	.53	(45)
MSMA-DSMA				
Xanthium stramonium	paraquat		.50	(46)
	bentazon		.65	(46)
Chlorsulfuron				
Datura stramonium[a]	imazaquin	FW	0.03	(47)
Paraquat				
Conyza canadensis	glufosinate	PS	.26	(17)

FW = fresh weight. PS = Photosynthetic CO_2 fixation. Thylakoids = photosystem II activity of isolated thylakoids. The I_{50} is the concentration lowering the parameter measured by 50%. R = resistant; S = susceptible. [a]Strain CSR2 only. [b]Where no I_{50} R/S ratio is given, there was a large degree of negative cross-resistance at a single herbicidal rate.

competitions should be considered at all stages. We have no information on differential resistant vs. susceptible survival in the seed bank. As no measurements have been made of the loss of resistant individuals during the off years, we can only generate a series of curves based on expected fitness differentials (Table I) or even higher fitness differentials based on negative cross-resistances (Table II) and the negative effects of the agronomic procedures used with other crops.

The Revised Model

The previous model kept track of the various influences affecting resistant and susceptible weeds. In particular, a seed bank was presumed to contain fully viable seeds for $\bar{n}$ years which then died. We neglected the loss of seeds from the seed bank due to germination. In the appendix to our earlier paper (3), an alternative model was developed based on the hypothesis that a constant fraction of seeds in the seed bank perish each year ($\delta^{(R)}$ for resistants and $\delta^{(S)}$ for susceptibles). In this alternative model the effects of germination on the seed bank are tallied. The two models agreed and yielded Eq. 1 when the factors f and α of Eq. 1 satisfied $f\alpha \gg 1$, i.e., when the herbicide selection pressure was sufficiently high. We now know that this will not be met in rotational years with fitness near 0.1 and $\alpha < 10$.

Our revised model starts with the model in the appendix of ref. (3), as it is somewhat more accurate and biologically realistic. Our previous discussion of the preferred model was terse, so we rederive the equations in Appendix 1 of this chapter. The somewhat more accurate counterpart to Eq. 1 is

$$N_n^{(R)} = N_0^{(R)} \left\{ \left(1 - \gamma^{(R)}\right)\left(1 - \delta^{(R)}\right) + f\alpha[1 - \left(1 - \gamma^{(S)}\right)\left(1 - \delta^{(S)}\right)] \right\}^n \quad \text{(Eq. 2)}$$

This equation explicitly depends on the fraction of susceptible ($\gamma^{(S)}$) and resistant ($\gamma^{(R)}$) seeds that germinate. It also depends on all other aspects, including relative germination ability of resistants in competition with susceptibles. See Eq. A1.20 for the exact definition of the fitness factor f. For simplicity, and in the absence of the detailed information that would permit use of Eq. 2, we shall generally assume that the fraction germinating in a given season is small ($\gamma^{(R)} \ll 1$, $\gamma^{(S)} \ll 1$) and that resistant and susceptible seeds perish to an equal extent in the seed bank:

$$\delta^{(R)} = \delta^{(S)} \equiv \delta \quad \text{(Eq. 3)}$$

With this, Eq. 2 becomes $N_n^{(R)} = N_0^{(R)} \{1 + \delta(f\alpha - 1)\}^n$ (Eq. 4a)

Another important special case of Eq. 2 corresponds to a "no till" situation. Here all seeds germinate. Thus $\gamma^{(R)} = \gamma^{(S)} = 1$ so that Eq. 2 reduces to

$$N_n^{(R)} = N_0^{(R)}[f\alpha]^n \quad \text{(Eq. 4b)}$$

As shown in Ref. (3), the average lifetime of seeds in the seedbank, $\bar{n}$, is related to the factor δ by $\bar{n} = \dfrac{1}{\delta}$ (Eq. 5)

Thus, Eq. 4a, reduces to Eq. 1, in cases when $f\alpha \gg 1$.

Perhaps more meaningful than the average lifetime, $\bar{n}$, of seeds in the seedbank is their half-life, $n_{\frac{1}{2}}$. If there are N_0 seeds in the bank initially, after n years $N_0(1-\delta)^n$ seeds remain. The half-life, $n = \frac{1}{2}$, is thus calculated from $N_0(1-\delta)^n = \frac{1}{2}N_0$, so that δ and $n_{\frac{1}{2}}$ are related by

$$n_{\frac{1}{2}} = \frac{\log(\frac{1}{2})}{\log(1-\delta)} = \frac{0.3}{\log[1/(1-\delta)]} \quad \text{and} \quad 1-\delta = \left(\frac{1}{2}\right)^{1/n_{\frac{1}{2}}} \quad \text{(Eq. 6)}$$

Table III. Relationship Between Seed Bank Survival Fraction $(1-\delta)$, Seed Bank Half-Lives, $n_{\frac{1}{2}}$, and Average Seed Bank Residence Time, $(\bar{n})$

δ	$1-\delta$	$n_{\frac{1}{2}}$	$\bar{n}$
0.50	0.50	1	2.0
0.29	0.71	2	3.4
0.21	0.79	3	4.8
0.16	0.84	4	6.3
0.13	0.87	5	7.7
0.07	0.93	10	14.3

(From Eqs. 5, 6).

For $\delta \ll 1$ $n_{\frac{1}{2}} \approx \dfrac{0.7}{\delta}$ (Eq. 7)

which should be compared to Eq. 5. Table III provides values of δ for several different half-lives.

We previously (3) made calculations with Eq. 1, the counterpart of Eq. 4, for a number of parameter values, showing the enrichment of resistance when various levels of herbicides with different selection pressures are applied annually. We also estimated the effects of periodically stopping herbicide use by setting $\alpha = 1$ in calculating the resistance enrichment factor (Fig. 1B).

We now realize that the fitness coefficient may differ in "on" years from its value in "off" years. Thus we will be working with the following enrichment factor (H), obtained from Eq. 4a, that gives the increase in resistance following a period of p "on" years of herbicide application and q "off" years without herbicide:

$$H_{p,q} = [1 + \delta(\alpha f_{on} - 1)]^p [1 - \delta(1 - f_{off})]^q \quad \text{(Eq. 8)}$$

Note that the p "on" years and the q "off" years can occur in any order during the $p+q$ year period that is under study. Note also that the "off" factor $1-\delta(1-f_{off}) = 1-\delta+\delta f_{off}$ can be approximated by $1-\delta$ if $f_{off} \ll 1$. There is very little effect of a small "off" fitness (e.g. $f_{off} < 0.3$), as there will be a significant loss of resistant seeds during the off years owing to the decimation of these seeds in the seed bank by natural causes (rotting, insects, etc.). This loss will not be compensated for by the small addition that stems from less fit resistant seeds. Note that the "off" factor is simply f_{off} in the absence of a seedbank ($\delta = 1$), as the only factor affecting seed number will be seed deposition, however small. If $\delta \approx 1$ then a strong influence of f_{off} can be expected.

If we compare the results of the original model concerning decline of a resistant population upon stoppage of the initially used herbicide (Fig. 1B) with the present model (Fig. 2), we see that the new model accounts for more rapid reduction of the proportion of resistant individuals in the off years. Plots of the various rotational schemes (Fig. 2) show that there are some situations where resistant individuals disappear in "off" years almost as rapidly as they are enriched for in "on" years. This gives a very slow rate of overall enrichment, showing that it will take very many years for resistance to become a major problem. These scenarios are summarized in Table IV, showing the (log) factor of enrichment at the end of a 30 year period. When this factor is compared with the initial frequency of resistance (N_0), we can estimate whether resistant populations should have evolved. If N_0 is 10^{-20}, (a guess for the N_0 for triazine resistance), fitness is 0.3 or less, and selection pressure is low, i.e. $\alpha < 20$, we see that resistant populations will not evolve under many of these scenarios. With chlorsulfuron, N_0 should be 10^{-6} to 10^{-8} (47–50), explaining why the appearance of resistance was so rapid. The selection pressure of 2,4-D is so low that no resistance has occurred in 36 years of monoculture wheat (51), as would be expected from Table IV.

The important predictive uses of this model are two-fold:
(a) to design rotational scenarios to delay resistance as much as possible yet still to obtain cost-effective weed control using herbicides such as chlorsulfuron and atrazine, which are among the least expensive and most active selective herbicides for wheat and corn, respectively;
(b) once resistance has occurred, to design strategies whereby herbicide usage is stopped for a number of years, until the level of resistance is below a certain proportion, and then resume limited use, during a certain proportion of the rotation cycle. Such strategies have been designed for insecticides where there already are predominantly resistant populations. These become diluted because of fitness, and the migration of susceptible individuals into the area (see other Chapters in this book). The treatment strategies are designed such that the maximum proportion of resistant individuals does not exceed a certain limit percentage.

With this model a kill percentage can be calculated that will give any (within reason) desired degree of resistance enrichment (E) after p "on" years and q

Table IV. (Log) Enrichment (log H, see Eq. 8) of Resistant Individuals in Weed Populations Over a 30 Year Period Under Different Herbicide Rotations

Rotation strategy	α (%)	EK	Fitness in "off" years			
			0.9	0.3	0.1	0.01
			Log_{10} of enrichment factor (Log H)			
No Rotation	2	50	3.4	3.4	3.4	3.4
	10	90	17.0	17.0	17.0	17.0
	20	95	24.8	24.8	24.8	24.8
	100	99	44.6	44.6	44.6	44.6
2 on; 1 off	2	50	2.1	1.3	.9	.7
	10	90	11.2	10.3	10.0	9.8
	20	95	16.4	15.5	15.2	15.0
	100	99	29.6	28.7	28.4	28.2
1 on; 1 off	2	50	1.5	0.2	-.3	-.6
	10	90	8.3	7.0	6.5	6.2
	20	95	12.2	10.9	10.3	10.1
	100	99	22.1	20.8	20.2	20.0
1 on; 2 off	2	50	0.9	-.9	-1.6	-1.9
	10	90	5.4	3.6	2.9	2.6
	20	95	8.0	6.2	5.5	5.2
	100	99	14.6	12.8	12.1	11.8

EK = effective kill of weeds for a whole season. The data were calculated assuming a seed bank release of $\delta = 0.3 (\bar{n} = 3.3)$. The table is best used to compare expected mutation frequencies for resistance (e.g. ca 10^{-6} for a dominant monogenic trait; 10^{-12} for recessive monogenic mutants) and determining if resistance would be expected with different management regimes. A minus sign indicates that there is a negative enrichment for resistance.

"off" years. To do this, set the expression $H_{p,q}$ of Eq. 8 equal to any given constant E and solve for α. We obtain from the equation $H_{p,q} = E$ for high persistence herbicides, when $f_{on} = 1$,

$$\alpha = 1 + \delta^{-1} \left\{ E^{1/p} [1 - \delta(1 - f_{off})]^{-q/p} - 1 \right\} \quad \text{(Eq. 9)}$$

We expect that f_{off} will be less than unity for low persistence herbicides. In such cases the right side of Eq. 9 must be divided by f_{on}. For example, if the relative lack of fitness in resistants during on years is expressed by $f_{on} = 0.5$, then the permitted value of α will be doubled.

An important special case of Eq. 9 is when *resistance stasis* is desired ($E = 1$), and resistants have low fitness in off years ($f_{off} \ll 1$). In this case, Eq. 9

simplifies to the approximate formula

$$\alpha \approx 1 + \delta^{-1}\left\{[1-\delta]^{-q/p} - 1\right\} \qquad \text{(Eq. 10)}$$

If duration in the seed bank is long enough so that $\delta \ll 1$, we may make a further approximation to obtain $\quad \alpha \approx q/p \qquad$ (Eq. 11)

Another important special case is that of no-till agricultural practices that effectively abolish the seed bank. Taking Eq. 4b into account in such instances, the value of the herbicide kill factor for resistance enrichment E is simply

$$\alpha = E^{1/(p+q)}(f_{off})^{-q/(p+q)} \qquad \text{(Eq. 12)}$$

Values of α, given by Eq. 12, that yield resistance stasis ($E = 1$) for various values of f_{off} in different treatment regimes are shown in Fig. 3. Given this information, weed control of strategies can be designed where there will be no enrichment of resistant individuals. It is clear that one cannot obtain stasis with continuous use of a herbicide, although one can still ensure that the rate of enrichment is low.

Returning to Eq. 9, we show in Fig. 4 the values of α that provide resistance stasis, as a function of the duration of the seedbank, for $f_{off} = 0.2$ (Fig. 4A) and for values of f_{off} that are so small compared to unity ($f_{off} < .05$) that f_{off} essentially has no effect (Fig. 4B). Three possible rotation strategies are examined. We see that there can be cases where there is no enrichment at all for resistance. If the effective kill of 2,4-D in wheat is only 50–60% due to late weed germination, then under low fitness and a 1:1 enrichment there is no enrichment (Fig. 4). Stasis can even be obtained with selection pressures above 90% if there is a 2 or more year interval between the treatments with the herbicide. Stasis is impossible with very high selection pressure herbicides in usual rotational sequences. Long duration in the seed bank is actually a deterrent to stasis, as resistant seeds act as a buffer for longer periods (Fig. 4).

For comparison, values of α are depicted in Fig. 5 that will just double resistance in three years if the "1-on: 2-off" strategy is employed. Note from Fig. 5 that at an intermediate value of δ, i.e., at an intermediate duration of the seeds in the seed bank, the doubling of resistance occurs very slowly at the lowest selection pressures. This means that if the frequency of resistance is 10^{-6}, then it will take almost 60 years for resistant populations to predominate. A parallel phenomenon is observed in Fig. 6 where the enrichment factor $H_{1,2}$ (one "on", two "off") for $\alpha = 10$ (90% effective kill) has a maximum as a function of δ.

To understand the qualitative behavior shown in Figs. 5 and 6 it is helpful to consider analytic formulae. Consider, for example, the maxima found in Fig. 6. From Eq. 8 the enrichment from a 1 on: 2 off strategy is given (with $f_{on} = 1$) by

$$H_{1,2} = [1 + \delta(\alpha - 1)][1 - \delta(1 - f_{off})]^2 \qquad \text{(Eq. 13)}$$

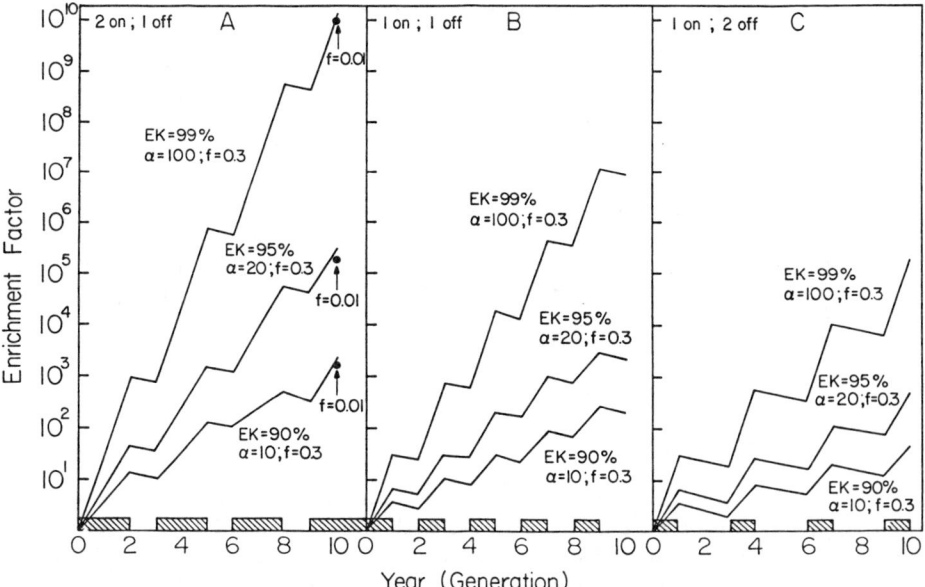

Figure 2. Prediction of the effect of herbicide rotation; revised model. Three rotational scenarios are shown for herbicides with different selection pressures. Drawn based on Eq. 8. The seed bank resistance time is assumed to be $\delta = 0.3$; $\bar{n} = 2$ years; $\bar{n} = 3.3$ years.

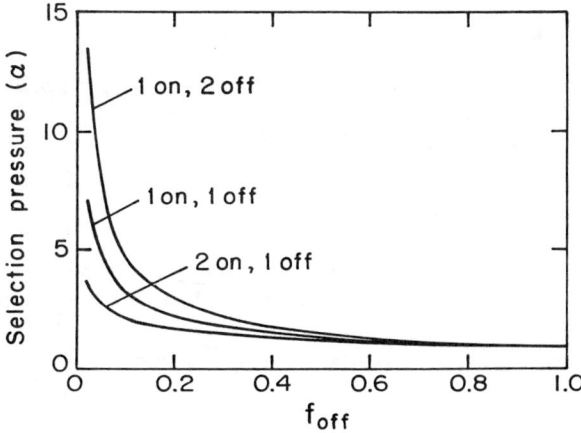

Figure 3. Resistance stasis for "no till" agriculture. Values of selection pressure (α) and fitness in "off" years that will allow no enrichment for resistance (stasis), when $f_{on} = 1$. This was drawn based on Eq. 12. The effective kills are based upon total lack of herbicidal effect on the resistant individuals.

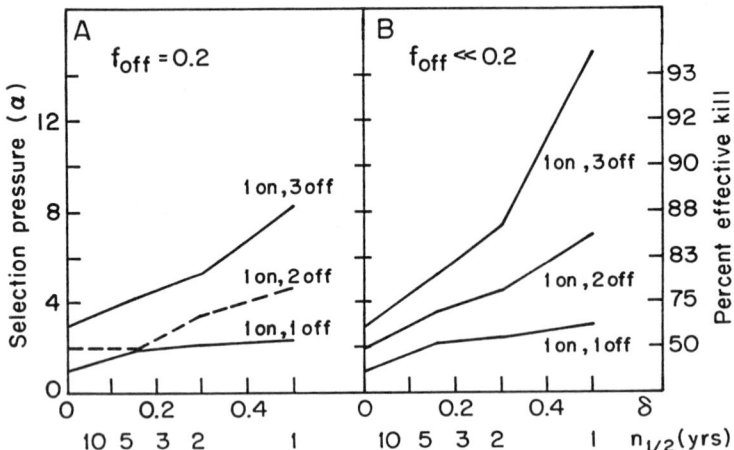

Figure 4. Lack of enrichment for resistance under various selection pressures (α), under different rotation strategies, with different weed seed dynamics in the seed bank. The selection pressure is also shown as "effective kill" (the percent reduction in sensitive propagules over a whole season) with the assumption that the rare resistant individuals are totally unaffected by the herbicide. Here $f_{on} = 1$ while f_{off} takes a relatively large value (A) and a (B) value so small that f_{off} has a negligible effect.

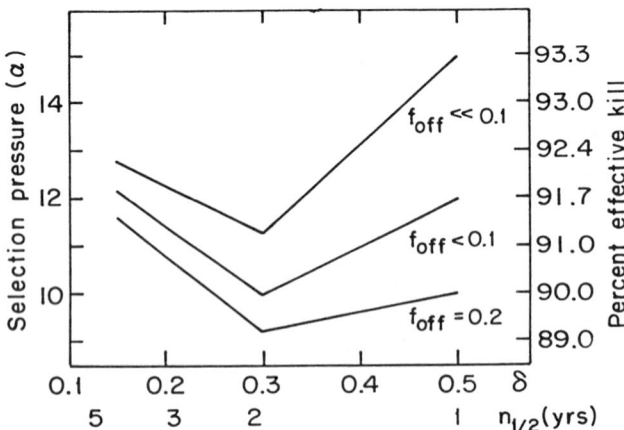

Figure 5. Selection pressures that cause a doubling of the proportion of resistant individuals in the population every 3 years. The data are given for a 1 on: 2 off rotational strategy, with different fitnesses in the off years ($f_{on} = 1$), and different seed bank dynamics.

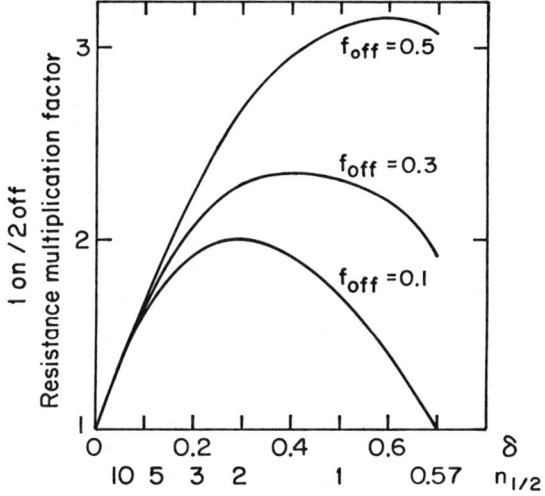

Figure 6. Amplification of resistance in a 3 year period as a function of fitness in the off years ($f_{on} = 1$) and seed bank dynamics. A 1 on: 2 off rotational scheme is used and the "on" herbicide has a fixed selection pressure of $\alpha = 10$ (effective kill = 90 %).

The derivative of $H_{1,2}$ with respect to δ vanishes at

$$\delta = \frac{\alpha - 1 - 2(1 - f_{off})}{3(1 - f_{off})(\alpha - 1)} \approx \frac{1}{3(1 - f_{off})} \qquad \text{(Eq. 14)}$$

As is required to make biological sense, this value of δ is less than unity, provided that

$$f_{off} < \frac{2\alpha}{3\alpha - 1} \approx \frac{2}{3} \qquad \text{(Eq. 15)}$$

where in Eqs. 14 and 15 the approximations are derived under the supposition that $\alpha \gg 1$.

It thus transpires that if Eq. 15 holds, then $H_{1,2}$ has a maximum at the value of δ given by Eq. 14. It is this maximum that is depicted in Fig. 6. The reason for this somewhat surprising behavior can be ascertained by examining Eq. 13. Decreasing δ (i.e. at seed banks with longer durations) decreases the first factor in Eq. 13 and increases the second, squared, factor. This reflects the fact that seed bank inertia slows both the enrichment in resistant individuals owing to herbicide use, and also slows the decrease in resistance when such use ceases. If $f_{off} \approx 1$ (as it may, with chlorsulfuron), the effect on the second factor in Eq. 13 is negligible, as the coefficient of δ almost vanishes. But if f satisfies Eq. 15, the effect of decreasing δ (increasing $\bar{n}$ and $n_{\frac{1}{2}}$) is predominantly to decrease the loss of resistance (in off years) when δ is sufficiently large, but strongly to decrease the enrichment of resistance (in on years) when δ is sufficiently small. We then expect maximum enrichment at an intermediate value of δ.

This new model describes the enrichment for resistant individuals in the population only when they are still only a very small proportion of the total population, and not when a population actually nears resistance. New considerations are thus needed for resistance management once there is a preponderance of resistant individuals. The consideration of the lower fitness in the off years can be further refined when actual information is available on fitness in different field situations with different species, as well as more information on the diminution of resistant populations under various agronomic and herbicide treatments. Such information will also be needed for many types of herbicide resistance to allow more accurate predictions on management strategies. What is clear is that all resistances so far have occurred in monoherbicide/monoculture and that weeds receiving the same number of treatments, but over a longer period in rotational situations, have not evolved resistant populations. Most resistant individuals are less fit (Table I) because of the nature of the target site mutation conferring resistance or of the genes required to detoxify the herbicide. In other cases resistant individuals may not be unfit, or because of their nuclear inherited nature, many of the deleterious co-mutations may be lost. There is evidence that chlorsulfuron resistant mutants do not have lower productivity (48), but the competitive fitness of these plants has not been measured. There are theoretical reasons, based on the site of the mutation on the gene to assume that these sulfonylurea resistant mutants need not be very unfit (52). Thus, rotating chlorsulfuron with other herbicides may not delay resistance beyond the proportion of off years, as it has with atrazine and trifluralin. The added value of rotation will only be where the fitness is very low in the off season.

In summary: the best tactics to prevent or delay the appearance of resistant populations are

(a) to use herbicide mixtures of compounds acting at different sites of action and having different modes of degradation, preferably with herbicides having heavy negative cross resistances.

(b) to use rotations of herbicides having different sites of action and different modes of degradation; preferably with herbicides having negative cross resistance.

(c) to employ mechanical cultivations in the rotations.

It is hard to meet these criteria with some monocultures, especially wheat. Wheat probably has only a single mode of degradation (2). In such situations it is necessary either to (a) rotate crops to allow herbicide rotation; (b) rotate with herbicides having placement selectivity (e.g. ref. 53).

It is clear that there will be problems with those high selection pressure herbicides having resistant mutants that are fit. The only alternative is to replace these with less persistent herbicides of the same group, having less selection pressure and thus partially offsetting the lack of fitness. The model that we have described, with minor modifications would be appropriate for use with other pesticides. The term "generation" would have to be used where we used seasons or years. Refuges and and immigration would replace the less mobile seed bank. Our models best describe the early enrichment of resistant individuals in populations. The simplifications make the model less accurate when populations already have a large percentage of resistant individuals. Models for resistance management, once resistance predominates, are found in earlier chapters.

Acknowledgments

The authors thank Drs. J. Dekker and J.S. Holt for their comments on this chapter. J.G. is the Gilbert de Botton Professor of Plant Sciences and L.A.S. is the Henry and Bertha Benson Professor of Mathematics.

APPENDIX 1
Deriving the Revised Model

The superscripts "S" and "R" denote "susceptible" and "resistant." Subscript n denotes the years or generations a herbicide is used.

$G_n^{(R)}$, $G_n^{(S)}$: The total number of seeds in the ground per unit area during early spring of nth year, ready to germinate.

$\gamma^{(R)}$, $\gamma^{(S)}$: Germination fraction.

$\delta^{(R)}$, $\delta^{(S)}$: Fraction of seeds in seed bank that perish over a year.

$\theta^{(R)}$, $\theta^{(S)}$: Fraction of seeds deposited in fall that are viable the next spring. In (3) we took $\theta^{(R)}$, $\theta^{(S)} = 1$.

$N_n^{(R)}$, $N_n^{(S)}$: Number of seeds deposited in the fall of year n.

D: Total number of seeds deposited in ground every year. Under any given cropping situation, the number of seeds deposited will be approximately the same each year, despite herbicides, owing to "Parkinsonian" plasticity.

$\rho_n^{(R)}$, $\rho_n^{(S)}$: Fraction of seeds deposited that are resistant, susceptible.

The various effects in the absence of herbicide are documented as follows. To calculate the number (per area) of viable seeds in the spring of next year, begin with this year's figures, subtract the seeds that have germinated, subtract the number that perish, and add new seeds that are deposited in the fall and remain viable over the winter. For resistants

$$G_{n+1}^{(R)} = G_n^{(R)} - \gamma^{(R)} G_n^{(R)} - \delta^{(R)} \left[G_n^{(R)} - \gamma^{(R)} G_n^{(R)} \right] + \theta^{(R)} N_n^{(R)} \quad \text{(Eq. A1.1)}$$

Equivalently $\quad G_{n+1}^{(R)} = G_n^{(R)} (1 - \gamma^{(R)})(1 - \delta^{(R)}) + \theta^{(R)} N_n^{(R)} \quad$ (Eq. A1.2)

Similarly, for susceptibles

$$G_{n+1}^{(S)} = G_n^{(S)} (1 - \gamma^{(S)})(1 - \delta^{(S)}) + \theta^{(S)} N_n^{(S)} \quad \text{(Eq. A1.3)}$$

The number of seeds deposited is found by taking the fraction of D, the total of seeds deposited, that are susceptible and resistant:

$$N_n^{(R)} = \rho_n^{(R)} D, \quad N_n^{(S)} = \rho_n^{(S)} D \quad \text{(Eq. A1.4a, b)}$$

Let us introduce more notations:
$\beta^{(R)}$, $\beta^{(S)}$: Fraction of germinated seeds that become established.

$\psi^{(R)}$, $\psi^{(S)}$: Fraction of established plants that survive to the fall, in the absence of herbicide.

$P_n^{(R)}$, $P_n^{(S)}$: Number of plants remaining in the fall

To calculate the number of plants of a given type (susceptible or resistant) in the fall of year n, start with the number of seeds at the beginning of the year, multiply by the germination percentage, by the fraction of germinated seeds that become established, and by the fraction that survives through the summer, obtaining

$$P_n^{(R)} = G_n^{(R)} \gamma^{(R)} \beta^{(R)} \psi^{(R)}, \quad P_n^{(S)} = G_n^{(S)} \gamma^{(S)} \beta^{(S)} \psi^{(S)} \quad \text{(Eq A1.5a, b)}$$

If resistants and susceptibles produced an equal number of seeds per plant, then the fraction of total seed production that was resistant would be found by determining the fraction of total plants in the fall that were resistant, giving

$$\rho_n^{(R)} = \frac{P_n^{(R)}}{P_n^{(R)} + P_n^{(S)}}, \quad \rho_n^{(S)} = \frac{P_n^{(S)}}{P_n^{(R)} + P_n^{(S)}} \quad \text{(Eq. A1.6)}$$

The factor ν is introduced, to give the relative contribution of resistant and susceptible plants to seed production:

$$\rho_n^{(R)} = \frac{\nu P_n^{(R)}}{\nu P_n^{(R)} + P_n^{(S)}}, \quad \rho_n^{(S)} = \frac{P_n^{(S)}}{\nu P_n^{(R)} + P_n^{(S)}} \quad \text{(Eq. A1.7)}$$

Note that $\rho_n^{(R)} + \rho_n^{(S)} = 1$ (Eq. A1.8)

which correctly shows that seeds are either susceptible or resistant. Note also from Eq. A1.7 that $\rho_n^{(R)} / \rho_n^{(S)} = \nu \left[P_n^{(R)} / P_n^{(S)} \right]$ (Eq. A1.9)

which illuminates the meaning of ν as a fitness factor converting plant ratio to seed ratio. Thus, for example, if $\nu = 0.5$, then resistant plants suffer a fitness deficiency that results in each resistant plant producing half as many seeds as its susceptible counterpart.

We can see the role of most of the other fitness factors by combining Eqs. Eq. A1.9 and Eq. A1.5 to give

$$\frac{\rho_n^{(R)}}{\rho_n^{(S)}} = \left[\chi \frac{\gamma^{(R)} \epsilon^{(R)} \psi^{(R)}}{\gamma^{(S)} \epsilon^{(S)} \psi^{(S)}} \right] \frac{G_n^{(R)}}{G_n^{(S)}}$$

The expression in the square brackets is a reproductive fitness that converts the ratio of resistant to susceptible (viable) seeds in the spring $\left(G_n^{(R)}/G_n^{(S)}\right)$ to the corresponding ratio of seeds deposited in the fall $\left(\rho_n^{(R)}/\rho_n^{(S)}\right)$.

This provides all the equations needed, namely Eq. A1.2 to Eq. A1.6, but they are complicated. They can be simplified under the warranted assumption that (until resistance is manifest) the proportion of resistants is small in the total population. If $\nu P_n^{(R)} \ll P_n^{(S)}$, then from Eq. A1.7

$$\rho_n^{(S)} \approx 1 \quad , \quad \rho_n^{(R)} \approx \nu P_n^{(R)}/P_n^{(S)} \qquad \text{(Eq. A1.10a, b)}$$

Hence, from Eq. A1.10a and Eq. A1.4a: $N_n^{(S)} \approx D$ (Eq. A1.11)

Eq. A1.11 merely says that if there are relatively few resistants, the number of susceptible seeds each year is roughly D, the invariant total number of seeds deposited. The number of resistants is relatively small, but we of course wish to keep track of the increase of this relatively small number.

If we substitute Eq. A1.11 into Eq. A1.3, we obtain

$$G_{n+1}^{(S)} = G_n^{(S)}\left(1-\gamma^{(S)}\right)\left(1-\delta^{(S)}\right) + \theta^{(S)} D \qquad \text{(Eq. A1.12)}$$

We claim that rather quickly (in the absence of herbicide) the number of susceptible seeds in the spring will approach a constant value, G:

$$G_n^{(S)} \to G \qquad \text{(Eq. A1.13)}$$

To find this constant, substitute $G_n^{(S)} = G$ and $G_{n+1}^{(S)} = G$ into Eq. A1.12.

This gives $G = G(1-\gamma^{(S)})\left(1-\delta^{(S)}\right) + \theta^{(S)} D$ (Eq. A1.14)

Solving this algebraic equation, we find that

$$G = \frac{\theta^{(S)} D}{1 - (1-\gamma^{(S)})(1-\delta^{(S)})} \qquad \text{(Eq. A1.15)}$$

It is assumed henceforth that the number of susceptible seeds in the spring is this constant value, G.

We are ready for the final formulation. From Eq. A1.4a and Eq. A1.10b

$$N_n^{(R)} = D\rho_n^{(R)} = D\nu P_n^{(R)}/P_n^{(S)} \qquad \text{(Eq. A.16)}$$

Employing Eq. A1.5 and Eq. A1.13, we obtain

$$N_n^{(R)} = D\nu \frac{\gamma^{(R)}\beta^{(R)}\psi^{(R)}}{\gamma^{(S)}\beta^{(S)}\psi^{(S)}} \frac{G_n^{(R)}}{G} \qquad \text{(Eq. A1.17)}$$

Substitution into Eq. A1.2 and employment of Eq. A1.15 yields

$$G_{n+1}^{(R)} = K G_n^{(R)} \qquad \text{(Eq. A1.18)}$$

where $K = \left(1 - \gamma^{(R)}\right)\left(1 - \delta^{(R)}\right) + f\left[1 - \left(1 - \gamma^{(S)}\right)\left(1 - \delta^{(S)}\right)\right]$ (Eq A1.19)

In Eq. A1.19, the fitness factor $f = \nu \dfrac{\gamma^{(R)}\beta^{(R)}\psi^{(R)}\theta^{(R)}}{\gamma^{(S)}\beta^{(S)}\psi^{(S)}\theta^{(S)}}$ (Eq. A1.20)

includes seed production (ν), germination (γ) establishment (β), survival during the growing season (ψ), and over-winter seed survival (θ). The other factors in Eq. A1.19 again refer to germination (γ) and also to loss of seeds from the seed bank (δ) to other causes. All these factors combine in Eq. A1.19 to form the overall *resistance multiplicative factor*, K.

Eq. A1.19 shows how the number of resistant seeds change from year to year. There are three main effects.

(i) There is a loss of resistant seeds in the ground by rotting and germination. As in seen in Eq. A1.2 this effect is quantitated by $(1 - \gamma^{(R)})(1 - \delta^{(R)})$. To this one must add the gain due to new seeds, which is described by (ii) and (iii).

(ii) By our Parkinsonian assumption, there is a constant total number of seeds deposited each fall. Those that are not susceptible are resistant. The factor $(1 - \gamma^{(S)})(1 - \delta^{(S)})$, as in (i) gives the loss of susceptibles. The factor $1 - (1 - \gamma^{(S)})(1 - \delta^{(S)})$ thus quantifies the fraction of seeds that are resistant.

(iii) Differentials in fitness translate the relative number of seeds in a given year to the relative number of seeds deposited next year. This is expressed in the fitness factor f.

We show in Appendix 2 that on the reasonable assumption that susceptibles are more fit than resistants, then $K < 1$. By Eq. A1.19, any given proportion of resistants will then decrease to zero by a fraction K per generation. If mutations were considered, then the decrease would be to a low level determined by the mutation frequency and the fitness.

Let us now take herbicide into account by defining

$\alpha^{(R)}$, $\alpha^{(S)}$: the fraction of (resistant, susceptible) plants that survive the herbicide treatment.

We again obtain an equation of the form Eq. A1.18, but the resistance decrement factor, K ($K < 1$), is replaced by a new constant that incorporates the effect of herbicide

$$G^{(R)}_{n+1} = H G^{(R)}_n$$ (Eq. A1.21)

The *resistance enrichment factor in the presence of herbicides*, H, differs from the factor K (Eq. A1.19) only in that the fitness factor f is replaced by $f\alpha$, where

$$\alpha = \alpha_R/\alpha_S$$ (Eq. A1.22)

This gives Eq. 2 of the main text, as $G^{(R)}_n$ is proportional to $N^{(R)}_n$, by Eq. A1.17. It can be shown that (with the approximation that resistants are far less numerous than susceptibles) the enrichment factor H of Eq. A1.21 is identical to the corresponding quantity A in the end of the Appendix to ref. (3) with the minor change that an additional viability factor, θ, has been considered here.

APPENDIX 2
Proof that $K \leq 1$

Consider K of Eq. A1.19. Write f of Eq. A1.20 as $f = g\gamma^{(R)}/\gamma^{(S)}$. Assume that the fitness of resistant individuals is less than or equal to that of susceptibles:

$$\gamma^{(R)} \leq \gamma^{(S)} \leq 1 \quad , \quad \delta^{(S)} \leq \delta^{(R)} \leq 1 \quad , \quad g \leq 1 \quad \text{(Eq. A2.1)}$$

As the factor multiplying it is non-negative, K is an increasing function of g. The largest permissible value is

$$K_{g=1} = (1 - \gamma^{(R)})(1 - \delta^{(R)}) + (\gamma^{(R)}/\gamma^{(S)})\left[1 - (1 - \gamma^{(S)})(1 - \delta^{(S)})\right]$$

However

$$K_{g=1} - 1 = \delta^{(R)} \left\{ \gamma^{(R)} - 1 + \left(\delta^{(S)}/\delta^{(R)}\right) \gamma^{(R)} \left[(1/\gamma^{(S)}) - 1\right] \right\}$$

The quantity in the curly brackets is less than

$$\gamma^{(R)} - 1 + \gamma^{(R)} \left[(1/\gamma^{(S)}) - 1\right] = -1 + \gamma^{(R)}/\gamma^{(S)} \leq 0$$

Thus $K_{g=1} \leq 1$ and $K \leq 1$, as required. Clearly, if any of the inequalities in Eq. A2.1 are strict, then $K < 1$.

Literature Cited

1. Gressel, J. In *Proc. Brit. Crop Protect. Conf.-Weeds*, 1987, p. 479.
2. Gressel, J. *Wheat Herbicides — The Challenge of Emerging Resistance.* Biotechnology Affiliates: Checkendon/Reading, U.K., 1988, p. 247.
3. Gressel, J.; Segel, L.A. *J. Theor. Biol.* 1978, **75**, 349–71.
4. Gressel, J.; Segel, L.A. In *Herbicide Resistance in Plants*, Le Baron, H.M.; Gressel, J., Eds.; Wiley: New York, 1982, p. 235.
5. Conard, S.G.; Radosevich, S.R. *J. Appl. Ecol.* 1979, **16**, 171–7.
6. Ahrens, W.H.; Stoller, E.W. *Weed Sci.* 1983, **31**, 438–44.
7. Holt, J.S. *J. Appl. Ecol.* 1988, **25**, 307–18.
8. Stowe, A.E.; Holt, J.S. *Plant Physiol.* 1988, **87**, 183–9.
9. Warwick, S.I.; Black, L. *Can. J. Bot.* 1981, **59**, 689–93.
10. Gressel, J.; Ben-Sinai, G. *Plant Sci.* 1985, **38**, 29–32.
11. Jacobs, B.F.; Duesing, J.H.; Antonovics, J.; Patterson, D.T. *Can. J. Bot.* 1988, **66**, 847–59.
12. Mudge, L.C.; Gossett, B.J.; Murphy, T.R. *Weed Sci.* 1984, **32**, 591–4.

13. Heap, I.; Knight, R. *Austral. J. Agric. Res.* 1988 (in press).
14. Watson, D.; Mortimer, A.M.; Putwain, P.D. In *Proc. Brit. Crop Protect. Conf.-Weeds*, 1987, p. 917.
15. Putwain, P.D.; Scott, K.R.; Holliday, R.J. In *Herbicide Resistance in Plants*, LeBaron, H.; Gressel, J., Eds.; Wiley: New York, 1982, p. 99.
16. Solymosi, P.; Lehoczki, E.; Laskay, G. *Weed Sci.* 1986, **34**, 175-80.
17. Pölös, E.; Mikulas, J.; Szigeti, Z.; Laskay, G.; Lehoczki, E. In *Proc. Brit. Crop. Protect. Conf.-Weeds*, 1987, p. 909.
18. Arntzen, C.J.; Duesing, J.H. In *Advances in Gene Technology*, Downey, K.; Voellmy, R.W.; Ahmand, F.; Schultz, J., Eds.; Academic: New York, 1983, p. 273.
19. Gressel, J. In *Pesticide Resistance: Strategies and Tactics for Management*, National Acad. Press: Washington, D.C., 1986, p. 54.
20. Heap, I.M. In *Proc. 8th Austral. Weeds Conf.*, 1987, p. 114.
21. Moss, S.R. In *Proc. Brit. Crop Protect. Conf.-Weeds*, 1987, p. 879.
22. Kemp, M.S.; Caseley, J.C. In *Proc. Brit. Crop. Protect. Conf.-Weeds*, 1987, p. 895.
23. Gressel, J. In *Plant Growth Substances 1988*, Pharis, R.P.; Rood, S., Eds.; Springer: Heidelberg, 1989 (in press).
24. Ireland, C.R.; Telfer, A.; Covello, P.S.; Baker, N.R.; Barber, J. *Planta* 1988, **173**, 459-67.
25. Lemoine, Y.; Dubacq, J.-P.; Zabulon, G. *Can. J. Bot.* 1986, **64**, 2999-3007.
26. Sinclair, J.; Macdonald, P. *Can. J. Bot.* 1987, **65**, 2147-51.
27. Yaacoby, T.; Schonfeld, M.; Rubin, B. *Weed Sci.* 1986, **34**, 181-4.
28. Gasquez, J.; Compoint, J.P. *Agro Ecosyst.* 1981, **7**, 1-10.
29. Darmency, H.; Gasquez, J. *New Phytol.* 1983, **95**, 299-304.
30. Warwick, S.I.; Black, L.P. *New Phytol.* 1986, **104**, 661-70.
31. Darmency, H.; Gasquez, J. *New Phytol.* 1983, **95**, 289-97.
32. Ducruet, J.-M.; Ort, D.R. *Plant Sci.* 1988, **56**, 39-48.
33. Vencill, W.K.; Foy, C.L.; Orcutt, D.M. *Envir. Exp. Bot.* 1987, **27**, 473-80.
34. Gasquez, J.; Darmency, H.; Compoint, J.P. *Weed Res.* 1981, **21**, 219-25.
35. Weaver, S.E.; Thomas, A.G. *Weed Sci.* 1986, **34**, 865-70.
36. Mapplebeck, L.R.; Souza-Machado, V.; Grodzinski, B. *Can. J. Plant Sci.* 1982, **62**, 733-9.
37. Beversdorf, W.D.; Hume, D.J.; Donnelly-Vanderloo, M.J. 1988. submitted.
38. Vaughn, K.C.; Marks, M.D., Weeks, D.P. *Plant Physiol.* 1987, **83**, 956-64.

39. Fuerst, E.P.; Nakatani, H.Y.; Dodge, A.D.; Penner, D.; Arntzen, C.J. *Plant Physiol.* 1985, **77**, 984–9.
40. Oettmeier, W.; Masson, K.; Fedtke, C.; Konze, J.; Schmidt, R.R. *Pestic. Biochem. Physiol.* 1982, **18**, 357–67.
41. Lehoczki, E.; Laskay, G.; Pölös, E.; Mikulas, J. *Weed Sci.* 1984, **32**, 669–74.
42. Salhoff, C.R.; Martin, A.R. *Weed Sci.* 1986, **34**, 40–2.
43. Clay, D.V. In *Proc. Brit. Crop Protec. Conf.-Weeds*, 1987, p. 925.
44. Bulcke, R.; Verstraete, F.; VanHimme, M.; Stryckers, J.; In *Weed Control on Vine and Soft Fruits*; Calvalloro, R.; Robinson, D.W., Eds.; Balkema: Rotterdam, 1987, p. 57.
45. Lutman, P.J.W.; Snow, H.S. In *Proc. Brit. Crop Protect. Conf.-Weeds*, 1987, p. 901.
46. Haigler, W.E.; Gossett, B.J.; Harris, J.R.; Toler, J.E. *Weed Sci.* 1988, **36**, 24–7.
47. Saxena, P.K.; King, J. *Plant Physiol.* 1988, **86**, 863–7.
48. Chaleff, R.S.; Ray, T.B. *Science* 1984, **223**, 1148–51.
49. Haughn, G.W.; Somerville, C. *Mol. Gen. Genet.* 1986, **204**, 430–4.
50. Stannard, M.E.; Fay, P.K. In *Abstracts-Weed Sci. Soc. Amer.* 1987, p. 61.
51. Hume, L. *Can. J. Bot.* 1987, **65**, 2530–6.
52. Schloss, J.V.; Ciskanik, L.M.; Van Dyk, D.E. *Nature* 1988, **331**, 360–2.
53. Glasgow, J.L.; Mojica, E.; Baker, D.R.; Tillis, H.; Gore, N.R.; Kurtz, P.G. In *Proc. Brit. Crop Protect. Conf.-Weeds*, 1987, p. 27.

RECEIVED June 27, 1989

Chapter 31

Herbicide-Resistant Plants Carrying Mutated Acetolactate Synthase Genes

Mary E. Hartnett, Chok-Fun Chui, C. Jeffry Mauvais, Raymond E. McDevitt, Susan Knowlton, Julie K. Smith, S. Carl Falco, and Barbara J. Mazur

Agricultural Products Department, E. I. du Pont de Nemours and Company, Experimental Station, P.O. Box 80402, Wilmington, DE 19880-0402

Acetolactate synthase (ALS) is the target enzyme for three unrelated classes of herbicides, the sulfonylureas, the imidazolinones, and the triazolopyrimidines. We have cloned the genes which specify acetolactate synthase from a variety of wild type plants, as well as from plants which are resistant to these herbicides. The molecular basis of herbicide resistance in these plants has been deduced by comparing the nucleotide sequences of the cloned sensitive and resistant ALS genes. By further comparing these sequences to ALS sequences obtained from herbicide-resistant yeast mutants, two patterns have become clear. First, the ALS sequences that can be mutated to cause resistance are in domains that are conserved between plants, yeast and bacteria. Second, identical molecular substitutions in ALS can confer herbicide resistance in both yeast and plants. These findings have been extended by oligonucleotide directed *in vitro* mutagenesis of plant ALS genes, followed by introduction of the mutated genes into sensitive plants. The herbicide-resistant transgenic plants so produced provide additional evidence for the commonality of mutations which specify herbicide resistance in ALS genes. Some implications of this work for predicting and addressing the problem of herbicide-resistant weeds are discussed.

Commercial herbicides are traditionally discovered by screening chemical compounds for toxicity to weeds and secondarily for lack of toxicity to particular crop species. Selective activity to weeds but not to crops is often due to metabolism of the herbicide to a non-toxic derivative by the crop. Development of one new herbicide often requires the expensive screening of thousands of compounds and generally is carried out only for the major crop species. To increase grower options by enabling the use of desirable herbicides on additional crop species, to increase the margin of safety in the use of selective herbicides, and to increase options for crop rotations, genetic modification of crops to express resistance to herbicides can be used. Herbicide resistance can be introduced into plants by three different strategies. One method is mutation breeding, in which mutations are introduced into the germ line through chemical mutagenesis of the seeds or pollen. A second method employs plant cell culture to select individual resistant cells *in vitro* followed by regeneration of plants from the resistant cells. A third method, which will be discussed in this paper, is genetic transformation. In this approach genes that code for altered

0097–6156/90/0421–0459$06.00/0
© 1990 American Chemical Society

herbicide target proteins that are not inhibited by the herbicide, or genes that code for herbicide detoxifying enzymes, are transferred to plant cells, and resistant plants are regenerated. We have successfully used the genetic transformation technique to produce crop plants that are resistant to the sulfonylurea herbicides.

BACKGROUND

The sulfonylurea herbicides are a new family of chemical compounds, some of which are selectively toxic to weeds but not to crops. The selectivity of the sulfonylureas results from their metabolism to non-toxic compounds by particular crops, but not by weeds. In addition to efficient weed control, the sulfonylurea herbicides provide environmentally desirable properties such as field use rates as low as two grams/hectare and very low toxicity to mammals. The high specificity of the herbicides for their molecular target contributes to both of these properties. In addition, the low toxicity to mammals results from their lack of the target enzyme for the herbicides. Sulfonylureas inhibit the enzyme acetolactate synthase (ALS), also known as acetohydroxyacid synthase (AHAS), which catalyzes the first common step in the biosynthesis of the branched chain amino acids leucine, isoleucine and valine. In mammals these are three of the essential amino acids which must be obtained through dietary intake because the biosynthetic pathway for the branched chain amino acids is not present. The prototype structure of a sulfonylurea herbicide is shown in Figure 1.

Prior to the identification of the site of action of the sulfonylurea herbicides, herbicide-resistant tobacco plants were selected *in vitro* using cell culture techniques. Genetic studies of the resistant plants indicated that the resistance phenotype was dominant or semi-dominant and segregated as a single nuclear gene in one of two linkage groups (1). The site of action of the sulfonylurea herbicides was identified through physiological studies in bacteria. Sulfonylurea herbicides inhibited the growth of some bacteria on minimal media, but not on rich media or on minimal media supplemented with the branched chain amino acids. Biochemical studies indicated that bacteria produced an ALS enzyme that was inhibited by the sulfonylurea herbicide sulfometuron methyl at nanomolar concentrations (2). Three isozymes of ALS are present in *Escherichia coli* and *Salmonella typhimurium*; ALS II and ALS III are sensitive to inhibition by the herbicide, whereas ALS I is insensitive to the herbicide (2,3). Immediately following the discovery of the site of action of the sulfonylurea herbicides in bacteria, the sensitivity of plant ALS to sulfonylurea herbicides was demonstrated (4) and biochemical, genetic, and physiological studies showed that resistant tobacco plants expressed a herbicide-insensitive form of the ALS enzyme that cosegregated with the herbicide-resistant whole plant phenotype (5). Herbicide-resistant mutants of bacteria, yeast, and algae have been isolated (2,6,7). In genetic studies of these organisms the herbicide-resistance phenotype has been shown to segregate with the herbicide-insensitive ALS enzyme. The genes coding for the resistant form of ALS in bacteria and yeast have been mapped to the *ilvG* and *ILV2* loci, respectively, and both sensitive and resistant forms of the genes have been sequenced.

ALS has also been shown to be inhibited by two other structurally unrelated classes of herbicides, the imidazolinones (8,9) and the triazolopyrimidines (10,11). It has been shown that the toxicity of the sulfonylurea herbicides to bacteria is due, in part, to the accumulation of an ALS substrate α-ketobutyrate, which is itself toxic. It has been suggested that the dual effects of the accumulation of a toxic substrate and the inability to synthesize isoleucine, leucine and valine make ALS a particularly good target for herbicides (12).

PLANT ALS GENES

The ALS gene from the yeast *Saccharomyces cerevisiae* was cloned using a high copy number plasmid library to rescue wild type yeast from minimally inhibitory concentrations of herbicide. The resistance was due to overproduction of the enzyme from the plasmid, which allowed cells to overcome the inhibition by the herbicide (6). Localization of the ALS gene on the plasmid was carried out by deletion analysis, by transposon mutagenesis (13), and by heterologous hybridization to a cloned *Salmonella typhimurium* ALS gene. The hybridization observed between the yeast and bacterial genes indicated sequence conservation between ALS genes from distantly related organisms Subsequent comparisons of the deduced amino acid sequences from the yeast and bacterial genes revealed three regions of extensive homology separated by four nonconserved regions (15). These results suggested that hybridization probes could be designed to isolate ALS genes from other organisms (14). To explore this possibilty, fragments of the yeast ALS gene were hybridized to genomic DNA from *Anabaena 7120* to determine the optimum fragment giving cross-hybridization. A fragment encompassing almost the entire ALS coding region of the yeast clone was chosen and used to screen genomic libraries of *Anabaena 7120*, *Arabidopsis thaliana* and *Nicotiana tabacum* (tobacco), under low stringency conditions. Hybridization was detected to genomic clones from all three species and phage carrying the presumptive ALS genes were isolated (16).

Complete sequence analyses of the *Arabidopsis* and tobacco ALS genes indicated that they code for proteins of 670 (2013 bp) and 667 (2004 bp) amino acids, respectively, with predicted molecular weights of about 73 kilodaltons. Neither of the two genes contain introns, as had been inferred from hybridization experiments with the yeast gene probe which indicated that the plant genes were coded by contiguous 2 Kb DNA fragments. The plant genes are highly conserved, with 75% sequence identity at the nucleotide level and 85% at the amino acid level (16). When compared to the deduced amino acid sequences of ALS from yeast and bacteria, both plant ALS enzymes maintain the three conserved regions previously noted in the microbial enzymes. The sequence similarity between the plant genes, however, extends into the four non-conserved regions from bacteria and yeast (Figure 2). One region of the plant genes, the 5' end of the coding sequence, is not conserved between *Arabidopsis* and tobacco. This portion of each gene is believed to code for a chloroplast transit sequence, since ALS is nuclear encoded (1,5) but the enzyme functions in the chloroplast (17,18). Because the degree of homology in this region is greater at the nucleotide level than at the amino acid level and the hydrophobicity plots of these transit peptides reveal regions of similar profiles, it was suggested that the hydrophobicity characteristics are more important for function than is the primary amino acid sequence (16).

Southern blot analyses of genomic DNA from *Arabidopsis* and tobacco were used to determine the number of ALS genes in these two species. When the genomic blots were hybridized with the isolated ALS genes as probes, two fragments of digested tobacco DNA, and one fragment of digested *Arabidopsis* DNA hybridized (16). This suggested that there are two ALS genes in tobacco and one in *Arabidopsis*. The molecular data correlated well with genetic data which had identified two distinct loci that could mutate to yield herbicide-resistant ALS enzymes in tobacco (1,5), which is an allotetraploid. Similar DNA hybridization analyses of soybean and maize indicated that each contains multiple ALS genes.

HERBICIDE-RESISTANT ACETOLACTATE SYNTHASE

Genomic DNA libraries were prepared from two herbicide-resistant lines of tobacco, the C3 and Hra lines, and probed with the cloned tobacco ALS gene. The C3 line is 100-fold more resistant to the sulfonylurea herbicide chlorsulfuron than is wild type

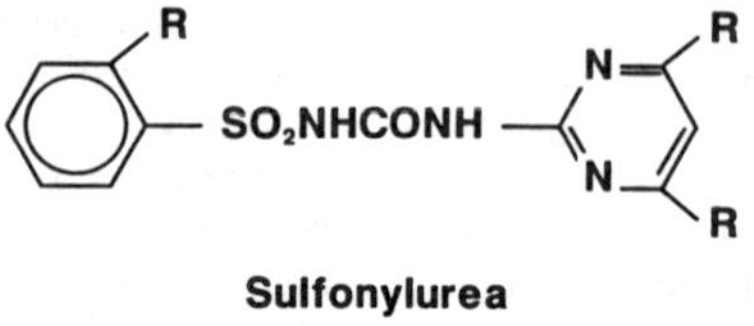

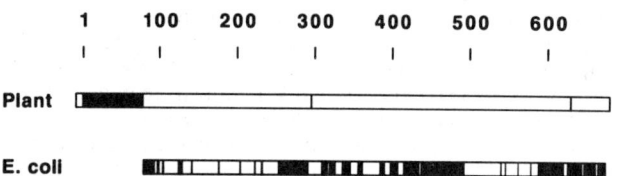

FIGURE 1   Prototype structure of the sulfonylurea herbicides.

FIGURE 2   Amino acid conservation between acetolactate synthase genes. The numbers indicate amino acid residues. The first bar represents a comparison of the deduced amino acid sequences of ALS from tobacco and *Arabidopsis*; the second bar represents a comparison between the amino acid sequences of the three *E. coli* ALS isozymes. Regions of conservation are shown in white.

tobacco and has been shown genetically to be mutated at the SuRA locus (1). The Hra tobacco line, which was isolated by two successive rounds of selection to obtain a highly resistant mutant, is 1000-fold more resistant and is mutated at the SuRB locus (19). Two different genes, representing the two genetic loci, SuRA and SuRB, were isolated from both mutant lines. Comparison of the DNA sequences of the genes from both loci indicated 97% identity at the nucleotide level and 99% identity at the amino acid level in the mature protein regions. The putative transit peptide regions differ considerably more, as a result of in-frame nucleotide duplications or deletions and 23 nucleotide substitutions (20).

The complete DNA sequence of the four genes revealed the identity of the mutations at each of the loci that confer herbicide resistance. The SuRA gene from the C3 line codes for a substitution of glutamine for proline at amino acid position 194. A mutation at the analogous site in the yeast ALS gene (amino acid position 192), which is within a conserved region of all ALS genes, had previously been shown to confer resistance (21). Interestingly, the SuRB gene from the Hra line codes for an alanine for proline substitution at the analogous position (amino acid 191), as well as a leucine for tryptophan substitution at amino acid position 568 (20). The two mutations found in this gene are consistent with the two cycles of selection used to isolate the Hra line. The tryptophan residue at position 568 is within another highly conserved region of ALS and an analogous mutation was found in the yeast ALS gene (see below). Similarly, the mutant ALS gene from a herbicide-resistant *Arabidopsis* line was isolated using the wild type *Arabidopsis* ALS gene as a probe and sequenced. The mutation in the resistance gene encoded a serine substitution for the analogous proline (amino acid position 197) in the *Arabidopsis* protein (22). The identification of mutations in diverse ALS genes coding for substitutions of the equivalent proline or tryptophan residues lead us to postulate that mutations resulting in herbicide resistance might generally be conserved. The results from three different experimental approaches are consistent with this postulate.

The first approach employed selection of mutations in the yeast ALS gene to uncover additional sites of herbicide-resistance mutations. Forty-one independently isolated spontaneous mutations in the yeast ALS gene were characterized by DNA sequencing. These mutations revealed 24 different amino acid substitutions that occur at 10 different sites ranging from the amino to the carboxy ends of the protein. The amino acids at these 10 sites are highly conserved among natural herbicide-sensitive ALS enzymes; the amino acid residues present in the wild type herbicide-sensitive yeast enzyme have been found in all wild type plant ALS enzymes that have been sequenced.

Site-directed mutagenesis was used to make additional amino acid substitutions at these sites in yeast ALS. At some of the sites, e.g. ala117, pro192, or trp586, nearly any substitution for the wild type amino acid that was tested resulted in a herbicide-resistant enzyme (Table I). Each of the mutant enzymes was characterized by enzyme assays to compare its activity, and its sensitivity to the sulfonylurea herbicide chlorimuron ethyl, to the wild type enzyme. These analyses have indicated that some of the mutations have little adverse effect on the activity of the enzyme, while decreasing sensitivity to the herbicide from three to greater than one thousand-fold. The characteristics of these mutant enzymes were further evaluated *in vivo* in order to investigate the utility of particular herbicide/mutant enzyme combinations (Falco *et al.*, manuscript in preparation).

In the second approach, herbicide-resistance mutations in the *Arabidopsis* ALS gene were studied in *E. coli*. To do this, wild type and mutant *Arabidopsis* genes were functionally expressed in *E. coli*, such that the plant genes complemented a branched chain amino acid auxotrophy in the bacteria (Smith *et al.* 1989, PNAS in press). ALS enzyme assays on extracts prepared from *E. coli* expressing the mutant *Arabidopsis* gene indicated that the mutant enzyme is resistant to sulfonylurea herbicides but is sensitive to the imidazolinone herbicide imazaquin. This selective resistance of the *Arabidopsis* mutation when expressed in *E. coli* is consistent with the activity measurements determined in plants (23). A simple in vivo assay employing filter paper disks impregnated with herbicides has allowed the screening of the mutant genes against a large number of herbicides. These two types of assays thus provide a rapid and facile means for determining the sensitivity of mutant plant enzymes to a variety of herbicides.

The third approach used oligonucleotide site-directed mutagenesis of the tobacco ALS gene to analyze mutations that confer herbicide resistance. Among the three mutant genes isolated from the herbicide-resistant tobacco and *Arabidopsis* plants, two mutation sites were identified. Mutations were made in the wild type gene from the SuRA locus at these sites and at additional sites predicted to confer resistance from the yeast experiments. DNA fragments carrying the mutations were subcloned into the wild type gene from the SuRB locus via a common restriction enzyme fragment creating chimeric ALS genes (Figure 3). This resulted in expression of the gene from the more active SuRB promoter and permitted evaluation of the mutations in a novel ALS enzyme to further test the generality of resistance mutations.

Tobacco protoplasts were transformed with mutant and wild type chimeric tobacco ALS genes by direct DNA uptake and analyzed for the ability to produce calli on minimally inhibitory concentrations of the herbicide. In the experiment shown in Table II, four novel herbicide-resistance mutations in the tobacco ALS gene were identified. When introduced into the chimeric tobacco ALS, a proline to serine change at amino acid residue 191, which had been shown to produce herbicide-resistant enzymes from yeast and *Arabidopsis*, allowed the formation of herbicide-resistant tobacco calli. The protoplast transformation experiment also revealed that substitution of leucine for tryptophan at position 568 or alanine for proline at position 191 resulted in herbicide resistance. While these substitutions had been shown to confer resistance individually in yeast ALS, they had previously been tested only in combination in a plant gene isolated from the tobacco Hra mutant. Finally, when alanine 199 was substituted with aspartic acid, resistant calli formed, revealing a site that had been previously identified in yeast ALS, but not plant ALS, to confer resistance (Hartnett *et al.* manuscript in preparation.). Tests of mutations coding for substitutions at two other sites, which were shown to produce herbicide-resistant yeast enzymes, were negative, but inconclusive (not shown). These results confirmed the utility of the microbial systems for predicting and evaluating herbicide-resistance mutations and support the postulate that mutations resulting in herbicide resistance are conserved.

ENGINEERING HERBICIDE-RESISTANT CROPS

The mutant genes isolated from the Hra and C3 lines of tobacco were reintroduced into tobacco and tested for the ability to confer herbicide resistance. These experiments indicated that a mutant gene from either the SuRA or SuRB locus can confer herbicide resistance in transgenic tobacco plants (20, 24). Since the introduction of the mutant gene from the Hra line resulted in higher levels of resistance, this gene was transferred to commercial lines of tobacco, and regenerated plants were assayed for levels of sulfonylurea resistance. Several different methods were used to determine the levels of resistance in the transformed plants, and to assess the feasibility of their use in the field. Resistance was measured by assaying leaf ALS activity, callus growth, and seed germination and growth in the presence of herbicide, and by monitoring plant phytotoxicity after foliar spray applications of the herbicide. The results of these tests were consistent, yet indicated the need to monitor resistance by several methods in order to identify those lines most suitable for crop breeding.

TABLE I. Herbicide Resistant Yeast ALS Mutants

Wild Type Amino Acid Residue		Amino Acid Substitutions Resulting in Resistance
116G	--->	S N
117A	--->	P S T I L V N Q D E K R H W F Y M
192P	--->	A S V Q E R W Y
200A	--->	T V D E R W Y C
251K	--->	P T N D E
354M	--->	V K C
379D	--->	G P S V N E W
583V	--->	A N Y C
586W	--->	G A S I L V N E K R H Y C
590F	--->	G L N R C

TABLE II. Herbicide Resistant Callus Following Co-Uptake of Kanamycin Resistance and Sulfonylurea Resistance Genes

ALS Gene	Protoplast Transformants %Chlorsulfuron^r/Kanamycin^r
none	0.3
wild type[1]	0.0
pro191-->ala	11.8
pro191-->ser	39.0
trp568-->leu	5.5
ala199-->asp	29.5
SuRB-Hra	29.3
pro191-->ala trp568-->leu	50.0
pro191-->ser trp568-->leu	27.2

[1]The wild type and mutant genes (except for SuRB-Hra) were chimeric tobacco ALS genes as described in the text and shown in Figure 3.

```
                    1810                      1830                      1850                      1870                      1890                      1910
AATCGAGTGAGGACTTGAGACATTCTGAGCTCACGGTATCCCGTGCAAGTACTTTGATGGTCTTGAGCTTTCCAACTGGGATGAGCTTTCCCTTTCAATGTTGGTATGC
  S   S   E   D   L   R   R   F   V   E   L   T   G   I   P   V   A   S   T   L   M   G   L   G   A   F   P   T   G   D   E   L   S   L   S   M   L   G   M   H
          1930                      1950                      1970                      1990                      2010                      2030
ATGTACTGTTTATGCTAATTATGCTGTGGACAGTAGTGATTTGTTGCTCGCATTTGGGGTGAGTTTGATAGAGTTACTGAAGCTTTGCTAGCCGAGCAAAAATTG
  G   T   V   Y   A   N   Y   A   V   D   S   S   D   L   L   L   A   F   G   V   R   F   D   D   R   V   T   G   K   L   E   A   F   A   S   R   A   K   I   V
          2050                      2070                      2090                      2110                      2130                      2150
TTCCATTGATATTGATTCAGCTGAGATTGGAAAGAACAAGCAGCCTCATGTTTCCATTTGTGCAGATATCAAGTTGCCGTTACAGGGTTTGAATTCGATCTGGAAGTAAGGAAGGTA
  H   I   D   S   A   E   I   G   K   N   K   Q   P   H   V   S   I   C   A   D   I   K   L   A   L   Q   G   L   N   S   I   L   E   S   K   E   G   K
          2170                      2190                      2210                      2230                      2250                      2270
AACTGAAGTTGGATTTTTCTGCTTGGGAGGCAGGAGTTGACGGAGCAGAAGTTGAAGCACCCATTGAACTTTAAAACTTTTGGTGATGCAATTCCTCCGCAATATGCTATCCAGGTTCTAG
  L   K   L   D   F   S   A   W   R   Q   E   L   T   E   Q   K   V   K   H   P   L   N   F   K   T   F   G   D   A   I   P   P   Q   Y   A   I   Q   V   L   D
          2290                      2310                      2330                      2350                      2370                      2390
ATGAGTTAACTAATGGAATGCTATTAAGTACTGGTGTGGGCAACACCAGATGGGCTTGCTCAATACTACTAAGTACAGAAGCCAATGGTTGACATCTGGTGGATTAGGAG
  E   L   T   N   G   N   A   I   I   S   T   G   V   G   Q   H   Q   M   W   A   A   Q   Y   Y   K   Y   R   K   P   R   Q   W   L   T   S   G   G   L   G   A
          2410                      2430                      2450                      2470                      2490                      2510
CAATGGGATTTGGTTTGCCCGCTATTGGTCGCGTGTTGGAAGACCGATGAAGTTGTTGTTGTTGACATTGATGGTGATGGCAGTTTCATCATGAATGTGCAGGAGCTTGCAACAATTA
  M   G   F   G   L   P   A   A   I   G   A   A   V   G   R   P   D   E   V   V   V   D   I   D   G   D   G   S   F   I   M   N   V   Q   E   L   A   T   I   K
          2530                      2550                      2570                      2590                      2610                      2630
AGTGGAGAATCTCCCAGTTAAGATTATGTTACTGAATAATCAACACCTTGGAATGGTGGTTCAATTGGAGGATCGGTTCTATAAGGCTAACAGAGCACACACATACTTGGGAATCCTT
  V   E   N   L   P   V   K   I   M   L   L   N   N   Q   H   L   G   M   V   V   Q   L   E   D   R   F   Y   K   A   N   R   A   H   T   Y   L   G   N   P   S
          2650                      2670                      2690                      2710                      2730                      2750
CTAATGAGGGGGAGATCTTTCCTAATATTGAAATTTGCAGAGGCTTGTGGCGTACCTGCGAGGTGCGAGGTGACACAGGATGATCTTAGGCGCCTATTCAAAAGATGTTAGACACTC
  N   E   A   E   I   F   P   N   M   L   K   F   A   E   A   C   G   V   P   A   A   R   V   T   H   R   D   D   L   R   A   A   I   Q   K   M   L   D   T   P
          2770                      2790                      2810                      2830                      2850                      2870
CTGGCCACTACTTGTTGGATGTGATTGATTGTACCCTCATCAGGAACATGTTCTACCTATGATTCCCAGTGCCGGGGGCTTTCAAAGATGTGATCACAGAGGGTGACGGAGAAGTTCTATTGAC
  G   P   Y   L   L   D   V   I   V   P   H   Q   E   H   V   L   P   M   I   P   S   G   G   A   F   K   D   V   I   T   E   G   D   G   R   S   S   Y   *
                    2890                      2910                      2930
TTTGAGGTGCTACAGAGCTAGTTCTAGCCCTTGTATTATCTAAAATAAAC
```

FIGURE 3 Nucleotide and amino acid sequences of a chimeric tobacco ALS gene. The SuRB portions extend from nucleotides 1-1236 and from 2656-2930. The SuRA portion extends from nucleotides 1237-2655 and contains two mutations; a C to G mutation at nucleotide 1455 results in a substitution of alanine for proline at amino acid position 191 and a G to T mutation at nucleotide 2587 results in a substitution of leucine for tryptophan at amino acid position 568.

Table III shows a comparison between the percentage of leaf ALS activity that is uninhibited by 10 parts per billion of the sulfonylurea herbicide chlorimuron ethyl, and progeny segregation ratios for the seed of the transgenic plants produced by self-fertilization. The level of herbicide-resistant enzyme activity ranges between 30 and 70 percent of the total ALS activity. The segregation analyses of the progeny plants produced from self-fertilization indicated that the majority of transformants contained the ALS gene integrated at one or two loci. The highest level of resistance was seen in a plant in which the gene had integrated at four loci. However, the number of loci at which resistant ALS genes were integrated was not the sole factor that affected the degree of resistance of the transformants. This is exemplified by plants 40 and 41, which grew equally well in the presence of herbicide, and which had equivalent levels of resistant enzyme. Yet, segregation analyses indicated that plant 40 has two resistant ALS loci, while plant 41 had one. Two possible explanations are that the ALS gene in plant 41 is integrated in a particularly favorable position for expression, or that tandem copies of the gene are present at the single locus. For plant breeding purposes, a high level of herbicide resistance originating from a single genetic locus is preferred.

As a measure of agronomically useful herbicide resistance, some of the transgenic tobacco lines were evaluated in 1987 and 1988 field tests conducted in North Carolina in collaboration with the Northrup King Co. Foliar sprays of the herbicide chlorimuron ethyl were applied to transplanted seedlings at rates of 0, 8, 16 and 32 grams of herbicide per hectare. Transformed plants showed no damage at the highest application rate tested, which is four times that of a typical field application rate of 8 grams of herbicide per hectare (Figure 4). Thus, transformation with the mutant gene from the Hra line provides an effective means of conferring sulfonylurea resistance in crop species.

The mutant gene from the Hra line was also used to transform a number of heterologous species to herbicide resistance. In one example, transgenic tomato plants were assayed for the level of herbicide-resistant ALS enzyme activity in the presence of 10 parts per billion of chlorimuron ethyl. The resistant enzyme activity ranged between 30 and 65 percent of the total ALS activity in the transformed plants. These values are in the same range as those found in the transgenic tobacco plants. Transgenic tomato lines were then tested in the field in Delaware, Florida and France in 1988. A photograph from one of the test plots is shown in Figure 5. These results demonstrated that the mutant tobacco gene can be effective in conferring herbicide resistance to a heterologous species (Figure 5). Similarly, the mutant *Arabidopsis* gene was introduced into tobacco and produced plants resistant to chlosulfuron, demonstrating that this ALS gene could also be functionally expressed in a heterologous plant (22).

THE OCCURENCE OF HERBICIDE-RESISTANT WEEDS

Many of the characteristics which combine to make ALS an excellent target for engineering beneficial herbicide resistance in crop plants may also lead to the proliferation of herbicide-resistant weeds. These characteristics include the following: sulfonylurea herbicide resistance is a semi-dominant trait that is carried on a nuclear gene(s); ALS is the single primary site of action; there are multiple positions in ALS that can be mutated to confer herbicide resistance; mutant ALS enzymes can possess full catalytic activity. The latter property results in engineered crop plants that are fit, but can equally well result in weed biotypes that are fit.

The properties of the herbicides that target ALS can also contribute to the emergence of resistant weeds. There are several classes of compounds which target ALS, including the sulfonylureas, the imidazolinones and the triazolopyrimidines. Within these classes are a number of herbicides that are used at rates that kill a high proportion of the weeds, thus increasing the likelihood that resistant biotypes will

TABLE III. ALS Activity and Inheritance of Herbicide
Resistance in Transformed Tobacco Lines

	% Uninhibited ALS Activity[1]	Segregation Ratio Resistant/Sensitive
NK326 (WT)	7	---
NK326 #1	36	3/1
NK326 #9c	47	3/1
NK326 #9d	37	3/1
NK326 #10	26	3/1
NK326 #10c	56	3/1
K14 (WT)	7	---
K14 #7	71	255/1
K14 #11	52	3/1
K14 #27	45	3/1
K14 #29	30	3/1
K24 #31	44	3/1
K14 #32c	32	3/1
K14 #40	41	15/1
K14 #41	40	3/1
K14 #42	29	3/1
K14 #53	37	3/1
K14 #54	34	3/1
K14 #54A	28	3/1

[1]The ALS activity in each line is related to the activity in the absence of a sulfonylurea herbicide which is taken as 100 percent.

FIGURE 4   Results from a field test of herbicide-resistant tobacco lines, conducted in North Carolina in 1987 in conjunction with Northrup King Co. The row of plants to the right of center contains elite tobacco lines that were transformed with the mutant tobacco ALS gene from the herbicide-resistant Hra line, while the row to the left of center has non-transformed control plants. After treatments with the sulfonylurea herbicide chlorimuron ethyl at 4X normal field application rate, the weeds were killed, the non-transformed controls were severely injured and remained stunted, and the transformed plants remained as vigorous as the unsprayed controls.

FIGURE 5   Results from a field test of herbicide-resistant tomatoes conducted in Delaware in 1988. The variety Herbst red cherry was transformed with the mutant tobacco ALS gene from the herbicide-resistant Hra line. Two different transformed lines flank the non-transformed control plants in the center row. After treatment with a sulfonylurea herbicide at 16X normal field application rate, the weeds and the non-transformed plants were killed while the transformed plants remained vigorous, and subsequently flowered and set fruit.

rapidly dominate the weed population. In addition, many of the compounds have extended residual activity in soil, which creates a continuous selection pressure for the emergence of herbicide-resistant weeds. Finally, crop management practices, such as the tendency to use these herbicides continually in mono-culture cropping systems, can accelerate the proliferation herbicide-resistant weeds.

Indications that herbicide-resistant weeds might arise came from the relative ease of isolation of resistant mutants of tobacco and *Arabidopsis* (1, 23) and from the fitness of the mutant plants. Recently, resistant biotypes of prickly lettuce, Russian thistle, chickweed, and Kochia, which exhibit herbicide-resistant ALS activity, have been identified in a number of locations, following prolonged use of long-residual sulfonylurea herbicides (Saari *et al.*, manuscript in preparation). In order to avoid further increases in the proportion of resistant biotypes, a number of control measures have been implemented. These include changes in the recommended herbicide use patterns, such as reduction in use rates as well as number of applications of the same herbicide in a growing season, and elimination of the use of the same herbicide in successive seasons. The replacement of long soil-residual herbicides with short soil-residual herbicides has also been recommended. Finally, the use of tank mixes of herbicides with differing modes of action has been advised. These changes in management practices should delay the proliferation of herbicide-resistant weeds and allow these environmentally favorable compounds to continue to be used for weed control. The engineering of herbicide-resistant crops will provide additional flexibility, for example by allowing the use of a short-residual herbicide on a crop where no such selective herbicide was available. In addition, the engineering of crops resistant to multiple herbicides will expand the opportunities for the use of tank mixes of herbicides with differing modes of action.

ACKNOWLEDGMENTS

We would like to acknowledge our many colleagues who have participated in these studies. Our co-workers at Du Pont include Tony Guida and Tim Ward on molecular analyses of the ALS genes, Todd Houser, Chris Kostow, Craig Sanders, and Carol Beaman on plant transformations and Gary Fader on green house spray tests. Joan Odell, Perry Caimi, Robert LaRossa, Naren Yadav, Roy Chaleff, Tom Ray, Tina Van Dyk, and Dana Smulski have made valuable contributions to this project in many ways. Finally, we have enjoyed collaborations with John Bedbrook, Jeff Townsend, Kathy Lee, and Pamela Dunsmuir at Advanced Genetic Sciences, Inc. and Chris Somerville and George Haughn at Michigan State University.

LITERATURE CITED

1. Chaleff, R. S.; Ray, T. B. Science 1984, 223, 1148-1151.
2. LaRossa, R. A.; Schloss, J. V. J. Biol. Chem. 1984, 259, 8753-8757.
3. LaRossa, R. A.; Smulski, D. R. J. Bacteriol. 1984, 160, 391-394.
4. Ray, T. B. Plant Physiol. 1984, 75, 827-831.
5. Chaleff, R. S.; Mauvais, C. J. Science 1984, 224, 1443-1445.
6. Falco, S. C.; Dumas, K. D. Genetics 1985, 109, 21-35.
7. Hartnett, M. E.; Newcomb, J. R.; Hodson, R. C. Plant Physiol. 1987, 85, 898-901.
8. Muhitch, M. J.; Shaner, D. L.; Stidham, M. A. Plant Physiol. 1987, 83, 451-456.
9. Shaner, D. L.; Anderson, P. C.; Stidham, M. A. Plant Physiol. 1984, 76, 545-546.
10. Hawkes, T. R.; Howard, J. L.; Pontin, S. E. In Herbicides and Plant Metabolism; Dodge, E. D., Ed.; Cambridge Academic: Cambridge, 1988.

11. Kleswick, W. A.; Ehr, R. J.; Gerwick, B. C.; Monte, W. T.; Pearson, N. R.; Costales, M. J.; Meikle, R. W. European Patent Application 0142152, 1984.
12. LaRossa, R. A.; Van Dyk, T. K.; Smulski, D. R. J. Bacteriol. 1987, 169, 1372-1377.
13. Van Dyk, T. K.; Falco, S. C.; LaRossa, R. A. Appl. Environ. Microbiol. 1986, 51, 206.
14. Mazur, B. J.; Chui, C. -F.; Falco, S. C.; Mauvais, C. J.; Chaleff, R. S. In The World Biotech Report 1985; Online International: New York, 1985.
15. Falco, S. C.; Dumas, K. D.; Livak, K. J. Nucleic Acids Res. 1985, 13, 4011-4027.
16. Mazur, B. J.; Chui, C. -F.; Smith, J. K. Plant Physiol. 1987, 85, 1110-1117.
17. Miflin, B. J. Plant Physiol. 1974, 54, 550-555.
18. Jones, A. V.; Young, R. M.; Leto, K. J. Plant. Physiol. 1985, 77, 5293-5297.
19. Chaleff, R. S.; Sebastian, S. A.; Creason, G. L.; Mazur, B. J.; Falco, S. C.; Ray, T. B.; Mauvais, C. J.; Yadav, N. S. In Molecular Strategies for Crop Protection; Alan R. Liss, Inc.: New York, 1987.
20. Lee, K. Y.; Townsend, J.; Tepperman, J.; Black, M.; Chui, C.-F.; Mazur, B.; Dunsmuir, P.; Bedbrook, J. EMBO J. 1988, 7, 1241-1248.
21. Yadav, N.; McDevitt, R. E.; Benard, S.; Falco, S. C. Proc. Natl. Acad. Sci. USA 1986, 83, 4418-4422.
22. Haughn, G. W.; Smith, J.; Mazur, B.; Somerville, C. Mol. Gen. Genet. 1988, 211, 266-271.
23. Haughn, G.; Somerville, C. Mol. Gen. Genet. 1986, 204, 430-434.
24. Mazur, B. J.; Falco, S. C.; Knowlton, S.; Smith, J. K. In Plant Molecular Biology; von Wettstein, D., Chua, N.-H., Eds.; Plenum Press: New York, 1987; Vol. 140.

RECEIVED June 27, 1989

Chapter 32

Genetic Modification of Crop Responses to Imidazolinone Herbicides

K. E. Newhouse[1], D. L. Shaner[1], T. Wang[1], and R. Fincher[2]

[1]Agricultural Research Division, American Cyanamid Company, P.O. Box 400, Princeton, NJ 08540
[2]Department of Plant Biotechnology, Pioneer Hi-Bred International, Inc., Johnston, IA 50131

>Resistance to imidazolinone herbicides is being introduced into commercial corn varieties. The program to create these hybrids has proceeded through the initiation, implementation and mutant characterization phases, and is currently in the final commercial development phase. Resistant corn was produced by in vitro selection and regeneration of resistant plants. Genetic analyses established that resistance is conferred by a single, semidominant mutant allele. Biochemical analyses demonstrated that resistance is due to altered forms of the enzyme acetohydroxyacid synthase. Commercial sales of imidazolinone-resistant hybrid corn are expected to begin in 1992. This paper tracks the imidazolinone-resistant corn project from its inception through development of the commercial product.

The development of herbicide-resistant crop varieties as undertaken by American Cyanamid is an evolutionary process that proceeds through a series of decision points. Initially, the market potential for herbicide resistance in the crop is evaluated. After a project is assessed as worthwhile, a scheme to make the crop resistant is established and implemented. Once the scheme proves successful and resistance is introduced into the crop of interest, the trait is characterized to assess commercial utility. A method for delivering resistant crops to the marketplace is then identified. Finally, through close cooperation between the seed company and the chemical company, the herbicide-resistant crop becomes available commercially.

This paper follows the evolution of the imidazolinone-resistant hybrid corn project at American Cyanamid. The initial considerations that provided the rationale for the project are discussed. The selection, regeneration, and characterization of the herbicide-resistant plants is described. Finally, options for commercializing the herbicide-hybrid package are discussed.

0097–6156/90/0421–0474$06.00/0
© 1990 American Chemical Society

Why make corn resistant to imidazolinone herbicides?

Selective herbicides are lethal to weeds, yet leave crop plants undamaged. Random screening of synthetic molecules has been the standard industrial approach for identifying selective herbicides. The past success of this approach is apparent when one considers the large number of excellent selective herbicides currently available (1).

However, ideal levels of crop selectivity are difficult to achieve. Several approaches to supplement natural crop selectivity have been used, including; development of chemical safeners that improve herbicide tolerance of corn and sorghum (2), restriction of herbicide use to tolerant cultivars (3), transfer of tolerance to crop cultivars by breeding methods (4), and optimization of the site, timing, and method of herbicide application.

In 1982, American Cyanamid began commercial development of the imidazolinone class of herbicides. These herbicides control a broad spectrum of economically important weeds in corn and soybeans. Although differences in tolerance of corn hybrids for imidazolinone herbicides exist (5), no hybrids with sufficient tolerance to allow application of current imidazolinone herbicides to corn fields have yet been identified. Other methods to supplement the inadequate levels of natural tolerance have likewise been ineffective.

An alternative to discovering a new herbicide for a crop is to change that crop's tolerance for an existing herbicide. This genetic approach has several advantages (6): 1) Adapting registered herbicides to new crops minimizes the huge expense associated with discovery, development, registration, and marketing of a new product. 2) Greater margins of crop tolerance may be achieved. 3) The familiarity of farmers with an existing product facilitates the introduction of new uses for that product. 4) Herbicide-resistant crops may improve flexibility for crop rotations, by alleviating follow crop concerns. 5) Finally, herbicide-resistant crops may create markets for new products that are superior to current products.

Development of imidazolinone-resistant corn provides new markets for products currently being sold by American Cyanamid. Resistant corn also offers farmers new weed control options. For these reasons, in 1982 American Cyanamid initiated a project to introduce imidazolinone-resistance into this important crop species.

Selection of imidazolinone-resistant corn

The procedure by which imidazolinone-resistant corn was selected has been described previously, and will be reviewed only briefly here (7). The program was directed by Dr. Paul Anderson of Molecular Genetics Incorporated (MGI), under a contract with American Cyanamid. Scientists at MGI were among the first to regenerate corn routinely from tissue culture.

Embryogenic corn tissue cultures were grown in media containing sublethal doses of imidazolinone herbicides, and sectors of rapidly growing tissue were identified visually. These rapidly growing sectors were then subjected to successive selection cycles of increasing herbicide concentrations. After several passages, homogeneous resistant cell lines were obtained.

Several independently selected resistant cell lines with distinct phenotypes were identified by this protocol.

Regeneration of plants from the imidazolinone-resistant cell cultures was invariably the most difficult part of the procedure. Cells in culture for long periods of time lose their morphogenic capacity. Imidazolinones may also affect morphogenic capacity. In addition, regenerated plants were often highly or completely sterile.

Fertile plants were obtained from five resistant cell lines. The properties of three of these five mutants are now described.

Characterization of imidazolinone-resistance mutations

Identification of the mode of action of the imidazolinones occurred while resistant cell lines were being isolated. Imidazolinones inhibit acetohydroxyacid synthase (AHAS; EC 4.1.3.18), the first enzyme in the pathway of branched chain amino acid synthesis (8). Imidazolinone-resistant cell lines provide proof that inhibition of AHAS is the site of action of the imidazolinones: AHAS activities in extracts from resistant corn cell lines are highly resistant to inhibition by imidazolinone herbicides (7).

Inheritance studies were conducted with plants regenerated from imidazolinone resistant cell lines. Plants from nonsegregating resistant progenies were crossed with susceptible inbred lines. The resultant F_1 hybrids were then selfed and crossed back to the susceptible parental inbred line. Progenies were treated with imidazolinone herbicides in order to monitor the segregation of resistance. The F_2 progenies (from selfed F_1 hybrids) segregated in a 3-resistant : 1-susceptible ratio. The progeny from the testcrosses (F_1 x susceptible parental inbred) segregated in a 1-resistant : 1-susceptible ratio. These results are consistent with the hypothesis that resistance is conferred by a single dominant allele. The phenotype is actually semidominant at higher rates of herbicide application; that is, homozygous resistant plants are resistant to higher levels of herbicide than are heterozygous plants. Each of the corn mutants tested displayed this same pattern of semidominant inheritance.

Imazethapyr

Imazaquin

Sulfometuron methyl

Other classes of herbicides in addition to the imidazolinones kill plants by inhibiting AHAS; e.g., the sulfonylureas (9,10). To characterize the level and spectrum of resistance for the three resistant corn lines, plants were treated with two imidazolinones, imazethapyr and imazaquin, and with the sulfonylurea sulfometuron methyl.

Table I. Increase in resistance to three AHAS-inhibiting herbicides provided by three imidazolinone-resistance mutants[1]

Herbicide	Imidazolinone-resistance mutant		
	XA17	XI12	QJ22
	(Increased level of resistance)		
Imazethapyr	>1000X	1000X	300X
Imazaquin	>1000X	60X	15X
Sulfometuron methyl	>1000X	<2X	<2X

[1]Resistance determinations are based on postemergence herbicide application rates that cause 50% growth inhibition.

The XA17 mutation confers a very high level of resistance to all three herbicides (Table I). In the homozygous state, the XA17 mutation increases resistance to each of the three herbicides tested by more than 1000-fold. The XI12 mutation confers a high degree of resistance to imazethapyr, somewhat less resistance to imazaquin, and little or no resistance to sulfometuron methyl. The QJ22 mutation provides a spectrum of resistance similar to that obtained with XI12; however, the level of resistance provided by QJ22 is only approximately one-fourth of that conferred by XI12.

Resistance at the whole plant level correlates closely with the herbicide insensitivity of AHAS activity in mutant extracts. For example, the effects of foliar applications of imazaquin on growth of B73, XA17, XI12, and QJ22 are displayed in Figure 1. Growth of the sensitive inbred B73 was inhibited at all rates applied. XA17 was highly resistant, XI12 was intermediate, and QJ22 had the least resistance to imazaquin. Imazaquin inhibition curves of AHAS activities from these genotypes display the same ranking of resistance (Figure 2).

The close correlation between seedling growth and AHAS activity resistance profiles is compatible with AHAS being the only site of action in corn for imidazolinone and sulfonylurea herbicides. A single mutation can offer either cross-resistance to both the imidazolinones and the sulfonylureas, or resistance only to the imidazolinones.

AHAS is feedback regulated by valine and leucine. The feedback regulation site is apparently distinct from the herbicide binding sites, because

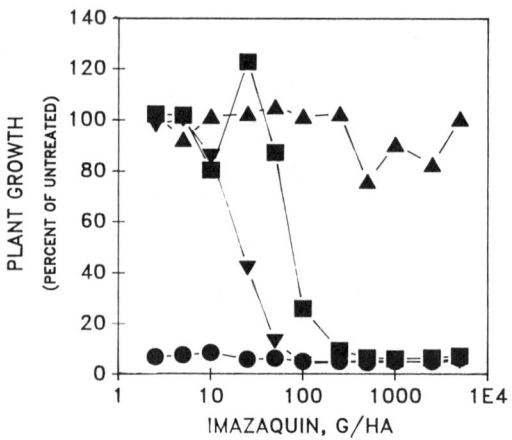

Figure 1. Growth of sensitive corn (B73 ●—●) and three imidazolinone-resistance mutants (XA17 ▲—▲, XI12 ■—■, and QJ22 ▼—▼) after treatment with imazaquin. Treatments were applied postemergence in a greenhouse trial.

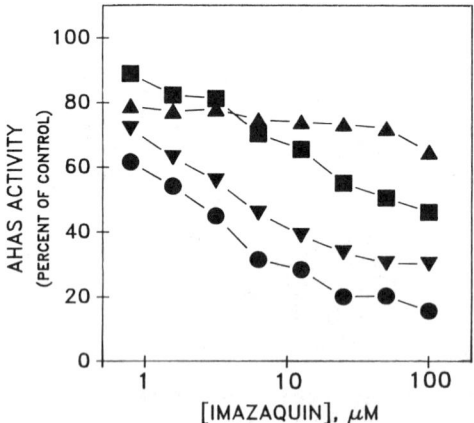

Figure 2. Inhibition of acetohydroxyacid synthase from sensitive corn (B73 ●—●) and three imidazolinone-resistance mutants (XA17 ▲—▲, XI12 ■—■, and QJ22 ▼—▼) by imazaquin.

plants with resistant AHAS appear normal in the absence of herbicide. Resistant mutants have normal levels of total AHAS activity. Imidazolinone-resistant corn plants appear to be as fit as their sensitive counterparts.

Bringing herbicide-resistant corn to the marketplace

The imidazolinone-resistant corn project successfully identified mutant alleles that confer greater than 1000-fold increased resistance to imidazolinone herbicides. Still, key questions must be answered before the resistant corn can be marketed.

In this case, imidazolinone-resistant corn was identified through efforts initiated by a chemical company. Thus, American Cyanamid needed a cooperating party to supply herbicide-resistant hybrid seed to the farmer. Decisions on collaborators, rights to exclusivity, royalties, obligations, and deadlines all entered into the objective of achieving maximum market penetration in the shortest possible time period.

American Cyanamid has an agreement with Pioneer Hi-Bred International, Inc., to bring resistant corn to the marketplace. Pioneer is introducing imidazolinone-resistance alleles into their elite proprietary inbred lines. A traditional backcrossing program with more than 100 inbreds is being used to convert the lines to imidazolinone-resistant forms. The selection of which inbreds to convert changes as newer inbreds are identified and inferior inbreds are discontinued, and as conversion is completed for inbreds currently in the program.

Six to eight years are required to convert the parental inbreds that will be used for the imidazolinone-resistant hybrids. Three generations of backcrossing are completed each year in Hawaii. Two years, or six backcross generations, are sufficient for the majority of the conversions. Another year is required to obtain the converted inbred in a homogeneous, homozygous resistant form. Three years of hybrid yield testing are typically conducted after conversion, to assure the absence of deleterious characters.

If different mutants are available, the seed company and the chemical company must agree on which mutations to commercialize. Level of resistance, cross tolerance spectra, performance as a heterozygote (for hybrid crops), and stability of resistance across genotypes and environments are factors that are considered in making this decision.

Also, the choice of resistance mutation(s) must be made in conjunction with the choice of herbicide(s). For the current example, corn without resistance alleles is highly sensitive to imazaquin and sulfometuron methyl, but has some natural tolerance for imazethapyr (Table II). A 7.5X increase in resistance would provide safety to postemergence applications of imazethapyr, whereas the corresponding increased resistance needed for safe applications of imazaquin would be 120X. Thus, the natural tolerance of a crop for a herbicide determines the level of resistance needed to use that herbicide on the crop.

The XA17 mutation provides sufficient resistance for commercial application rates of imazethapyr, imazaquin, or sulfometuron methyl. The XI12 mutation allows use of either of the two imidazolinones, but does not provide safety to sulfometuron methyl. QJ22 permits use of imazethapyr, but not of imazaquin or sulfometuron methyl. Thus, each allele offers unique marketing opportunities.

Close coordination between the seed company and the chemical company is essential to assure that yield trials, efficacy trials, hybrid seed production, and the required herbicide label extensions proceed in a concerted fashion. The imidazolinone herbicide must be ready to be used when imidazolinone-resistant corn is ready for sale. Joint launch of the resistant hybrids and the herbicide label extension that includes resistant corn will be coordinated by American Cyanamid and Pioneer Hi-Bred.

Table II. Natural tolerance levels of sensitive corn inbred B73 to three AHAS inhibiting herbicides

Herbicide	Natural tolerance rate[1], (g/ha)	Field usage rate[2], (g/ha)	Resistance increase providing a 1X safety margin[3]
Imazethapyr	10	75	7.5X
Imazaquin	1.25	150	120X
Sulfometuron	0.125	10	80X

[1] Herbicides were applied postemergence in a greenhouse study
[2] These rates provide approximately equal levels of weed control
[3] Field usage rate ÷ natural tolerance rate

Sales of a corn herbicide to be applied only to specifically labelled corn hybrids has never before been attempted. Marketing of this hybrid-herbicide package requires novel strategies for both American Cyanamid and Pioneer Hi-Bred. Farmer and dealer education about the package will be an integral part of insuring proper usage and customer satisfaction with product performance.

Summary

The project to develop corn hybrids resistant to imidazolinone herbicides has been very successful. Corn with excellent resistance characteristics was selected, and semidominant resistance alleles are being introduced into proprietary inbred lines of Pioneer Hi-Bred International, Inc. Commercial corn hybrids resistant to imidazolinone herbicides will be the first major biotechnology product in cereals to enter the marketplace.

The cooperation between American Cyanamid and Pioneer Hi-Bred provides an excellent opportunity to investigate the introduction of value added traits into crops. Even though the program has been successful, 10 years will have elapsed from initiation to first commercial release. In the future, newer technologies will allow for these programs to move more quickly, and be more responsive to rapid changes in the marketplace.

Acknowledgments

We thank Roy Chaleff for his critical review of this manuscript, and helpful suggestions for its improvement.

Literature Cited

1. Herbicide Handbook of the Weed Science Society of America, Fifth Edition; Weed Science Society of America: Champaign, Illinois, 1983.
2. Hatzios, K.K. Adv. in Agron. 1983 36, 265-316.
3. Tottman, D.R. Ann. Appl. Biol. 1984 104, 151-159.
4. Edwards, J.R.; Barrentine, W.L.; Kilen, T.C. Crop Science 1976 16, 199-120.
5. Renner, K.A.; Meggitt, W.F.; Penner, D. Weed Science 1988 36, 625-629.
6. Chaleff, R.S. In Biotechnology and Crop Improvement and Protection; Day, P.R., Ed.; British Crop Protection Council Publications: Croyden, England, 1986; pp 111-121.
7. Shaner, D.L.; Anderson, P.C. in Biotechnology in Plant Science; Zaitlin, M.; Day, P.R.; Hollaender, A., Eds.: Academic Press: Orlando, Fla, 1985; pp 287-299.
8. Shaner, D.L.; Anderson, P.C.; Stidham, M.A. Plant Physiol. 1984, 76, 545-546.
9. Chaleff, R.S.; Ray, T.B. Science 1984, 223, 1148-1151.
10. Chaleff, R.S.; Mauvais, C.J. Science 1984, 224, 1443-1445.

RECEIVED June 27, 1989

Author Index

Annamalai, P., 249
Arneson, Phil A., 264
Berg, D., 184
Böger, Peter, 407
Brattsten, L. B., 42
Brent, K. J., 303
Brown, Thomas M., 61
Büchel, K.-H. 184
Bull, D. L., 118
Butters, J. A., 199
Chui, Chok-Fun, 459
Croft, B. A., 149
Darmency, H., 353
Davidse, L. C., 215
Delp, C. J., 320
Denholm, I., 92
Dennehy, T. J., 77
Eckert, Joseph W., 286
Falco, S. Carl, 459
Farnham, A. W., 92
Fincher, R., 474
Fry, W. E., 275
Fujimura, Makoto, 224
Gasquez, J., 353
Gelernter, Wendy D., 105
Georghiou, George P., 18
Green, Maurice B., ix
Gressel, J., 430
Hargreaves, J. A., 199
Hartnett, Mary E., 459
Hollomon, D. W., 199, 303
Holmwood, G., 184
Holt, Jodie S., 419
Holtum, J. A. M., 394
Inoue, Hirokazu, 224
Inoue, I., 237
Ishii, H., 237
Iwasaki, S., 237

Kato, Toshiro, 224
Kemp, Malcolm S., 376
Knowlton, Susan, 459
Krämer, W., 184
Lalithakumari, D., 249
LeBaron, Homer M., 336
Liljegren, D. R., 394
Martinson, T. E., 77
Matthews, J. M., 394
Mauvais, C. Jeffry, 459
Mazur, Barbara J., 459
McDevitt, Raymond E., 459
McFarland, Janis, 336
Menn, J. J., 118
Milgroom, M. G., 275
Moberg, William K., 1
Morton, H. V., 170
Moss, Stephen R., 376
Newhouse, K. E., 474
Nyrop, J. P., 77
Oeda, Kenji, 224
Pontzen, R., 184
Powles, S. B., 394
Riley, S. L., 135
Rowland, M., 92
Sandmann, Gerhard, 407
Sato, Z., 237
Sawicki, R. M., 92
Schwinn, F. J., 170
Segel, L. A., 430
Shaner, D. L., 474
Shaw, M. W., 303
Smith, Julie K., 459
Thomas, Tudor H., 376
Vaughan, Martin A., 364
Vaughn, Kevin C., 364
Wade, M., 320
Wang, T., 474

INDEX

Affiliation Index

AFRC Institute of Arable Crops Research, 92, 199, 303, 376
American Cyanamid Company, 474
Bayer AG, 184
Ciba-Geigy Corporation, 170, 336
Clemson University, 61
Cornell University, 77, 264, 275
E. I. du Pont de Nemours and Company, 1, 135, 459
Fruit Tree Research Station, 237
Institut National de la Recharche Agronomique, 353
Mycogen Corporation, 105
National Institute of Agro-Environmental Sciences, 237
Oregon State University, 149
Pioneer Hi-Bred International, Inc., 474
Rutgers University, 42
Saitama University, 224
Shell Research, Ltd., 320
Sumitomo Chemical Company, Ltd., 224
U.S. Department of Agriculture, 118, 364
Universität Konstanz, Federal Republic of Germany, 407
University of Adelaide, Australia, 394
University of Bristol, United Kingdom, 199, 303, 376
University of California–Riverside, 18, 286, 419
University of Madras, India, 249
University of Tokyo, 237
Wageningen Agricultural University, Netherlands, 215
Weizmann Institute of Science, 430

Subject Index

A

Abscisic acid, regulation of accumulation, 383,387
Acaricides, resistance by spider mites, 77–89
Acetolactate synthase
  inhibition by imidazolinones and triazolopyrimidines, 460
  sensitivity to sulfonylurea herbicides, 460
Acetolactate synthase gene
  activity and inheritance of resistance in tobacco, 461
  amino acid conservation between genes, 461,462f
  cloning, 461
  field test resistant crop results, 468,471–472f
  herbicidal resistance formation, 461,463
  hybridization, 461
  number of genes in *Arabidopsis* and tobacco, 461
  sequence analyses, 461
  use in engineering herbicide-resistant crops, 464,468,469t,470–471f
Acetolactate synthase inhibitors
  resistance mechanism, 344
  weed resistance, 344

Acetylcholinesterase
  bimolecular rate constants for inhibition by organophosphates and carbamate, 53t
  consequences of existence of isoenzymes in insects, 54
  effect of *N*-alkyl groups of insecticide on resistance, 70t
  inhibition by insecticides, 69–70
  inhibition by organophosphates and carbamates, 52
  phosphorylated vs. phosphinylated recovery, 70,71f
  reaction, 52
  reaction mechanism, 51
  spontaneous reactivation from tobacco budworm larvae, 70,71f
  subsites, 51–52
Acetylcholinesterase-inhibiting insecticides
  mechanism, 69
  occurrence of resistance alleles, 69–70
Activated oxygen, initiation of herbicide-induced degradations in cells, 407,409
Agricultural fungicides, chromosomal control of resistance, 260–261
Agricultural pests, resistance to chemical control, 1
Agrochemical(s)
  areas, 2
  importance, 1–2

Agrochemical resistance
  challenge, 7
  characteristics, 10–11
  consequences, 4–5
  current literature, 3–4
  definition, 7–8
  historical perspective, 9–11
  industry involvement, 3
  nature, 6–7
  need for communication and cooperation, 12
  phenomenon, 6
  previous literature, 3
  prospects, 5–6
  role of chemists, 2–3
  scope, 4
  stepwise progression, 9–10
Alleles, definition, 9
Alopecurus myosuroides
  control, 376
  herbicidal resistance, 377–389
α-Amanitin, effect on endogenous RNA polymerase activity, 218t
1-Aminobenzotriazole
  physiological effects, 383,387
  synergism with phenylurea herbicides, 382–387
γ-Aminobutyric acid receptor, target for insecticides, 72
Anti-kaurene-oxidase plant growth regulators, synergism with herbicides, 387–388,389f
Antiresistance strategies
  communication in fungicidal resistance management, 178,179t
  comparative merits of two-use strategies, 175t
  computer-based models, 174–175
  design, 171–177
  elements of phenylamide strategy, 175,176t
  elements of strategies utilizing integrated crop production practices, 177t
  implementation, 177–178,179–180t
  importance, 170
  integration of technical parameters and management elements, 174,175–177t
  key elements, 173t
  long-term aspects, 178,181,182t
  management of fungicidal resistance in practice, 178,180t
  monitoring techniques, 174
  problems and prospects, 178,181t
  problems from industrial viewpoint, 178,181t
  risk evaluation, 171,172t,173
  strategies for major groups of systemic fungicides, 176t
  See also Resistance management
Aphids, insecticidal resistance, 22–23

Apple scab, use of benzimidazole, 328
Aryl ester hydrolase, role in insecticidal resistance, 69
Aryloxyphenoxypropionate herbicides, reduced sensitivity of target sites, 397,398t,399f,400
Assay methods for fungicidal response, disadvantages, 308,309t
Atrazine resistance, mutation, 355t
Australia, pyrethroid resistance, 139–140
Avermectins, disadvantages as insecticides, 43
Azidometalaxyl
  binding properties for phenylamide receptor, 221,222t
  structure, 221,222f
Azoles
  characterization, 184
  description, 184
  detoxification, 192,194f
  lipid composition, 192,195t,196
  mode of action studies in model systems, 185,186–187f
  mode of action studies with powdery mildew, 188,189–190f,191
  model studies on resistance, 191–192,193f
  target mutation, 192,194f
  uptake kinetics, 192,193f

B

Bacillus thuringiensis
  benefits, 105
  biology, 106
  commercial products, 108t
  control of resistant insects, 114–116f
  development, 108
  host range specificity, 107
  limitations of living rDNA products, 113
  MCap delivery system, 113–114
  mode of action, 106–107
  modification through genetic engineering, 112,113
  natural occurrence, 109
  potential for resistance, 113
  taxonomy, 106,107t
Bacillus thuringiensis based insecticides
  advantages, 109,112
  degradation curves, 110f,112
  delivery to target pests, 112
  economics of production, 111
  flexibility of use, 111
  limitations, 110f,112
  mode of action, 111
  narrow host range, 111
  residual activity, 110f,112

INDEX

Bacillus thuringiensis based insecticides—
  Continued
  safety, 109
  streamlined regulatory review, 109–111
Bacillus thuringiensis variety *san diego*
  discovery, 109
  role as insecticide, 109
Barley mildew, fungicidal resistance, 314
Benomyl
  fungicidal activity, 225
  resistance development, 289–290
Benzimidazole(s)
  accomplishments of FRAC Working
    Group, 322
  resistance development, 289–290
  risk of resistance development, 170–171
Benzimidazole fungicides
  resistance development, 224–225,237–238
  structures, 224,226*f*
Benzimidazole-resistance research and
    management goals
  to determine optimal benzimidazole use
    strategies, 327
  to implement optimal fungicidal use strategies
    through cooperation and education, 328
  to use resistance monitoring to study
    resistant populations, 328
Benzimidazole-resistant *Neurospora* strains
  cross resistance, 227*t*
  genetic resistance, 228*t*,229
Benzimidazole-resistant strains, negatively
    correlated cross resistance to
    N-phenylcarbamates, 225
Biochemical characterization of
    edifenphos-resistant mutants
  determination of DNA base composition, 251
  electrophoretic pattern of proteins,
    251,255*t*
  estimation of total DNA, RNA, and
    protein, 251
  estimation of total lipids and
    phospholipids, 251
  extraction of undegraded DNA, 251
  rate of efflux of electrolytes, 251,254*t*
Biomass production, indicator of
  fitness, 420
Biopesticidal resistance
  development of transgenic plants, 31
  Indian meal moth, 30
  rates and levels for houseflies and
    mosquitoes, 30,33*f*
  tobacco budworm, 30
Biopesticides, description, 30
Blackgrass, herbicidal resistance, 376–389
Blackgrass plants, morphological variation,
  377–378
Blue mold
  effect on decay of citrus fruits, 286–287
  spore production, 287

Bollworm, *See Heliothis zea*
sec-Butylamine, resistance development, 289

C

Calcium, effect on dinitroaniline herbicidal
  resistance, 366,367*f*
Carbamates
  factors influencing evolution of insecticidal
    resistance, 54–57
  metabolic resistance, 45–48
  structures, 44,46*f*
  target site interactions, 51–52
Case histories, fungicidal resistance, 284
Cattle tick, resistance management, 162
Cellular level, effect on resistance
  development, 151
Cerulenin
  effect on azoles, 195*t*,196
  structure, 196
Characterization of insecticides
  efficacy of insecticidal deposits, 95,96*f*
  persistence of deposits, 95,97,99*f*
Chlorotoluron
  phytotoxicity in liquid medium of
    blackgrass, 378,379*f*
  resistance in blackgrass, 377
  resistance mechanism, 382
  synergism with 1-aminobenzotriazole,
    383,384–385*f*
  synergism with triadimenol and
    triadimefon, 387–388,389*f*
  uptake, translocation, and degradation in
    blackgrass, 378,380*f*,381–382
Chlorotoluron-resistant blackgrass
  occurrence, 377
  resistance for herbicides, 377,379*t*
Chromosomes, definition, 9
Citrus fruit
  decay control strategies, 287,288*f*
  effect of molds on storage life and
    quality, 287
Citrus fruit molds, fungicidal resistance, 289
Cockroaches, pyrethroid resistance, 27,29
Colombia, pyrethroid resistance, 140
Colorado potato beetle, insecticidal
  resistance, 24,26,108
Commercial herbicides, discovery, 459
Corn
  characterization of imidazolinone-
    resistant mutations, 476,477*t*,478*f*,479
  importance of resistance to imidazolinone
    herbicides, 475
  selection of imidazolinone-resistant
    varieties, 475–476
Cotton
  insecticidal usage, 120
  interaction with *Heliothis*, 120
  susceptibility to insect damage, 119

Cotton bollworm, resistance management, 157
Cotton industry, potential effect of pyrethroid resistance, 135
Cotton pests, pyrethroid resistance, 26–27,28f
Cross resistance
 definition, 8,395
 Heliothis, 123–124
 mechanisms in annual ryegrass, 395–405
 occurrence, 261
 prediction of fungicidal resistance evolution, 307,308t
 subclasses, 8
Cucurbit powdery mildew, resistance to DMI fungicides, 202
Curative sprays, effect on fungicidal resistance, 282,284
Cyclohexanedione herbicides, reduced sensitivity of target sites, 397,398t,399f,400
Cytochrome P–450 oxidase inhibitors, synergistic effects, 382–389

D

2,4-D, weed resistance, 339
Dalapon, weed resistance, 339
Demethylation inhibitor(s) (DMI)
 accomplishments of FRAC Working Group, 323–324
 changes in sterol 14α-demethylase, 206,208t
 cross resistance, 202
 decreased fungicide uptake, 204
 detoxification, 206,208
 fungicidal activity, 200–201
 fungicidal transfer to barley powdery mildew, 204,205t
 genetics of resistance, 203–204
 resistance, 201,202t,203
 resistance mechanisms, 204–208
 sterol changes and resistance, 205–206,207f
 See also Sterol biosynthesis inhibitors
Demethylation inhibitor resistance, research and management goals, 332
Detection of esterases, approaches, 34–35
Diamondback moth
 insecticidal resistance, 22
 resistance management, 162
Dicarboximide(s)
 accomplishments of FRAC Working Group, 322–323
 risk of resistance development, 171
Dicarboximide resistance research and management goals
 to characterize field populations of dicarboximide-resistant strains of *Botrytis* and *Monilinia*, 329

Dicarboximide resistance research and management goals—*Continued*
 to establish comprehensive monitoring programs, 329
 to identify most effective use strategies to delay or prevent resistance development, 328–329
 to incorporate nonfungicidal disease control measures with dicarboximide use, 330
 to promote interindustry cooperation and communication among industry, institutions, regulatory agencies, and academia, 330
Diclofop-methyl
 effect on growth of annual ryegrass, 404f,405
 resistance mechanism, 344
Dicofol
 factors influencing resistance, 80
 frequencies of resistance, 80
 hypotheses for changes in susceptibility, 83–85
 regression of mean mortality for spider mites, 82–83,84f
Diethofencarb
 fungicidal activity, 225
 use in control of fungicide resistance, 244
Diethofencarb-resistant mutants of *Neurospora*
 examples, 229t
 genetic analysis, 229,230t
Diethofencarb-supersensitive mutants of *Neurospora*
 characterization, 227
 isolation, 227t
Dinitroaniline herbicides
 applications, 364
 effect of calcium on resistance, 366,367f
 intermediate-resistant goosegrass biotype, 374
 mechanisms of tolerance and resistance, 365–366,367f,368
 metabolism, 365–366
 occurrence of resistant biotype, 364–365
 resistance mechanism, 369,372,373f,374,421
 resistance susceptibility vs. lipid content, 365
 site of action, 366
 structural and cross-resistance studies, 366,368
 translocation, 365
Directional selection, definition, 191
Disruptive selection, definition, 191
Dominant gene, definition, 9
Drechslera oryzae protoplasts
 agarose gel electrophoresis, 252
 isolation, 251–252,256,257f
 isolation of plasmid DNA, 252

INDEX

E

Early-season tactics for resistance management, description, 128–129
Ecological level, effect on resistance development, 153–154
Edifenphos resistance, 259
Edifenphos resistance in *Pyricularia oryzae* and *Drechslera oryzae*
  adapted mutants, 250
  biochemical characterization of mutants, 251,254–258
  countermeasure, 252
  field mutants, 250
  fungal strains, 250
  isolation of plasmid from protoplasts, 251–252
  morphological characterization of mutants, 251
  mutagenesis, 250
  sensitivity, 259
  stable and virulent mutants, 250
  test for stability and pathogenicity, 250
Edifenphos-resistant mutants
  biochemical characterization, 251,254–255t
  growth, 252
  macromolecular content, 259–260
  macromolecular synthesis, 254t
  morphological characterization, 251
  morphology, 252,253f
Eleusine indica, See Goosegrass
Empirical modeling of resistance
  development and role, 100,101f
  simulations, 100,101f
Environment, effects on fitness, 425,427
Enzymes catalyzing insecticidal biotransformation
  aryl ester hydrolase, 69
  diagnostic synergists, 62–63,64f
  genetic control of monooxygenases, 63,66,67f
  glutathione *S*-transferases, 69
  hydrolases, 66,68t
  insecticidal sequestration, 68–69
  monooxygenase(s), 62
  monooxygenase inhibitors as synergists, 62,63t,64–65f
  parathion hydrolase, 69
Ergosterol biosynthesis inhibitors
  risk of resistance development, 171
  synergism with herbicides, 387–388,389f
Esterases
  classifications, 32
  detection approaches, 34–35
  gene amplification, 32,34
  role in insecticidal resistance, 32,33f
Evolution of resistance in field, influencing factors, 419

F

Feedback regulation site vs. herbicide binding site, 477,479
Fitness
  characteristics of plant used for determination, 420
  components, 419–420
  definition, 419
  effect on amplification of resistance, 446,449f,450
  environmental effects, 425,427
  genetic basis of differences, 424–425,426f
  of herbicide-resistant plants, effect of resistance mechanism, 421t
  measurement, 432
  mechanism of reduced productivity, 422,424,426t
  of resistant weeds
    changing of seed bank characters, 439
    density dependence, 439
    earliest and classical studies, 438
    effect of environment, 439
    from germination, 438
    narrow genetic base, 439
    resistant vs. susceptible biotypes, 438
    role in fungicidal resistance, 282
Fungicidal resistance
  case histories, 284
  effect of curative sprays, 282,284
  effect of growth rates, 281–282,283f
  effect of isolate fitness, 282
  evolution, 309,310f
  factors affecting development, 293–295
  heritability, 315
  interaction of resistant and sensitive biotypes during disease development, 294–295
  population size vs. probability of resistance occurrence, 277,278f,279
  predicted relative effects of strategies on resistance and disease suppression, 279,280t
  prediction, 305–316
  rationale for calculating probabilities for resistance occurrence, 277t
  role of migration, 282
  role of mutation, 276,277t,278f,279
  role of selection, 279–283
  selection and dispersal of resistant biotypes, 294
  strategies, 279t
Fungicidal resistance management
  communication, 178,179t
  dual goals, 275,276t
  management in practice, 178,180t
  monitoring methods, 264

Fungicidal resistance management—*Continued*
  predicted relative influence of mixtures and alternations on resistance and disease suppression, 280,281*t*
  predictive models, 264–265
  Resistan model, 265–269
  Sigatoka model, 269–273
Fungicidal resistance management of *Penicillium*
  delay in buildup of resistant biotypes, 299
  dispersal prevention of resistant biotypes, 300
  influencing factors, 296,297*f*
  reduction of fruit infection, 300
  sanitation, 296
  selectivity minimization, 296,298–299
  strategies, 295–300
Fungicidal resistance of molds in citrus fruits
  benzimidazoles, 289–290
  early fungicides, 289
  imazalil, 290–291,292*t*,293
Fungicidal sensitivity
  effect on fitness, 314,316*f*
  vs. stabilizing selection, 315,316*f*
Fungicide(s)
  control of citrus fruit decay, 287,288*f*
  discovery and development, 304
  prediction of resistance behavior, 304
  stepwise progression of resistance, 10
Fungicide Resistance Action Committee
  accomplishments, 324–325
  description, 320–321
  North American initiatives, 325–333
  objectives, 321
  principal functions, 321

G

Gene, definition, 9
Gene amplification, effect on biochemical mechanisms, 72,74*t*
Gene amplification in insects, role in insecticidal resistance, 32,34
Genetic experimentation for prediction of fungicidal resistance
  mutagenesis, 305–306
  recombination, 306
  selection experiments, 306–307
Genetics of insecticidal resistance, examples, 61–62
Genetic transformation, description, 459–460
Glasshouse crops, recommended use of dicarboximides, 323
Glutathione, effect of levels on peroxidative ethane formation, 414,415*f*,416
Glutathione *S*-transferase
  activity, 402,403*f*
  role in insecticidal resistance, 69
  role in organophosphate metabolism, 49,50*f*

Goals of North American Fungicide Workshop
  to anticipate impact of new candidate fungicides on resistance problems, 327
  to avoid establishment of resistant strains in new areas, 326–327
  to coordinate industrial, governmental, and academic activities, 327
  to discover and develop fungicides with new modes of action, 327
  to improve communication of resistance strategies at user level, 326
  to improve understanding and cooperation of EPA to implement FRAC resistance management strategies, 327
  to provide educational and training aids, 326
  to strengthen basic research on fungicidal resistance, 326
Goosegrass
  comparisons of tubulin and microtubules in R and S biotypes, 368–369,370–371*f*
  intermediate-resistant biotype, 374
  mechanism for dinitroaniline herbicidal resistance, 369,372,373*f*,374
Grapes, recommended use of dicarboximides, 323
Green mold
  effect on decay of citrus fruit, 286–287
  spore production, 287
Growth rates, effect on fungicidal resistance, 281–282,283*f*

H

Heliothis
  chronology of resistance development to insecticides, 120–122
  cross tolerance and resistance, 123–124
  development of resistance, 118
  economic impact, 118
  field resistance, 134
  interaction with cotton, 120
  resistance management, 125–126,127*f*, 128*t*,161–162
  resistance mechanisms, 122–123
  resistance monitoring, 124–125
Heliothis virescens
  economic impact, 118–119
  See also Tobacco budworm
Heliothis zea
  economic impact, 118–119
Herbicidal resistance
  in blackgrass
    mechanisms, 377–382
    synergistic effects of cytochrome P–450 oxidase inhibitors, 382–389
  delay strategies, 430–456
  prediction, 345,346–347*t*
  triazine resistance, 339,340–343*t*

INDEX

Herbicidal resistance—*Continued*
  in weeds
    genetic control, 354
    model, 354
    problem, 354
Herbicidal resistance formation
  genetic transformation, 459–472
  mutation breeding, 459
  plant regeneration from resistant cells, 459
Herbicidal resistance management, strategy rules for success, 351
Herbicide(s)
  appearance and proliferation of resistant weed biotypes, 394–395
  detoxification, 400,402,403–404f,405
  development, 459
  evolution of resistance in weeds, 430
  examples with high risk for weed resistance, 345,347t
  examples with low risk for weed resistance, 345,346t
  impact on modern agricultural technology, 337
  mechanisms of action, 337,339
  need for resistance prevention and management strategies, 349,350t,351
  need for variety, 345,348–349
  plant selectivity, 337
  reduced sensitivity of target sites, 397–401
  reduced translocation to active sites, 397
  resistance mechanisms, 344–345,421t
  stepwise progression of resistance, 10
  trends in first occurrence of herbicide-resistant biotypes, 349,350t
  variation of peroxidative response, 409,412f,413
Herbicide-induced degradations in cells, initiation by radicals or activated oxygen, 407,409
Herbicide-resistant acetolactate synthase
  Escherichia coli mutants, 463–464
  formation, 461
  mutations conferring resistance, 461
  mutations in tobacco gene, 464,465t
  nucleotide and amino acid sequences, 464,466–467f
  yeast mutants, 463,465t
Herbicide-resistant biotypes
  fitness and ecological adaptability, 422–427
  negative cross resistance, 440,441t
Herbicide-resistant corn
  marketing, 479–480
  natural tolerance levels to herbicides, 479,480t
Herbicide-resistant crop varieties
  advantages of changing crop tolerance for existing herbicide, 475
  development from acetolactate synthase genes, 464,468–471

Herbicide-resistant genes in weeds
  population structure, 358–359
  regulation of resistance, 359–360
  spontaneous mutants, 355,356t,357f,358
  spread of resistances from crops, 360–361
Herbicide-resistant plants, fitness, 421t
Herbicide-resistant weed(s)
  appearance and proliferation, 394–395
  genetic control of resistance, 354
  mechanisms for cross resistance in annual ryegrass, 395–405
  occurrence, 353,468,472
Heterozygous, definition, 9
High selection pressure herbicides, rate of resistance development, 430–431
Horn fly, pyrethroid resistance, 29–30
Houseflies
  resistance management, 157,160
  resistance mechanisms, 93,94t
Hydrolases
  insecticide sequestration, 68–69
  role in insecticidal resistance, 66,67t

I

Imazalil
  control of resistance, 290–291
  effect on lemon decay, 292t
  residue tolerance, 292–293
  resistance development, 290–291
Imazaquin
  structure, 476
  use in characterization of corn resistance, 477t,478f
Imazethapyr
  structure, 476
  use in characterization of corn resistance, 477t,478f
Imidazolinone(s), mode of action, 476
Imidazolinone herbicides
  commercial development, 475
  importance of corn resistance, 475
Imidazolinone-resistant corn
  marketing, 479–480
  natural tolerance levels to herbicides, 479,480t
  selection, 475–476
Imidazolinone-resistant mutations in corn
  characterization, 476,477t,478f,479
  inheritance studies, 476
India, pyrethroid resistance, 140–141
Indian meal moth, biopesticidal resistance, 30
Industry resistance groups, organization, 136,137f,138
Insecticidal biotransformation, catalysis by enzymes, 62–69
Insecticidal mixtures, evaluation, 98,100,101f

Insecticidal resistance
  appearance, 18
  chemical structure vs. resistance
    evolution, 56–57
  chemical usage vs. resistance evolution,
    55–56
  chronological increase in species resistant
    to insecticides, 19t,20f
  Colorado potato beetle, 24,26
  costs, 54–55,135
  critical cases, 21–26
  detection and monitoring tests, 34–35
  development(s)
    in biopesticidal resistance, 30–31,33f
    of mathematical model to describe
      evolution, 92
    in pyrethroid resistance, 26–30
  diamondback moth, 22
  documented cases of pyrethroid resistance,
    138,139t,140–141
  evolution, 55–56
  gene amplification, 72
  genetic basis, 61–62
  green peach aphid, 22–23
  leafminer, 23–24
  limitations of monitoring, 138
  linkage mapping of resistance genes in
    pests, 74
  mechanisms, 61,62t
  mosquitoes, 24,25f
  occurrence vs. economic importance, 21t
  occurrence vs. pesticide chemical group, 21t
  progress on mechanisms, 31–32,33f,34
  status, 19t,20f,21t
  whitefly, 22
Insecticidal resistance mechanisms
  categories, 61
  principal physiological mechanisms, 61,62t
Insecticide(s)
  acetylcholinesterase as target, 69,70t,71f
  appraisal of resistance mechanisms,
    97–98,99f
  characterization, 95,97
  chronology of resistance development in
    Heliothis species, 120
  ion channels as target, 70,72,73f
  stepwise progression of resistance, 9
  use on cotton, 120
Insecticide–synergist mixtures, role in insect
  control, 47–48
Insecticide Resistance Action Committee,
  description, 136,137f
Instability of resistance, definition, 79
Integrated pest management, 150
Ion channels, target for pyrethroid
  insecticides, 70,72,73f
Isoproturon
  phytotoxicity in liquid medium of
    blackgrass, 378,379f

Isoproturon—*Continued*
  resistance mechanism, 382
  synergism with 1-aminobenzotriazole,
    383,385f
  synergism with triadimefon, 388
  uptake, translocation, and degradation in
    blackgrass, 378,380f,381–382

L

Late-season tactics for resistance
  management, description, 130
Leafminers, insecticidal resistance, 23–24
Lemon molds, effect of imazalil, 291,292t
Life history level, effect on resistance
  development, 153
Lipids, effect on resistance susceptibility, 365

M

Malaria, insecticides used in control
  program, 24,25f
Malathion, enzyme hydrolysis, 66,67t
Malathionase, role in malathion resistance, 48
Management of fungicidal resistance, *See*
  Fungicidal resistance management
MCap delivery system
  advantages, 115
  description, 113–114
  EPA approval, 114
Mechanisms for cross resistance in *Lolium*
  rigidum
  enhanced detoxification of herbicides,
    400,402–405
  rapid repair of damage, 395
  reduced sensitivity of herbicidal target
    sites, 397–401f
  reduced translocation of herbicides to
    their active sites, 397
  reduced uptake of herbicides, 395
Metabolic resistance to carbamates
  hydrolysis, 47
  mixtures of insecticides, 47–48
  oxidation, 45,46f
Metabolic resistance to organophosphates
  detoxification, 48
  detoxification and activation by
    cytochrome P–450, 48,50f
  evolution, 49,51
  hydrolysis, 48–49
  hypothetical mechanism for P=S to P=O
    conversion, 48,50f
  oxidation, 48,50f
Metalaxyl
  effect of phenylamides on binding, 219,221t
  effect on endogenous RNA polymerase
    activity, 218t

Metalaxyl—*Continued*
  effect on uridine incorporation, 218,219*t*
  Scatchard plot of binding data, 218,219*t*
Metalaxyl resistance, case history, 284
N-Methylcarbamates
  development, 43
  target site interactions, 51–52
Methyl parathion
  acute toxicity, 43
  bioactivation and detoxication, 62,64*f*
  putative mechanisms of resistance, 72,74*t*
  recovery from tobacco budworms, 63,65*f*
  synergism in tobacco budworm, 63*t*
  use as insecticide, 43
Microtubules, effect on dinitroaniline herbicidal resistance, 366
Midseason tactics for resistance management, description, 130
Mid-South, U.S. cotton belt management programs, 142
Migration, role in fungicidal resistance, 282
Molecular analysis of sterol 14α-demethylase gene
  discussion, 208–209,210*f*
  DNA isolation, 209
  hybridization, 209,211
Monogenic cross resistance, definition, 8
Monooxygenases
  function, 62
  inducers of activity, 66,67*f*
  inhibitors as synergists, 62,63*t*,64–65*f*
Morpholines, fungicidal activity, 201
Mosquitoes, insecticidal resistance, 24,25*f*
Multifactorial analysis, prediction of fungicidal resistance evolution, 311,312*t*
Multiple resistances to herbicides, inadequacy of original population genetics model, 436
Mutagenesis, use in prediction of fungicidal resistance, 305–306
Mutation, role in fungicidal resistance, 276,277*t*,278*f*,279
Mutation breeding, description, 459

N

Negatively correlated cross resistance
  benefits in resistance management, 225
  benzimidazoles and *N*-phenylcarbamates
    binding model, 233,234*f*
    genetic studies, 228*t*,229
    molecular biology, 230,232*f*
    revertant analysis, 229–230*t*
    validity of *Neurospora* as model organism, 233,235
    binding site, 246–247bp

Negatively correlated cross resistance—*Continued*
  effect of genetic changes, 246
  herbicide-resistant biotypes, 440,441*t*
  sensitivity, 244,246
  strategy for control of fungicidal resistance, 244
Neurospora, validity as model organism for resistance mechanism determination, 233
Nicotine, target, 72
Nontriazine herbicidal resistance, occurrence and distribution of resistant weed biotypes, 339,342–343*t*
North American Fungicide Resistance Workshop
  general goals, 326–327
  research and management goals, 325
  benzimidazole resistance, 327–328
  dicarboximide resistance, 328–330
  DMI resistance, 332
  fungicidal resistance, 332–333
  phenylamide resistance, 330–331
  proceedings, 333
Nuclear isogenicity, formation, 438

O

Off-years during rotations, inadequacy of original population genetics model, 436–437
Organophosphates
  acute toxicities, 43*t*
  factors influencing evolution of insecticidal resistance, 54–57
  historical perspective as insecticide, 42,43*t*
  metabolic resistance, 48–49,50*f*,51
  metabolic resistance mechanisms, 44
  metabolism by glutathione transferases, 49,50*f*
  resistance, 44–45
  structures, 44,46*f*
  target site interactions, 51–52
  target site resistance, 44–45,51
Organophosphorus ester insecticides, resistance development, 121

P

Paraoxon, enzyme hydrolysis, 67*t*
Parathion
  acute toxicity, 43*t*
  resistance in spider mites, 80
  use as insecticide, 43
Parathion hydrolase, role in insecticidal resistance, 69
Pear psylla, resistance management, 160,162

Pendimethalin
  applications, 364
  occurrence of resistant biotype, 364–365
  See also Dinitroaniline herbicides
Penicillium digitatum, resistance to
  imazalil, 202
Penicillium molds
  control strategies, 287
  factors affecting resistance development, 293–295
  fungicidal resistance management strategies, 295–300
  interaction of resistant and sensitive biotypes during disease development, 294–295
  selection and dispersal of resistant biotypes, 294
Permethrin, structure, 95,96f
trans-Permethrin, enzyme hydrolysis, 67t
Peroxidizing herbicides
  effect of glutathione levels on ethane formation, 414,415f,416
  effect of α-tocopherol and ascorbate on tolerance, 414,415t
  initiation of peroxidation by protoporphyrin IX, 414
  light-dependent mechanism for starting radical formation, 409
  quantitation by ethane formation measurement, 407
  structures, 407,408f
Pesticidal resistance management
  applied science aspects, 154–155,156f
  basic science aspects, 151,153–154
  case histories of implementation, 157–162
  components, 155,156f,161
  definition, 149
  discipline needs, 151,152f
  issues areas of policy, 163t,164–165
  less successful management, 161–162
  progress in research, 150–151
  successful management, 157,160–161
  tactics, 155
Pesticides
  amounts used on major crops, 337,338t
  U.S. sales, 337,338t
Pests, genetic studies, 74
Phenylamide(s)
  accomplishments of FRAC Working Group, 322
  binding at phenylamide receptor, 219,220f,221t
  biological properties of photoaffinity probe for receptor, 221,222f,t
  effect of inhibitors on endogenous RNA polymerase activity, 218–219t

Phenylamide(s)—*Continued*
  effect of metalaxyl on incorporation of labeled precursors, 216t
  effect of metalaxyl on uridine incorporation, 218,219t
  effect on binding by cell-free mycelial extract, 219,221t
  effect on biosynthetic processes, 216t,217f,218–219t
  electrophoretic analysis of total RNA, 216,217f,218
  examples, 215
  mechanism of action, 215–221
  resistance problems, 215
  risk of resistance development, 171
  Scatchard plot of binding data, 218,219t
  structures, 215,217f
Phenylamide resistance research and management goals
  to determine base-line sensitivity to DMIs of selected pathogens, 332
  to determine mechanisms of resistance to DMIs, 332
  to implement use of management strategies developed by FRAC, 331
  to prioritize attention to two diseases, 330–331
  to standardize monitoring programs, 331
N-Phenylcarbamates
  antifungal activity against benzimidazole-resistant strains, 238
  mechanism of action to benzimidazole-resistant strains, 224–235,238
  structures, 225,226f
N-Phenylformamidoximes
  antifungal activity against benzimidazole-resistant strains, 238
  mechanism of action to benzimidazole-resistant strains, 238
3-Phenylpropylamines, fungicidal activity, 202
Phenylurea herbicides, degradation and detoxification in plants, 381
Photosystem II herbicides usage against atrazine-resistant weeds, 435–436
Phytophagous arthropod pests of apple, resistance management, 160
Plant pathologists, expertise, 275
Plasmid DNA, agarose gel electrophoresis, 256,258f
Policy
  definition, 163
  effect on resistance management approaches, 164–165
  issues for resistance management, 163t,164
Polygenic cross resistance, definition, 8–9

INDEX

493

Polygenic models, prediction of fungicidal resistance evolution, 314–315,316f
Polyunsaturated fatty acids, peroxidation, 407,410–411f
Pome fruit, recommended use of dicarboximides, 323
Population genetics model
  fitness differential, 437
  fitness of resistant weeds, 437–439
  negative cross resistance, 440,441t
  problems to be considered, 437
Population genetics model, original
  advantages and disadvantages, 431
  affirmation of model, 434–435
  cases of possible inadequacy, 435–437
  importance of selection pressure, 434–435
  prediction of effect of herbicidal rotation, 432,433f
  proportion of resistant individuals of weed species, 431–432
  seed bank life, 435
Population genetics model, revised
  advantages and disadvantages, 451
  amplification of resistance vs. fitness, 446,449f,450
  calculation of kill percentage, 444–445
  derivation, 452–455
  discussion, 442–444
  effect of herbicide rotation, 444,447f
  enrichment factor, 443–444
  enrichment under different herbicidal rotations, 444,445t
  lack of enrichment for resistance vs. selection pressures, 446,448f
  modifications, 440–441
  predictive uses, 444
  resistance status for no-till agriculture, 446,447f
  seed bank survival fraction vs. half-lives vs. residence time, 443t
  selection pressures vs. proportion of resistant individuals, 446,448f
Population structure of triazine-resistant biotypes
  monomorphism, 358
  origin, 359
  polymorphism, 358–359
Powdery mildew
  fungicidal resistance, 314
  fungicidal transfer, 204,205t
Prediction of fungicidal resistance evolution
  cross resistance, 307,308t
  detection and monitoring, 308,309t,310f,311
  genetic experimentation, 305–307
  mode of action, 307
  multifactorial analysis, 311,312t
  polygenic models, 314–315,316f
  single-gene models, 311,313t

Protoporphyrin IX
  formation, 413,415t
  initiation of peroxidation, 414
Pyrethroid(s)
  advantages, 134
  disadvantages as insecticides, 43
  resistance development, 121–122
Pyrethroid resistance
  1988 monitoring program results, 143,144–145f
  1989 monitoring program, 143
  Australia, 139–140
  cockroaches, 27,29
  Colombia, 140
  costs, 135
  documented cases, 138,139t
  early concern, 136
  horn fly, 29–30
  India, 140–141
  management guidelines, 142,143t
  pests of cotton, 26–27,28f
  Thailand, 140
  Turkey, 140
  U.S. cotton belt, 141
Pyrethroids Efficacy Group, description, 136,137f
Pyricularia oryzae protoplasts
  agarose gel electrophoresis, 252
  isolation, 251–252,256,257f
  isolation of plasmid DNA, 252
Pyrimidines, risk of resistance development, 170–171

R

R biotypes of goosegrass
  comparisons of tubulin and microtubules, 368–369,370–371f
  effects of taxol, 372,373f
  electron micrographs of microtubules, 369,370f
  Western blots of extracts, 369,371f
Recessive gene, definition, 9
Recombination, use in prediction of fungicidal resistance, 306
Regulation of resistance of triazine
  lamarkism for resistance, 359–360
  search for resistant plants, 359
Relative fitness
  definition, 85
  estimation, 87
Reproductive success, *See* Fitness
Resistance
  biochemical basis for phenylamide fungicides, 215–222
  definition, 7–8
  DMI fungicides, 201,202t

Resistance-countering strategies
  apparatus and techniques, 94
  appraisal of resistance mechanisms,
    97–98,99f
  characterization of insecticides,
    95,96f,97,99f
  development and role of empirical
    modeling, 100,101f
  evaluation of insecticidal mixtures,
    98,100,101f
  objectives of housefly work, 94–95
  resistance mechanisms isolated in
    houseflies, 93,94t
  work on tobacco whitefly, 102,103f
Resistance enrichment factor in presence of
  herbicides, calculation, 455
Resistance management
  cross tolerance and resistance, 123–124
  early-season tactics, 128–129
  importance of early planting and
    maturation, 126,127f,128t
  late-season tactics, 130
  midseason tactics, 130
  programs, 125–126
  progress in research, 150–151
  resistance mechanisms, 122–123
  resistance monitoring, 124–125
  simulation of magnitude and timing of
    Heliothis species generations, 126,127f,128t
  simulation of phenological information,
    126,127f
  use of negatively correlated cross resistance
    244,246–247
  See also Antiresistance strategies, Pesticidal
    resistance management
Resistance mechanism(s)
  cellular and organismal level, 151
  ecological and population genetics level,
    153–154
  effect of metabolic factors, 122
  influencing factors, 122
  life history level, 153
  sensitivity of target sites, 122–123
Resistance mechanism appraisal
  persistence of resistant genes in
    untreated populations, 98,99f
  phenotypic expression of resistance genes,
    97–98,99f
Resistance monitoring, *Heliothis*, 124
Resistance multiplicative factor
  calculation, 455
  proof of value less than or equal to
    1, 456
Resistance of mildew toward azoles,
  developmental models, 191–192,193f
Resistance to agrochemicals, *See*
  Agrochemical resistance
Resistance to biopesticides, *See*
  Biopesticidal resistance
Resistance to insecticides, *See*
  Insecticidal resistance
Resistance to pyrethroids, *See* Pyrethroid
  resistance
Resistan model of fungicidal management
  description, 265–266
  diagrammatic representation, 266,267f
  objectives, 265
  parameters, 266,268–269
  structure of simulation,
    266,267f,268–269
  use as management tool, 269
Resistant biotypes, lower productivity and
  competitive fitness, 432,434t
Resistant genotypes, effect of
  chemical-selecting agent, 420
Resistant populations, tactics to
  prevent or delay appearance, 451
Resistant weeds, fitness, 437–439
Rhizoxin
  antifungal activity, 247
  structure, 238,239f
  use in resistance control, 238–247

S

Sanitation, effect on fungicidal resistance
  management, 296
S biotypes of goosegrass
  comparisons of tubulin and microtubules,
    368–369,370–371f
  effects of taxol, 372,373f
  electron micrographs of microtubules,
    369,370f
  Western blots of extracts, 369,371f
Seed bank life, effect on herbicide
  resistance, 435
Selection, role in fungicidal resistance,
  279–283
Selection pressure
  definition, 432
  effect on herbicidal resistance, 434
Selective herbicides
  description, 475
  importance, 475
Selectivity of fungicides, effect on
  fungicidal resistance management,
  296,298–299
Sigatoka model of fungicidal management
  advantages, 270,273
  diagrammatic representation of banana leaf
    submodel, 270,271f
  diagrammatic representation of fungus
    development submodel, 270,272f
  management decision making, 269–270
  modules, 270
Single-gene models, prediction of fungicidal
  resistance evolution, 311,313t

INDEX

Spider mite(s)
  carbamate acaricide resistance, 78
  organophosphate resistance, 77
  organosulfur resistance, 78
  organotin resistance, 77–78
  ovicidal acaricide resistance, 78
  resistance to organic acaricides, 77
Spider mite resistance to acaricides
  assembling of management program, 79
  development of methods for resistance frequency estimation, 78–79
  experimental procedures, 81–82,85–87
  hypotheses for changes in susceptibility to dicofol, 83,85
  magnitude of potential fitness differences between resistant and susceptible genotypes, 87,88f,89
  regression of mean mortality, 82–83,84f
  stability, 79–81
  susceptibility of field population treatment history, 81–85
  system for studying resistance dynamics, 78–79
Spontaneous mutants of triazine-resistant biotypes
  fluorescence curve, 356,357f
  mutation for atrazine resistance, 355t
  mutator system, 356,358
  occurrence of mutant precursor plants in populations, 356t
  precursor plants, 356t,357f
  randomness, 358
Spread of resistances from crops
  introgressions, 360
  limitation, 360–361
Stability of resistance, definition, 79
Stable resistance, definition, 79
Sterol(s)
  biosynthesis, 200
  functions, 200
Sterol biosynthesis inhibiting fungicides
  current status of practical resistance, 201,202t,203
  examples, 200–201
  genetics of resistance, 203–204
  resistance mechanisms, 204–208
  sites of action, 200t
Sterol biosynthesis inhibitors
  cooperative effect of cerulenin and triadimenol, 195t
  description, 185
  detoxification, 192,194f
  effect of triadimenol and terbuconazole, 185,186f
  fatty acid analysis, 195t
  influence of oleic acid on efficacy, 195t
  lipid composition, 192,195t,196
  mechanism of action, 185,187f,188,190f

Sterol biosynthesis inhibitors—Continued
  mode of action studies in model systems, 185,186–187f
  mode of action studies with powdery mildew, 188,189–190f,191
  model studies on resistance, 191–192,193f
  sterol distribution in barley powdery mildew, 188,189f
  target mutation, 192,194f
  uptake kinetics, 192,193f
Sterol 14α-demethylase
  changes as cause for DMI fungicidal resistance, 206
  molecular analysis of gene from powdery mildew, 208–209,210f,211
Stone fruit brown rot, use of benzimidazole, 328
Stone and pome fruit, recommended use of dicarboximides, 323
Strawberries, recommended use of dicarboximides, 323
Sulfometuron methyl
  structure, 476
  use in characterization of corn resistance, 477t,478f
Sulfonylurea herbicides
  environmentally desirable properties, 460
  inhibition of acetolactate synthase, 460
  prototype structure, 460,462f
  reduced sensitivity of target sites, 400,401f
  resistance mechanism, 344
  selectivity, 460
  site of action, 460
  weed resistance, 344
Synthetic organic insecticides, resistance development, 120–121
Systemic fungicides, control of resistance, 249–250

T

Target site interactions, organophosphates and carbamates, 51–52
Target site resistance
  diagnosis, 52–54
  discovery, 52
  importance, 54
Tebuconazole, resistance behavior, 305
Tetradifon, resistance in spider mites, 80
Tetraethyl pyrophosphate, use as insecticide, 43
Texas, U.S. cotton belt management programs, 141–142
Thailand, pyrethroid resistance, 140
Theoretical population biologists, expertise, 275

Thiabendazole, resistance development, 289–290
Tobacco budworm
  biopesticidal resistance, 30–31
  See also Heliothis virescens
Tobacco hornworm, insecticidal resistance mechanism, 55–56
Transgenic plants, development as insecticides, 31
Tree fruit production, effect of acaricide resistance, 77
Triadimefon
  activity under semifield conditions, 388
  synergism with herbicides, 388,389f
Triadimenol, synergism with herbicides, 388,389f
Triazine resistance
  discovery, 339
  distribution of resistant dicot weeds, 339,340t
  distribution of resistant monocot weeds, 339,341t
  mechanism, 354–355,422
  occurrence, 422
  population structure, 358–359
  regulation of resistance, 359–360
  spontaneous mutants, 355,356t,357f,358
  spread of resistances from crops, 360–361
  whole plant differences between triazine-resistant and -susceptible biotypes, 422,423t
Triazine-resistant biotypes
  environmental effects on fitness, 425,427
  genetic basis of fitness differences, 424–425,426f
  mechanism of reduced productivity, 422,424,426t
  physiological differences between resistant and susceptible biotypes, 422,424,425t
  whole plants difference between resistant and susceptible biotypes, 422,423t
Trichlorphon, structure, 95,96f
Trifluralin
  applications, 364
  occurrence of resistant biotype, 364–365
  resistance mechanism, 344,421
  See also Dinitroaniline herbicides
Tubulin, site of action of dinitroaniline herbicides, 366
Tubulin genes
  amino acid substitutions, 231,233t

Tubulin genes—*Continued*
  characterization of plasmids for resistance site determination, 231,232f
  model for negatively correlated cross resistance, 233,234f
  mutations, 231
Turkey, pyrethroid resistance, 140

U

Undegraded DNA, characterization, 255t,256,257–258f
Unstable resistance, definition, 79
U.S. branch of Pyrethroids Efficacy Group
  description, 137f,138
  resistance monitoring efforts, 142,143t,144–145f
U.S. cotton belt
  pyrethroid resistance, 141
  resistance management programs, 141–142

V

Venturia nashicola, isolate preparation, 238,240
Venturia nashicola cellular protein–carbendazim binding
  binding assay procedure, 241
  binding assays with [^{14}C]carbendazim, 241,244,245f
  experimental materials, 240
  preparation of cell-free mycelial extracts, 240–241
  sensitivity of isolates to carbendazim, N-phenylcarbamate, N-phenyl formamidoxime, and rhizoxin, 240–241,242–243t
  use of negatively correlated cross resistance for resistance management, 244,246–247

W

Weed(s), evolution of resistant populations, 430
Weed control techniques in Europe, features, 353
Weed resistance, herbicides, 339,344
Weed scientists, contributions to agriculture, 337
Whitefly, insecticidal resistance, 22

Production: Peggy D. Smith
Indexing: Deborah H. Steiner
Acquisition: Cheryl Shanks

Elements typeset by Hot Type Ltd., Washington, DC
Printed and bound by Maple Press, York, PA

Paper meets minimum requirements of American National Standard for Information Sciences—Permanence of Paper for Printed Library Materials, ANSI Z39.48–1984 ∞

Other ACS Books

Chemical Structure Software for Personal Computers
Edited by Daniel E. Meyer, Wendy A. Warr, and Richard A. Love
ACS Professional Reference Book; 107 pp;
clothbound, ISBN 0–8412–1538–3; paperback, ISBN 0–8412–1539–1

Personal Computers for Scientists: A Byte at a Time
By Glenn I. Ouchi
276 pp; clothbound, ISBN 0–8412–1000–4; paperback, ISBN 0–8412–1001–2

Biotechnology and Materials Science: Chemistry for the Future
Edited by Mary L. Good
160 pp; clothbound, ISBN 0–8412–1472–7; paperback, ISBN 0–8412–1473–5

Polymeric Materials: Chemistry for the Future
By Joseph Alper and Gordon L. Nelson
110 pp; clothbound, ISBN 0–8412–1622–3; paperback, ISBN 0–8412–1613–4

The Language of Biotechnology: A Dictionary of Terms
By John M. Walker and Michael Cox
ACS Professional Reference Book; 256 pp;
clothbound, ISBN 0–8412–1489–1; paperback, ISBN 0–8412–1490–5

Cancer: The Outlaw Cell, Second Edition
Edited by Richard E. LaFond
274 pp; clothbound, ISBN 0–8412–1419–0; paperback, ISBN 0–8412–1420–4

Practical Statistics for the Physical Sciences
By Larry L. Havlicek
ACS Professional Reference Book; 198 pp; clothbound; ISBN 0–8412–1453–0

The Basics of Technical Communicating
By B. Edward Cain
ACS Professional Reference Book; 198 pp;
clothbound, ISBN 0–8412–1451–4; paperback, ISBN 0–8412–1452–2

The ACS Style Guide: A Manual for Authors and Editors
Edited by Janet S. Dodd
264 pp; clothbound, ISBN 0–8412–0917–0; paperback, ISBN 0–8412–0943–X

Chemistry and Crime: From Sherlock Holmes to Today's Courtroom
Edited by Samuel M. Gerber
135 pp; clothbound, ISBN 0–8412–0784–4; paperback, ISBN 0–8412–0785–2

For further information and a free catalog of ACS books, contact:
American Chemical Society
Distribution Office, Department 225
1155 16th Street, NW, Washington, DC 20036
Telephone 800–227–5558